Lecture Notes in Networks and Systems

Volume 59

Series editor

Janusz Kacprzyk, Polish Academy of Sciences, Warsaw, Poland
e-mail: kacprzyk@ibspan.waw.pl

The series "Lecture Notes in Networks and Systems" publishes the latest developments in Networks and Systems—quickly, informally and with high quality. Original research reported in proceedings and post-proceedings represents the core of LNNS.

Volumes published in LNNS embrace all aspects and subfields of, as well as new challenges in, Networks and Systems.

The series contains proceedings and edited volumes in systems and networks, spanning the areas of Cyber-Physical Systems, Autonomous Systems, Sensor Networks, Control Systems, Energy Systems, Automotive Systems, Biological Systems, Vehicular Networking and Connected Vehicles, Aerospace Systems, Automation, Manufacturing, Smart Grids, Nonlinear Systems, Power Systems, Robotics, Social Systems, Economic Systems and other. Of particular value to both the contributors and the readership are the short publication timeframe and the world-wide distribution and exposure which enable both a wide and rapid dissemination of research output.

The series covers the theory, applications, and perspectives on the state of the art and future developments relevant to systems and networks, decision making, control, complex processes and related areas, as embedded in the fields of interdisciplinary and applied sciences, engineering, computer science, physics, economics, social, and life sciences, as well as the paradigms and methodologies behind them.

Advisory Board

More information about this series at http://www.springer.com/series/15179

Samir Avdaković
Editor

Advanced Technologies, Systems, and Applications III

Proceedings of the International Symposium on Innovative and Interdisciplinary Applications of Advanced Technologies (IAT), Volume 1

Editor
Samir Avdaković
Faculty of Electrical Engineering
University of Sarajevo
Sarajevo, Bosnia and Herzegovina

ISSN 2367-3370 ISSN 2367-3389 (electronic)
Lecture Notes in Networks and Systems
ISBN 978-3-030-02573-1 ISBN 978-3-030-02574-8 (eBook)
https://doi.org/10.1007/978-3-030-02574-8

Library of Congress Control Number: 2016954521

This Springer imprint is published by the registered company Springer Nature Switzerland AG
The registered company address is: Gewerbestrasse 11, 6330 Cham, Switzerland

Contents

Mechatronics, Robotics and Embedded Systems

Information and Communication Technologies

Applied Mathematics

Detecting Functional States of the Rat Brain with Topological Data Analysis

Nianqiao Ju[1(✉)], Ismar Volić[2], and Michael Wiest[3]

[1] Department of Statistics, Harvard University, Cambridge, MA 02138, USA
nju@g.harvard.edu
[2] Mathematics Department, Wellesley College, Wellesley, MA 02481, USA
ivolic@wellesley.edu
[3] Neuroscience Program, Wellesley College, Wellesley, MA 02481, USA
mwiest@wellesley.edu

Abstract. One of the cutting-edge methods for analyzing large sets of data involves looking at their "shape", namely their geometry and topology. In this paper, we apply topological analysis to data arising from a neuroscience experiment involving multichannel voltage measurements of brain activity in awake rats. Data points are viewed as a point cloud, with distance defined using channel correlations or a Euclidean metric. Exploratory data analysis reveals that the topological structure defined in terms of a Euclidean metric can distinguish between a coherent oscillatory brain state and the desynchronized awake state, by associating different Betti numbers to the different brain states.

Keywords: Topological data analysis · mu rhythm · alpha rhythm
Rat brain · Persistent homology · Betti numbers · Local field potentials
Spike-and-wave

1 Introduction

Multi-channel neurophysiological recordings from the brain produce rich high-dimensional time series data from which neuroscientists attempt to distinguish different functional states and relate them to an animal or a person's behavioral capacities on the one hand and to underlying neural mechanisms on the other. Our goal is to explore whether topological data analysis, a new technique that has in recent years proved to be extremely fruitful in many fields, including in neuroscience (see [2] for a compilation of references), can reveal higher geometric structure in multichannel neural "local field potential" (LFP) voltage data and ultimately reveal information about functional states of the brain, or patterns of functional connectivity, that traditional methods cannot see. LFPs are analogous to electroencephalographic (EEG) recordings from the scalp, in that they reflect the electrical activities of many neurons acting in concert, but they are "depth EEGs" recorded using electrode arrays surgically implanted into selected brain areas to better discern the sources of neurologically important "brain waves".

In this paper we focus on a test case comparing the topological structure of two known distinct states of the awake rat brain as measured by multisite LFP recordings. One is a state which can appear in immobile but awake rats, in which the LFP at

S. Avdaković (Ed.): IAT 2018, LNNS 59, pp. 3–12, 2019.
https://doi.org/10.1007/978-3-030-02574-8_1

multiple cortical and subcortical sites in the rat brain oscillates in a coherent high-amplitude rhythm with a frequency around 10 Hz [4, 9, 16]. This state has been referred to as "high voltage spike and wave discharges" [11–13, 15] or informally as "mu rhythm" by analogy with a human brain rhythm in the same frequency range. For brevity in this study we will refer to this brain state as *mu*. We will compare episodes of this brain state to episodes of *non-mu* in which the brain is relatively "desynchronized", such that LFP fluctuations are smaller in amplitude and more broadband. Aside from being readily distinguishable in the LFP, these brain states have been shown to correspond to distinct modes of sensory processing [10].

The goal in this work is to apply topological analysis to the mu and non-mu data in hope that it can distinguish these states. This would support the possibility that topology might detect more subtle patterns that relate LFPs to behavioral and cognitive states.

Topology studies intrinsic geometric properties of objects, namely properties of the shape that remain unchanged after a continuous deformation. The most effective way of measuring and comparing such properties is to look at topological invariants of the space. A topological invariant is mathematical object, such as a polynomial or a group, that remains unchanged after the space is deformed. One of the most basic and effective class of invariants are homology groups. We will not define them precisely here since this is not needed for our purposes, but will say something about them in Sect. 2. For a precise definition, see [8] or [5]. Intuitively, homology groups keep track of the holes in a topological space. For example, the circle S^1 has a one-dimensional hole, while the sphere S^2 has a two-dimensional hole. Higher-dimensional topological objects might have higher-dimensional holes (in fact, the k-dimensional sphere S^k has a k-dimensional hole).

In topological data analysis, we view data as point clouds endowed with a certain geometry that in turn gives them the structure of a topological space. The points are intended to be thought of as finite samples taken from a geometric object, perhaps with noise. The geometry is provided by a distance function on the data, namely a notion of a distance between any two data points. The distance is defined using correlations between signals recorded from different parts of the rat's brain. From this distance function, one builds the topological space by means of a *Vietoris-Rips complex*. Finally, since we now have a topological space, we can compute its homology groups, thereby learning something about the shape of the data cloud from the information about its holes.

The paper is organized as follows: Some mathematical preliminaries, including basic background on homology and the Vietoris-Rips complex, are provided in Sect. 2. In Sect. 3 we describe the neurophysiological recording experiments and data set. Results are presented in Sect. 4 and we summarize our conclusion in Sect. 5.

2 Mathematical Background

Informally, a homology of a topological space X is the family of *homology groups*

$$H_0(X), H_1(X), H_2(X), \ldots \tag{2.1}$$

Each of them is a topological invariant that essentially counts the k-dimensional holes in X. The first homology group, $H_0(X)$, counts the number of connected components of the topological space, $H_1(X)$ counts the number of 1-dimensional holes, $H_2(X)$ counts the number of 2-dimensional holes, etc. For example, the homology groups of the circle S^1 are:

$$\begin{aligned} H_n(S^1) &= Z, \text{for } n = 0, 1; \\ H_n(S^1\} &= \{0\}, \text{for } n \geq 2. \end{aligned} \tag{2.2}$$

Here Z stands for the group of integers and $\{0\}$ for the trivial group. More generally, for a k-dimensional sphere S^k we have:

$$\begin{aligned} H_n(S^k) &= Z, \text{ for } n = 0, k; \\ H_n(S^k\} &= \{0\}, \text{for all other } n. \end{aligned} \tag{2.3}$$

What we mostly care about is the *rank*, namely the number of copies of Z, of each homology group, since this number essentially captures all the information about the group. The rank of the kth homology group is called the kth *Betti number*, denoted by

$$\beta_k = Rank(H_k(X)). \tag{2.4}$$

Thus β_k counts the number of kth dimensional holes. If $\beta_0(X) = 1$, then X consists of a single connected component; if $\beta_1(X) = 1$, then X has a single one-dimensional hole. A way to capture the number of holes is to see how many loops there are on the space that cannot be shrunk to a point (counting loops that can be deformed into one another as the same). An example that illustrates this is the torus $T^2 = S^1 \times S^1$, the Cartesian product of two circles (a hollow doughnut). It has one connected component, and so the 0th Betti number is 1; it has two holes because there are two essential loops (as shown in pink and red in the left panel of Fig. 2) that cannot be shrunk to points on the torus, so the 1-st Betti number is 2; and the space in the interior of T^2 is a two dimensional hole, so $\beta_2(T^2) = 1$.

In order to make a topological space out of a data set, one first defines a notion of a distance on it. Namely, to any two points x_i and x_j in the data set, we associate a non-negative number $d(x_i, x_j)$ satisfying the usual properties of a distance function, i.e. of a metric. Then one endows the data set with the structure of a Vietoris-Rips complex, the standard way to make a topological space out of the metric in the context of topological data analysis. Briefly, the Vietoris-Rips complex of a data cloud X, attached to the parameter $\varepsilon > 0$, and denoted by $VR(X,\varepsilon)$, is the simplicial complex (a space built out of triangles, tetrahedra, and their generalizations) whose vertex set is X and where $\{x_1,x_2, \ldots, x_k\}$ spans a k-simplex if and only if $d(x_i, x_j) \leq \varepsilon$ for all $0 \leq i, j \leq k$. For an overview of the Vietoris-Rips complex and the idea of topological data analysis in general, see [1] or [3]. Figure 1 illustrates the Vietoris-Rips complex of a simple data cloud for various values of ε.

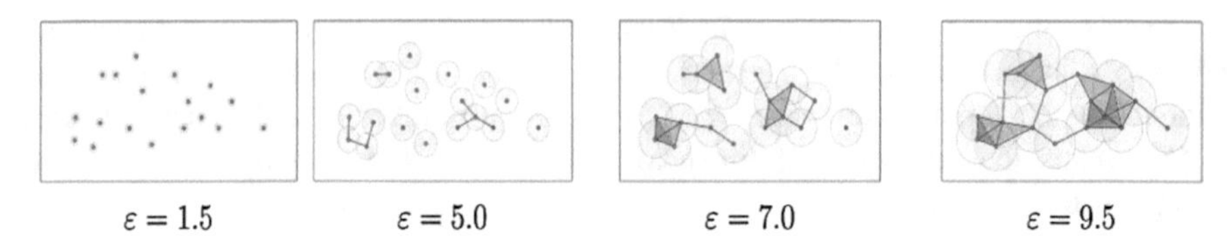

Fig. 1. Example of Vietoris-Rips complexes at different ε (figure is taken from the Javaplex documentation). Connected components are constructed so that data points within ε of each other belong to the same component.

Once the data cloud has been given the structure of a topological space like this, we can compute its homology groups $H_k(X)$, $k \geq 0$. This can be done algorithmically through linear algebra using various online data analysis packages.

The one used here was Javaplex [14]. Javaplex produces a *persistence barcode* for each homology group, with the number of bars that "survive" being the Betti number for that homology group. Figure 2 gives an example of the persistence barcodes for the torus. The interpretation is that the long bars are holes in the data cloud that appear for various values of ε, i.e. they are persistent, and this means that those holes are essential to the data cloud.

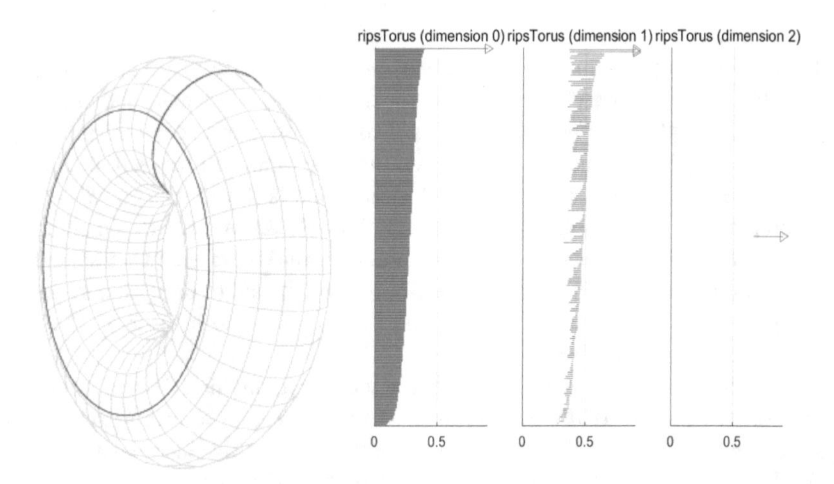

Fig. 2. A torus T^2 with β_0 (T^2) = 1, β_1 (T^2) = 2 and β_2 (T^2) = 1. We see that the barcode plot shows exactly these the Betti numbers. To read the Betti numbers, we count the number of arrows in the barcode plots associated with each dimension.

Note that all that is necessary to perform topological data analysis on a data cloud is the metric, i.e. the distance function; the rest is essentially automatically done by a computational tool such as Javaplex.

3 Materials and Methods

Local field potentials (LFPs) were recorded at 16 parietal and 16 frontal sites in the cortex of a male Long-Evans rat while the rat passively listened to 100 ms duration tones of two different pitches, presented with equal probability in random order. The sample rate was 1000 Hz. Trials were defined as segments of LFP from 0.5 s before each tone until 1.5 s after the tone. For the present study to avoid confounds due to the two pitches we only analyzed trials in which the lower pitched tone (1500 Hz) was presented to the rat. We first rejected artifact trials automatically using a 1.5 mV threshold.

During the passive recording session the rat spontaneously went in and out of the synchronized $\sim$10 Hz oscillatory state we are referring to as a mu-rhythm. Our goal is to compare the topology of mu and non-mu trials to see whether it can capture the difference in brain states. To identify mu and non-mu trials for the purposes of this comparison, one of us (MCW) with experience studying this brain state selected 126 *mu* trials and 136 *non-mu* trials based on visual inspection of one frontal LFP channel. The selected mu trials exhibited characteristic "spike-and-wave" patterns for the whole 2-second trial. Conversely, the trials selected as representative non-mu trials were free

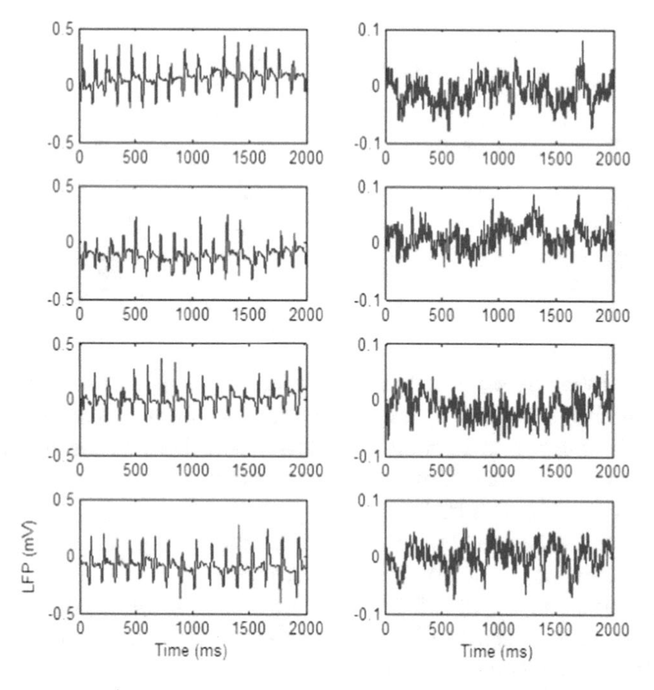

Fig. 3. The *left column* shows four examples of one local field potential (LFP) channel recorded from frontal cortex of an awake rat during episodes of an oscillatory brain state we refer to as mu. The *right column* shows four example trials recorded in the same rat during the same session, but while the rat's brain was in a relatively desynchronized state we refer to as non-mu. In every trial a brief tone stimulus was presented to the rat at 500 ms.

of the spike-and-wave oscillation for the entire trial. This procedure resulted in a set of 126 mu trials and 136 non-mu trials. Four examples of LFP recordings in each state are shown in Fig. 3. We chose these two brain states as a test case for our topological analysis because they are clearly distinct in the LFP, even to an untrained eye.

Thus the total data set comprised a

$$27 \times 126 \times 2001 = (\text{number of LFP channels}) \times (\text{number of trials}) \times (\text{number of time points})$$

3-dimensional grid for the mu trials plus a 27 × 136 × 2001 grid for the non-mu trial data. Further details about electrode implantation, recording coordinates, preprocessing, and other experimental procedures may be found in [6]. All procedures involving rats were approved by the Wellesley College Institutional Animal Care and Use Committee.

As a possible way to learn about functional connectivity between various parts of the brain, we analyze the data using a persistent brain network homology. For each trial, denote the data set as a string $C = (c_1, c_2, \ldots, c_n)$ consisting of n nodes where n is number of channels and each c_i is a 2001-dimensional vector whose coordinates are the LFPs at each ms of a 2.0 s trial. Inspired by an earlier paper [7], we calculate the distance matrix D based on correlation between channels, defined as

$$D_{ij} = \sqrt{1 - corr(c_i, c_j)} \tag{3.1}$$

where

$$\bar{c}_i = \frac{1}{2001}\sum_{t=1}^{2001} c_{it} \quad \text{and} \quad corr(c_i, c_j) = \frac{\sum_{t=1}^{2001} (c_{i,t} - \bar{c}_i)(c_{j,t} - \bar{c}_j)}{\sqrt{\sum_{t=1}^{2001} (c_{it} - \bar{c}_j)^2 \sum_{t=1}^{2001} (c_{jt} - \bar{c}_j)^2}} \tag{3.2}$$

is the sample correlation between signals from the i th and j th channel. The correlation, which is a number between −1 and 1, captures the linear relationship between the channels. If the correlation is close to 1, this would indicate the two channels are positively linearly related and "functionally connected". Figure 4 gives an example of a distance matrix for a sample trial.

With the metric now defined, we can associate the topological space $VR(C,\varepsilon)$ to our data, and then compute its homology using Javaplex.

In addition to the metric described above, we also implemented the naive Euclidean metric, treating each trial as 2001 points collected from a 27 dimensional space, endowed with the standard Euclidean distance.

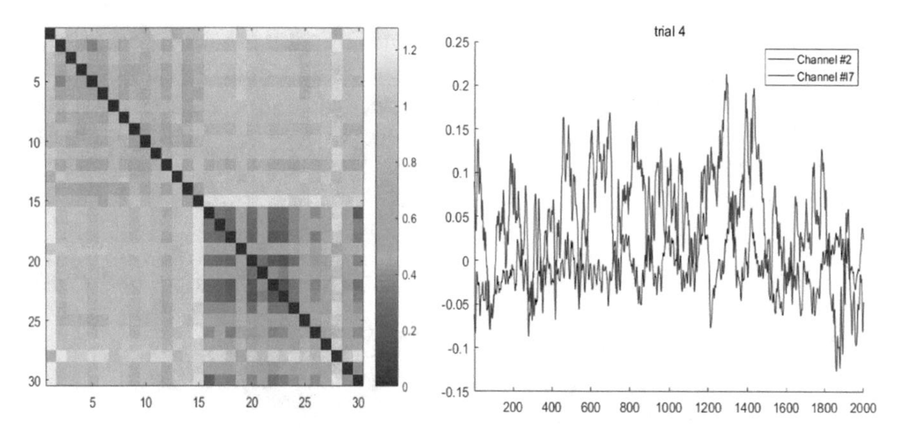

Fig. 4. *Left panel*: The distance matrix D for trial No. 4 - a single trial in a session where the rat sat passively while listening to 2 different beeps played in random order. Channels 1–15 are frontal channels, and channels 16–30 are parietal channels. *Right panel*: Signals from two channels in trial No. 4, a frontal channel #2 and a parietal channel #7. The horizontal axis shows time in milliseconds and the vertical axis shows the LFP voltage in millivolts.

4 Results

In order to test the potential of topological data analysis for understanding multi-channel LFP neural data, we compared the topology of mu trials, exhibiting a high-amplitude rhythmic 10 Hz oscillation, to the topology of relatively desynchronized non-mu trials. Examples of the two LFP states are shown in Fig. 3.

We take Trial 4, whose distance matrix and channels #2 and #7 are shown in Fig. 4, as an example to illustrate our correlation-based topological analysis. We obtained $\beta_0 = 2$ and $\beta_1 = 1$ as the only nontrivial Betti numbers for this trial. Topologically, this means that the data has two connected components and that one of the components has a 1-dimensional hole, or an essential circle that cannot be shrunk within the data cloud. Because the distance we defined arises from channel correlations, we believe the two connected components correspond to the two brain areas - the frontal and the parietal area.

We first used the correlation distance to analyze all 262 trials, and examine the resulting β_0 from the two groups. Unfortunately this metric turned out to be not revealing in distinguishing between mu and non-mu trials. We ran a Wilcoxon rank-sum test on the β_0's from the 262 trials to test the hypothesis that the two populations has the same distribution. This nonparametric test has a p-value of 0.0025, which means we can reject the null hypothesis at the 95% confidence level. We also ran a Student-t test (dof = 261) comparing the mean β_0 in each group. It returned a p-value of 0.001, supporting that the means are significantly different.

Although these differences are statistically significant due to the large number of trials, the differences are subtle. For example, Fig. 5 shows that knowing a trial's β_0 would not be sufficient to reliably predict whether it was a mu or non-mu trial. The distance based on correlations reduces size of the data from 27×2001 to a 27 by 27 distance matrix. Namely, the distance is summarizing all the information from time series data with rich structures into pairwise correlations, and this is possibly one reason why we observed only low-dimensional topological structure from the resulting Vietoris-Rips complex. This compression of the LFP information appears to be obscuring all the potential topological insight, and this is why we also tried the Euclidean metric.

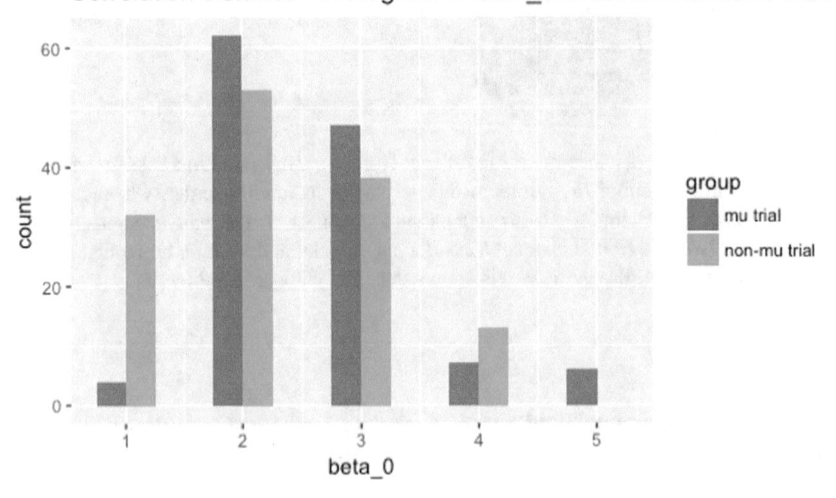

Fig. 5. Histogram of β_0 based on the correlation-metric defined in [7]. The *red bars* show the mu trials, and the zero Betti numbers have a mean β_0 of 2.60 with standard deviation 0.84. The *green bars* show the non-mu trials, and they have a mean β_0 of 2.24 with standard deviation of 0.92. The Wilcoxon rank-sum test has p-value equal to 0.0025, and the Student-t test comparing the two means has p-value equal to 0.001.

With the Euclidean metric, both trials in mu and non-mu group show larger β_0, which corresponds to number of connected components in the data cloud representing a trial. The histogram of these β_0's is shown in Fig. 6. The mu group has an average β_0 of 8.40 and standard deviation 2.32. The non-mu group has an average β_0 of 19.71 and standard deviation 5.67. The Wilcoxon rank-sum test has p-value equal to 9.5×10^{-36}, which suggests the two populations have different distribution and that the Euclidean metric can indeed be used as a way to detect difference in topological structures in mu and non-mu trials. The Student-t test has p-value 1.5×10^{-51}, so we can clearly reject the null hypothesis of equal means. Our findings suggest that the data from the mu trials "clusters" more, in the sense that it forms fewer separate connected components.

We also calculated β_1 for each trial, which is number of essential holes in the point-cloud data. Unfortunately this is not as illuminating as the β_0 data in terms of detecting mu trials: 19 out of 126 mu trials have β_1 equal to 1 and, for the non-mu trials, 2 out of 136 have $\beta_1 = 1$ and one has $\beta_1 = 2$.

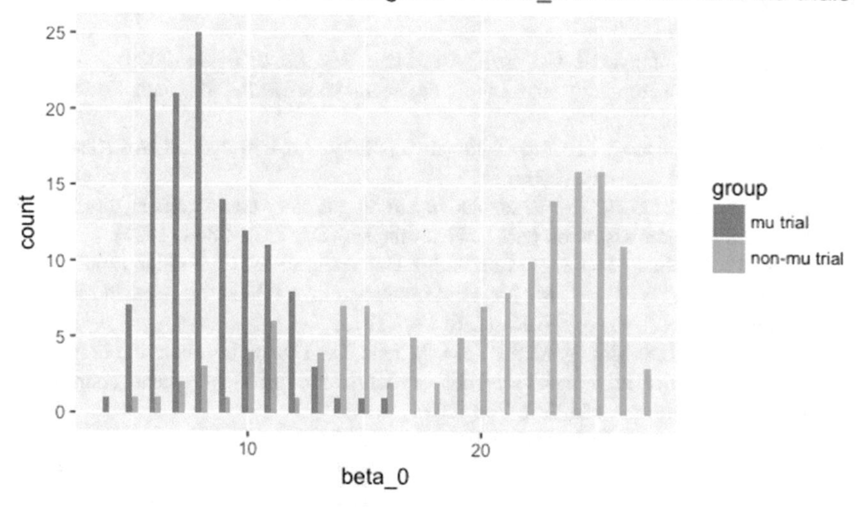

Fig. 6. Histogram of β_0 based on Euclidean distance. The *red bars* show the mu trials, with mean 8.40 and standard deviation 2.33. The *green bars* show the non-mu trials, with mean 19.71 and standard deviation 5.67. The Wilcoxon rank-sum test comparing the two groups has a p-value of 9.5×10^{-36}, which mean we can safely reject the hypothesis that the two populations are from the same distribution. The Student-t test comparing the means returns the p-value of 1.5×10^{-51}.

5 Conclusion

In order to test whether a topological analysis can capture differences between distinct brain states as measured by LFPs in awake rats, we compared Betti numbers for segments of multichannel LFP data recorded during an oscillatory "mu" state and a relatively desynchronized "non-mu" state. A Euclidean-based analysis found Betti-zero numbers in the mu state less than half their values in the non-mu state (Fig. 6), reflecting greater clustering of the data cloud in the non-mu state, and supporting that topological analysis can detect functional states of the brain in multichannel LFP data.

In the future, we would like to apply topological data analysis to more sessions and explore other metrics for defining simplicial complexes. It will be interesting to see whether the Betti numbers can capture more subtle functional differences in brain state than those we examined in this study, and whether higher-order Betti numbers can also be useful for distinguishing functional brain states.

Acknowledgments. The authors would like to thank the Wellesley College Science Center Summer Research Program and the Brachman-Hoffman Fellowship. Ismar Volić would also like to thank the Simons Foundation for its support. Michael Wiest's work was supported by National Science Foundation Integrative Organismal Systems grants 1121689 and 1353571.

References

1. Carlsson, G.: Topology and data. Bull. Am. Math. Soc. **46**, 255–308 (2009)
2. Curto, C.: What can topology tell us about the neural code? Bull. Am. Math. Soc. **54**, 63–78 (2016)
3. Edelsbrunner, H., Harer, J.: Computational Topology: An Introduction. American Mathematical Society, Providence (2009)
4. Fontanini, A., Katz, D.B.: 7 to 12 Hz activity in rat gustatory cortex reflects disengagement from a fluid self-administration task. J. Neurophysiol. **93**, 2832–2840 (2005)
5. Hatcher, A.: Algebraic Topology. Cambridge University Press, Cambridge (2001)
6. Imada, A., Morris, A., Wiest, M.: Deviance detection by a P3-like response in rat posterior parietal cortex. Front. Integr. Neurosci. **6**, 127 (2013)
7. Khalid, A., Kim, B.S., Chung, M.K., Ye, J.C., Jeon, D.: Tracing the evolution of multi-scale functional networks in a mouse model of depression using persistent brain network homology. NeuroImage. **101**, 351–363 (2014)
8. Munkres, J.: Topology, 2nd edn. Pearson, London (2000)
9. Nicolelis, M.A., Baccala, L.A., Lin, R.C., Chapin, J.K.: Sensorimotor encoding by synchronous neural ensemble activity at multiple levels of the somatosensory system. Science **268**, 1353–1358 (1995)
10. Nicolelis, M.A., Fanselow, E.E.: Thalamocortical [correction of Thalamcortical] optimization of tactile processing according to behavioral state. Nat. Neurosci. **5**, 517–523 (2002)
11. Polack, P.O., Charpier, S.: Intracellular activity of cortical and thalamic neurons during high-voltage rhythmic spike discharge in Long-Evans rats in vivo. J. Physiol. **571**, 461–476 (2006)
12. Rodgers, K.M., Dudek, F.E., Barth, D.S.: Progressive, seizure-like, spike-wave discharges are common in both injured and uninjured sprague-dawley rats: implications for the fluid percussion injury model of post-traumatic epilepsy. J. Neurosci. **35**, 9194–9204 (2015)
13. Shaw, F.Z.: 7–12 Hz high-voltage rhythmic spike discharges in rats evaluated by antiepileptic drugs and flicker stimulation. J. Neurophysiol. **97**, 238–247 (2007)
14. Tausz, A., Vejdemo-Johansson, M., Adams, H.: JavaPlex: a research software package for persistent (Co)homology. Software (2011). http://code.google.com/javaplex
15. Vergnes, M., Marescaux, C., Depaulis, A., Micheletti, G., Warter, J.M.: Spontaneous spike and wave discharges in thalamus and cortex in a rat model of genetic petit mal-like seizures. Exp. Neurol. **96**, 127–136 (1987)
16. Wiest, M.C., Nicolelis, M.A.: Behavioral detection of tactile stimuli during 7–12 Hz cortical oscillations in awake rats. Nat. Neurosci. **6**, 913–914 (2003)

Benford's Law and Sum Invariance Testing

Zoran Jasak(✉)

NLB Banka d.d., Sarajevo, Bosnia and Herzegovina
zoran.jasak@nlb.ba

Abstract. Benford's law is logarithmic law for distribution of leading digits formulated by P[D = d] = log(1 + 1/d) where d is leading digit or group of digits. It's named by Frank Albert Benford (1938) who formulated mathematical model of this probability. Before him, the same observation was made by Simon Newcomb. This law has changed usual preasumption of equal probability of each digit on each position in number. One of main characteristic properties of this law is sum invariance. Sum invariance means that sums of significand are the same for any leading digit or group of digits. Term 'significand' is used instead of term 'mantissa' to avoid terminological confusion with logarithmic mantissa.

1 Introduction

In article Note on the Frequency of use of different digits in natural numbers (Am J Math **4**(1):39–40, 1881) Simon Newcomb asserted *That the ten digits do not occur with equal frequency must be evident to any one making much use of logarithmic tables, and noticing how much faster the first pages wear out than the last ones. The first significant figure is oftener 1 than any other digit, and the frequency diminishes up to 9.* Newcomb did not give mathematical explanation of this observation, just relative frequencies which were verified later [1].

The same phenomenon was re-discovered by Benford (1938) [2] who gave the mathematical formulation:

$$P[D = d] = \log_{10}\left(1 + \frac{1}{d}\right) \tag{1}$$

This law is presented on Fig. 1.

He named this phenomenon by "Law of Anomalous number" because he asserted that "…*An analysis of the numbers from different sources shows that the numbers taken from unrelated subjects, such as a group of newspaper items, show a much better agreement with a logarithmic distribution than do numbers from mathematical tabulations or other formal data. There is here the peculiar fact that numbers that individually are without relationship are, when considered in large groups, in good agreement with a distribution law*".

For a long time this was treated just as curiosity. This law is a theoretical challenge from many theoretical and practical aspects and considered as unsolved problem [3].

S. Avdaković (Ed.): IAT 2018, LNNS 59, pp. 13–21, 2019.
https://doi.org/10.1007/978-3-030-02574-8_2

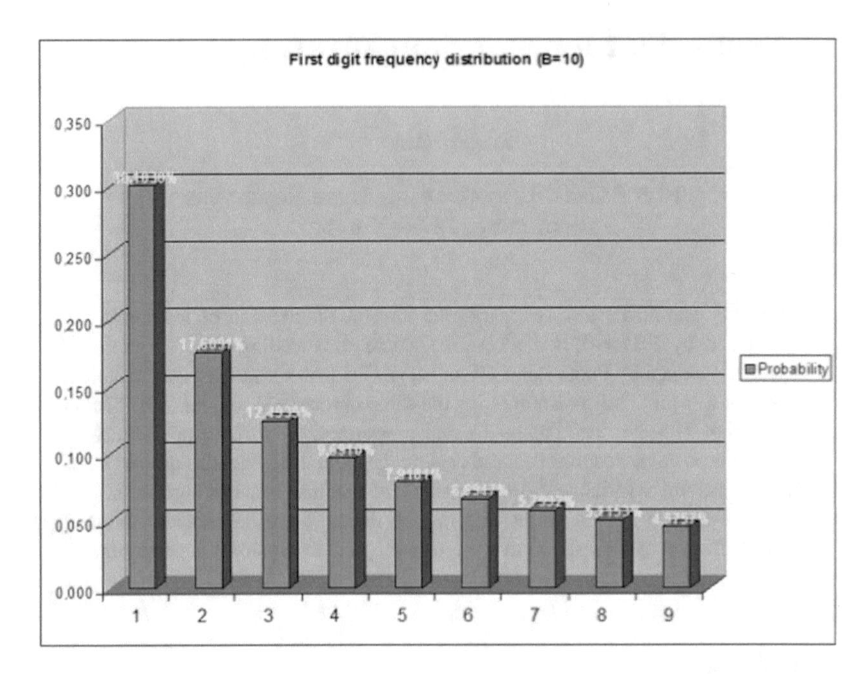

Fig. 1. Probabilities of leading digits for base 10

It's difficult to find an area in which this law cannot be applied. One of the most frequent use of this law is fraud detection. Basic premise is that is difficult to simulate numbers in ordinary unmanipulated processes which follows Benford's law exactly.

Exponential form of real number x in base B is:

$$x = S(x) \cdot b^m, m \in Z \tag{2}$$

The original word used to describe the coefficient $S(x)$ of floating-point numbers is *mantissa*. This usage remains common in computing and among computer scientists. However, this use of the word *mantissa* is discouraged by the IEEE floating point standard committee and by some professionals such as W. Kahan and D. Knuth because it conflicts with the pre-existing usage of mantissa for the fractional part of a logarithm. New term is *significand*.

Formal definition of significand is formulated by Berger and Hill [4].

Definition. The (decimal) significand function $S : R \to [1, 10)$ is defined as follows: if $x \neq 0$ then $S(x) = t$ where t is the unique number in $[1, 10)$ with $|x| = 10^k t$ for some (necessarily) unique $k \in Z$; if $x = 0$ then $S(x) = 0$.

2 Invariances

One of the most interesting properties of Benford's law are base, scale and sum invariance.

Base invariance means that the probabilities of leading digits have logarithmic law in any base $b \geq 2$. Mathematical formulaton of this property is:

$$P[D_1 = d|b] = \log_b\left(1 + \frac{1}{d}\right) = \frac{\log\left(1 + \frac{1}{d}\right)}{\log b} \quad (3)$$

In Fig. 2 the theoretical probabilities for bases 2 to 10 are presented.

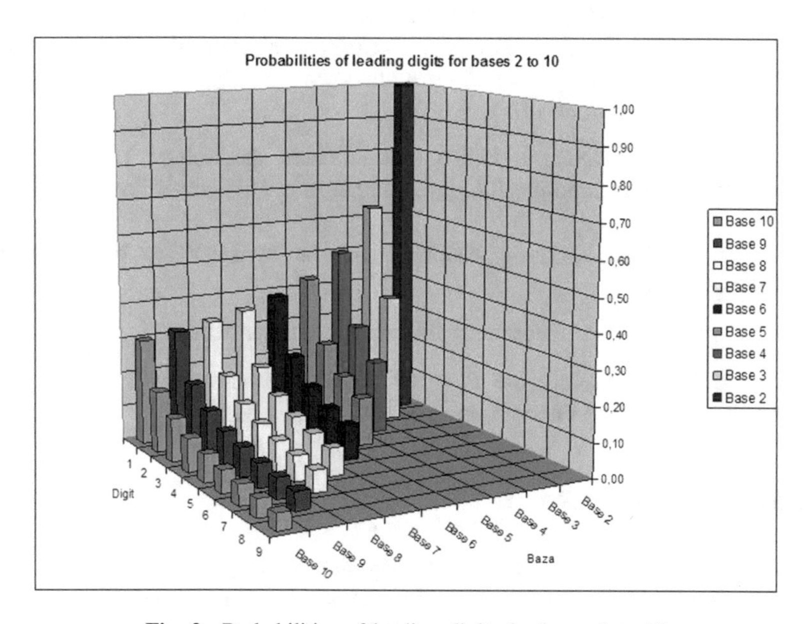

Fig. 2. Probabilities of leading digits for bases 2 to 10

Hill proved [5] that base invariance implies Benford's law.

Scale invariance means that probabilities of leading digits have the same probabilities if whole sample is multiplied by one positive number $\alpha \neq b^k, k \in Z$. This property has practical importance. It's possible to check, for example, do numerical values come from (un)manipulated source.

Hill [5] proved that scale invariance implies base invariance.

Idea of **sum invariance** is presented by Mark Nigrini who asserted in his Ph.D thesis (1992) that tables of unmanipulated accounting data closely follow Benford's Law and that sufficient long list of data for which BL holds the sum of all entries with leading digit d is constant for various d. Nigrini, in his book [6], calculated integral for $a \cdot r^{x-1}$ between leading digits ft and $ft + 1$; result doesn't depend of leading digits and he concluded that sum must be equal.

Extension of this observation can be stated for k-tuples of leading digits, which is called *sum invariance* property of Benford's law.

Formal definition of sum invariance is given by Berger and Hill [4, p. 61].

Definition. *A sequence* $\{x_n\}$ *of real numbers has sum invariant significant digits if, for every* $m \in N$, *the limit*

$$\lim_{N\to\infty} \frac{\sum_{n=1}^{N} S_{d_1,\ldots,d_m(x_n)}}{N}$$

exists and is independent of $d_1, \ldots, d_m$.

Here $S_{d_1,\ldots,d_m(x_n)}$ is significand with $d_1, \ldots, d_m$ as a leading digits.

Analytical tools relying on Benford's law are primarly oriented to analysis of frequencies. Sum invariance is interesting from practical point of view because it can be very efficient additional tool in all such analyses.

Some facts are important [7]:

- Significands of numbers in tables, not numbers themselves, must be added. Otherwise, single astronomically large number in a table would dominate all other sums;
- word 'constant' in Nigrini's statement can be translated to be 'constant in expectation'.

Pieter C. Allart proved theorem of this empirical observation.

Theorem. A probability measure P on (R^+, B) is sum-invariant if and only if its corresponding significand distribution P_S is Benford's law.

3 Results and Discussion

In my research idea is to investigate sum invariance not only for leading digits but for any k-tuple of consecutive digits inside the number and to propose testing method.

Sum invariance property can be extended for second, third, ... digit. In another words, in sample which follows Benford's law *sums of significands having same digits (or group of digits) on the same positions are the same*. There is no limitations on leading digits only. Another interpretation is that sum of significands for first digit d is $1/9$ of total sum of significands in sample.

Null hypothesis is: H_0: *Sum of significands for groups of consecutive digits are the same.*

Main problem in testing is to estimate expected sums of significands.

If we have sample in size of N elements, theoretical frequency for every digit d is given by

$$n_d = N \cdot \log_{10}\left(1 + \frac{1}{d}\right) \tag{4}$$

Sum invariance means that there is number T_d which is the sum of significands beginning with digit:

$$T_d = \sum_{i=1}^{n_d} S_i(d)$$

Here $S_i(d)$ is the significand of i-th numerical value having d as leading digit. Dividing this relation by n_d we have the average significand (arithmetic mean) for group of significands, denoted by $\overline{S(d)}$. This is analogue of the actual mean defined by Dumas and Devine [8, p. 16]:

$$AM = \frac{1}{N}\sum X_{collapsed}$$

where $X_{collapsed}$ is defined by

$$X_{collapsed} = \frac{10 \cdot X}{10^{int(\log_{10} X)}}$$

With accuracy of five digits the smallest and the biggest average significands for numbers beginning by digit 9 are 9.00000 and 9.99999 respectively. It's possible from this, by reccurence, to get smallest and biggest significands for other leading digits, denoted by S_{min} and S_{max} in Table 1. The same calculation can be conducted for groups of leading digits of any size.

Table 1. Theoretical minimal, average and maximal significands for one leading digit

Digit	S_Min	Average	S_Max
1	1.36803	1.44270	1.52003
2	2.33866	2.46630	2.59891
3	3.29615	3.47606	3.66239
4	4.24948	4.48142	4.72164
5	5.20095	5.48481	5.77882
6	6.15141	6.48716	6.83490
7	7.10129	7.48888	7.89031
8	8.05077	8.49019	8.94530
9	9.00000	9.49122	9.99999

Sum invariance is based on one interesting property of logaritmic curve [10]. If interval $[1, 10)$ is divided in subintervals of equal size, areas of curvilinear rectangle bounded by lines:

$$y = \log_{10} x,\ x = 0,\ l_3 = \log_{10}\left(1 + \frac{1}{d}\right),\ l_4 = \log_{10}\left(1 + \frac{1}{d+1}\right),$$

are equal. Lines l_3 and l_4 are Benford's probabilities and d are digits 1, 2, …, 9. Next theorem is very important [10].

Theorem. *A probabilistic measure P for Benford's law is sum invariant if and only if $[B^{k-1}, B^k)$ is divided on n subintervals of equal size.*

Digits 1 to 9 are one of ways in which interval *[1,9)* can be divided on subintervals of equal size. We can do it with any other interval $[B^{k-1}, B^k)$, where *B* is base.

Natural idea for sum invariance is to use average significands. They can be easely calculated by [10]:

$$x = \frac{\log_{10} e}{\log_{10}\left(1 + \frac{1}{d}\right)}$$

Where *d* is digit 1, 2, …, 9. This formula we got by use of mean value theorem for logarithmic curve on intervals *[d, d + 1)*. Average significands for leading digits are in Table 1. It can be easly verified that average significands are harmonic averages of minimal and maximal significands.

Theoretical sum of significands is proposed by use of formula [10]:

$$T_1 = 9 \cdot N \cdot \left(\sum_{d=1}^{9} \frac{1}{\bar{S}(d)}\right)^{-1}$$

Main reason for such proposal is that is not regular to use arithmetic but harmonic means. Adequacy of such approach is verified in [10].

This formula means that the sum of significands for one leading digit $T_1/9$ can be found if the sample size is multiplied by the harmonic mean of average significands, denoted here by $\bar{S}(d)$.

Expected sum of significands having the same leading digits can be found if we multiply average significand for this group by number of such significands. This formula is for 9 leding digits but it's can be easy extended for 90 two first digits, 10 s digits, 100 digits on the second and third position etc. By use of maximal and minimal

Table 2. Calculation of values from sample

Dig	Counts	Sam_Per	Sums	Av_Sig
1	4.047	0,35234	5.577,88735	1,37828
2	1.747	0,15210	4.019,78036	2,30096
3	1.222	0,10639	4.047,61105	**3,31228**
4	997	0,08680	4.282,58452	4,29547
5	921	0,08018	4.765,26876	5,17402
6	721	0,06277	4.520,78251	6,27016
7	623	0,05424	4.531,63929	7,27390
8	639	0,05563	5.326,60358	8,33584
9	569	0,04954	5.309,84965	9,33190

average significands from Table 2 we have lower and upper limit for sums. Same formula is used to calculate sums of significands, which is needed for testing purposes.

Expected sum of significands having the same digits on second position is 9/10 of sum on first position, namely [10]:

$$T_2 = \frac{9}{10} \cdot T_1$$

In this way we have adequate tools for testing of sum invariance property.

4 Sum Invariance Testing

Main goal for practicioners is to test sum invariance property. In other words, it's task is to investigate if there is any discrepancy between theoretical and sample sums of significands.

My proposal is to use f-divergence for testing sum invariance property. Divergence measures play an important role in statistical theory, especially in large sample theories of estimation and testing [9]. The underlying reason is that they are indices of statistical distance between probability distributions P and Q; the smaller these indices are, the harder is to discriminate between P and Q. Many divergence measures have been proposed since the publication of the paper of Kullback and Leibler [12].

In order to conduct a unified study of statistical properties of divergence measuers, Salicru, Morales, Prado and Menendez [9] proposed a generalized divergence which includes as particular cases other divergence measures. They proposed unified expression, called $(\underline{h}, \underline{\emptyset})$-divergence, as follows [9]:

$$D_{\underline{\emptyset}}^{\underline{h}}(\theta_1, \theta_2) = \int_{\Lambda} h_\alpha \left\{ \int_X f_{\theta_2}(x) \cdot \emptyset_\alpha \left(\frac{f_{\theta_1(x)}}{f_{\theta_2}(x)} \right) d\mu(x) - \emptyset_\alpha(1) \right\} d\eta(\alpha) \tag{5}$$

where $\underline{h} = (h_\alpha)_{\alpha \in \Lambda}$, $\underline{\emptyset} = (\emptyset_\alpha)_{\alpha \in \Lambda}$, $\emptyset_\alpha$ and h_α are real valued C^2 functions with $h_\alpha(0) = 0$ and η is σ-finite measure on the measurable space (Λ, β).

Let X be a random variable denoting the quotient between sum of significands for one leading digit and total sum of significands so we test the hypothesis that X has a uniform discrete distribution with probabilities $1/9$. Test statistic derived from (5) is [11]:

$$T_2 = 36 \cdot \left[9 \cdot \left(\sum_{i=1}^{9} \widehat{p_i}^{0,5} \right)^{-2} - 1 \right]$$

This statistic is used for first leading digits. Here $\widehat{p_i}$ denotes sample quotient between sum of significands for one digit and total sum of digits. Analogue statistic are derived for first two digits and for second digits [10].

This statistic, for $n = 9$ digits, has χ_8^2-distribution, what is described in [9], with appropriate statistical tables.

Advantage of this procedure is additivity of statistic T_2. We can make choice of groups of digits we want to test if we want intentionally exclude some digits or we are dealing with process which produces numbers with specific leading digits. The only condition is that we need at least two different digits in our sample.

This method is demonstrated on a sample of size of 11,486 elements. Minimal sample value is 10, maximal value is 176,932.50, average is 3,606.00, standard deviation is 7,793.29, total sum of all values is 41,418,526.12. All calculations are made on $\alpha = 0.05$ significance level. Table 2 presents these calculations.

In column DIG are leading digits, in column COUNTS are sample frequencies for every digit, in column SAM_PER are sample relative frequencies, in column SUMS are sample sums of significands for every digit and in column AV_SIG are average significands for every digit. Total sample sums of significands are 42382.00706 for first digits.

In Table 3 calculation of test statistic is presented.

Table 3. Calculation of test statistic

Rat_Th	Rat_Sa	pi*qi	Sqrt(AE)
0,11111	0,13161	0,01462	0,12092688
0,11111	0,09485	0,01054	0,10265714
0,11111	0,09550	0,01061	0,10301189
0,11111	0,10105	0,01123	0,10595976
0,11111	0,11244	0,01249	0,11177166
0,11111	0,10667	0,01185	0,10886663
0,11111	0,10692	0,01188	0,10899728
0,11111	0,12568	0,01396	0,11817162
0,11111	0,12529	0,01392	0,11798563

In column RAT_TH are quotients of theoretical sums for leading digits and total theoretical sum of digits. As it's expected, all quotients are 1/9. In column RAT_SA are quotients of sample sums for leading digits and total sample sum of digits. In column $p_i \cdot q_i$ are products of quotients RAT_TH and RAT_SA. In next column, SQRT(*), are square roots of product $p_i \cdot q_i$.

Value of statistics T_2 in this case is $T_2 = 0.11920$. Critical region corresponds to probability

$$P\left[|T_2| \geq \chi^2_{\frac{\alpha}{2};8}\right] = \alpha$$

For $\alpha = 0.05$ we have intervals (0; 2.1797307) and (17.53454614; $+\infty$). According to this we have no reason to accept hypothesis. It means that sums of significands in sample are not equaly distributed.

5 Conclusions

In this text testing of sum invariance of Benford's law is presented. My proposal is to use average significands and additional method for calculating of expected sums of significands. Using of f-divergence as a test procedure has some big advantages like additivity proerty.

Acknowledgments. I wish to thank to Mr. Wilhelm Schappacher for great support in my work.

References

1. Newcomb, S.: Note on the frequency of use of different digits in natural numbers. Am. J. Math. **4**, 39–40 (1881)
2. Benford, F.A.: The law of anomalous numbers. Proc. Am. Philos. Soc. **78**, 551–572 (1938)
3. Strauch, O.: Unsolved problems. Tatra Mt. Math. Publ. **56**(3), 175–178 (2013)
4. Berger, A., Hill, T.P.: Theory of Benford's Law. Probab. Surv. **8**, 1–126 (2011). https://doi.org/10.1214/11-ps175. ISSN 1549-5787
5. Hill, T.P.: Base invariance implies Benford's Law. Proc. Am. Math. Soc. **123**(3), 887–895 (1995)
6. Nigrini, M.: Forensic Analytics – Methods and Techniques for Forensic Accounting Investigations, pp. 144–146. Wiley, Hoboken (2011)
7. Allart, P.C.: A Sum-invariant Charcterization of Benford's Law. AMS (1990)
8. Dumas, C., Devine, J.S.: Detecting evidence of non-compliance in self-reported pollution emissions data: an application of Benford's law. Selected Paper American Agricultural Economics Association Annual Meeting Tampa, Fl, 30 July–2 August 2000
9. Salicru, M., Morales, D., Menendez, M.L., Pardo, L.: On the Application of Divergence Type Measures in Testing Statistical Hypotheses. J. Multivar. Anal. **51**, 372–391 (1994)
10. Jasak, Z.: Sum invariance testing and some new properties of Benford's law, Doctorial dissertation. University in Tuzla, Bosnia and Herzegovina (2017)
11. Jasak, Z.: Benford's law and invariances. J. Math. Syst. Sci. **1**(1), 1–6 (2011). (Serial No.1). ISSN 2159-5291
12. Kullback, S., Leibler, R.A.: On information and sufficiency. Ann. Math. Stat. **22**(1), 79–86 (1951)

Using Partial Least Squares Structural Equation Modeling to Predict Entrepreneurial Capacity in Transition Economies

Matea Zlatković(✉)

Faculty of Economics, University of Banja Luka, Banja Luka, Bosnia and Herzegovina
matea.zlatkovic@ef.unibl.org

Abstract. Many theoretical and empirical studies indicate the significant influence of environmental challenges and characteristics on entrepreneurship. Drawing insights from this research, this paper defines the structural model to analyze synergistic influences of certain elements of Entrepreneurial Factor Conditions on the entrepreneurial capacity in Slovenia and Bosnia and Herzegovina. The analyzed structural model consists of three environmental dimensions – entrepreneurial education and training, cultural and social norms and research and development, and higher-order construct entrepreneurial capacity as a final target dependent variable. Partial Least Squares Structural Equation Modeling analyzed relationships between chosen variables. The obtained results indicate the highest significance of the cultural and social norms of entrepreneurial capacity in both countries. Entrepreneurial education and training does not have the direct effect on entrepreneurial capacity in factor-driven Bosnia and Herzegovina's economy which suggests that education programs are insufficiently extended with necessary tools for starting and managing the new business. Research and development has an important role in entrepreneurial capacity in both countries because as it yields innovation as a generator of ideas for new business and technological changes creating new opportunities for entrepreneurship activities.

1 Introduction

Countries of varying degrees of development differ in terms of overall social, political and cultural trends reflected in the entrepreneurial behavior of the population as well as on the scale and structure of entrepreneurial endeavors. The level of economic development directly influences entrepreneurial conditions and the environment as the basic preconditions of entrepreneurial behavior. In addition to the personal traits, skills and motivations of individuals, entrepreneurial behavior depends on the availability of entrepreneurial capital, government programs, and policies, physical infrastructure, entrepreneurship education etc.

The conceptual model of the entrepreneurial environment presented in the Global Entrepreneurship Monitor (GEM) is in all segments supported by the views of the classical Austrian economic school. The model encompasses general national conditions affecting business activities such as institutions, macroeconomic stability,

S. Avdaković (Ed.): IAT 2018, LNNS 59, pp. 22–35, 2019.
https://doi.org/10.1007/978-3-030-02574-8_3

infrastructure, education, technology availability, the market as well as specific entrepreneurial conditions that encompass government policy, programs, funding, and market openness.

In addition to affecting business activities, institutions also influence the creation of a motivational structure in society and decision-making related to entrepreneurial endeavors. Formal institutions include politics, judiciary and bureaucracy, and informal expectations, norms and social networks [21].

This research aims to carry out an empirical review of the impact of the size of the entrepreneurial framework on entrepreneurial capacity, based on the GEM data for two transition economies of varying degrees of development, Bosnia and Herzegovina and Slovenia, using the PLS structural equation model.

The contribution of this research is the application of the PLS model of structural equations for GEM data analysis on the example of transition countries. Similar studies in this context, according to the author's knowledge, have not been made. There are not too many researchers that used a similar methodology for examples of developed countries [16, 22, 28].

2 Literature Review and Development of Hypothesis

The entrepreneurial activity varies considerably between countries with a lower level of development, higher degree of inequality and high unemployment rates on one hand, and highly developed countries on the other [1, 2]. According to some authors, developed countries need to improve their supportive policies for start-ups and the promotion of commercial exploitation of scientific achievements and, in particular, the link between higher education and industry, while developing countries should use the economy of scale, encourage direct foreign investment, improve education and reduce corruption [3, 30].

In this paper, the variables used in the previous researches [16, 23, 28] are used in the study of the influence of the entrepreneurial framework on entrepreneurial capacity (ENTCAP) in selected economies. One of the variables that describe the entrepreneurial framework is entrepreneurial education and training (EET) that differs from general education [16]. Entrepreneurial education and training has the role of educating the population for entrepreneurial endeavors. The second variable is the transfer of research and development (R&D). According to many studies, the development of new technologies contributes to the improvement of business processes and the creation of new business opportunities [7, 29]. The third variable of the entrepreneurial framework refers to cultural and social norms (CSN). It assesses the impact of social values and attitudes on the entrepreneurial commitment to the individual. Entrepreneurial capacity is represented by two variables: motivation (MOT) and skills (SKILLS), motivating individuals to launch business ventures and their skills to implement entrepreneurial initiatives.

There are a large number of studies on entrepreneurial education [15], some of which deal with program issues and how to implement this education [26]. Entrepreneurial knowledge and skills make them more capable of recognizing business opportunities and launching new jobs, increasing their productivity, and improving their productivity [8, 16]. Based on the exposition, the first hypothesis is set:

H1: Entrepreneurial education and training has a direct positive impact on entrepreneurial capacity in Slovenia and Bosnia and Herzegovina.

Entrepreneurial studies show that entrepreneurial initiative depends on the characteristics of the entrepreneurial character, its innovativeness, readiness to take risks and self-efficacy, skills and competencies, as well as family environment and growth [12]. In the culture of society, the perceptions that arise from collective learning reflect the cumulative experiences of societies that have been incorporated in all its pores over time [19]. The collective historical knowledge and experience of the society have a significant influence on the formation of present knowledge and formation of values and norms, through various forms of educational systems. Previous research is the basis for the next hypothesis:

H2: Entrepreneurial education and training has a direct positive impact on cultural and social norms in Slovenia and Bosnia and Herzegovina.

Knowledge-based economies attach great importance to linking the scientific and entrepreneurial sector, and this connection is considered to have a significant impact on their innovative potential. Innovative possibilities are considered from different perspectives: through the change of production function [25], the process of information [29], success of product innovation strategy [7], application of innovations. Innovation is considered to be a complex phenomenon that leads to generating new ideas [4]. The transfer of knowledge and technology between the scientific sector and the economy is a process through which scientific research and discoveries find practical application [25]. Based on the aforementioned research hypothesis is defined:

H3: Transfer of research and development has a direct positive impact on entrepreneurial capacity in Slovenia and Bosnia and Herzegovina.

Research into the impact of cultural and social norms on the entrepreneurship development indicate differences in the cultural models of different countries. Based on the observation of various institutional mechanisms, Hofstede identified four dimensions of culture that he considered important for entrepreneurial activity: distance from power, avoidance of uncertainty, individualism and masculinity [14]. Characteristics of the period and the place where the entrepreneur lives, the fact that some cultures encourage entrepreneurship more than others and the fact that business creation is largely a regional phenomenon play a significant role [18]. Some authors emphasize the positive relationship between three cultural dimensions: cognitive, regulators and normative and entrepreneurship. The degree of social acceptability, admiration, and respect of entrepreneurial activities is often the most important predictor of entrepreneurship [5, 27]. The following hypothesis emerges from the above-outlined:

H4: Cultural and social norms have a direct positive impact on entrepreneurial capacity in Slovenia and Bosnia and Herzegovina.

Based on the literature research and hypothesis set out in the empirical part of the paper, the model shown in the next picture will be considered (Fig. 1).

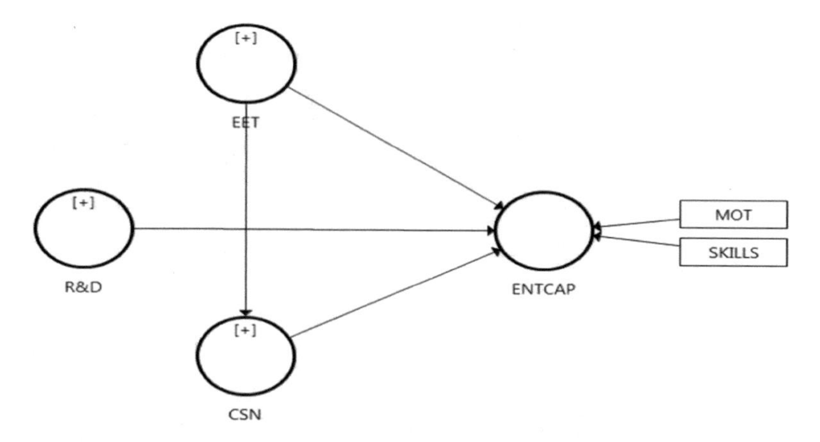

Fig. 1. Proposed research model

3 Research Methodology

To test structural model, data for Bosnia and Herzegovina and Slovenia are used from Global Entrepreneurship Monitor (GEM), particularly from the National Expert Survey (NES), for 145 and 143 experts from Slovenia and Bosnia and Herzegovina, respectively. These surveys consist of opinions of experts measured on the 5 points Likert scale where 1 equals totally false and 5 equals totally true (GEM, 2013). These data covered several years in the range from 2010 to 2013 for both countries. The full sample size is 280 cases. The size of the sample for each year is small so the data are polled across several years which provided the stability of different measures. The one-way ANOVA is employed to examine the existence of the differences in mean values of different measures in different years. The Leven's test of equality of variances, which was used for the assessment of homogeneity of variances across groups, indicated that mean values of examined constructs in the structural model did not differ significantly in observed period for analysis in two countries. Examined variables are not directly measurable so Partial Least Squares Structural Equation Modeling (PLS-SEM) is applied. The assessment of the structural model is performed in SmartPLS 3.2.7.

3.1 Theoretical Background of the PLS-SEM

Partial least squares present diagram of relationships between variables which have to be estimated in structural equation model analysis. Figure 2 displays an example of path model of latent constructs (Y_1, $Y_{2,}$ and Y_3 presented by circles) and its indicators called manifest variables or items (x_1, x_2, x_3 and etc. presented by rectangles). Path models consist of two models: structural and measurement model. Structural and measurement models are also called inner and outer models [17].

Original PLS algorithm [31] has three stages which consist of the estimation of weights and latent constructs scores through four steps in first stage, estimation of outer loadings/weights and path coefficients in the second stage, and estimation of location parameters in third stage. These four steps in the first stage are initialization, inner

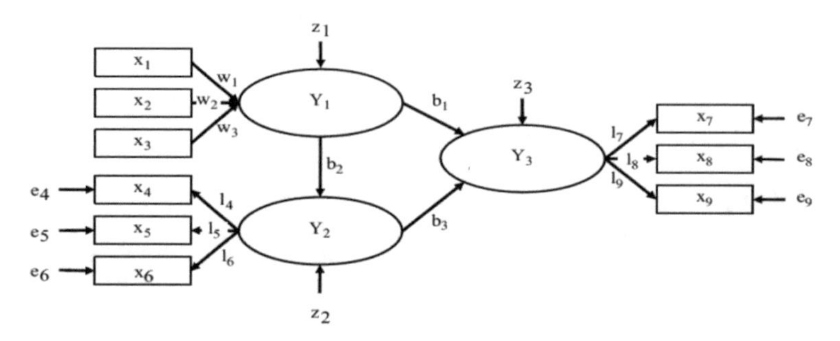

Fig. 2. An example of path model (Partial Least Squares Structural Equation Modeling) [24]

weights estimation, inside approximation, outer weights estimation and outside approximation. PLS algorithm starts with an initialization stage where preliminary latent variable scores are established, typically using unit weights for all indicators in the measurement models. This is performed in step 4, stage 1, called outside approximation [11]. Outside approximation calculates initial variable score in the following manner:

$$Y_{jn} := \sum\nolimits_{k_j} \tilde{w}_{k_j} x_{k_j n} \tag{1}$$

Stage 1 of the PLS algorithm through iterations determines inner weights and latent variable scores using four step procedure. The first step, in stage 1, uses obtained latent variables scores from initialization of the PLS algorithm to determine the inner weights b_{ji} (path coefficients) between adjacent dependent latent variables Y_j and independent latent variable Y_i. There are three possible ways to estimate inner weights [6, 17]: centroid, factor, and path weighting scheme. If the covariance between Y_j and Y_i *(cov* $(Y_{j,}\ Y_i))$ is positive, inner weights are set at +1 in centroid scheme and if the $cov(Y_{j,}\ Y_i)$ is negative then the inner weights are set at −1. In the factor weighting scheme, $cov(Y_{j,}\ Y_i)$ corresponds to inner weights and is set at 0 if the two latent variables are not connected. The direction of the relationships in the inner model is taken into consideration in case of path weighting scheme.

In the second step, inside approximation calculates proxies for latent variables in the following manner:

$$\tilde{Y}_j := \sum\nolimits_i b_{ji} Y_i \tag{2}$$

Further, in the third step, the outer weights estimation defines new outer weights/loadings which indicate the strength of the relation between every latent construct and its corresponding indicator. Therefore, there are two types of measurement models: mode A (reflective model) in which bivariate correlation between each indicator and corresponding latent construct is determined in the following manner:

$$\tilde{Y}_{jn} = \sum\nolimits_{k_j} \tilde{w}_{k_j} x_{k_j n} + d_{jn} \tag{3}$$

and mode B (formative model) in which indicators weights are computed by regressing every latent construct on its indicator:

$$x_{kjn} = \tilde{w}_{k_j} \tilde{Y}_{jn} + e_{k_j n} \quad (4)$$

Here, x_{kjn} presents raw data for indicator k $(k = 1, \ldots, K)$ of the latent construct j $(j = 1, \ldots, J)$ and observation n $(n = 1, \ldots, N)$, $\tilde{Y}_{jn}$ are scores of latent variables from the inside approximation in step 2, $\tilde{w}_{k_j}$ outer weights from step 3, d_{jn} is error term from bivariate regressions and $e_{k_j n}$ error term from multiple regression. The linear combination of the $\tilde{w}_{k_j}$ updated from step 3 and x_{kjn} is used to update latent variable scores in fourth step – outside approximation. PLS-SEM algorithm uses standardized data as input and standardizes latent variable scores in second and fourth steps. After step 4, following iteration begins and algorithm terminates when weights from step 3 change marginally from one to next iteration (until convergence is achieved $1 * 10^{-7}$) or when the maximum number of iterations is achieved (usually 300).

3.1.1 Measurement and Structural Model Evaluation

The outer model represents the measurement model of latent constructs. There exist two different ways to measure latent constructs called the reflective and formative way of measurement. Evaluation of the reflective model includes assessment of indicator reliability, internal consistency reliability, convergent validity and discriminant validity. Composite reliability (ρ_c) is used as a measure of the internal consistency reliability defined as:

$$\rho_c = \frac{\left(\sum_{k=1}^{K} l_k\right)^2}{\left(\sum_{k=1}^{K} l_k\right)^2 + \sum_{k=1}^{K} \mathrm{var}(e_k)} \quad (5)$$

where l_k presents standardized outer loadings of indicator k of the latent constructs which is measured by K indicators, e_k is a measurement error of the indicator k and $var(e_k)$ is a variance of the measurement error defined as $1-l_k^2$. Another measure of the internal consistency that is used is Cronbach's α.

Convergent validity represents the measure of the extent to which construct's variance is explained by its indicators. It is also called communality because it is assessed by average variance extracted (AVE) across all items associated with the observed latent construct. AVE is calculated by following formula:

$$AVE = \frac{\left(\sum_{k=1}^{K} l_k^2\right)}{K} \quad (6)$$

Discriminant validity determines whether observed construct is really, and in which extent, distinct from other constructs. A new criterion for discriminant validity assessment is heterotrait-monotrait ratio ($HTMT$) [13], defined by following formula:

$$HTMT_{ij} = \frac{1}{K_i K_j} \sum_{g=1}^{K_i} \sum_{h=1}^{K_j} r_{ig,jh} \div \left(\frac{2}{K_i(K_j - 1)} * \sum_{g=1}^{K_i-1} \sum_{h=g+1}^{K_i} r_{ig,ih} * \frac{2}{K_j(K_j - 1)} * \sum_{g=1}^{K_j-1} \sum_{h=g+1}^{K_j} r_{jg,jh} \right)^{\frac{1}{2}} \quad (7)$$

where $r_{ig,\, jh}$ indicate correlations of the indicators. *HTMT* values close to 1 indicate a lack of discriminant validity. Using bootstrapping procedure it can be tested whether *HTMT* is significantly different from 1 ($HTMT_{inference}$).

Assessment of the formative model includes: convergent validity, assessment of the collinearity issues and assessment of the significance and relevance of the formative indicators weights.

Convergent validity is determined by redundancy analysis [6] which uses a formative measured construct as exogenous latent variable predicting the same constructs measured by a reflective indicator or a single item. Colinearity assessment uses the variance inflation factor (*VIF*) which can be computed by the formula:

$$VIF_k = \frac{1}{1 - R^2_{x_k}} \quad (8)$$

where $R^2_{x_k}$ is the coefficient of determination of the *k*-th regression. The following step in the formative model assessment is to identify statistical significance and relevance of indicators' weights. The bootstrapping procedure must be conducted. If the weight's confidence interval does not include 0, it is not significant, which means that the indicator should be considered for removal (also check loading of indicator).

The following step is to estimate the structural model. First, to assess collinearity tolerance, *VIF* measures need to be applied. The primary criterion for assessment of the structural model is the coefficient of determination (R^2). The change in the R^2 values is presented by f^2 effect size. Next, blindfolding procedure is used to assess model's predictive relevance and Q^2 is obtained. The change of the Q^2 value is presented by q^2 effect size. The PLS predict procedure uses training and holdout sample to generate and evaluate predictions from PLS model estimations. The significance of the path coefficient is determined in similar ways as the significance of the formative indicators' weights by bootstrapping procedure.

4 Results and Discussion

4.1 Assessment of Measurement Model

Assessing the measurement model for reflective indicators in PLS is based on individual item reliability, construct reliability and discriminant validity in both countries, Bosnia and Herzegovina and Slovenia. The proposed measurement models rely on reflective indicators. In this study, most of the reflective indicators have loadings above or very near 0.7 (Table 1), although very few individual items are below the accepted threshold and, despite that, they are retained because values of composite reliability and average variance extracted for all reflective constructs are above the threshold.

Table 1. Indicator reliability for Bosnia and Herzegovina and Slovenia, respectively

EET	B&H	Slovenia	R&D	B&H	Slovenia	CSN	B&H	Slovenia
D01	0.615	0.729	E01	0.799	0.624	I01	0.845	0.824
D02	0.7	0.847	E02	0.722	-	I02	0.909	0.839
D03	0.705	0.804	E03	0.717	-	I03	-	0.847
D04	0.834	0.759	E04	0.673	0.654	I04	0.834	0.849
D05	0.842	0.695	E05	0.74	0.729	I05	0.697	0.721
D06	0.678	0.713	E06	0.764	0.846			

In order to assess construct validity of reflective measurement models both convergent validity and discriminant validity are used. Construct reliability is measured by the composite reliability and Cronbach's α which are above recommended acceptable threshold 0.7 [10, 20] (Table 2). The average variance extracted (*AVE*) is used as a criterion of convergent validity. Values of *AVE* are above threshold of 0.5 so the convergent validity of latent constructs is established (Table 2).

Table 2. Internal consistency – construct reliability and convergent validity for Bosnia and Herzegovina and Slovenia, respectively

	Cronbach's α	rho_A	Composite reliability	Average variance extracted (AVE)
Bosnia and Herzegovina				
CSN	0.892	0.900	0.921	0.702
EET	0.842	0.898	0.874	0.538
ENTCAP	-	1	-	-
R&D	0.831	0.838	0.877	0.543
Slovenia				
CSN	0.875	0.876	0.910	0.669
EET	0.853	0.861	0.891	0.577
ENTCAP	-	1	-	-
R&D	0.683	0.706	0.808	0.516

In order to establish discriminant validity, the heterotrait-monotrait (*HTMT*) ratio is used. The obtained values of *HTMT* are below predefined threshold (0.85 as conservative criterion and 0.9 as liberal criterion) and discriminant validity is established. In addition, bootstrapping procedure with 5.000 sub-samples is used to determine $HTMT_{inference}$. *HTMT* values close to 1 indicate a lack of discriminant validity (Table 3).

Table 3. Heterotrait-monotrait ratio and bootstrapping procedure for assessing $HTMT_{inference}$ for Bosnia and Herzegovina and Slovenia, respectively

Bosnia and Herzegovina									
	CSN	EET	R&D		Original sample (O)	Sample mean (M)	Bias	2.50%	97.50%
CSN				EET → CSN	0.272	0.296	0.023	0.125	0.465
EET	0.272			R&D → CSN	0.472	0.475	0.003	0.307	0.616
R&D	0.472	0.371		R&D → EET	0.371	0.388	0.017	0.195	0.551
Slovenia									
CSN				EET → CSN	0.556	0.557	0.001	0.379	0.698
EET	0.556			R&D → CSN	0.39	0.415	0.025	0.224	0.533
R&D	0.39	0.52		R&D → EET	0.52	0.542	0.022	0.362	0.655

4.2 Assessment of Structural Model

Further, structural models for both countries (Bosnia and Herzegovina and Slovenia) were assessed. The first step in the assessment of structural models is to examine the presence of the collinearity issues. *VIF* values are in the range from 1.154 to 1.278 for Bosnia and Herzegovina, and from 1.068 to 1.443 for Slovenia, which is lower than threshold *VIF* value of 5. Path coefficients and explained variance of endogenous constructs for Bosnia and Herzegovina are shown in Fig. 3.

The proposed path model for Bosnia and Herzegovina explains: 7.4% of cultural and social norms' variance ($R^2 = 0.074$) and 39.4% ($R^2 = 0.394$) of entrepreneurial capacity's variance. According to path coefficients, it can be concluded that the strongest effect on final target endogenous construct entrepreneurial capacity have cultural and social norms (0.473, $p = 0.000$), followed by research and development (0.201, $p = 0.022$) and entrepreneurial education and training (0.107, $p = 0.244$). The significance of direct effects of these constructs is examined by bootstrapping procedure (5.000 sub-samples) which established significance at level 5% of all constructs except for entrepreneurial education and training. The effect of entrepreneurial education and training on cultural and social norms is 0.273 ($p = 0.008$). According to obtained path coefficients and their significance, it can be concluded that majority of the hypothesis are supported in case of factor-driven economy Bosnia and Herzegovina with exception of hypothesis 1. Even though path coefficients provide information about the size of direct effects, in this model there is mediation present and hence some indirect effects need to be considered. Total effect analysis of higher order construct entrepreneurial capacity and its significance are presented in Table 4. As it can be seen, all total effects of constructs on final target construct entrepreneurial capacity are significant. Cultural and social norms have the strongest effect on entrepreneurial capacity (0.473), followed by entrepreneurial education and training (0.236) and research and development (0.201). Finally, taking into account outer weights of two dimensions, motivation and skills, of higher order reflective-formative construct entrepreneurial capacity in Bosnia and Herzegovina, it can be observed that skills (0.763) are much more important than motivation (0.409) in constructing entrepreneurial capacity.

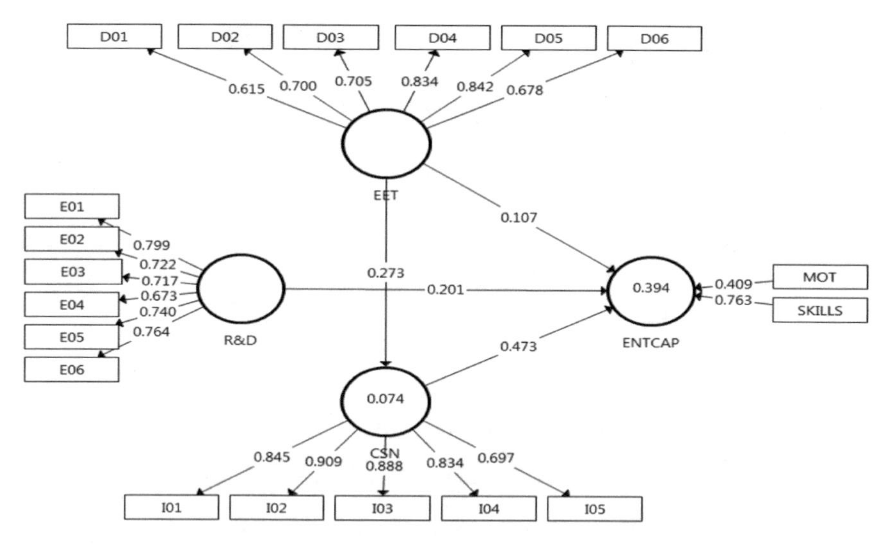

Fig. 3. Structural model results for Bosnia and Herzegovina

Table 4. Total effect analysis and bootstrapping results in Bosnia and Herzegovina

	Original sample (O)	Sample mean (M)	Standard deviation (STDEV)	T Statistics (O/STDEV|)	P values
CSN → ENTCAP	0.473	0.471	0.081	5.841	0
EET → CSN	0.273	0.294	0.099	2.747	0.006
EET → ENTCAP	0.236	0.255	0.108	2.173	0.03
R&D → ENTCAP	0.201	0.205	0.088	2.276	0.023

Concerning the Slovenia case, results of structural models show that 23.6% of cultural and social norms' variance and 45.6% of entrepreneurial capacity's variance are explained by other influential constructs. According to path coefficients, it can be concluded that the strongest effect on final target endogenous construct entrepreneurial capacity have cultural and social norms (0.469, p = 0.000), followed by entrepreneurial education and training (0.223, p = 0.012) and research and development (0.139, p = 0.031). The significance of direct effects of these constructs on final target construct entrepreneurial capacity is examined by bootstrapping procedure (5.000 subsamples) which established significance at level 5% of all of them. The effect of entrepreneurial education and training on cultural and social norms is stronger than in case of Bosnia and Herzegovina (0.488). According to the obtained path coefficients and their significance, it can be concluded that all hypothesis are supported in the case of innovation-driven economy Slovenia (Fig. 4).

As in the case of Bosnia and Herzegovina, mediation is present in this structural model for Slovenia and therefore it is important to examine direct and possible indirect effects of certain construct on final target construct entrepreneurial capacity. Beside

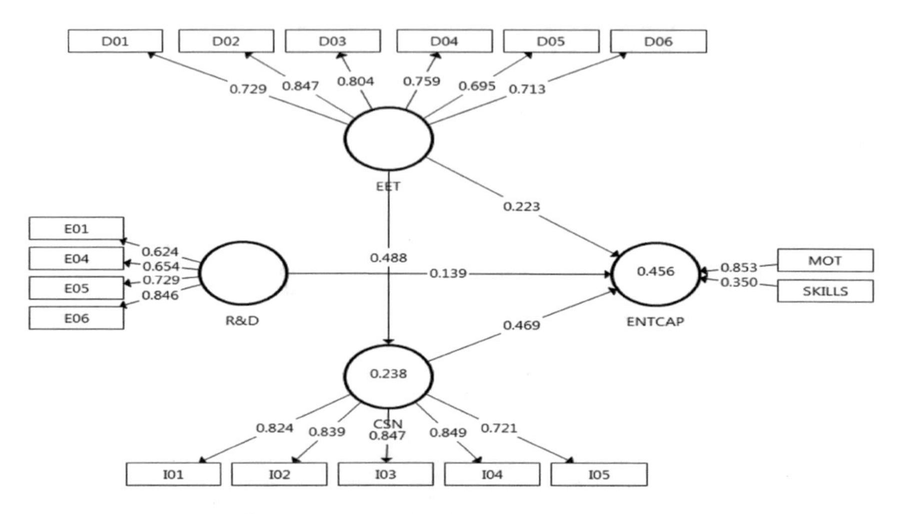

Fig. 4. Structural model results for Slovenia

Table 5. Total effect analysis and bootstrapping results in Slovenia

	Original sample (O)	Sample mean (M)	Standard deviation (STDEV)	T Statistics (|O/STDEV|)	P values
CSN → ENTCAP	0.469	0.466	0.085	5.517	0
EET → CSN	0.488	0.495	0.07	6.978	0
EET → ENTCAP	0.451	0.452	0.069	6.538	0
R&D → ENTCAP	0.139	0.151	0.065	2.147	0.032

direct effect that entrepreneurial education and training has on final target construct entrepreneurial capacity, its indirect effect through cultural and social norms is also present. Total effect analysis of higher order construct entrepreneurial capacity and its significance obtained by bootstrapping procedure are presented in Table 5.

As it can be seen, all total effects of constructs on final target construct entrepreneurial capacity are significant. Cultural and social norms have the strongest effect on entrepreneurial capacity (0.469), followed by entrepreneurial education and training (0.451), and research and development (0.139). Finally, taking into account outer weights of two dimensions, motivation and skills, of higher order reflective-formative construct entrepreneurial capacity in Slovenia, it can be observed that motivation (0.853) is much more important than skills (0.350) in constructing entrepreneurial capacity.

In order to compare results of Bosnia and Herzegovina and Slovenia, the total effects have been analyzed of entrepreneurial education and training and cultural and social norms on final target higher-order construct entrepreneurial capacity. In both countries, cultural and social norms have the strongest effect on entrepreneurial capacity. The total effect of entrepreneurial education and training on entrepreneurial capacity is more pronounced in Slovenia (0.451) than in Bosnia and Herzegovina (0.236). In Slovenia, research and development have a slightly lower effect on entrepreneurial capacity (0.139) than in Bosnia and Herzegovina (0.201). Entrepreneurial education and training have nearly two times stronger effect on cultural and social norms (0.488) in Slovenia than in Bosnia and Herzegovina. Finally, when analyzing the higher order reflective-formative construct entrepreneurial capacity, it can be concluded that in case of Slovenia motivation is much more important dimension than skills opposite to Bosnia and Herzegovina.

In order to assess the relative predictive relevance of a predictor construct on endogenous construct through out-of-sample prediction, blindfolding procedure for obtaining Q^2 values is employed [9, 10]. The blindfolding procedure yields Q^2 values above 0 for both structural models that indicate models' satisfactory predictive relevance.

5 Conclusion

The aim of this research was to analyze how several entrepreneurial framework conditions affect the entrepreneurial capacity of the society of two countries in different stages of development [16, 23, 28]. The results obtained through this research support the fact that cultural and social norms in both countries, regardless of their development phase, have the strongest positive contribution to the entrepreneurial capacity of society. In this way, hypothesis 4 is supported. In this research, entrepreneurial capacity is modeled as a second-order construct measured through two formative dimensions: motivation and skills. According to the obtained results, education and training for entrepreneurship is not positively associated with entrepreneurial capacity in Bosnia and Herzegovina. Hence, hypothesis 1 is rejected. This is the reason why we cannot say that the different educational and training systems in Bosnia and Herzegovina have included into their programs necessary tools for training in establishing and managing new enterprises in sufficient way to influence entrepreneurial capacity. This concerns both dimensions of entrepreneurial capacity, motivation and skills, in the way that there is not sufficient motivation in society to take advantage of business opportunities and no necessary skills to convert them into successful enterprises. Nevertheless, this kind of educational programs and system in Bosnia and Herzegovina do influence entrepreneurial capacity not directly but through cultural and social norms, and values which they create in the population of Bosnia and Herzegovina. This means that hypothesis 2 is supported in both countries. In conclusion, they have an indirect effect on motivation and skills of potential entrepreneurs to seek opportunities and start a new business. These social and cultural norms have the strongest effect on the entrepreneurial capacity of society in both countries and so hypothesis 4 is supported. They stimulate individuals to continuously search and acquire knowledge and essential skills to fully exploit business opportunities and convert them into new enterprises. On the other hand, technological changes yield new business and create new opportunities through the combination of resources in another form or through creating production functions. Technological innovation and the research and development contribute to the entrepreneurial capacity of the environment in both analyzed countries, which indicates that hypothesis 3 is supported.

References

1. Acemoglu, D.: Modeling inefficient institutions. In: Blundell, R., Newey, W., Persson, T. (eds.) Advances in Economic Theory, Proceedings of 2005 World Congress, pp. 341–380. Cambridge University Press, New York (2006)
2. Aidis, R., Estrin, S., Mickiewicz, T.: Institutions and entrepreneurship development in Russia: a comparative perspective. J. Bus. Ventur. **23**(6), 656–672 (2008)
3. Amorós, J., Bosma, N.S.: Global entrepreneurship monitor 2013 global report: fifteen years of assessing entrepreneurship across the globe (2014). http://www.gemconsortium.org/docs/3106/gem-2013-global-report/. Accessed 10 Jan 2018
4. Burgelman, R., Maidique, M., Wheelwright, S.: Strategic Management of Technology and Innovation. McGraw-Hill Higher Education, New York (2001)
5. Busenitz, L.W., Gómez, C., Spencer, J.W.: Country institutional profiles: unlocking entrepreneurial phenomena. Acad. Manag. J. **43**(5), 994–1003 (2000)
6. Chin, W.W.: The partial least squares approach to structural equation modeling. In: Marcoulides, G.A. (ed.) Modern Methods for Business Research, pp. 295–336. Lawrence Erlbaum Associates, Mahwah (1998)
7. Cooper, S.Y., Park, J.S.: The impact of incubator organizations on opportunity recognition and technology innovation in new, entrepreneurial high-technology ventures. Int. Small Bus. J. **26**(1), 27–56 (2008)
8. Davidsson, P.: Nascent entrepreneurship: empirical studies and developments. Found. Trends Entrep. Res. **2**(1), 1–76 (2006)
9. Hair, J.F., Ringle, C.M., Sarstedt, M.: Partial least squares structural equation modeling: rigorous applications, better results and higher acceptance (2013)
10. Hair, J.F., Hult, G.T., Ringle, C.M., Sarstedt, M.: A Primer on Partial Least Squares Structural Equation Modeling (PLS-SEM). Sage, Thousand Oaks (2014)
11. Hair, J.F., Hult, G.T., Ringle, C.M., Sarstedt, M.: A Primer on Partial Least Squares Structural Equation Modeling (PLS-SEM), 2nd edn. Sage, Thousand Oaks (2017)
12. Henry, C., Hill, F., Leitch, C.: Entrepreneurship Education and Training. Ashgate, Aldershot (2003)
13. Henseler, J., Ringle, C.M., Sarstedt, M.: A new criterion for assessing discriminant validity in variance-based structural equation modeling. J. Acad. Mark. Sci. **43**(1), 115–135 (2015)
14. Hofstede, G.: Culture's Consequences: International Differences in Work Related Values. Sage, Beverly Hills (1984)
15. Katz, J.A.: The chronology and intellectual trajectory of American entrepreneurship education. J. Bus. Ventur. **18**(2), 283–300 (2003)
16. Levie, J., Autio, E.: A theoretical grounding and test of the GEM model. Small Bus. Econ. **31**(3), 235–263 (2008)
17. Lohmöller, J.B.: Predictive vs. structural modeling: Pls vs. ml. In: Latent Variable Path Modeling with Partial Least Squares, pp. 199–226 (1989). Physica, Heidelberg
18. Mueller, S.L., Thomas, A.S.: Culture and entrepreneurial potential: a nine country study of locus of control and innovativeness. J. Bus. Ventur. **16**(1), 51–75 (2000)
19. North, D.C.: Understanding the Process of Economic Change. Princeton University Press, Princeton (2005)
20. Nunnally, J.: Psychometric Theory, 2nd edn. McGraw-Hill, New York (1978)

21. Nyström, K.: The institutions of economic freedom and entrepreneurship: evidence from panel data. Public Choice **136**(3–4), 269–282 (2008)
22. Reynolds, P., Storey, D.J., Westhead, P.: Cross-national comparisons of the variation in new firm formation rates. Reg. Stud. **28**(4), 443–456 (1994)
23. Reynolds, P., Bosma, N.S., Autio, E., et al.: Global entrepreneurship monitor: data collection design and implementation 1998–2003. Small Bus. Econ. **24**(3), 205–231 (2005)
24. Sarstedt, M., Henseler, J., Ringle, C.M.: Multigroup analysis in partial least squares (PLS) path modeling: alternative methods and empirical results. In: Sarstedt, M., Schwaiger, M., Taylor, C.R. (eds.) Measurement and Research Methods in International Marketing (Advances in International Marketing), vol. 22, pp. 195–218. Emerald Group Publishing Limited, Bingley (2011)
25. Schumpeter, J.A.: Business Cycles: A Historical and Statistical Analysis of the Capitalist Process. McGraw-Hill, New York (1939)
26. Solomon, G.T., Duffy, S., Tarabishy, A.: The state of entrepreneurship education in the United States: a nationwide survey and analysis. Int. J. Entrep. Educ. **1**(1), 65–86 (2002)
27. Spencer, J.W., Gómez, C.: The relationship among national institutional structures, economic factors and domestic entrepreneurial activity: a multi-country study. J. Bus. Res. **57**(10), 1098–1107 (2004)
28. Sternberg, R., Wennekers, S.: Determinants and effects of new business creation using global entrepreneurship monitor data. Small Bus. Econ. **24**(3), 193–203 (2005)
29. Tushman, M.L., Anderson, P.: Technological discontinuities and organizational environments. Adm. Sci. Q. **31**(3), 439–465 (1986)
30. Wennekers, S., Van Wennekers, A., Thurik, R., Reynolds, P.: Nascent entrepreneurship and the level of economic development. Small Bus. Econ. **24**(3), 293–309 (2005)
31. Wold, H.: Path models with latent variables: the NIPALS approach. In: Quantitative Sociology, pp. 307–357 (1975)

Mathematical Modeling and Statistical Representation of Experimental Access

Amina Delić-Zimić(✉) and Fatih Destović

Faculty of Educational Sciences, University of Sarajevo,
Skenderija 72, Sarajevo, Bosnia and Herzegovina
minchysarajevo@hotmail.com, fatih_d@msn.com

Abstract. In this paper, the research on the application of mathematical modeling in the problem teaching mathematics in analyzed schools was carried out in the form of an experiment, with the aim of demystifying its role and importance in increasing the educational and functional effect. In the period of one school semester, in five primary schools it has been shown that mathematical modeling can be implemented and can achieve significant effects. The statistical overview of selected groups demonstrated the application of mathematical modeling in experimental classes, but gave an overview of the results of applying mathematics and modeling in control groups. Tests were performed using Kolmogorov-Smirnov, Shapiro-Wilkov, Mann-Whitney and Wilcoxon test. The analysis and comparison of the results before and after the experimental program indicates a significant improvement in student knowledge and mathematical competences using the methodology of mathematical modeling which was carried out using the experiment.

Keywords: Mathematical modeling · Experiment · Research
Statistical display · Analysis · Improvement of student knowledge

1 Introduction

Recognizing the application of mathematical modeling of problem teaching mathematics and respecting its essential characteristics in elementary school are the first steps towards quality teaching mathematics. The process of student problem solving, as well as various ways of modeling, contribute to the creative experience of students. In recent years, the approach to solving mathematical problems by modeling is intensively analyzed [1].

The report of the European Commission "Mathematics in Education in Europe" (2015) expressed concern about the poor students achievement in mathematics, proposing a reform of the curriculum of mathematics, the use of innovative teaching methods, and the improvement of teacher education. In the world reports, they also advocate for better mathematical education and a focus on teacher support, emphasizing the application of mathematical knowledge and problem-solving skills that will contribute to improving student competencies in mathematics [2]. In this paper we investigated the presence of problem teaching mathematics in our schools by experimental research on the application of mathematical modeling, effects and comparisons.

S. Avdaković (Ed.): IAT 2018, LNNS 59, pp. 36–48, 2019.
https://doi.org/10.1007/978-3-030-02574-8_4

The research was conducted in five primary schools in Sarajevo Canton (in different municipalities) in two (2) experimental and ten (10) control classes. The students in experimental classes, with pleasure and excitement, took part in the work, awaited for a new game, a new problematic situation, a teaching poster or a new idea that we implemented on school tablets and computers. Setting up mathematical problems, selecting the way of presenting mathematical situations and expressing one's own opinion on questions in an experiment is very important to present to students in an interesting way. By developing the application of modeling, the application of mathematical education involving concrete skills is achieved [3]. In this paper the views of teachers, parents and students were examined. The results related to the students and their progress in applying modeling in the problem teaching mathematics will be presented.

By modernizing schools and opening them up to the newspapers in education, engaging in various projects, contemporary and advanced programs (Step by Step Method, RWCT), teachers have become more sensible and more prepared for modern approaches [4]. Application of modern methods, forms and means of work, new educational technologies (computer connectivity, mathematical softwares, smart board), placing students at the center of the educational process, applying self-assessment, self-evaluation, etc. contributed to modernizing teaching methods. They are more willing to be educated, and in some schools to actively participate in education related to mathematics education, thus complement problem teaching by modeling, and achieve mathematical competence among students [4].

2 Application of Experimental Models

By correct selection of mathematical content, the teacher will improve the success of his students and interest them in mathematics. "The quantity and especially the quality of students' knowledge is a reliable measure of the teacher's work, that is, the success of the teaching" [5]. In experimental classes (52 students), as the basic teaching system problem teaching mathematics by models was carried out, combining interactive, programmed, project, individual and group heuristic teaching. Based on their own empirical research, the significance and value of problem teaching mathematics by modeling were determined, as well as the effects of its application in comparison with the traditional mathematics education in lower grades of elementary school, with appropriate statistical data processing. The results presented in this paper represent the continuation of the research presented in Refs. [Implementation of ICT in Education, Modern teaching approaches, Application of Problem Teaching in Lower Grades of Primary Schools and its Impact on Improving Mathematical Knowledge of Students]. Many papers and research have confirmed the influence of teachers on the value and progress of students in mathematics. "The research by Rosenthal and Jacobsen showed that students from whom teachers expected intellectual advancement, have really made that progress" [6].[1]

[1] This effect is known as the "Pigmalion effect" (Vizek-Vidović and associates, 2003)

Application of Model 1. - Experimental Application of Interactive Learning Through Group Work in Problem Teaching (Logical Tasks) - We have resolved in problem-solving groups, thus creating models of solutions that are acceptable. Together we selected the best and the shortest solution. We tried each proposed idea and combination together, and then the students in the groups (the 1st group drew in the notebook, the second group worked on the board, and the 3rd group on the poster). The students freely asked questions, talked in the group, and wanted to come to a solution as soon as possible. All the students with visible interest participated in the work and presented the group solution (Fig. 1).

Fig. 1. Realized activities according to Model 1

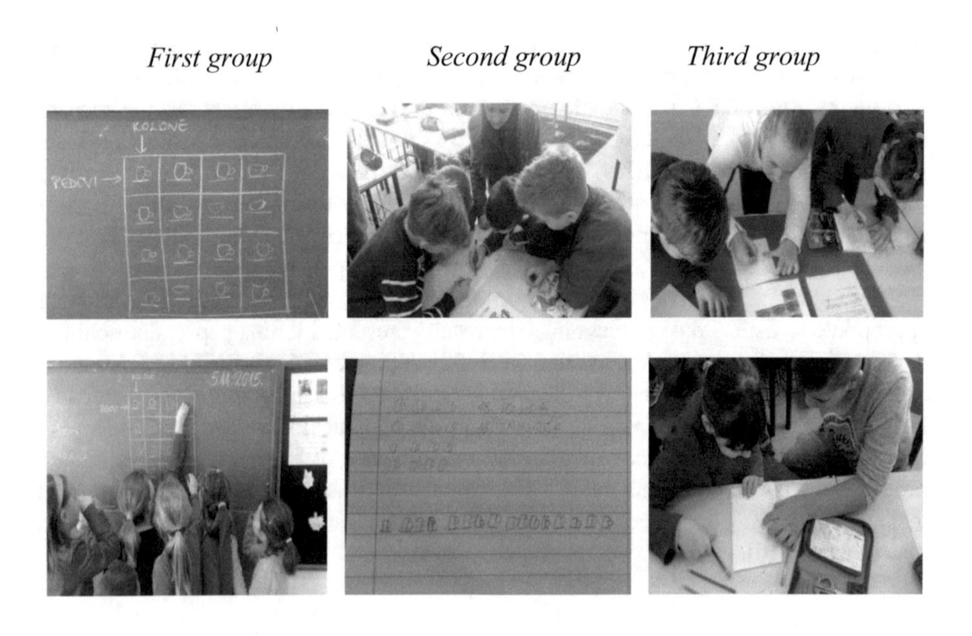

Application of Model 2. - Experimental Application of ICT in Problem Teaching - Working in Pairs - The "Tandem Group" provides great opportunities for different teacher participation and teaching process control [7]. In particular, if teaching is organized with ICT, teachers have a certain number of computers and tablets for realizing modeling through solving the problematic tasks assigned to the link that a student has to take: Calculate the following tasks independently in pairs using the table of local values. Exercise 1 (students open link to tasks) (Fig. 2).

Fig. 2. Realized activities according to Model 2

Application of Model 3. - Experimental Application of Problem Teaching Through the Project "Triangle in Geogebra" - Project Based Learning (PBL) - By applying this model, the teacher carefully uses selected educational activities that are long-lasting, interdisciplinary, focused on a student and current issues, problems and practice. PBL allows students to work together to solve specific problems and eventually present their work. "The final project result can be a multimedia presentation, a performance, a written report, a website or a constructed product" [8], as the students of the experimental departments did (Fig. 3).

Fig. 3. Realized activities according to Model 3

Application of Model 4 - Experimental Application of Individualization of Teaching Sheets in Problem Teaching Mathematics by Level of Progression (Differentiated Tasks) - Implementation of Model 4 at three complexity levels was realized in order to adapt the teaching to different possibilities of students. Students should be given the opportunity to choose their tasks in accordance with their abilities, since only then the intellectual capacities of the students will be maximally exploited, and during the classe everyone will be equally engaged. [9] We resolved interesting textual and Sudoku tasks, and we tried to engage all students (Fig. 4).

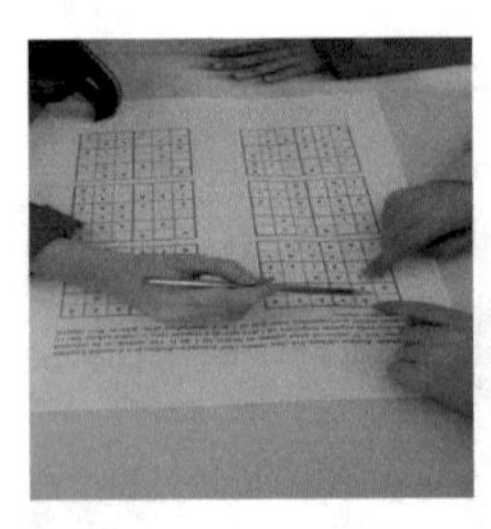
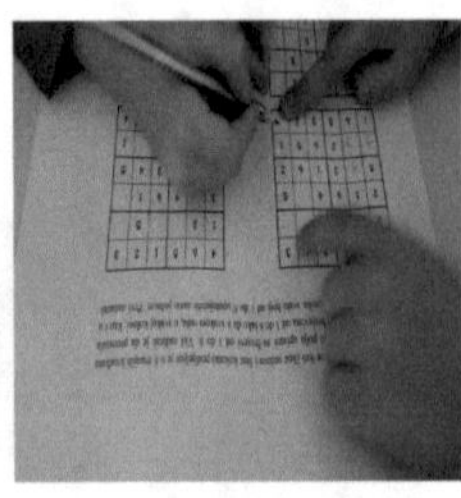
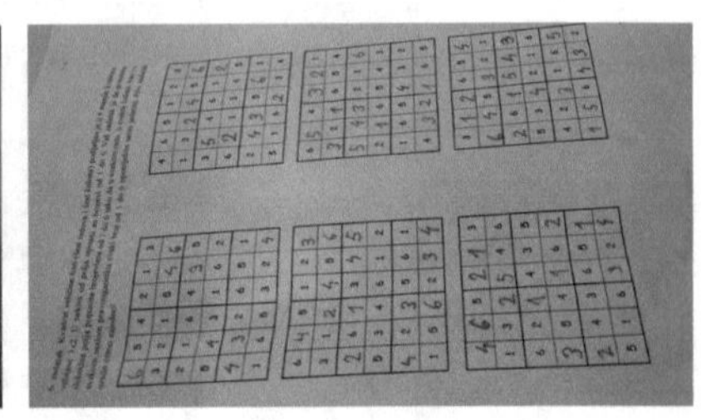

Fig. 4. Realized activities according to Model 4

3 Methodology and Data

At the beginning of the research, an **initial general knowledge test** was conducted in 12 classroom of fourth grade students, in logical problem assignments, tasks of change, comparison and combining, on the basis of which student grouping was performed. Two experimental classes of the IV grade (52 students) at the Elementary School "Aleksa Šantić" applied problem teaching mathematics according to the models which included: the choice of tasks (when adopting, repeating and practicing), a high-quality methodological approach to the problem teaching mathematics (presentations, the use of ICT through group, tandem and individual work), interactive approach, modeling, task analysis and tracking of their learning in content that is in accordance with the curriculum of Class IV.

The research was carried out using the general methods: causal and descriptive, with the experiment occupying a special place within the causal method. The analysis was performed by qualitative and quantitative methods of the statistical method. The following were used: Servej method, Theory of theoretical analysis of content, Kausal-descriptive method, Experimental and Statistical method. Data processing and analysis was performed using the IBM SPSS statistics program. 20, for which we needed mathematical approaches in the statistical formulation of the sought. Statistical tests of Kolmogorov-Smirnov, Shapiro-Wilkov, Mann-Whitney and Wilcoxon were used. In K-S and S-W tests, the **distribution of the frequencies** of some variables is done by sorting the results by size from the smallest to the largest, and if the number of individual results is determined, the frequency (f) of that variable is given [10]. If the number is large, as in our research, classes are created and grouping of results into classes is carried out. The smallest result in the Xmin variable and the largest Xmax is found, the result range R = Xmax − Xmin is found, *k - the number of classes*, and *i - the width of the class* is determined, and it is calculated by the formula for k. On the basis of absolute frequencies, **the cumulative and relative frequencies** are calculated, and on the basis of the boundaries of the class, the averages are calculated. For example. absolute, cumulative, relative frequencies, mid-range variables:

x - results, SR - average of class, F - frequencies, FR - relative frequencies, FK - cumulative frequencies, (FK) R - coupling relative frequencies

Frequency tags: cf - cumulative frequency, f(Md) - frequency of the class in which the median is requested (medial interval).

The sum of the square of the deviation of each result from the arithmetic mean divided by the number of these results is the variance. It is calculated according to the formula:

$$\sigma^2 = \frac{1}{n}\sum_{i-1}^{n}(x_i - \bar{x})^2 = \frac{1}{n}\sum_{i-1}^{n} x_i - \bar{x}^2$$

After the K-S and S-W tests, the following ranks were used: **Test of Sum of Ranks** (Mann-Witney's U test, Wilcoxon T-test). The M-W test uses more information, ranks and therefore can be considered better and "stronger" than others. As with the median test, with the test of the sum of rank we tested whether two samples belong to the population with the same median. In comparison to the E and K groups, we separately summed up all ranks in our research. Due to the control, we checked the sum of the ranks with the formula:

$T1 + T2 = \frac{N(N+1)}{2}$, after that, we calculated the value $z = \frac{|2Ti - Ni(N+1)| - 2}{2\sqrt{\frac{N1N2(N+1)}{3}}}$

Ti - any sum of ranks; $|\ |$ - absolute value; Ni - number of subjects in the group from which we took Ti.; z - limit value; we compared the U value [10].

4 III Tests Analysis (Initial T1 and After Conducting the Experiment by Modeling T2)

The main data source in our research is an assessment of initial knowledge and an experiment in which deliberately a change was induced, in order to examine the effectiveness of the impact of the application of work by models. An experimental model was applied, with even parallel groups in which the process is carried out simultaneously with the scheme K - classical work and E - experimental work. This basic model of experiment has parallel groups, each of which is the carrier of its experimental factor. In studying the influence of teaching methods, it is recommended to equate pre-knowledge and general mathematical ability, since they are mostly contributing to the realization of materials [11] (Table 1).

Of the total number of 266 students surveyed, according to the division into three categories, 71 (26.69%) below the average, 98 averagc (36.84%) and 97 above-average students (36.47%) were examined. Students were divided according to the grade criteria from 1 to 3 IP, 4 P and 5 NP. in the K group of 214 students. By the same norm, we divided students of the E group of 52 students.

Table 1. Total division of students into three categories

		Group				
		Control		Experimental		
		Number of students N	Percentage N %	Number of students N	Percentage N %	Distribution
General success, score in mathematics Test 01 - total number of points	0	2	.9%	0	0.0%	
	2	3	1.4%	0	0.0%	
	3	3	1.4%	0	0.0%	
	4	6	2.8%	1	1.9%	
	5	4	1.9%	0	0.0%	
	6	9	4.2%	2	3.8%	
	7	5	2.3%	4	7.7%	
	8	11	5.1%	1	1.9%	
	9	10	4.7%	6	11.5%	**26.69%**
	10	**18**	**8.4%**	**5**	**9.6%**	
	11	**13**	**6.1%**	**6**	**11.5%**	
	12	**30**	**14.0%**	**8**	**15.4%**	**36.84%**
	13	20	9.3%	5	9.6%	
	14	29	13.6%	4	7.7%	
	15	31	14.5%	4	7.7%	
	16	20	9.3%	6	11.5%	**36.47%**

From Table 2 we see the results by classes on Test 1. In the control class they are ranging from ($\bar{X} = 3.16$) to ($\bar{X} = 4.2$). Two experimental compartments have ($\bar{X} = 3.52$) and ($\bar{X} = 3.59$), which in the overall comparison corresponds to $\bar{X}$: E - 3.55, K - 3.53, according to which the groups are equal. In order to ensure that the E and K group are equal, after the analysis of the success of the classes, we have statistically analyzed the overall results of Test 1 in order to determine whether there is a statistical difference between E and K groups. Both tests showed that there was no statistically significant difference between the E and K groups at the initial test. In order to investigate the distribution normality, Kolmogorov-Smirnov (hereinafter K-S) and/or Shapiro-Wilk's test (hereinafter S-W test) should be carried out (Table 3).

The two tables above show us the results of the Mann-Whitney test between the control and experimental group of students for the total number of points in Test 1. At a value of $U = 5218{,}500$ and a value of $p > 0.05$ ($p = 0.485$), we conclude that there is no statistically significant difference between these two groups of students when it comes to the total number of points on Test 1.

Table 2. Results of initial test T1

School	Groups	Class	Number of students	Results by classes
Elementary school "Aleksa Šantić"	**Experimental**	**IV 1** **IV 2**	**25** **27**	**3. 52** **3. 59**
	Control	IV 3 IV 4	20 20	3. 35 4. 2
Elementary school "Behaudin Selmanović"	Control	IV 1 IV 2	23 23	3. 17 3. 22
Elementary school "Skender Kulenović"	Control	IV 1 IV 2	24 22	3. 83 3. 5
		IV 3 IV 4	25 22	3. 4 3. 73
Elementary school "Velešićki heroji"	Control	IV 1 IV 2	17 18	3. 70 3. 16
Total of 1 school	2 E	2 classes	52	E - 3.55
Total of 4 school	10 K	102 classes	214	K - 3.53
Total of 4 school	2 E i 10 K	122 classes	266	**E - 3.55** **K - 3.53**

Table 3. (a) Results of Test 1 - comparison of K and E groups for the total number of points in Test 1

Ranks

group		N	$\bar{x}$ of ranks	Σ Of ranks
Test 01 – Total number of points	control	214	135.11	28914.50
	experimental	52	126.86	6596.50
	total	266		

Test Statistics[a]

	Test 01 – Total number of points
Mann-Whitney U	5218.500
Wilcoxon W	6596.500
Z	-.698
Statistically sig. difference p	.485

5 T2 Test Results After Conducting the Experiment by Modeling

The knowledge test as a check of overall activities, after the experimental modeling program which was conducted in 5 schools, as agreed with the management and teachers of the departments in which it was studied, was done by 266 students. Out of which 52 students in 2 classes made an experimental group, and 214 deployed in 10

Table 4. Results of students on Test 2 by classes

School	Groups	Class	Number of students	Results of T2 by classes
Elementary school "Aleksa Šantić"	Experimental	**IV 1** **IV 2**	**25** **27**	**4. 64** **4. 62**
	Control	IV 3 IV 4	20 20	4. 2 4. 6
Elementary school "Behaudin Selmanović"	Control	IV 1 IV 2	23 23	3. 48 3. 52
Elementary school "Skender Kulenović"	Control	IV 1 IV 2	24 22	4. 16 3. 86
		IV 3 IV 4	25 22	3. 64 4
Elementary school "Velešićki heroji"	Control	IV 1 IV 2	17 18	3. 71 3. 44
Total of 1 school	2 E	2 classes	52	E = 4.63
Total of 4 school	10 K	10 classes	214	K = 3.86
Total of 4 school	2 E i 10 K	12 classes	266	**E = 4.63** **K = 3.86**

departments consisted of a control group. Table 4 shows the results achieved by the students of these two groups by classes:

Experimental departments have achieved significantly better results. After applying the experimental program of mathematical modeling during a single semester, students of the E department showed better knowledge in the tasks of changing, combining and comparing. Also, we notice that the students of grade IV4 MSA achieved a score of 4.6, which is closest to the success of the E department. After the analysis of the success of the departments, the total results of the E and K groups were statistically compared.

The difference was shown between the E and K groups by the arithmetic mean of the total number of points on Test 2 of $\bar{x} =$ 174, 98 (E group) and $\bar{x} =$ 123, 42 (K group), then based on the sum of the ranges $\Sigma = 9099{,}00$ (E groups) and $\Sigma = 26412.00$ (K group), which should be further analyzed by MW test, notice values U difference and W test, see Z values.

In Table 5, the total number of points on Test 2 was compared between the control and experimental group of students. The value of the Mann-Whitney test is U = 3407 and the test is significant at $p < 0.01$. This tells us that there is a statistically significant difference between the control and the experimental group in the total number of points on Test 2. In order to determine in which direction there is a difference, we will see a table with sum of ranks and determine that students from the experimental group achieve a statistically significant higher score on Test 2 of students from the control group students.

Table 5. (a) Razlika M i Σ - E i K grupe Test 2 (b) M-W testiranje E i K

Ranks				
group		N	$\overline{X}$ of ranks	Σ of ranks
Test 02 – Total number of points	control	214	123.42	26412.00
	experimental	52	174.98	9099.00
	Total	266		

M-W I W Test Statistics[a]	
	Test 02 – Total number of points
Mann-Whitney U	3407.000
Wilcoxon W	26412.000
Z	-4.372
Statistically sig. difference p	.000

6 The Difference in the Achieved Results of Test 1 and Test 2

In the table we see the results of both tests that students worked in E and K groups. Comparing them, it can be seen that each group has made progress in solving mathematical problems in relation to the first test. The experimental group of students achieved better results, but results also show a visible individual improvement in relation to the first test. Each department achieved progress in Test 2 versus Test 1, as shown in Table 6.

Table 6. Results on Test 1 and Test 2 by classes

Number of students – Test 1	Results by classes – Test 1	Number of students – Test 2	Results by classes – Test 2
25 27	3. 52 3. 59	25 27	4. 64 4. 63
20 20	3. 35 4. 2	20 20	4. 2 4. 6
23 23	3. 17 3. 22	23 23	3. 48 3. 52
24 22	3. 83 3. 5	24 22	4. 12 3. 86
25 22	3. 4 3. 73	25 22	3. 64 4
17 18	3. 70 3. 16	17 18	3. 71 3. 44
E (52)	E = 3.55	52	E = 4.63
K (214)	K = 3.53	214	K = 3.86
Total 266	**E = 3.55** **K = 3.53**	266	**E = 4.63** **K = 3.86**

From the Table 6. on the average of the classes in the E and K groups, the progression of students in the E departments was noticed. Individual progress can be seen from 3.52 to 4.64, and from 3.59 to 4.63, which is extremely successful. Specific results have been achieved in K4 IV of 4.6 in the final test, which is a consequence of more regular use of PNM.

The difference in the results of Test 1 and Test 2 graphically shows the relationship between the median values of the total number of points scored in the tests (Fig. 5).

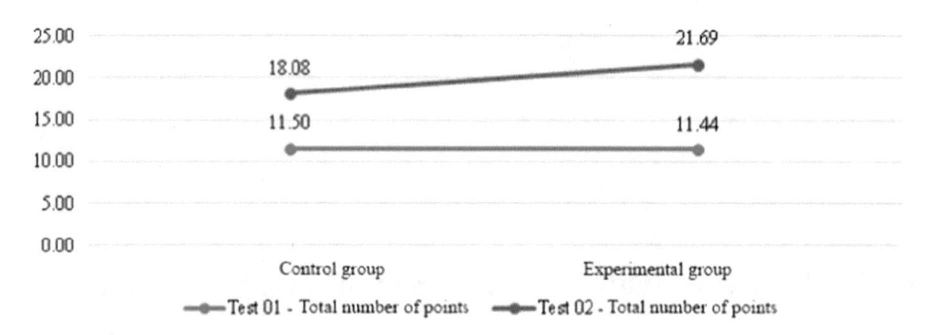

Fig. 5. Graphic representation of the difference in mean values on Test 1 and Test 2 (y axis - sum of total points)

The chart shows that the results of the initial Test 1 between groups E and K were equalized by mean values of 11.50 and 11.44. However, after regular application of problem teaching mathematics, a model which contributed to student progress, more modern methods, means and forms of work, the results of Test 2 are significantly different between these two groups. The mean value of the K group on Test 2 is 18.08, and E groups 21.69, which is the mean difference of 3.61 between these two groups, which shows that "regular application of modeling contributes to much better qualitative and quantitative results of mathematical knowledge, i.e. the abilities of our students in mathematics" [12]. Students who often do problematic tasks and mathematics by modeling achieve better results in mastering content from mathematics, i.e. there is a statistically significant difference of $p < 0.01$ in the adoption of mathematical contents in classes where this teaching is carried out by conventional (traditional) methods and procedures.

7 Conclusion

After analyzing and researching in our schools it has been shown that mathematical modeling can be successfully applied in mathematics classes in elementary school, and that this teaching as an efficient and productive form of learning, influences the increase in overall educational performance in mathematics teaching. It has been noted that students of all departments showed interest and progress in mathematics. By researching and monitoring students' responses, it was noticed that mathematics by models was more interesting for students and they were more motivated to work and learn in relation to traditional teaching.

The teacher is there to help, instruct and enable the fulfillment and achievement of goals, both to students and to mathematics in general with a higher quality mathematical and creative approach to solving mathematical problems and tasks, as well as understanding and adopting mathematical concepts and modeling. In some schools, depending on the management, teachers have more modern teaching tools, technology and creative didactic materials, and in other teachers themselves buy materials that will enable them to better teaching.

Not in one of the schools the problem teaching mathematics and mathematical modeling is emphasized, but it is "implicit" in GPP for all grades, which means that it is up to mathematics teachers if will it be represented and realized.

The announcement of the research encouraged teachers to apply mathematical modeling, and approached the selection of more modern methods and procedures, working methods and quality problematic tasks that would influence access to mathematics in general. The research and presentation of the results will give at least a small contribution to the improvement of the quality of this subject, which always has the worst results on tests and finals.

Through the experimental application of the model of problem teaching, the teaching of mathematics was promoted, thereby increasing the interests and results of students in mathematics. Progress in mathematics can be fully achieved through quality teaching activities, thought-out and well-planned and guided mathematical contents.

The teacher must encourage creativity in the student, find opportunities and ways for the student to think and create.

From this paper, we expect to improve the teaching approach to mathematics as something ubiquitous around us, around students, and as a teaching subject that is in daily use, that students recognize that "Mathematics is everything that surrounds us".

Suggestions for further research are an analysis of the entire population of students from the first to the fifth grade of elementary school, in order to determine the state of mathematics teaching, and draw conclusions about the plan and programme, and the possible curriculum of mathematical content. Furthermore, students' mathematical abilities could be evaluated by the schools, and the results should be compared in order to give instructions to schools that do not achieve significant results. In order to improve the teaching of mathematics, PPZ and MONKS can arrange competitions at the class level, so that everyone would try to achieve the best possible results.

References

1. Bikić, N.: Efekti postignuti rješavanjem problemskih zadataka u nastavi. Univerzitet u Zenici, Pedagoški fakultet, Zenica (2012)
2. Azevedo, F., Sessa, A., Sherin, B.: An evolving framework for describing student engagement in classroom activities. J. Math. Behav. **31**(2), 270–289 (2012)
3. Christy, F.E.: Teaching problem solving in mathematics: cognitively guided instruction in kindergarten. Dissertations, Hamline University (2016)

4. Bogdanović, Z.: Strategije rešavanja matematičkih zadataka u nižim razredima osnovne škole. Istraživanje matematičkog obrazovanja **8**, 67–74 (2013)
5. Pejić, M.: Neuspjeh u nastavi matematike osnovne i srednje škole i njegovi glavni uzroci. Naša škola. **49**(25), 3–14 (2010)
6. Vizek-Vidović, V., Rijavec, M., Vlahović-Štetić, V., Miljković, D.: Psihologija učenja i poučavanja matematike. Psihologija obrazovanja. IEP, d.o.o. – Vern, Zagreb (2003)
7. Muminoviæ, H.: Osnovi didaktike. DES doo: Centar za napredne studije, Sarajevo (2013)
8. Delić-Zimić, A.: Modern Teaching Approaches. Springer, Heidelberg (2017). International Perspectives on the Teaching and Learning of Mathematical
9. Zlokapa, B.: Kritički prikaz - komentar na tekst: (Na)učiti kako se uči (matematika), autora prof. Milana Matijevića. Istraživanje matematičkog obrazovanja **IV**(6), 5–7 (2012)
10. Brkić, M., Kundačina, M.: *Statistika u istraživanju odgoja i obrazovanja*. Jela educa, Sarajevo (2003). Brkić, M. (2013)
11. Rešić, S.: Matematika i metodika početne nastave matematike. Papir karton, Tuzla (2013)
12. Presmeg, N.: Visualization in High School Mathematics - For the Learning of Mathematics Modelling. Springer, Heidelberg (2006). International Perspectives on the Teaching and Learning of Mathematical

Advanced Electrical Power Systems (Planning, Operation and Control)

Comparison of Different Techniques for Power System State Estimation

Dženana Tomašević[1(✉)], Samir Avdaković[1], Zijad Bajramović[1], and Izet Džananović[2]

[1] Faculty of Electrical Engineering, University of Sarajevo, Sarajevo, Bosnia and Herzegovina
dzenana.tomasevic@hotmail.com
[2] PE Elektroprivreda B&H, Sarajevo, Bosnia and Herzegovina

Abstract. A power system state estimation is one of the most important applications in the power system control center. Power systems are continuously monitored in order to determine a reliable data of the state variables. Conventional power system state estimators, which are iterative and less accurate, are subject to important changes because of the extensive use of Phasor Measurement Units (PMUs). Although there are many advantages related to their use, PMUs are highly expensive and cannot be implemented at all buses. Therefore, it is very important to find an optimal number of PMUs and their locations in power system to achieve full observability. This paper presents an overview of different techniques for power system state estimation. It also investigates the outcomes of including PMUs in conventional state estimation process. The effectiveness of the proposed state estimation techniques is validated using IEEE test system with 14 buses.

1 Introduction

State estimation is a critical component of the contemporary electric power system. It provides a reliable real-time data of the voltage magnitudes and relative phase angles of the system. The idea of state estimation in power systems was first suggested by Fred Schweppe [1–3]. Computational developments after this pioneering work made the state estimation a standard function in every control centre [4] and led to the establishment of the first Supervisory Control and Data Acquisition (SCADA) Systems and, with further advances of the SCADA system computers, the Energy Management Systems (EMS). Power system state estimator includes the following functions: measurement prefiltering, topology processor, observability analysis, state estimation and bad data processing. It serves as a filter between the raw measurements received from the system and all the applications that constitute the EMS [5].

Early state estimation algorithms used measurements of line flows to obtain state variables. Bus voltage magnitudes and angles could not be measured directly. With the introduction of PMUs, it became possible to measure the power system state directly. A PMU device provides precise information of system state in synchronized way. All measurements are assigned with a time stamp to synchronize the data during transmission via communication link [6]. The use of phasor measurements for state

S. Avdaković (Ed.): IAT 2018, LNNS 59, pp. 51–61, 2019.
https://doi.org/10.1007/978-3-030-02574-8_5

estimation increases both speed and accuracy of the state estimation technique. Although PMUs provide precise measurements of system state, they cannot be implemented at all buses and the average cost per PMU is very high. Considering small adjustments, phasor measurements from PMUs may be included in the traditional state estimation. Therefore, finding an optimal number of PMUs as well as their locations is very important and has always been a subject to optimization problem. An overview of solution methods for this problem is discussed in [7, 8].

Unlike classic estimation technique which has to deal with iterative solution of non-linear equations, the PMU measurements are linear functions of state variables. There are two general techniques that are used to combine PMU measurements with traditional SCADA data. The first method integrates PMU measurements into classic state estimator and processes them in the same iterative procedure. Another algorithm utilizes the estimate obtained in classic state estimator through a post-processing step. The state vector is converted to rectangular form, comprising real and imaginary parts of bus voltages. Then it is fed into the linear estimator along with voltage and current measurements. Both methods are identical but it is preferable to use linear estimator as less complicated and non-iterative algorithm [9].

In this paper the basic principle of classic state estimation (CSE), the concept of non-linear weighted least squares (WLS) is briefly explained since it is used as a part of the hybrid state estimator. The concept of linear weighted least squares is discussed and applied to linear state estimation (LSE) for two cases: utilizing PMU measurements exclusively and combined with traditional estimate. Generalized integer linear programming formulation for optimal PMU placement (OPP) problem for full observability is adopted and presented. The proposed power system state estimation techniques were applied to IEEE test system with 14 buses. The obtained results were compared with the results of WLS estimator with reliable bad data processing proposed in [10] and the fast Weighted Least Absolute Value (WLAV) estimator proposed in [11].

This paper is organized as follows. Theoretical principles of the proposed power system techniques with the OPP problem are presented in the second, third, fourth and fifth part. The obtained results are presented in sixth part. Seventh part concludes the paper and suggests future work.

2 Classic State Estimation

The classical state estimators use SCADA measurements and network topology information to calculate the state vector x, which comprises voltage magnitudes and phase angles at the system nodes. As for the classic power system state estimator, this is performed with non-linear WLS. Measurement vector is formulated as follows [12]:

$$z = \begin{bmatrix} z_1 \\ z_2 \\ \vdots \\ z_m \end{bmatrix} = \begin{bmatrix} h_1(x_1, x_2, \ldots, x_n) \\ h_2(x_1, x_2, \ldots, x_n) \\ \vdots \\ h_m(x_1, x_2, \ldots, x_n) \end{bmatrix} + \begin{bmatrix} e_1 \\ e_2 \\ \vdots \\ e_m \end{bmatrix} = h(x) + e \tag{1.1}$$

where $h(x)$ is vector of non-linear functions relating measurements to the state vector x and e is the vector of measurements errors which is assumed to have zero mean and constant variance σ_i^2. The standard deviation σ_i of each measurement i is calculated to reflect the expected accuracy of the corresponding meter used [12].

According to the WLS method, following objective function has to be minimized:

$$J(x) = \sum_{i=1}^{m} \frac{(z_i - h_i(x))^2}{R_{ii}} = [z - h(x)]^T R^{-1} [z - h(x)] \tag{1.2}$$

where R is the measurement covariance matrix, while R^{-1} is the diagonal weighting matrix of measurement variance.

The minimization of the objective function is achieved by satisfying the first-order optimality conditions:

$$g(x) = \frac{\partial J(x)}{\partial x} = -H^T(x) R^{-1} [z - h(x)] = 0 \tag{1.3}$$

where $H(x) = \partial h(x)/\partial x$ is Jacobian matrix and contains first partial derivatives of $h(x)$ with respect to the state variables. The estimation problem can be solved using Gauss-Seidel iterative method. The iteration procedure finishes when the specified error convergence limit is reached. The state vector in the kth iteration is computed as:

$$\left|G(x^k)\right| \Delta x^{k+1} = H^T(x^k) R^{-1} \left[z - h(x^k)\right] \tag{1.4}$$

$$\Delta x^{k+1} = x^{k+1} - x^k \tag{1.5}$$

where $G(x^k) = \partial g(x^k)/\partial x$ is the gain matrix which is sparse, positive definite and symmetric provided that the system is fully observable [12].

If the power system has N buses, then the state vector will have $(2N - 1)$ elements, N bus voltage magnitudes and $(N - 1)$ phase angles where the angle is zero at reference bus. The relevant measurement functions that relate conventional measurements to state variables, assuming the general π-model for the network branches, are [13]:

- Real and reactive power injection at bus i:

$$P_i = V_i \sum_{j \in N_i} V_j (G_{ij} \cos\theta_{ij} + B_{ij} \sin\theta_{ij}) \tag{1.6}$$

$$Q_i = V_i \sum_{j \in N_i} V_j (G_{ij} \sin\theta_{ij} - B_{ij} \cos\theta_{ij}) \tag{1.7}$$

- Real and reactive power flow from bus i to bus j:

$$P_{ij} = V_i^2 (g_{si} + g_{ij}) - V_i V_j (g_{ij} \cos\theta_{ij} + b_{ij} \sin\theta_{ij}) \tag{1.8}$$

$$Q_{ij} = -V_i^2 \left(b_{si} + b_{ij} \right) - V_i V_j \left(g_{ij} \sin \theta_{ij} - b_{ij} \cos \theta_{ij} \right) \tag{1.9}$$

- Line current flow magnitude from bus i to bus j:

$$I_{ij} = \frac{\sqrt{P_{ij}^2 + Q_{ij}^2}}{V_i} \tag{1.10}$$

where V_i, θ_i is the voltage magnitude and phase angle at bus i, $\theta_{ij} = \theta_i - \theta_j$, g_{ij}, b_{ij} correspond respectively to series conductance and susceptance of the line connecting buses i and j, g_{si}, b_{si} are shunt conductance and susceptance of the line and G_{ij}, B_{ij} are respectively real and imaginary parts of ij-th element of admittance matrix. N_i refers to the set of bus numbers that are directly connected to bus i. The state vector obtained by classic state estimator is composed of bus voltage magnitudes and phase angles:

$$x^T = [\theta_1, \theta_2, \ldots, \theta_n, V_1, V_2, .., V_n] \tag{1.11}$$

3 Linear State Estimation

PMUs provide direct measurements of power system state variables. As a result, the relation between state vector and measurements vector is linear. Although installation of PMUs at every bus has many advantages, there are several issues that must be considered.

The objective function that must be minimized, is the same as (1.2). The only difference is that measurement functions $h(x)$ are linear. Equation (1.1) can be written as [14]:

$$z = h(x) + e = Bx + e \tag{1.12}$$

where B is the system matrix.

The state vector obtained by linear state estimator is calculated as follows:

$$x = \left[B^T R^{-1} B \right]^{-1} B^T R^{-1} z = Mz \tag{1.13}$$

where the matrix M is constant as long as the bus structure does not change. It can be computed offline and stored for real-time use [14]. This equation is linear and no iterations are needed. Therefore, the computation process can be significantly simplified.

4 Hybrid Linear State Estimation

The proposed hybrid linear state estimator uses both conventional and PMU measurements for estimating the state variables of power system. In the first step, only conventional measurements are processed. In order to formulate the post-processing step in a linear form, the state vector obtained in the first step is converted to rectangular form and is fed into the linear estimator in the second step. PMU measurements vector is augmented by the estimate from the classic state estimator in the following form:

$$z_H = \begin{bmatrix} V_{CSE} \\ V_{PMU} \\ I_{PMU} \end{bmatrix} \quad (1.14)$$

The state vector in this case is calculated as follows:

$$x_H = \left[B_H^T R_H^{-1} B_H\right]^{-1} B_H^T R_H^{-1} z_H \quad (1.15)$$

where B_H is the system matrix augmented by the earlier derived system matrix B, R_H is the hybrid covariance matrix. The covariance matrix of linear estimator is diagonally concatenated with the covariance matrix of classic state estimator.

5 Optimal Placement Problem

PMUs are power system devices that provide direct measurements of the system state. A PMU placed at a given bus is capable of measuring the voltage phasor of the bus as well as the phase currents for all lines incident to that bus [15]. The number of PMUs that can be installed in power system is limited due to their cost. Therefore, the objective of the OPP problem is to find minimum number of PMUs such that the entire system is observable [16].

The OPP problem in terms of integer linear programming (ILP) for N-bus system is formulated as follows [17]:

$$\begin{aligned} &\text{Min} \sum_{k=1}^{N} x_k \\ &\text{subject to } T_{PMU} X \geq b_{PMU} \\ &X = [x_1, x_2, \ldots, x_N]^T \\ &x_i \in \{0, 1\} \end{aligned} \quad (1.16)$$

where T_{PMU} is incidence matrix, x_i is PMU placement vector of binary values and b_{PMU} is the inequality constraints vector.

If the conventional measurements are considered in OPP problem, then some constraints must be considered to ILP algorithm [17]. These constraints form a matrix T_{meas} where each column represents a bus associated to conventional measurements. Buses that are not associated with conventional measurements are added together with matrix T_{meas} in the matrix T_{con}.

The OPP problem in terms of integer linear programming (ILP), considering the conventional measurements, is formulated as follows [17]:

$$\begin{aligned} & \text{Min} \sum_{k=1}^{N} x_k \\ & \text{subject to } T_{con} P T_{PMU} X \geq b_{con} \\ & X = [x_1, x_2, \ldots, x_N]^T \\ & x_i \in \{0, 1\} \end{aligned} \quad (1.17)$$

where P is a permutation matrix and b_{con} is the inequality constraints vector.

The generalized form of OPP problem for the cases of redundant PMU placement and full observability is given as follows [17]:

$$\begin{aligned} & \text{Min} \sum_{k=1}^{N} x_k \\ & \text{subject to } \; G T_{PMU} X \geq B_G \\ & X = [x_1, x_2, \ldots, x_N]^T \\ & x_i \in \{0, 1\} \end{aligned} \quad (1.18)$$

where matrix G is the transformation matrix and the column vector B_G indicates the redundancy requirements for all buses.

The OPP algorithm was tested on IEEE 14 bus test system. Integer programming problem was solved using MATLAB Optimization Toolbox. Results show that in case of complete observability and without conventional measurements number of PMUs for exclusive PMU measurements is 4. The number of PMUs for case with inclusion of conventional measurements is 3. It should be noted that the number of conventional measurements and their location in the system can change obtained results.

6 Results and Discussion

The power system state estimation techniques were applied to IEEE test system with 14 buses (Fig. 1). The conventional measurements were placed to obtain complete observability of the proposed system. The OPP algorithm was used to determine the locations of the PMUs.

Table 1 provides the placement of the conventional measurements and the PMUs in the test system. The measurement error variance is assigned to each measurement type according to [18]. Results were compared to true values obtained from power flow calculation output.

In order to compare effectiveness of different estimation techniques, obtained values of voltage magnitudes and angles are shown graphically for all 14 buses (Figs. 2 and 3).

When comparing the results with true values, it can be concluded that the estimation process is enhanced with the inclusion of PMUs. The CSE technique shows large deviation of estimated value from the true value while the LSE technique with PMUs located at the optimal positions shows very good precision and little estimation error. The inclusion of conventional measurements in linear estimator does not affect much the overall result in hybrid estimator. The results of the fast WLAV estimator show the improvement of computational efficiency and accuracy. If a measurement is

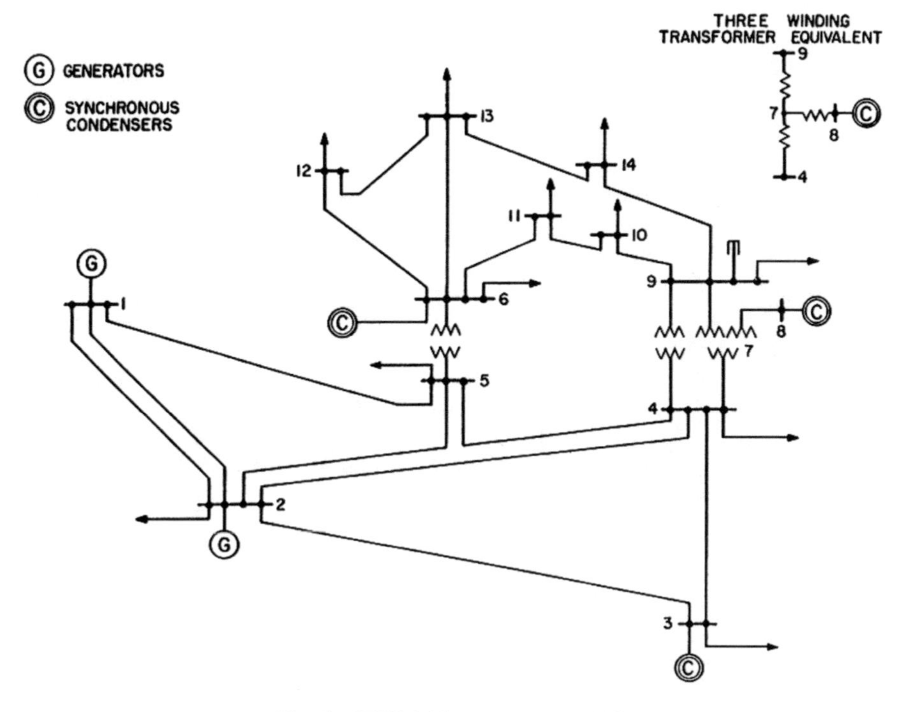

Fig. 1. IEEE 14 bus test system [18]

Table 1. Measurement locations for the IEEE 14 test system

Measurement type	Measurement location
Voltage magnitude	1
Power injection	2, 3, 6, 8, 10, 11, 12, 14
Power flow	1-2, 2-3, 4-2, 4-7, 4-9, 5-2, 5-4, 5-6, 6-13, 7-9, 11-6, 13-14

bad, it should be detected, identified and removed from the estimator calculations [19]. The proposed WLS estimator with reliable bad data processing uses largest normalized residual test to detect bad data. Once all bad data are eliminated, the final estimate can be evaluated by reducing the tolerance [20]. The obtained results show more accurate states in comparison with the classical WLS state estimator.

In order to compare proposed techniques for power system state estimation, several error criteria were adopted. State estimation results for the IEEE 14 test system are presented in Table 2. Figures 4 and 5 give the estimation errors of voltage magnitudes and angles.

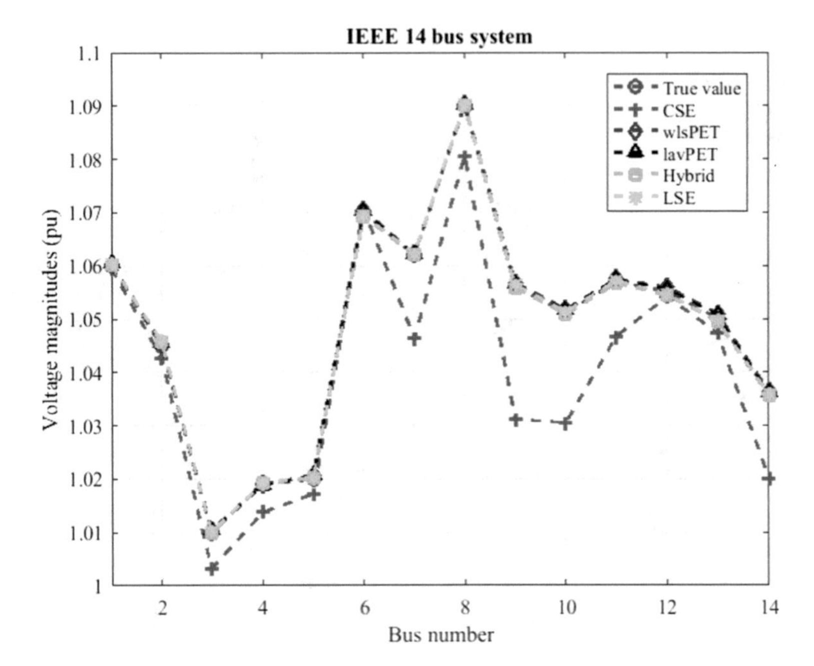

Fig. 2. Simulation result of voltage magnitudes for IEEE 14 test system

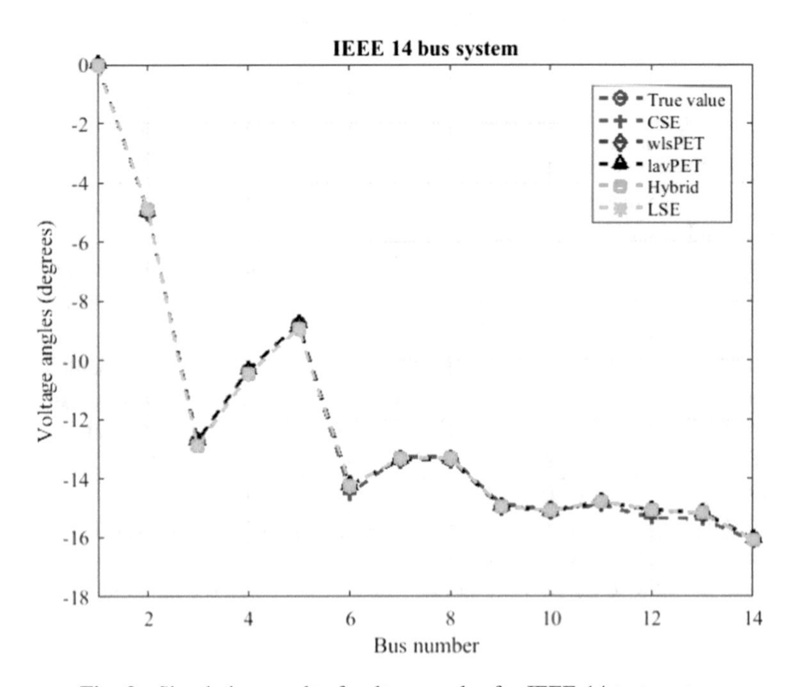

Fig. 3. Simulation result of voltage angles for IEEE 14 test system

Table 2. State estimation results for the IEEE 14 test system

Error criteria	CSE	LSE	Hybrid	wlsPET	lavPET
Maximum magnitude estimation error	0.0236	$8.1700 \cdot 10^{-4}$	$6.4603 \cdot 10^{-4}$	$8.5714 \cdot 10^{-4}$	$6.6667 \cdot 10^{-4}$
Maximum angle estimation error	0.2490	0.1860	0.1822	0.0099	0.0099
Mean magnitude estimation error	0.0085	$3.5650 \cdot 10^{-4}$	$3.1187 \cdot 10^{-4}$	$5.1429 \cdot 10^{-4}$	$3.7143 \cdot 10^{-4}$
Mean angle estimation error	0.0935	0.0558	0.0560	0.0072	0.0036

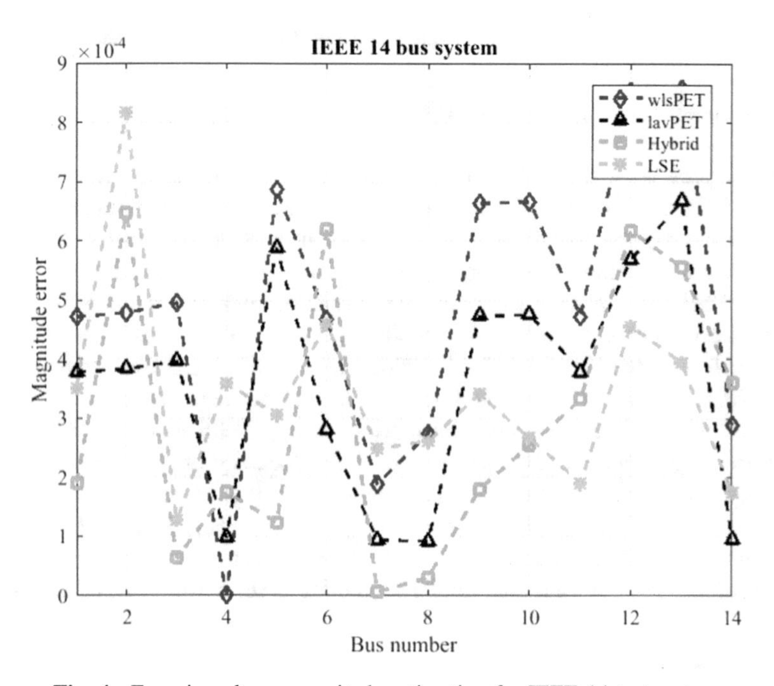

Fig. 4. Error in voltage magnitude estimation for IEEE 14 test system

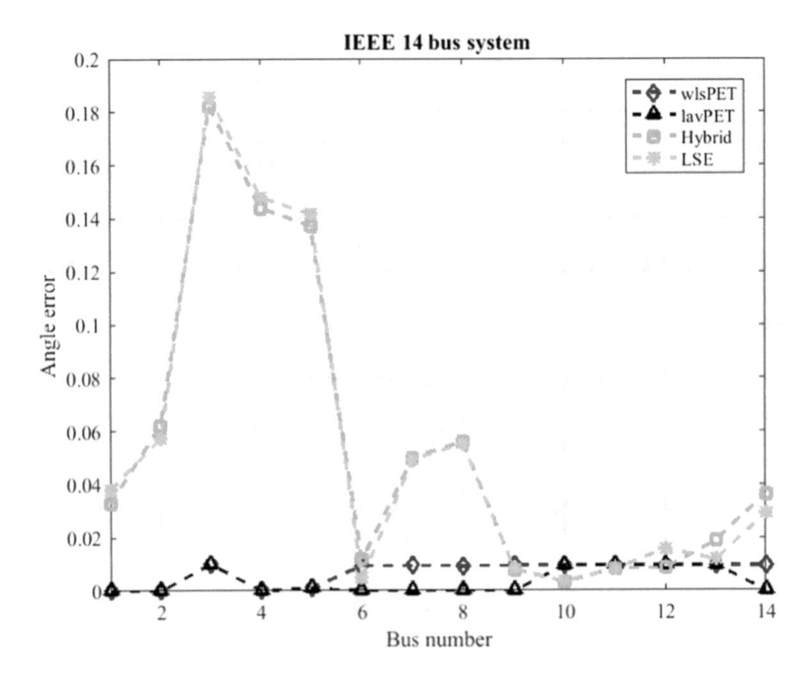

Fig. 5. Error in voltage angle estimation for IEEE 14 test system

7 Conclusions and Future Work

This paper presents a comparison of different techniques for power system state estimation. WLS estimation is a widely used and well-investigated technique. However, it is not robust and requires bad data processor. PMUs become more attractive because they provide direct measurements of state variables. In order to show the importance of including PMUs in classical estimator, linear state estimation technique incorporating phasor measurements as well as conventional measurements was discussed. Linear approach shows time benefits for simulation process with PMU measurements exclusively. The generalized formulation shows that the OPP problem can be modelled linearly and solved effectively using integer linear programming. WLS estimator with a new bad data processing is more reliable and improves the accuracy of the estimation process. WLAV estimator uses a fast algorithm that is simple to implement and computationally more efficient. The future work can include comparison of other hybrid state estimators to one proposed in this paper as well as other state estimation techniques, bad-data and redundancy analysis for all models and estimate for larger power system test cases.

References

1. Schweppe, F.C., Wildes, J.: Power system static-state estimation, part I: exact model. IEEE Trans. Power Appar. Syst. **pas-89**(1), 120–125 (1970)
2. Schweppe, F.C., Rom, D.B.: Power system static-state estimation, part II: approximate model. IEEE Trans. Power Appar. Syst. **pas-89**(1), 125–130 (1970)

3. Schweppe, F.C.: Power system static-state estimation, part III: implementation. IEEE Trans. Power Appar. Syst. **pas-89**(1), 130–135 (1970)
4. Wood, A.J., Wollenberg, B.F.: Power Generation, Operation, and Control. Wiley, Hoboken (1996)
5. Gomez-Exposito, A., Conejo, A.J., Canizares, C.: Electric Energy Systems – Analysis and Operation. CRC Press, Boca Raton (2009)
6. De La Ree, J., Centeno, V., Thorp, J.S., Phadke, A.G.: Synchronized phasor measurement applications in power systems. IEEE Trans. Smart Grid **1**, 20–27 (2010)
7. Nuqui, R.F., Phadke, A.G.: Phasor measurement unit placement techniques for complete and incomplete observability. IEEE Trans. Power Deliv. **20**, 2381–2388 (2005)
8. Nazari-Heris, M., Mohammadi-Ivatloo, B.: Application of heuristic algorithms to optimal PMU placement in electric power systems: an updated review. Renew. Sustain. Energy Rev. **50**, 214–228 (2015)
9. Zhou, M., Centeno, V.A., Thorp, J.S., Phadke, A.G.: An alternative for including phasor measurements in state estimators. IEEE Trans. Power Syst. **21**, 1930–1937 (2006)
10. Monticelli, A., Garcia, A.: Reliable bad data processing for real-time state estimation. IEEE Trans. Power Appar. Syst. **pas-102**(5), 1126–1139 (1983)
11. Abur, A., Celik, M.K.: A fast algorithm for the weighted least absolute value state estimation. IEEE Trans. Power Syst. **6**, 1–8 (1991)
12. Abur, A., Gomez-Exposito, A.: Power System State Estimation – Theory and Implementation. Marcel Dekker, New York City (2004)
13. Glover, J.D., Sarma, M.S., Overbye, T.J.: Power System Analysis and Design. Cengage Learning, Boston (2010)
14. Phadke, A.G., Thorp, J.S.: Synchronized Phasor Measurements and Their Applications. Springer, Heidelberg (2008)
15. Xu, B., Abur, A.: Observability analysis and measurement placement for systems with PMUs. In: Power Systems Conference and Exposition (2004)
16. Gou, B.: Optimal placement of PMUs by integer linear programming. IEEE Trans. Power Syst. **23**, 1525–1526 (2008)
17. Gou, B.: Generalized integer linear programming formulation for optimal PMU placement. IEEE Trans. Power Syst. **23**, 1099–1104 (2008)
18. Majdoub, M., Sabri, O., Cheddadi, B., Belfqih, A., Boukherouaa, J., El Mariami, F., Cherkaoui, N.: Study of state estimation using weighted least squares method. Int. J. Adv. Eng. Res. Sci. **3**(8), 55–63 (2016)
19. Soni, S., Bhil, S., Mehta, D., Wagh, S.: Linear state estimation model using phasor measurement unit (PMU) technology. In: IEEE International Conference on Electrical Engineering, Computing Science and Automatic Control (2012)
20. Vishnu, T.P., Viswan, V., Vipin, A.M.: Power system state estimation and bad data analysis using weighted least square method. In: IEEE International Conference on Power, Instrumentation, Control and Computing (2015)

Fuzzy Multicriteria Decision Making Model for HPP Alternative Selection

Zedina Lavić(✉) and Sabina Dacić-Lepara

Department of Capital Investments, EPC Elektroprivreda B&H D.D. Sarajevo, Sarajevo, Bosnia and Herzegovina
z.lavic@epbih.ba

Abstract. In decision making models developed using the Analytic Hierarchy Process, expert judgments expressed verbally are quantified using the Saaty's scale in the way that they were assigned strict values (crisp numbers). As expert judgments are not really strict, there is a need for these judgments to be more properly quantified. Instead of strict values, they need to be assigned fuzzy numbers. In this paper a fuzzy model for multicriteria decision making on the selection of the best alternative from the Pareto set of technical solutions for the hydroelectric power plant is developed using a Fuzzy Analytic Hierarchy Process. The model is tested on a concrete example and the results are compared with the results of a model developed using the classic Analytic Hierarchy Process.

Keywords: Sustainable development · Hydropower plant Multicriteria decision making · Fuzzy analytic hierarchy process Economic · Technical · Social and environmental criteria

1 Introduction

When comparing matrix elements in the classical Analytic Hierarchy Process [1], evaluations given by experts are expressed verbally. They are in the terms of the Saaty's scale represented by crisp numbers. Expert evaluations can be more realistically represented using the theory of fuzzy sets and fuzzy logic [2]. Unlike classical sets, fuzzy sets have no sharp boundaries, and therefore, in the form of fuzzy numbers, are applicable for the quantification of linguistic variables with vague, imprecise values. In the decision-making model on the selection of the best alternative from the Pareto set of technical solutions for the hydro power plant [3], the criteria are selected respecting the goals of sustainable development and they are classified in the three groups:

- economic and technical criteria
- environmental criteria
- social criteria.

Among criteria there are both qualitative and quantitative ones. In this paper, the mentioned model is fuzzified using a Fuzzy Analytic Hierarchy Process (FAHP) [4]. Applied FAHP, based on Chang's fuzzy extent analysis [5–9], is very appropriate for

S. Avdaković (Ed.): IAT 2018, LNNS 59, pp. 62–69, 2019.
https://doi.org/10.1007/978-3-030-02574-8_6

the selection of the best alternative from the Pareto set of technical solutions for the hydro power plant.

2 Fuzzy Analytic Hierarchy Process

The basic steps of the fuzzy analytic hierarchy process are consistent with the steps of the classic analytic process. There is a difference in the presentation of expert evaluations and the synthesis of results, i.e. the calculation of the ultimate priority of alternatives. In the synthesis of the results, Chang's fuzzy extent analysis was applied.

The development of the model is implemented through a series of steps:

1. Selection of decision making criteria,
2. Selection of methods for multicriteria decision aid,
3. Structuring of the decision making hierarchy,
4. Application of the Fuzzy Analytic Hierarchy Process method,
5. Testing of the model.

3 Hierarchy of Decision Making and Model Parameters

The problem of decision making is structured in a hierarchy that (as in [3]) has 4 hierarchical levels:

1. Objective: selection of the best alternative from the Pareto set of HPP alternatives
2. Criteria groups
3. Criteria (listed in the criteria groups)
4. Alternatives (Pareto set of alternatives).

The number of alternatives is a variable in the model and the specific parameters of the model are (as in [3]):

1. Number of levels in a hierarchical structure (3),
2. Number of criteria groups (3),
3. Total number of criteria (18),
4. Number of economic and technical criteria (5),
5. Number of environmental aspect criteria (6),
6. Number of social aspect criteria (7).

4 Input/Output

Expert's assessments of criteria importance and priorities of alternatives are the inputs of the model. The expert's assessments are fuzzy and they are entered into pairwise comparison matrices:

1. Pairwise comparison fuzzy matrix of criteria groups with respect to the objective

2. Pairwise comparison fuzzy matrix of economic and technical criteria with respect to the corresponding group
3. Pairwise comparison fuzzy matrix of environmental criteria with respect to the corresponding group
4. Pairwise comparison fuzzy matrix of social criteria with respect to the corresponding group
5. Pairwise comparison fuzzy matrices of alternatives with respect to each of the criterion.

The output of the model are priority values for alternatives (crisp numbers), sorted from largest to smallest. The best alternative has the highest value of priority.

5 Consistency Check

Pairwise comparison matrices are positive, square and reciprocal. Elements of matrices are fuzzy numbers from fuzzified Saaty's scale and consistency check is made according [1]. The fuzzy scale is shown in Table 1. If classical matrix is consistent, than correspondent fuzzy matrix is consistent too.

Table 1. Fuzzified Saaty's scale

Definition	Importance intensity	Reciprocals
Equal importance	(1, 1, 1)	(1, 1, 1)
Moderate importance	(2, 3, 4)	(1/4, 1/3, 1/2)
Strong importance	(4, 5, 6)	(1/6, 1/5, 1/4)
Very strong importance	(6, 7, 8)	(1/8, 1/7, 1/6)
Extreme importance	(9, 9, 9)	(1/9, 1/9, 1/9)
Intermediate values	(1, 2, 3), (3, 4, 5), (5, 6, 7) and (7, 8, 9)	(1/3, 1/2, 1), (1/5, 1/4, 1/3), (1/7, 1/6, 1/5) and (1/9, 1/8, 1/7)

6 Algorithm Model

The model can be realized through the algorithm whose flowchart is presented in Fig. 1.

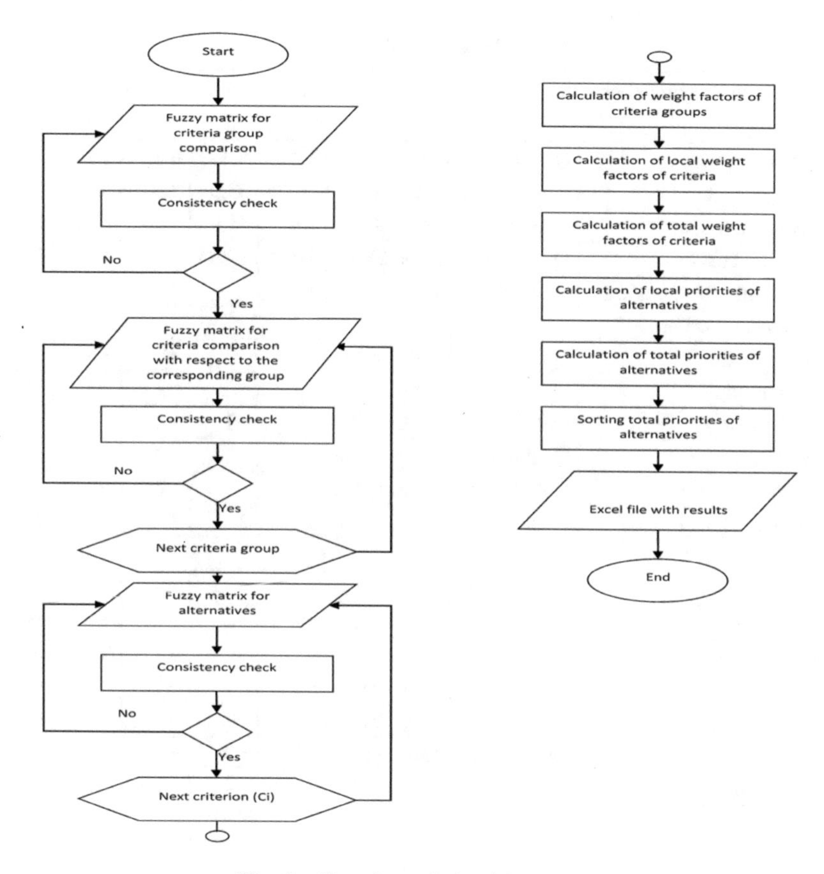

Fig. 1. Flowchart of algorithm model

7 Testing Model on a Concrete Example and Results

The developed model is tested on a concrete example of the selection problem for the best alternative of 5 HPP scheme alternatives, using a Java application made on the basis of the algorithm of the developed model. The alternatives are given in the Table 2. The evaluation scale for qualitative criteria (attributes) is given in the Table 3. To allow comparison of results obtained by applying FAHP to s AHP based model results, the test was performed on the same set of alternatives (in the same example) by the same criteria as in [3].

The decision maker made the pairwise comparisons to give assessment of importance of criteria groups with respect to the objective (matrix 1 in Table 4). The decision maker's assessment of criteria importance with respect to the corresponding criteria group are given in matrices 2–4 in Table 4 and the decision maker's assessment of the priorities of alternatives with respect to each of the criteria are given in matrices 5–22 in Table 4. The results (priorities of alternatives) are given in Table 5.

Table 2. The alternatives

Economic and technical criteria				Alternatives				
				A1	A2	A3	A4	A5
1.	Installed capacity	Quantitative	MW	5.12	6.45	6.35	10.45	10.18
2.	Energy generation	Quantitative	GWh/a	26.04	33.94	32.30	56.27	53.81
3.	Investment costs	Quantitative	Mio EUR	35.5	41.9	41.2	72.7	66.4
4.	Specific costs	Quantitative	EUR/kWh	1.365	1.234	1.277	1.291	1.233
5.	Flexibility in operation	Qualitative	Ling. value	1	2	2	3	3
Environmental criteria								
1.	Aquatic life	Qualitative	Ling. value	−2	−3	−3	−3	−3
2.	Fauna/Flora	Qualitative	Ling. value	−2	−2	−2	−3	−3
3.	Water quality	Qualitative	Ling. value	−1	−1	−1	−2	−2
4.	Air quality	Qualitative	Ling. value	−1	−2	−2	−2	−2
5.	Noise and vibration	Qualitative	Ling. value	−1	−2	−2	−2	−2
6.	Climate change	Qualitative	Ling. value	2	2	2	2	2
Social criteria								
1.	Resettlement	Qualitative	Ling. value	−1	−3	−3	−4	−4
2.	Loss of land	Qualitative	Ling. value	−1	−3	−3	−4	−4
3.	Cultural heritage	Qualitative	Ling. value	0	0	0	0	0
4.	Traffic and infrastructure	Qualitative	Ling. value	−1	−3	−3	−3	−3
5.	Labour and employment	Qualitative	Ling. value	1	−2	−2	−2	−2
6.	Landscape	Qualitative	Ling. value	−2	−3	−3	−4	−4
7.	Community acceptance	Qualitative	Ling. value	−2	−4	−4	−4	−4

Table 3. The evaluation scale for qualitative criteria

	Definition of impact (linguistic values)
3	Strongly positive: highly beneficial effect, affecting a wide area and/or an important parameter
2	Positive: beneficial effect
1	Small positive: beneficial effect of lesser importance
0	None: no or negligible impact
−1	Small negative: negative impact of limited duration
−2	Negative: undesirable or harmful effect of limited concern
−3	Strongly negative: mitigation possible
−4	Strongly negative: mitigation not possible

Table 4. Pairwise comparisons fuzzy matrices

Matrix	No.	CR
[(1, 1, 1) (1, 1, 1) (1, 1, 1) | (1, 1, 1) (1, 1, 1) (1, 1, 1) | (1, 1, 1) (1, 1, 1) (1, 1, 1) |]	1	0.0%
[(1, 1, 1) (1, 2, 3) (2, 3, 4) (2, 3, 4) (1, 2, 3) | (1/3, 1/2, 1) (1, 1, 1) (1, 1, 1) (1, 1, 1) (1, 1, 1) | (1/4, 1/3, 1/2) (1, 1, 1) (1, 1, 1) (1, 1, 1) (1, 1, 1) | (1/4, 1/3, 1/2) (1, 1, 1) (1, 1, 1) (1, 1, 1) (1, 1, 1) | (1/3, 1/2, 1) (1, 1, 1) (1, 1, 1) (1, 1, 1) (1, 1, 1) |]	2	0.71%
[(1, 1, 1) (1, 1, 1) (1, 1, 1) (1, 1, 1) (1, 1, 1) (1, 1, 1) | (1, 1, 1) (1, 1, 1) (1, 1, 1) (1, 1, 1) (1, 1, 1) (1, 1, 1) | (1, 1, 1) (1, 1, 1) (1, 1, 1) (1, 1, 1) (1, 1, 1) (1, 1, 1) | (1, 1, 1) (1, 1, 1) (1, 1, 1) (1, 1, 1) (1, 1, 1) (1, 1, 1) | (1, 1, 1) (1, 1, 1) (1, 1, 1) (1, 1, 1) (1, 1, 1) (1, 1, 1) | (1, 1, 1) (1, 1, 1) (1, 1, 1) (1, 1, 1) (1, 1, 1) (1, 1, 1) |]	3	0.0%
[(1, 1, 1) (1, 2, 3) (2, 3, 4) (3, 4, 5) (3, 4, 5) (3, 4, 5) (3, 4, 5) | (1/3, 1/2, 1) (1, 1, 1) (1, 2, 3) (2, 3, 4) (2, 3, 4) (2, 3, 4) (2, 3, 4) | (1/4, 1/3, 1/2) (1/3, 1/2, 1) (1, 1, 1) (1, 2, 3) (1, 1, 1) (1, 1, 1) (1, 1, 1) | (1/5, 1/4, 1/3) (1/4, 1/3, 1/2) (1/3, 1/2, 1) (1, 1, 1) (1, 1, 1) (1, 1, 1) (1, 1, 1) | (1/5, 1/4, 1/3) (1/4, 1/3, 1/2) (1, 1, 1) (1, 1, 1) (1, 1, 1) (1, 1, 1) (1, 1, 1) | (1/5, 1/4, 1/3) (1/4, 1/3, 1/2) (1, 1, 1) (1, 1, 1) (1, 1, 1) (1, 1, 1) (1, 1, 1) | (1/5, 1/4, 1/3) (1/4, 1/3, 1/2) (1, 1, 1) (1, 1, 1) (1, 1, 1) (1, 1, 1) (1, 1, 1) |]	4	0.93%
[(1, 1, 1) (1/5, 1/4, 1/3) (1/4, 1/3, 1/2) (1/9, 1/9, 1/9) (1/9, 1/8, 1/7) | (3, 4, 5) (1, 1, 1) (1/3, 1/2, 1) (1/8, 1/7, 1/6) (1/7, 1/6, 1/5) | (2, 3, 4) (1, 2, 3) (1, 1, 1) (1/7, 1/6, 1/5) (1/7, 1/6, 1/5) | (9, 9, 9) (6, 7, 8) (5, 6, 7) (1, 1, 1) (1, 1, 1) | (7, 8, 9) (5, 6, 7) (5, 6, 7) (1, 1, 1) (1, 1, 1) |]	5	7.28%
[(1, 1, 1) (1/5, 1/4, 1/3) (1/4, 1/3, 1/2) (1/9, 1/9, 1/9) (1/9, 1/8, 1/7) | (3, 4, 5) (1, 1, 1) (1/3, 1/2, 1) (1/8, 1/7, 1/6) (1/7, 1/6, 1/5) | (2, 3, 4) (1, 2, 3) (1, 1, 1) (1/7, 1/6, 1/5) (1/7, 1/6, 1/5) | (9, 9, 9) (6, 7, 8) (5, 6, 7) (1, 1, 1) (1, 1, 1) | (7, 8, 9) (5, 6, 7) (5, 6, 7) (1, 1, 1) (1, 1, 1) |]	6	7.28%
[(1, 1, 1) (1/5, 1/4, 1/3) (1/4, 1/3, 1/2) (1/9, 1/9, 1/9) (1/9, 1/8, 1/7) | (3, 4, 5) (1, 1, 1) (1/3, 1/2, 1) (1/8, 1/7, 1/6) (1/7, 1/6, 1/5) | (2, 3, 4) (1, 2, 3) (1, 1, 1) (1/7, 1/6, 1/5) (1/7, 1/6, 1/5) | (9, 9, 9) (6, 7, 8) (5, 6, 7) (1, 1, 1) (1, 1, 1) | (7, 8, 9) (5, 6, 7) (5, 6, 7) (1, 1, 1) (1, 1, 1) |]	7	7.28%
[(1, 1, 1) (1/6, 1/5, 1/4) (1/5, 1/4, 1/3) (1/4, 1/3, 1/2) (1/7, 1/6, 1/5) | (4, 5, 6) (1, 1, 1) (3, 4, 5) (3, 4, 5) (1/3, 1/2, 1) | (3, 4, 5) (1/5, 1/4, 1/3) (1, 1, 1) (2, 3, 4) (1/5, 1/4, 1/3) | (2, 3, 4) (1/5, 1/4, 1/3) (1/4, 1/3, 1/2) (1, 1, 1) (1/6, 1/5, 1/4) | (5, 6, 7) (1, 2, 3) (3, 4, 5) (4, 5, 6) (1, 1, 1) |]	8	9.68%
[1 1/2 1/2 1/3 1/3 | 2 1 1 1/2 1/2 | 2 1 1 1/2 1/2 | 3 2 2 1 1 | 3 2 2 1 1 |] [(1, 1, 1) (1/3, 1/2, 1) (1/3, 1/2, 1) (1/4, 1/3, 1/2) (1/4, 1/3, 1/2) | (1, 2, 3) (1, 1, 1) (1, 1, 1) (1/3, 1/2, 1) (1/3, 1/2, 1) | (1, 2, 3) (1, 1, 1) (1, 1, 1) (1/3, 1/2, 1) (1/3, 1/2, 1) | (2, 3, 4) (1, 2, 3) (1, 2, 3) (1, 1, 1) (1, 1, 1) | (2, 3, 4) (1, 2, 3) (1, 2, 3) (1, 1, 1) (1, 1, 1) |]	9	0.34%
[(1, 1, 1) (1, 2, 3) (1, 2, 3) (1, 2, 3) (1, 2, 3) | (1/3, 1/2, 1) (1, 1, 1) (1, 1, 1) (1, 1, 1) (1, 1, 1) | (1/3, 1/2, 1) (1, 1, 1) (1, 1, 1) (1, 1, 1) (1, 1, 1) | (1/3, 1/2, 1) (1, 1, 1) (1, 1, 1) (1, 1, 1) (1, 1, 1) | (1/3, 1/2, 1) (1, 1, 1) (1, 1, 1) (1, 1, 1) (1, 1, 1) |]	10	0.0%
[(1, 1, 1) (1, 1, 1) (1, 1, 1) (1, 2, 3) (1, 2, 3) | (1, 1, 1) (1, 1, 1) (1, 1, 1) (1, 2, 3) (1, 2, 3) | (1, 1, 1) (1, 1, 1) (1, 1, 1) (1, 2, 3) (1, 2, 3) | (1/3, 1/2, 1) (1/3, 1/2, 1) (1/3, 1/2, 1) (1, 1, 1) (1, 1, 1) | (1/3, 1/2, 1) (1/3, 1/2, 1) (1/3, 1/2, 1) (1, 1, 1) (1, 1, 1) |]	11	0.0%

(*continued*)

Table 4. (*continued*)

Matrix	No.	CR
[(1, 1, 1) (1, 1, 1) (1, 1, 1) (1, 2, 3) (1, 2, 3) | (1, 1, 1) (1, 1, 1) (1, 1, 1) (1, 2, 3) (1, 2, 3) | (1, 1, 1) (1, 1, 1) (1, 1, 1) (1, 2, 3) (1, 2, 3) | (1/3, 1/2, 1) (1/3, 1/2, 1) (1/3, 1/2, 1) (1, 1, 1) (1, 1, 1) | (1/3, 1/2, 1) (1/3, 1/2, 1) (1/3, 1/2, 1) (1, 1, 1) (1, 1, 1) |]	12	0.0%
[(1, 1, 1) (1, 2, 3) (1, 2, 3) (1, 2, 3) (1, 2, 3) | (1/3, 1/2, 1) (1, 1, 1) (1, 1, 1) (1, 1, 1) (1, 1, 1) | (1/3, 1/2, 1) (1, 1, 1) (1, 1, 1) (1, 1, 1) (1, 1, 1) | (1/3, 1/2, 1) (1, 1, 1) (1, 1, 1) (1, 1, 1) (1, 1, 1) | (1/3, 1/2, 1) (1, 1, 1) (1, 1, 1) (1, 1, 1) (1, 1, 1) |]	13	0.0%
[(1, 1, 1) (1, 2, 3) (1, 2, 3) (1, 2, 3) (1, 2, 3) | (1/3, 1/2, 1) (1, 1, 1) (1, 1, 1) (1, 1, 1) (1, 1, 1) | (1/3, 1/2, 1) (1, 1, 1) (1, 1, 1) (1, 1, 1) (1, 1, 1) | (1/3, 1/2, 1) (1, 1, 1) (1, 1, 1) (1, 1, 1) (1, 1, 1) | (1/3, 1/2, 1) (1, 1, 1) (1, 1, 1) (1, 1, 1) (1, 1, 1) |]	14	0.0%
[(1, 1, 1) (1, 1, 1) (1, 1, 1) (1, 1, 1) (1, 1, 1) | (1, 1, 1) (1, 1, 1) (1, 1, 1) (1, 1, 1) (1, 1, 1) | (1, 1, 1) (1, 1, 1) (1, 1, 1) (1, 1, 1) (1, 1, 1) | (1, 1, 1) (1, 1, 1) (1, 1, 1) (1, 1, 1) (1, 1, 1) | (1, 1, 1) (1, 1, 1) (1, 1, 1) (1, 1, 1) (1, 1, 1) |]	15	0.0%
[(1, 1, 1) (2, 3, 4) (2, 3, 4) (3, 4, 5) (3, 4, 5) | (1/4, 1/3, 1/2) (1, 1, 1) (1, 1, 1) (1, 2, 3) (1, 2, 3) | (1/4, 1/3, 1/2) (1, 1, 1) (1, 1, 1) (1, 2, 3) (1, 2, 3) | (1/5, 1/4, 1/3) (1/3, 1/2, 1) (1/3, 1/2, 1) (1, 1, 1) (1, 1, 1) | (1/5, 1/4, 1/3) (1/3, 1/2, 1) (1/3, 1/2, 1) (1, 1, 1) (1, 1, 1) |]	16	0.84%
[(1, 1, 1) (2, 3, 4) (2, 3, 4) (3, 4, 5) (3, 4, 5) | (1/4, 1/3, 1/2) (1, 1, 1) (1, 1, 1) (1, 2, 3) (1, 2, 3) | (1/4, 1/3, 1/2) (1, 1, 1) (1, 1, 1) (1, 2, 3) (1, 2, 3) | (1/5, 1/4, 1/3) (1/3, 1/2, 1) (1/3, 1/2, 1) (1, 1, 1) (1, 1, 1) | (1/5, 1/4, 1/3) (1/3, 1/2, 1) (1/3, 1/2, 1) (1, 1, 1) (1, 1, 1) |]	17	0.84%
[(1, 1, 1) (1, 1, 1) (1, 1, 1) (1, 1, 1) (1, 1, 1) | (1, 1, 1) (1, 1, 1) (1, 1, 1) (1, 1, 1) (1, 1, 1) | (1, 1, 1) (1, 1, 1) (1, 1, 1) (1, 1, 1) (1, 1, 1) | (1, 1, 1) (1, 1, 1) (1, 1, 1) (1, 1, 1) (1, 1, 1) | (1, 1, 1) (1, 1, 1) (1, 1, 1) (1, 1, 1) (1, 1, 1) |]	18	0.0%
[(1, 1, 1) (2, 3, 4) (2, 3, 4) (2, 3, 4) (2, 3, 4) | (1/4, 1/3, 1/2) (1, 1, 1) (1, 1, 1) (1, 1, 1) (1, 1, 1) | (1/4, 1/3, 1/2) (1, 1, 1) (1, 1, 1) (1, 1, 1) (1, 1, 1) | (1/4, 1/3, 1/2) (1, 1, 1) (1, 1, 1) (1, 1, 1) (1, 1, 1) | (1/4, 1/3, 1/2) (1, 1, 1) (1, 1, 1) (1, 1, 1) (1, 1, 1) |]	19	0.0%
[(1, 1, 1) (3, 4, 5) (3, 4, 5) (3, 4, 5) (3, 4, 5) | (1/5, 1/4, 1/3) (1, 1, 1) (1, 1, 1) (1, 1, 1) (1, 1, 1) | (1/5, 1/4, 1/3) (1, 1, 1) (1, 1, 1) (1, 1, 1) (1, 1, 1) | (1/5, 1/4, 1/3) (1, 1, 1) (1, 1, 1) (1, 1, 1) (1, 1, 1) | (1/5, 1/4, 1/3) (1, 1, 1) (1, 1, 1) (1, 1, 1) (1, 1, 1) |]	20	0.0%
[(1, 1, 1) (1, 2, 3) (1, 2, 3) (2, 3, 4) (2, 3, 4) | (1/3, 1/2, 1) (1, 1, 1) (1, 1, 1) (2, 3, 4) (2, 3, 4) | (1/3, 1/2, 1) (1, 1, 1) (1, 1, 1) (1, 2, 3) (1, 2, 3) | (1/4, 1/3, 1/2) (1/4, 1/3, 1/2) (1/3, 1/2, 1) (1, 1, 1) (1, 1, 1) | (1/4, 1/3, 1/2) (1/4, 1/3, 1/2) (1/3, 1/2, 1) (1, 1, 1) (1, 1, 1) |]	21	1.59%
[(1, 1, 1) (2, 3, 4) (2, 3, 4) (2, 3, 4) (2, 3, 4) | (1/4, 1/3, 1/2) (1, 1, 1) (1, 1, 1) (1, 1, 1) (1, 1, 1) | (1/4, 1/3, 1/2) (1, 1, 1) (1, 1, 1) (1, 1, 1) (1, 1, 1) | (1/4, 1/3, 1/2) (1, 1, 1) (1, 1, 1) (1, 1, 1) (1, 1, 1) | (1/4, 1/3, 1/2) (1, 1, 1) (1, 1, 1) (1, 1, 1) (1, 1, 1) |]	22	0.0%

Table 5. Priorities of alternatives

Alternative	Priority (AHP) [3]	Priority (FAHP)
A1	0.2491	0.3334
A5	0.2182	0.1979
A4	0.2061	0.2283
A2	0.1659	0.1201
A3	0.1603	0.1201

8 Conclusion

In this paper the FAHP model for HPP scheme selection from Pareto set of solutions (HPP scheme alternatives) is developed. This model was developed by the fuzzification of the existing model based on classical AHP in such way that expert's assessments (linguistic variables) was represented by triangular fuzzy numbers and Chang's fuzzy extent analysis was applied. The fuzzy scale was obtained by the appropriate fuzzification of the Saaty's scale so that the consistency of the classical matrix implies the consistency of the fuzzy matrix. The model was tested on concrete example (in Java application, made according to the algorithm). The best alternative in the FAHP model was also the best in the classical AHP model. The FAHP model is more realistic than classical AHP model though expert's assessments (linguistic variables) instead of by crisp are represented by fuzzy numbers. Future research can be conducted in the direction of the FAHP model extensions for group multicriteria decision making.

References

1. Saaty, T.L.: Analytic Hierarchy Process. McGraw-Hill, New York (1980)
2. Zadeh, L.A.: Fuzzy set. Inf. Control **8**, 338–353 (1965)
3. Lavic, Z., Dacić-Lepara, S.: Multicriteria decision making model for HPP alternative selection. In: Hadžikadić, M., Avdaković, S. (eds.) Advanced Technologies, Systems, and Applications II. AG 2018. Lecture Notes in Networks and Systems, vol. 28, pp 170–177. Springer, Heidelberg (2018) https://doi.org/10.1007/978-3-319-71321-2_14
4. Kahraman, C.: Fuzzy Multi-criteria Decision Making Theory and Applications with Recent Developments. Springer, New York (2008)
5. Kabir, G., Akhtar, H.A.: Comparative analysis of AHP and fuzzy AHP models for multicriteria inventory classification. Int. J. Fuzzy Log. Syst. (IJFLS) **1**(1), 1–16 (2011)
6. Chang, D.Y.: Applications of the extent analysis method on fuzzy AHP. Eur. J. Oper. Res. **95** (1996), 649–655 (1996)
7. Zyoud, S., Kaufmann, L., Shaheen, H., Samhan, S., Fuchs-Hanusch, D.: A framework for water loss management in developing countries under fuzzy environment: integration of fuzzy AHP with fuzzy TOPSIS. Expert Syst. Appl. **61**, 86–105 (2016)
8. Bozbura, F.T., Beskese, A., Kahraman, C.: Prioritization of human capital measurement indicators using fuzzy AHP. Expert Syst. Appl. **32**(4), 1100–1112 (2007). https://doi.org/10.1016/j.eswa.2006.02.006
9. Chen, J., Hsieh, H., Do, Q.: Evaluating teaching performance based on fuzzy AHP and comprehensive evaluation approach. Appl. Soft Comput. **28**, 100–108 (2015)

The Valuation of Kron Reduction Application in Load Flow Methods

Tarik Hubana[1(✉)], Sidik Hodzic[2], Emir Alihodzic[3], and Ajdin Mulaosmanovic[4]

[1] Public Enterprise Elektroprivreda of Bosnia and Herzegovina Mostar, Mostar, Bosnia and Herzegovina
t.hubana@epbih.ba

[2] Raible + Partner GmbH & Co., Munich, Germany
s.hodzic@raible.de

[3] Public Enterprise Elektroprivreda of Bosnia and Herzegovina, Zenica, Zenica, Bosnia and Herzegovina
e.alihodzic@epbih.ba

[4] Public Enterprise Elektroprivreda of Bosnia and Herzegovina, Jablanica, Jablanica, Bosnia and Herzegovina
a.mulaosmanovic@epbih.ba

Abstract. The presence of new smart monitoring devices in the modern power system eases the operation of the system, and as a part of this, computational load flow calculations are becoming more present. Load flow calculations as a part of the smart grid monitoring systems have an important role in the real time calculations. Thus, their optimization is crucial, by making the calculation time as fast as possible, by using less hardware resources. One way to achieve this is to use the Kron reduction for node elimination in the load flow process. This paper analyses and compares the efficiency of the Kron reduction in different load flow methods. Results obtained from the developed real 10-bus, 17-bus and 50-bus systems show the efficiency of the Kron reduction applied on five different load flow methods. It is demonstrated that the proposed method is capable to significantly reduce the computational time and the number of load flow's iterations. This paper makes a contribution to the existing body of knowledge by testing and comparing the Kron reduction effect on multiple load flow methods, whose application represents an improvement when compared to the operation of the standard load flow algorithms.

1 Introduction

Current power systems must be able to meet ever-growing demands for electrical energy, deal with constant additions and expansion of power facilities, increasing number of renewable energy sources, power electronics devices, etc. Power systems analysis need to be carried out constantly in modern power systems to make them efficient, functional and well-planned. Power flow analysis is often a first step in that process. It is an essential part of planning, managing and monitoring processes in power systems. Because of that, it is crucial to make power flow analysis fast as possible in order to obtain fast and valid solutions to control those processes in

S. Avdaković (Ed.): IAT 2018, LNNS 59, pp. 70–85, 2019.
https://doi.org/10.1007/978-3-030-02574-8_7

real-time. Power flow analysis methods are based on an iterative solution of a large nonlinear equations systems. Gauss-Seidel and Newton-Raphson (NR) methods are commonly used to solve the power flow model. Both of these methods in their design include large matrices with a lot of interconnected data both depending on a system scale. Hence the reason why power flow analysis in large power systems is in most cases very slow and does not meet the requirements for efficient power analysis in terms of time needed to obtain results. In spite of ever-growing computer speed and memory performance, it is very important to find a way to simplify those large systems of nonlinear equations and consequently reduce the calculating time. One of the methods that is widely used in this purpose is Kron reduction. Kron reduction is basically a method for eliminating unnecessary data from large matrices in power flow analysis calculations. The outcome of that elimination is reduced calculating time.

2 Literature Review

From the moment power flow analysis problem started to emerge, engineers conducted a lot of attempts to solve this problem and make it applicable in the real-time studies. The starting point for all of those attempts is of course formulation of the basic power flow problem which is nicely presented in [1]. Furthermore, this basic form of power flow analysis started to evolve and develop. One of the proposed improvements is Fast Decoupled Load Flow model which is presented in [2] and this method itself evolved in terms of speed as described in [3] and in terms of widening its applicable area as described in [4]. As mentioned before Gauss-Seidel and Newton-Raphson methods are commonly used to solve the power flow model and their basic mathematical formulations will be described in this paper.

There are a lot of papers describing how to develop these basic methods. In [5], an algorithm that is used for the solution of three- phase (or unsymmetrical) power flow analysis of both transmission and distribution systems under unsymmetrical operating conditions is described. One of the further developments of the Newton-Raphson method is described in [6] where elements of the Jacobian matrix are obtained considering the power flows in the network elements. In that manner, in [7], an adjustments incorporated within the Newton–Raphson method are used to obtain a faster rate of convergence, to provide a larger region of convergence or both. Furthermore, [8] proposes an entire family of numerically efficient algorithms and shows that Newton–Raphson's method and most robust power flow techniques proposed in the literature are particular cases of the given formulation. As far as Gauss–Seidel method is concerned, the parallelization and implementations of Gauss–Seidel algorithms for power flow analysis have been investigated in [9]. Of course, there are a lot of papers that give comparisons between these two methods and one of them is [10] where an operational comparison between Gauss–Seidel and Newton Raphson method is depicted using MATLAB simulation software for a 4–Bus system. Also, there are certain techniques proposed for solving power flow model and in [11] a general power flow algorithm for three-phase four-wire radial distribution networks, considering neutral grounding, based on backward–forward technique, is proposed.

As mentioned before, one of the methods that is widely used for power flow model reduction is Kron reduction. Authors in [12] provide a comprehensive and detailed graph-theoretic analysis of the Kron reduction process and physical insights in the application domains of Kron reduction. In domain of electrical networks, papers [13] and [14] describe how to perform Kron reduction for a class of electrical networks called homogeneous electrical networks without steady state assumptions. Generally there are different techniques for power flow model reduction, which is described in many papers. Some of the proposed techniques use Kron reduction as [15] which analyzes the reduction process relating the two power network models where the reduced admittance matrix is obtained by a Schur complement. Also, [16] proposes model-reduction methods based on singular perturbation and Kron reduction to reduce large-signal dynamic models of inverter-based islanded microgrids. Both for the linear and the nonlinear model of the network using Kron reduction, [17] proposes explicit reduced order models which are expressed in terms of ordinary differential equations. Of course, there are other methods for power flow model reduction and in purpose of comparison, [18] explores some of the commonly used static network reduction techniques, such as Ward reduction, Kron reduction, Dimo's method, and Zhukov's method, and the performances of the reduced networks are evaluated in terms of their ability to follow the busbar voltages of the original network, with changes in operating conditions. This section shows that there are a lot of researches in this area. The aim is always the same – to improve traditional power flow models. This paper will show to what extent it is possible in different power flow models by using Kron reduction.

3 Methodology

In this chapter an overview of the traditional commonly used load flow methods is given, as well as the mathematical formulation of the Kron reduction process.

3.1 Y-Matrix Method

This iterative process for solving the power flow problem is applied by direct implementation of the iterative scheme on certain conductivity matrix, leaving unknown voltages on the left side of a system of nonlinear equations. For solving the system with this method, any of three processes can be applied (Jacobi, Gauss-Seidel, successive overreaction). For example, if Gauss-Seidel (YGS) iterative process is applied, the corresponding iterative scheme appears in following form [19]:

$$\begin{gathered} U_1^{k+1} = \frac{1}{Y_{11}}\left(\frac{P_1 - jQ_1}{U_1^{k*}} - \mathrm{Y}_{10}\mathrm{U}_0 - \mathrm{Y}_{11}U_1^k - \mathrm{Y}_{12}U_2^k - \mathrm{Y}_{13}U_3^k \ldots - Y_{1N}U_N^k\right) \\ U_2^{k+1} = \frac{1}{Y_{11}}\left(\frac{P_2 - jQ_2}{U_2^{k*}} - \mathrm{Y}_{20}\mathrm{U}_0 - \mathrm{Y}_{21}U_1^{k+1} - \mathrm{Y}_{23}U_3^k \ldots - Y_{2N}U_N^k\right) \\ \vdots \\ U_N^{k+1} = \frac{1}{Y_{NN}}\left(\frac{P_N - jQ_N}{U_N^{k*}} - \mathrm{Y}_{\mathrm{N}0}\mathrm{U}_0 - \mathrm{Y}_{\mathrm{N}1}U_1^{k+1} - \mathrm{Y}_{\mathrm{N}2}U_2^k - \mathrm{Y}_{\mathrm{N}3}U_3^{k+1} \ldots - Y_{N,N-1}U_{N-1}^{k+1}\right). \end{gathered} \tag{1}$$

With known members of matrix [Y] and known voltage of reference busbar (e.g. $U_0 = 1, 1 + j0$), iterative process starts with definition of initial busbar voltage vector [19]:

$$[U]^0 = \begin{bmatrix} U_1^0 \\ U_2^0 \\ \vdots \\ U_N^0 \end{bmatrix} \tag{2}$$

Members of initial busbar voltage vector can all be equal (e.g. $U_i^0 = 1 + j0$). After each iterative cycle, solution convergence is being controlled [19]:

$$\left| [U]^{k+1} - [U]^k \right| \leq \varepsilon \tag{3}$$

An iterative process is terminated at the moment when the module of voltage difference of all busbars between two adjacent iterations is less or equal to a defined value of variable **ε**. Iterative procedure in case of P|U| busbars is implemented as follows [19]:

- Based on known voltages of all busbars in iterative cycle **k**, reactive power of busbar **m** can be determined as follows:

$$Q_m^k = \mathrm{Im}\left\{ -U_m^{k*}(Y_{m0}U_0 + \sum_{l=1}^{N} Y_{ml}U_l^k) \right\} \tag{4}$$

- Obtained value of reactive power enables determination of all voltages in iterative cycle **k + 1**. On this way, the voltage of busbar **m** can also be obtained. If the obtained voltage of busbar **m** is expressed through its real and imaginary part, i.e.:

$$U_m^{k+1} = e_m^{k+1} + jf_m^{k+1} \tag{5}$$

Corresponding voltage angle of busbar **m** is:

$$\theta_m^{k+1} = arctg(\frac{f_m^{k+1}}{e_m^{k+1}}) \tag{6}$$

- Considering that busbar **m** has defined voltage module |Um|, new voltage value of busbar **m** is determined by:

$$U_m^{k+1} = |U_m|\cos(\theta_m^{k+1}) + j|U_m|\sin(\theta_m^{k+1}) \tag{7}$$

- Based on this corrected voltage of busbar **m**, reactive power is determined by using the relation (4).

Devices that produce or consume reactive power, have limits for their power values. These limits for reactive power of busbar **m** are defined with [19]:

$$Q_m^{\min} \le Q_m \le Q_m^{\max} \tag{8}$$

During the previously described iterative process, it may occur that reactive power defined by the relation (4) comes out of the limits defined in (8). In this case, two following rules have to be applied [19]:

- If $Q_m > Q_m^{\max}$ then $Q_m = Q_m^{\max}$,
- If $Q_m < Q_m^{\min}$ then $Q_m = Q_m^{\min}$.

When one of previously described cases occurs, voltage correction of busbar **m** will not be done, to ensure the given voltage module for this busbar. Voltage correction is implemented again when the value of reactive power Q_m comes back between limits given in relation (8).

3.2 Z-Matrix Method

This method for solving the power flow problem is similar to Y-Matrix Method. The difference is that before the beginning of iterative process, inversion of Conductivity Matrix [Y] must be performed. Nodal currents, determined by the voltages of the iteration cycle **k**, have the following form [19]:

$$I_i^k = \frac{P_i - jQ_i}{U_i^{k*}} - Y_{i0}U_0 \tag{9}$$

System of equations arranged for Z Gauss-Seidel (ZGS) iterative process is [19]:

$$\begin{aligned} U_1^{k+1} &= Z_{11}I_1^k + Z_{12}I_2^k + Z_{13}I_3^k + \ldots + Z_{1N}I_N^k \\ U_2^{k+1} &= Z_{21}I_1^k + Z_{22}I_2^k + Z_{23}I_3^k + \ldots + Z_{2N}I_N^k \\ &\vdots \\ U_N^{k+1} &= Z_{N1}I_1^k + Z_{N2}I_2^k + Z_{N3}I_3^k + \ldots + Z_{NN}I_N^k \end{aligned} \tag{10}$$

Iterative process begins with determination of busbar voltages starting vector. On this basis, nodal currents can be defined according to the relation (9). Furthermore, new voltages are determined by the Eq. (10). Described process is to be reiterated until the solution convergence is achieved.

Busbars with defined voltage module (P|U| type busbar) are taken into consideration similarly as the procedure described by Y-Matrix Method. An iterative procedure is applied on following relation [19]:

$$Q_m = \mathrm{Im}\left\{-U_m^* Y_{m0} U_0 - \frac{U_m^*}{Z_{mm}}\left(U_m - \sum_{l=1, l \neq m}^{N} Z_{ml} I_l\right)\right\} \tag{11}$$

The rest of the process is the same as Y-Matrix Method. In previous considerations, matrix [Y] and [Z] are set in relation to the ground as reference busbar, marked with index **0**.

3.3 Newton-Raphson Process-Mathematical Formulation

For this method, the reference busbar is marked with **0** and it has defined voltage module |U| and angle θ_0. The remaining busbars are P|U| and PQ busbars. **N** is the total number of system busbars, not including the reference busbar. Nodal power of the equivalent generator for each node is given by active and reactive power, creating a system of nonlinear algebraic equations that is used for execution of the final equations of Newton-Raphson (NR) process [19]:

$$P_i = \sum_{k=0}^{N} |U_i||U_k|(g_{ik} \cos \Theta_{ik} + b_{ik} \sin \Theta_{ik}),\ \mathrm{i} = 0,\ 1,\ 2,\ \ldots,\ \mathrm{N} \tag{12}$$

$$Q_i = \sum_{k=0}^{N} |U_i||U_k|(g_{ik} \sin \Theta_{ik} - b_{ik} \cos \Theta_{ik}),\ \mathrm{i} = 0,\ 1,\ 2,\ \ldots,\ \mathrm{N}$$

Considering that the busbar **0** was declared as the reference busbar (|U| and θ_0 are defined), it is necessary to determine N system busbars (N voltage modules and N voltage angles) [19]:

$$[\theta] = \begin{bmatrix} \theta_1 \\ \vdots \\ \theta_N \end{bmatrix} \qquad [|U|] = \begin{bmatrix} |U|_1 \\ \vdots \\ |U|_N \end{bmatrix} \tag{13}$$

Vectors [θ] and [|U|] (13) represent the solution vector for the system of Eq. (12). The mathematical formulation of the NR process begins with a basic equation for power flow calculation with the standard NR process [19]:

$$\begin{bmatrix} [J]_{11} & [J]_{12} \\ [J]_{21} & [J]_{22} \end{bmatrix} \begin{bmatrix} \Delta\theta \\ \Delta|U| \end{bmatrix} = -\begin{bmatrix} \Delta P \\ \Delta Q \end{bmatrix} \tag{14}$$

3.3.1 Decoupled Newton-Raphson Process-Mathematical Formulation

In real electrical power systems, differences between voltage angles of particular (adjacent) system elements are very small, so that corresponding sinuses of these angles can be ignored. Besides this, the active conductance of submatrix members of Jacobian matrix can also be ignored, which is a case in the transmission grid. With previously mentioned limits, members of submatrices $[J]_{12}$ i $[J]_{21}$, that represent derivations of active power by voltage modules, are as follows [19]:

$$\begin{bmatrix} [J]_{11} & 0 \\ 0 & [J]_{22} \end{bmatrix} \begin{bmatrix} \Delta\theta \\ \Delta|U| \end{bmatrix} = - \begin{bmatrix} \Delta P \\ \Delta Q \end{bmatrix} \tag{15}$$

Previous equation can be divided into two separate equations [19]:

$$\begin{aligned} [J]_{11}\Delta\theta &= -\Delta P \\ [J]_{22}\Delta|U| &= -\Delta Q \end{aligned} \tag{16}$$

Equations (16) represent basic equations of Decoupled Newton-Raphson (DNR) process, developed from a standard NR process.

3.3.2 Fast Decoupled Newton-Raphson Process-Mathematical Formulation

Assuming that p.u. values of voltage modules are approximately equal (which is a case for real electrical Power Systems) and if great island conductivities are excluded from consideration (great capacitors, great electric dumpers, etc.), then the other island conductivities in real electrical Power Systems are relatively small and they can be ignored. Furthermore, if Eq. (16) are written in developed form and if it is assumed that all remaining voltages are approximately equal to zero (what is a case in real Power Systems if values are expressed in p.u.), then by using simple mathematical operations following equations in matrix form are obtained [19]:

$$[B]^{'}[\Delta\theta] = \left[\frac{\Delta P}{|U|}\right] \tag{18}$$

$$[B]^{''}[\Delta|U|] = \left[\frac{\Delta Q}{|U|}\right] \tag{19}$$

Equations (18) and (19) represent basic equations of Fast-Decoupled Newton-Raphson (FDNR) procedure used in iterative scheme.

Values of voltage modules and voltage angles in each iteration are given [19]:

$$\begin{aligned} [\Delta\theta]^{k+1} &= [\theta]^{k} + [\Delta\theta]^{k} \\ [\Delta|U|]^{k+1} &= [|U|]^{k} + [\Delta|U|]^{k} \end{aligned} \tag{20}$$

Iterative process continues until the convergence criteria is achieved [19]:

$$\max\left|\begin{bmatrix} [\Delta\theta]^{k+1} - [\theta]^{k} \\ [\Delta|U|]^{k+1} - [|U|]^{k} \end{bmatrix}\right| < \varepsilon \tag{21}$$

Where ε is convergence parameter.

3.4 Math Formulation of Kron Reduction Method

A widely used method for reducing the number of nodes in electrical power systems is invested by Gabriel Kron, and it is known as Kron reduction. This method is used to eliminate busbars which are not of interest for detailed analysis. It is mostly used for power system analysis where the focus is on voltages at some selected busbars. It ensures the size reduction of large power system matrix, and therefore it allows the reduction of the computing time. To explain the Kron reduction method it is required to start from linear system of algebraic equations as [19]:

$$[S] \cdot [X] = [B] \tag{22}$$

Where:

[S] – coefficient matrix N x N
[X] – required solution vector
[B] – right – hand side of equation

Equation (22), where vector $[X]_1$ is part of a vector [X] that is necessary for the analysis, and vector $[X]_2$ the eliminating vector, can be represented as [19]:

$$\begin{bmatrix} [S]_{11} [S]_{12} \\ [S]_{21} [S]_{22} \end{bmatrix} \cdot \begin{bmatrix} [X]_1 \\ [X]_2 \end{bmatrix} = \begin{bmatrix} [B]_1 \\ [B]_2 \end{bmatrix} \tag{23}$$

To eliminate vector $[X]_2$, supposing that $[S]_{22}$ is not singular matrix, relation (23) can be written as [19]:

$$[S]_{11} \cdot [X]_1 + [S]_{12} \cdot [X]_2 = [B]_1 \tag{24}$$

$$[S]_{21} \cdot [X]_1 + [S]_{22} \cdot [X]_2 = [B]_2 \tag{25}$$

Moving the $[S]_{21} \cdot [X]_1$ to right hand side of equation [19]:

$$[S]_{22} \cdot [X]_2 = [B]_2 - [S]_{21} \cdot [X]_1 / \cdot [S]_{22}^{-1} \tag{26}$$

To find vector $[X]_2$, multiplying last relation by inverse $[S]_{22}$ gives [19]:

$$[X]_2 = [S]_{22}^{-1} \cdot ([B]_2 - [S]_{21} \cdot [X]_1) \tag{27}$$

Equation (27) can be written as follows [19]:

$$[S]_{11} \cdot [X]_1 + [S]_{12} \cdot [S]_{22}^{-1} \cdot ([B]_2 - [S]_{21} \cdot [X]_1) = [B]_1 \tag{28}$$

$$([S]_{11} - [S]_{12} \cdot [S]_{22}^{-1} \cdot [S]_{21}) \cdot [X]_1 = [B]_1 - [S]_{12} \cdot [S]_{22}^{-1} \cdot [B]_2 \tag{29}$$

If the $[S]^k$ and $[B]^k$ matrix are involved, then the last relations can be represented as [19]:

$$[S]^k = [S]_{11} - [S]_{12} \cdot [S]_{22}^{-1} \cdot [S]_{21} \quad (30)$$

$$[B]^k = [B]_1 - [S]_{12} \cdot [S]_{22}^{-1} \cdot [B]_2 \quad (31)$$

$$[S]^k \cdot [X]_1 = [B]^k \quad (32)$$

To get solution for required $[X]_1$ vector, it can be written [19]:

$$[X]_1 = ([S]^k)^{-1} \cdot [B]^k \quad (33)$$

This method is known as a Kron reduction method for eliminating a certain number of unknown variables.

3.5 Test Systems

For the purpose of testing the impact of the Kron reduction process applied to the load flow methods, new three test systems are developed. These test systems present the parts of the real electric power system of Bosnia and Herzegovina. First one is a simple 10-bus system. At busbars 6, 7 and 9 there is power generation and busbar 0 is taken as slack busbar. At busbar 4 there is no power generation or power consumption, so this busbar is not of interest. This means that this busbar should be eliminated in analysis, so busbar 4 will be eliminated using the Kron reduction method. The electrical scheme of 10-bus test system is shown in Fig. 1.

The second test system that is developed is a 17-bus test system. Three generators are connected via 10/110 kV transformers and three generators via 15/220 kV transformers. The load is connected to busbar 6 over a 220 kV line from one side and through a 110/220 kV transformer from the other side. The 10 kV generators are connected to busbar 7 and then over a 110 kV line to 110/220 kV transformer and busbar 6, where the load is connected. At busbars 0, 1, 2, 3, 4 and 5 power generation is present, and only busbar 6 is a load busbar. The slack busbar is busbar 0. Ten busbars 7, 8, 9, 10, 11, 12, 13, 14, 15 and 16 are not of interest for analysis, because these busbars have neither power generation nor power consumption. These busbars will be eliminated using the Kron reduction process. The electrical scheme of this system is shown at Fig. 2.

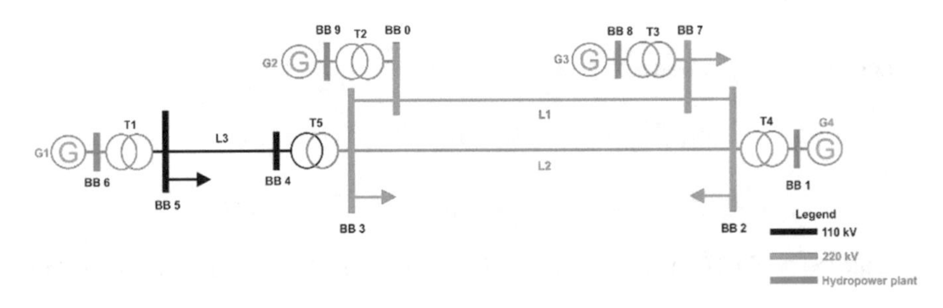

Fig. 1. 10-bus test system

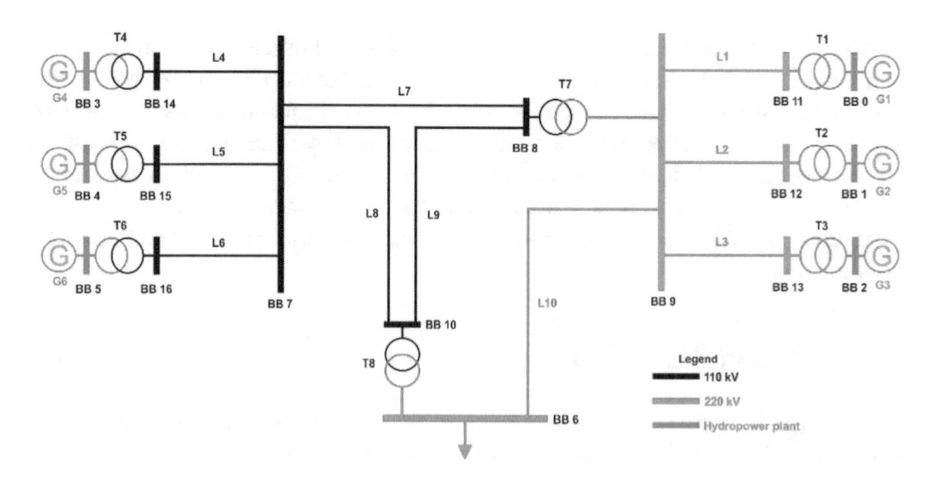

Fig. 2. 17-bus test system

The largest developed test system in this paper is a 50-bus system and it used for power flow analysis as third example. This system is more complicated than other two. This system represents a part of a real Bosnia and Herzegovina electric power system with interconnections to Montenegrin and Croatian power systems, and it is composed of eight hydroelectric power plants, one pumped - storage hydroelectric power plant, one thermal power plant, 110 kV lines, 220 kV lines and 400 kV lines. Total installed power generation capacity is 1726.6 MW. At 21 busbars there is neither power generation nor power consumption, so these busbars are out of interest, and will be eliminated using the Kron reduction process. These busbars are numbered from 29 to 50. The 50-bus test system is shown in Fig. 3.

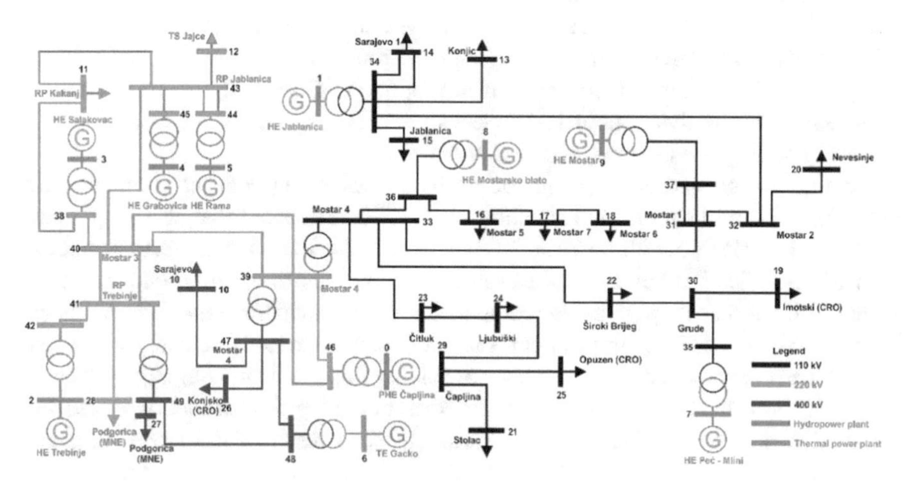

Fig. 3. 50-bus test system

All simulations are carried out in MATLAB software. For matrix reduction process, the appropriate MATLAB code is developed. Firstly, the calculations for power flow analysis have been carried out without using the Kron reduction process for different calculation methods. After eliminating the busbars, these calculations are repeated. All calculations are carried out on different computer hardware configurations, and are repeated multiple times, to reduce the possible errors. Computational time and number of iterations for all power flow methods in each test system are recorded and then used for further analysis.

4 Results and Discussion

In the following section the results of the load flow simulations on three different test systems are presented. Firstly, the computing time reduction will be analyzed, for the each developed test system. Results are presented in Figs. 4, 5, 6, 7 and 8, and present average reduced computing time for each load flow method. In the case of 10-bus system load flow, as shown in Fig. 4, the computation time reduction by applying Kron reduction is highest in the case of YGS method and it is 55.05%. After that, the highest time reduction occurred in case of DNR with 23.63%, FDNR with 14.61%, ZGS with 13.38%, and finally in the case of the NR method with 12.96%. These results occurred because the test system is quite small and unconnected (radial), which results in a large number of zero elements in Y matrix of the system. Because of that, by applying the Kron reduction principle, the load flow computational time has reduced significantly in the YGS method.

In the case of the 17-bus test system load flow, the computing time reduction with Kron reduction is highest in the case of YGS method and it is 90.25%. After that, the highest time reduction occurred in case of DNR with 74.58%, NR method with 65.68%, FDNR method with 35.74% and ZGS method with 10.85%. The 17-bus test system is quite radial, which results with a large number of zero elements in Y matrix of the system. Because of that, by applying the Kron reduction principle, the load flow computational time has reduced significantly in the YGS method, and in this case the reduction is even higher than in 10-bus test system. It is a result of the large number of busbars, and larger number of eliminated busbars.

In the case of load flow calculation in the 50-bus test system the computing time reduction is highest for the DNR method and it is 70.18%. After that, the biggest saves are in the NR method with 66.78%, YGS with 63.08%, FDNR with 53.19% and ZGS with 20.96%. The 50-bus system also has many busbars that can be eliminated, and therefore Kron reduction gives big computation time reductions, especially in the case of the YGS method. Since this system is larger than the previous two analyzed systems, the NR method is the dominant method for solving this problem, with large computing time reduction. In case of this test system, the computing time reduction is generally higher than in 10 and 17-bus test systems. This is a result of the larger number of eliminated busbars. The results demonstrated the effect of Kron reduction process, and it is clear that the computational time and iteration number reductions rise with the rise of the busbars number and system size.

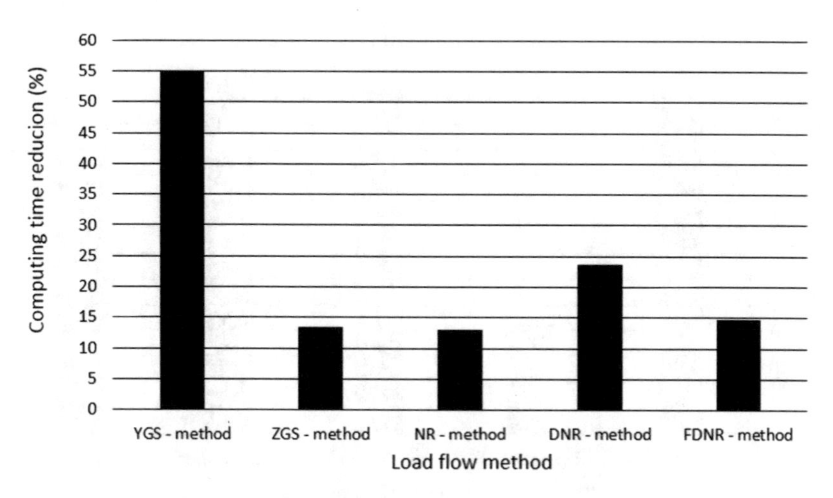

Fig. 4. Load flow computing time reduction for 10-bus test system

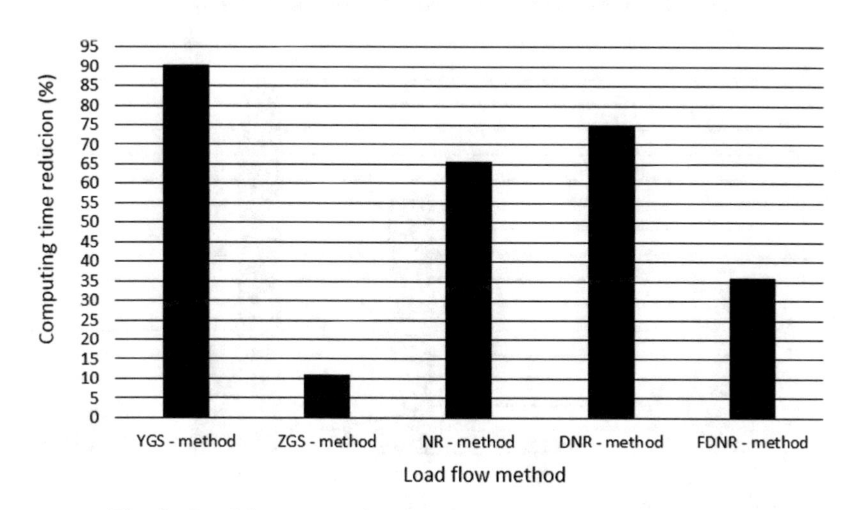

Fig. 5. Load flow computing time reduction for 17-bus test system

Since the 50-bus test system has the largest computer requirements for the load flow calculation, the computing time is more representative for this test system case. Figure 7. Shows the load flow computing time comparison for 50-bus test system for different load flow calculation methods. The time reduction is highest in the case of NR method and DNR method, and smallest in the case of ZGS method. The ZGS method is the fastest method in this test system case, but however the process of the Y matrix inversion, that is a preparatory part of this method, is long and requires many computer resources. Thus, this method is not common in real time load flow calculations.

If the load flow iteration number is analyzed, all methods except YGS method have a small iteration number reduction. However, the YGS method iteration reduction is significant and it is 48.23% in the 10-bus system, 86.92% in the 17-bus system and

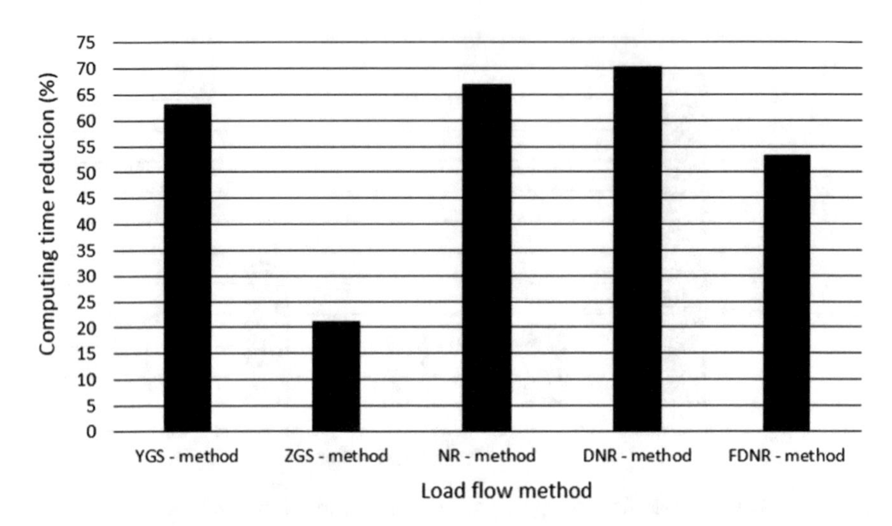

Fig. 6. Load flow computing time reduction for 50-bus test system

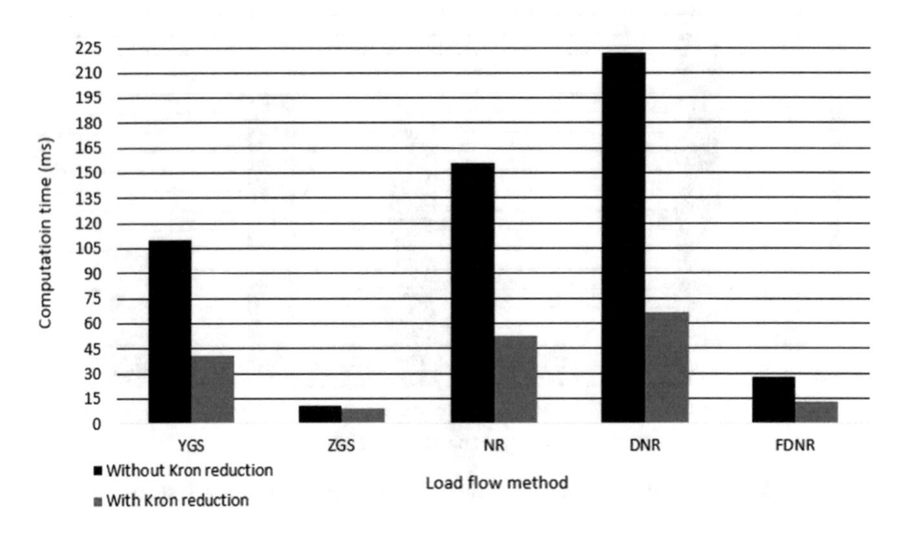

Fig. 7. Load flow computing time comparison for 50-bus test system

46.04% in the 50-bus test system, as shown in Fig. 5. Kron matrix reduction significantly reduces the iteration number in YGS load flow method.

It is well known that this large iteration number is main disadvantage of this method. The main reason for this are the relatively unconnected system busbars (Y-matrix has many zero elements). Thus, the voltage change on the one busbar only affects directly connected and near busbars. Since all busbars voltages are being calculated in the same iterative procedure, the final solution needs a large number of iterations. This is where Kron reduction takes part and eliminates unnecessary busbars, and consequently results in large iteration number reduction.

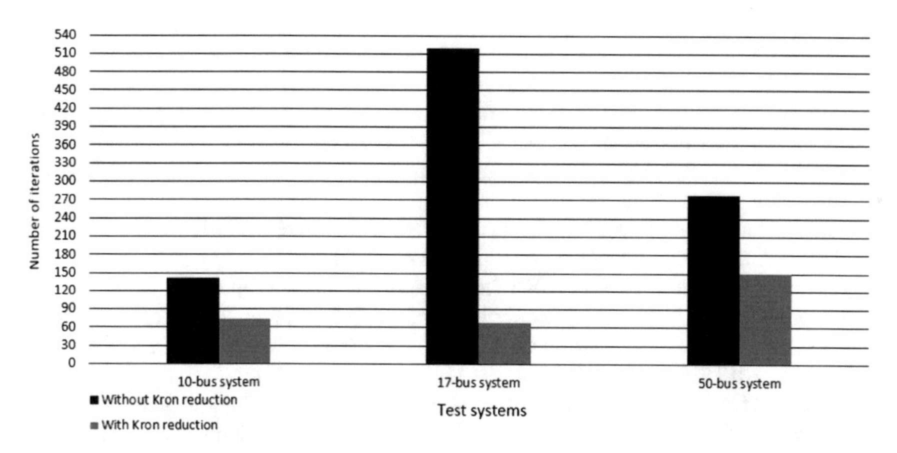

Fig. 8. YGS load flow method number of iterations

5 Conclusions

This paper analyzed the impact of the Kron reduction process to the calculation time and iteration number improvements in the different load flow methods. The Kron reduction process is validated on the 3 three developed power systems. The multiple simulations carried out on the three different computer hardware configurations showed significant improvements in all load flow methods.

The impact of Kron reduction process in every method depends mostly from the method iterative process and topology of the power system. The best results are obtained in the 17-bus test system, where more than half of the busbars are eliminated, as busbars that are not of interest. Despite the different computer hardware configurations, the Kron reduction process has a similar effect in all configurations. The calculation time reduction is less expressed in the ZGS method because the ZGS algorithm doesn't consider currents in the busbars that are not of interest. The larger the system is, the more busbars that are not of interest are present, and the improvements in the calculation time and iteration number are higher. In the terms of iteration number reduction, the Kron reduction process has a significant impact on the YGS method. It is well known that this method needs a large number of iterations for the solution, especially when the system busbars are poorly connected.

The results demonstrated that the Kron reduction process can have significant impact in the load flow calculation process, and the improvements could help the real-time load flow calculations and simulation to become more accepted in the future smart grid monitoring technologies. This paper is a part of an ongoing research, whose objective is to make the power system operation and development easier by improving the traditional approaches for load flow calculations. Moreover, important future research direction in this area should extend the number of system elements and valuate the integration in real-time load flow calculations.

References

1. Thomas, A.: An overview of power flow analysis. Paper presented at American Control Conference, San Francisco, CA, USA, 22–24 June 1983 (1983)
2. Van Ameronge, R.A.M.: A general-purpose version of the fast decoupled load flow. IEEE Trans. Power Syst. **4**(2), 760–770 (1989)
3. Behnam-Guilani, K.: Fast decoupled load flow: the hybrid model. IEEE Trans. Power Syst. **3** (2), 734–742 (1988). https://doi.org/10.1109/59.192929
4. Takamichi, O., Daiki, Y., Kaoru, K.: The development and the application of fast decoupled load flow method for distribution systems with high R/X ratios lines. Paper presented at 2013 IEEE PES Innovative Smart Grid Technologies (ISGT), Washington, DC, USA, 24–27 February 2013 (2013). https://doi.org/10.1109/isgt.2013.6497842
5. Nguyen, H.L.: Newton-Raphson method in complex form [power system load flow analysis]. IEEE Trans. Power Syst. **12**(3), 1355–1359 (1997). https://doi.org/10.1109/59.630481
6. Nursyarizal, M.N., Ramiah, J., Perumal, N.: Newton-Raphson state estimation solution employing systematically constructed jacobian matrix. In: Kankesu, J. (ed.) Advanced Technologies. InTech (2009). https://doi.org/10.5772/8200
7. Lagace, P.J.: Power flow methods for improving convergence. Paper presented at IECON 2012 – 38th Annual Conference on IEEE Industrial Electronics Society, Montreal, QC, Canada, 25–28 October 2012 (2012). https://doi.org/10.1109/iecon.2012.6388538
8. Federico, M.: Continuous Newton's method for power flow analysis. IEEE Trans. Power Syst. **24**(1), 50–57 (2009). https://doi.org/10.1109/tpwrs.2008.2004820
9. Huang, G., Ongsakul, W.: Managing the bottlenecks of a parallel Gauss-Seidel algorithm for power flow analysis. Paper presented at Proceedings of Seventh International Parallel Processing Symposium, Newport, CA, USA, 13–16 April 1993 (1993). https://doi.org/10.1109/ipps.1993.262781
10. Sreemoyee, C., Suprovab, M.: A novel comparison of Gauss–Seidel and Newton–Raphson methods for load flow analysis. Paper presented at 2017 International Conference Power and Embedded Drive Control (ICPEDC), Chennai, India, 16–18 March 2017 (2017). https://doi.org/10.1109/icpedc.2017.8081050
11. Ciric, R.M., Feltrin, A.P., Ochoa, L.F.: Power flow in four-wire distribution networks-general approach. IEEE Trans. Power Syst. **18**(4), 1283–1290 (2003). https://doi.org/10.1109/tpwrs.2003.818597
12. Florian, D., Francesco, B.: Kron reduction of graphs with applications to electrical networks. IEEE Trans. Circuits Syst. I: Regul. Pap. **60**(1), 150–163 (2013)
13. Caliskan, S.Y., Tabuada, P.: Towards Kron reduction of generalized electrical networks. J. Autom. (J. IFAC) **50**(10), 2586–2590 (2014)
14. Caliskan, S.Y., Tabuada, P.: Kron reduction of power networks with lossy and dynamic transmission lines. Paper presented at 2012 IEEE 51st Annual Conference on Decision and Control (CDC), Maui, HI, USA, 10–13 December 2012 (2012). https://doi.org/10.1109/cdc.2012.6426580
15. Dorfler, F., Bullo, F.: Synchronization of power networks: network reduction and effective resistance. In: IFAC Proceedings Volumes, vol. 43, no. 19, pp. 197–202 (2010). https://doi.org/10.3182/20100913-2-fr-4014.00048
16. Luo, L., Dhople, S.V.: Spatiotemporal model reduction of inverter-based islanded microgrids. IEEE Trans. Energy Conversion **29**(4), 823–832 (2014). https://doi.org/10.1109/tec.2014.2348716

17. Monshizadeh, N., De Persis, C., Van der Schaft, A.J., Scherpen, J.M.A.: A networked reduced model for electrical networks with constant power loads. Paper presented at American Control Conference (ACC), Boston, MA, USA, 6–8 July 2016 (2016). https://doi.org/10.1109/acc.2016.7525479
18. Ashraf, S.M., Rathore, B., Chakrabarti, S.: Performance analysis of static network reduction methods commonly used in power systems. Paper presented at 2014 Eighteenth National Power Systems Conference (NPSC), Guwahati, India, 18–20 December 2014 (2014). https://doi.org/10.1109/npsc.2014.7103837
19. Sadovic, S.: Analiza Elektroenergetskih Sistema. Faculty of Electrical Engineering, University of Sarajevo (2011)

Application of Artificial Neural Network and Empirical Mode Decomposition for Predications of Hourly Values of Active Power Consumption

Maja Muftić Dedović(✉), Nedis Dautbašić, and Adnan Mujezinović

Faculty of Electrical Engineering, Sarajevo, Bosnia and Herzegovina
maja.muftic-dedovic@etf.unsa.ba

Abstract. The precision of load forecasting is of great importance for power distribution systems planning and management. As load data are highly non-linear and nonstationary time series, ordinary methods of linear prediction seem insufficient. In this paper, for the active power consumption forecasting, two methods are used. A method using artificial neural network (ANN) based technique is developed for short-term and mid-term load forecasting of power distribution system. Aiming to increase the accuracy of load prediction, method using artificial neural network and Empirical Mode Decomposition (EMD) technique for short-term and mid-term load forecast is developed. Two cases are used to validate the prediction methods.

Keywords: Load forecast · Artificial neural network (ANN) Empirical mode decomposition (EMD)

1 Introduction

The prediction and precision of hourly values of active power consumption is of great importance for power distribution systems planning and management. Modern power companies rely on accurate electricity load forecasting to minimize financial risk and optimize operational efficiency and reliability. Forecasting of electricity consumption can be divided into three categories: short-term predictions, from one hour to one week, medium-term predictions, from one week to one year, and long-term predictions, usually longer than one year. This paper addresses the short-term prediction and medium-term prediction of hourly values of active power consumption. Their biggest difference is exactly in precision. An accurate forecast of medium-term load (MTLF) is essential for studying expansion plans for distribution systems [1]. Also, the medium-term electric load as a function of time has complex non-linear behaviour that makes ordinary methods of linear prediction seem insufficient.

In literature can be found different approaches for studying load forecasting. In [2] an investigation for the short term (up to 24 h) load forecasting of the demand is presented, using a Multiple Linear Regression (MLR) method. Conventional methods for load forecasting are regression models, time series (for example ARMA (autoregressive moving average), ARIMA (autoregressive integrated moving average)),

S. Avdaković (Ed.): IAT 2018, LNNS 59, pp. 86–97, 2019.
https://doi.org/10.1007/978-3-030-02574-8_8

pattern recognition, Kalman filters, trend analysis (trending), end-use analysis is that the demand for electricity depends on what it is used for (the end-use), econometric [3]. But when variables are nonstationary it is better to obtain forecast on the basis of artificial neural networks [4]. Also in order to achieve more accuracy, in literature can be found hybrid models that are combination of ANN models and Hilbert-Hung Transform [5]. Model based on combination of HHT and ANN is proposed in the paper [6] and simulation results indicate that accuracy of the forecasting model is higher than the traditional linear combination model. Also in [7] is proposed the short-time and monthly load forecasting of power system based on HHT. Because the EMD has some serious disadvantages including border effect and mode mixing, decomposition of EMD is improved with first-order difference algorithm. This method is proposed in [8] as a new short-term load forecasting model of power system based on HHT and ANN. The new two-stage adaptive approach that combines the effective technique for time series analysis HHT and ANN forecasting technology is discussed in papers [9, 10].

This paper is organized as follow: In Sect. 2 will be presented ANN technique and EMD technique for temporal predictions, separately, and in the end of this chapter will be explained model that combines these two techniques. Then in Sect. 3 are presented input data and experimental calculations, and this chapter is divided in two parts according to two tasted cases. Results and conclusion are analysed in Sect. 4.

2 Model for Load Forecasting

The precision of load forecasting is of great importance for power distribution systems planning and management. As load data are highly nonlinear and nonstationary time series, ordinary methods of linear prediction seem insufficient. Furthermore, in this paper for short-term and mid-term load forecasting will be used ANN technique. Aiming to increase the accuracy of load prediction, method using ANN will be combined with EMD technique for short-term and mid-term load forecast.

2.1 ANN Technique

Basic approach of ANN technique is presented [11]:

Step 1. An artificial neuron, receive raw data or the output from another neuron, process them and produce one output.
Step 2. Network – input layer, output layer with one or more hidden layers. Network topology may be different depending on the problem to be solved, the type of input layer, output layers and other factors.
Step 3. Artificial neuron model:

- Input variable to the ANN (load, temperature…etc.)
- Number of classes (weekday, weekend, season…etc.)
- What to forecast: (hourly loads, next day peak load, next day total load…etc.)
- Neural network model (number of hidden layer, number of neuron in the hidden layer…etc.)
- Training method and stopping criterion

- Activation functions
- Size of the training and test data [3].

In this paper is used backpropagation learning algorithm and Levenberg-Marquardt algorithm is learning method applied for numerical optimization. Equations (1) to (6) describe simplified learning algorithm.

$$o_{pi} = f_i\left(net_{pi}\right) \tag{1}$$

$$net_{pi} = \sum\nolimits_k w_{ik}\, o_{pk}, \tag{2}$$

where: o_{pi} is the i-th component of the actual output with input pattern p, f_i is differentiable and nondecreasing sigmoid type function and w_{ik} is weight between i^{th} input and k^{th} neuron.

After weight initialization, it is necessary to initial the error at the output, so it can be written:

$$E_p = \frac{1}{2}\sum\nolimits_i \left(t_{pi} - o_{pi}\right)^2 \tag{3}$$

According to weights adjusting, the gradient of E_p with respect to w_{ij} is used, so it can be written:

$$-\frac{\partial \mathrm{E}_p}{\partial w_{ij}} = \delta_{pi}\, o_{pj}, \tag{4}$$

where: E_p is error on pattern p and δ_{pi} is defined as:

$$\delta_{pi} = f_i'\left(net_{pi}\right)\sum\nolimits_k \delta_{pk}\, w_{ki}. \tag{5}$$

After calculation of actual output and corresponding error, weight are modified to minimize error:

$$\Delta w_{ij}(n+1) = \eta\delta_{pi}o_{pi} + \alpha\Delta w_{ij}(n), \tag{6}$$

where: η is the learning rate parameter and α is the momentum constant to determine the effect of past weight changes [12].

2.2 EMD Approach

The Empirical Mode Decomposition is a technique to decompose a given signal into a set of elemental signals called Intrinsic Mode Functions (IMF). The Empirical Mode Decomposition is the part of the Hilbert-Huang Transform that comprises also a Hilbert Spectral Analysis and an instantaneous frequency computation.

First, every IMF must satisfy the following two conditions:

(a) Extreme number (minima and maxima) of zero-crossing points should be equal or differ only by one.
(b) The median value of envelopes (obtained by the local maxima and the local minima) at any point for intrinsic mode functions is equal to zero.

Algorithm of EMD can be presented as follows:

Step 1. Detect all the local extrema and using a cubic spine line generates its upper and lower envelopes ($e_{up}(t)$ and $e_{low}(t)$, respectively);
Step 2. Calculate the mean value of the $e_{up}(t)$ and $e_{low}(t)$:

$$m_1(t) = \left(e_{up}(t) + e_{low}(t)\right)/2. \tag{7}$$

And find $h_1(t)$ as follows:

$$h_1(t) = x(t) - m_1(t). \tag{8}$$

Repeat steps until h_1 satisfies a set of predetermined stopping criteria for the IMF function as expressed:

$$c_1 = h_{1k}. \tag{9}$$

Step 3. If $h_1(t)$ does not meet the criteria for the existence of IMF, the previous procedure is repeated and determined $h_{11}(t)$ as a new signal:

$$h_{11}(t) = x(t) - m_{11}(t), \tag{10}$$

where $m_{11}(t)$ is mean value of the upper and lower envelope of the signal $h_1(t)$.
Step 4. Compute residue and repeat step 2 k times to obtain n IMFs and residual, then the new signal is as follow:

$$x(t) = \sum\nolimits_{i=1}^{n} c_i(t) + r_n(t) = \sum\nolimits_{i=1}^{q} c_i(t) + \sum\nolimits_{j=q+1}^{p} c_j(t) + \sum\nolimits_{k=p+1}^{n} c_k(t) + r_n(t), \tag{11}$$

where $q < p < n$, $c_i(t)$ are *ith* IMF, $c_j(t)$ are the components representing properties of the series and $c_k(t)$ and $r_n(t)$ are the final residual, trend non-sinusoidal components [13, 14].

2.3 Two-Stage Adaptive Model for Load Forecasting

In this paper the two-stage adaptive load forecasting model is proposed. The first stage involves the decomposition of the initial time of analysis of nonstationary time series - empirical mode decomposition and the technology of artificial neural network forecasting. This combined model is proposed in order to improve the accuracy of the short-term and mid-term load forecasting.

Two-stage adaptive model for load forecasting construction can be presented as follows:

Step 1. According to the procedure described in the Sect. 2.2, EMD approach is applied on load time series and this non stationary signal is decomposed into the several IMFs and the residual.
Step 2. This computed values of the IMFs and the residual are used as input variables for the artificial neural network model.
Step 3. Obtaining results according to the selection (based on input data, computed values of the IMFs and the residual) of the best neural network prediction structure.

3 Input Data and Experimental Calculations

The data on hourly active power consumption in Bosnia and Herzegovina for the year 2011 are plotted on Fig. 1. From these data, input variables are extracted and used for the two tested cases. The first test case represents mid-term load forecasting and input variables are one month, precisely 720 h of active power consumption. The second test case represents short-term load forecasting but input variables are three days, precisely 72 h of active power consumption.

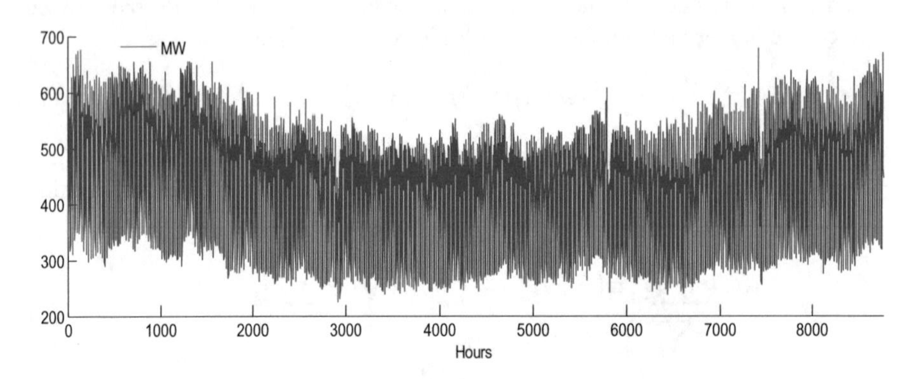

Fig. 1. Hourly values of active power consumption in Bosnia and Herzegovina for 2011

3.1 First Test Case - Mid-Term Load Forecasting

The input data for the first test case are plotted on Fig. 2. The first test case denotes the mid-term load forecasting and is separated in two parts. In first, ANN model is introduced and the second part of the first test case is application of the two stage adaptive model based on joint usage of the ANN and EMD approach.

For the applied ANN model all input date are time series and weekdays and hours are normalized and generated as sinusoidal and cosinusoidal variables. Other input data are the active power consumption for 720 h. Forecasting methodology is performed using Levenberg-Marquardt as learning method for training weights and bias values. Results are evaluated using statistical descriptors and two hidden layers with 15

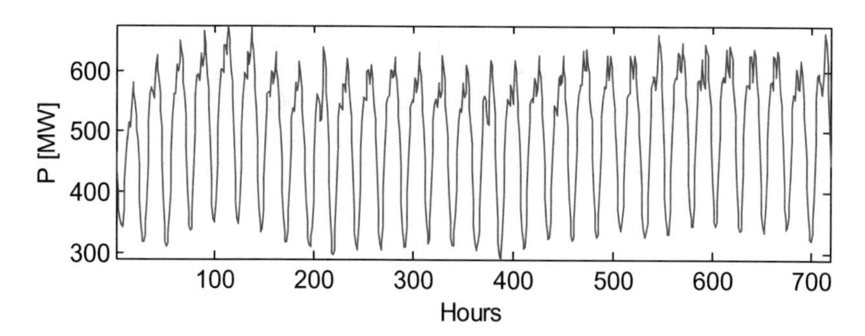

Fig. 2. 720 h of active power consumption, input signal for the first test case

neurons with 5 input layers. Nodes are selected according to minimum error. Transfer function log-sigmoid and tan-sigmoid are performed and for output neuron is used linear transfer function.

First of all dataset are divided into train, train target, test, test target and network output. The ANN model has three layers. These layers are input data or train dataset, hidden layer and output layer. ANN are trained with following input data: weekday, hours and 720 h values of active power consumption from 2011. Nodes in input layer are connected to the nodes in hidden layer and activation function is obtained over signals and then sent to the output layer reaching the target. Algorithm repeats until minimum error between outputs and targets are achieved [11]. Target data are second 720 h of active power consumption of Bosnia and Herzegovina in 2011. Because predictions are for the 720 h or 30 days this test case represents mid-term load forecasting.

The results obtained only applied ANN model are plotted on Fig. 3.

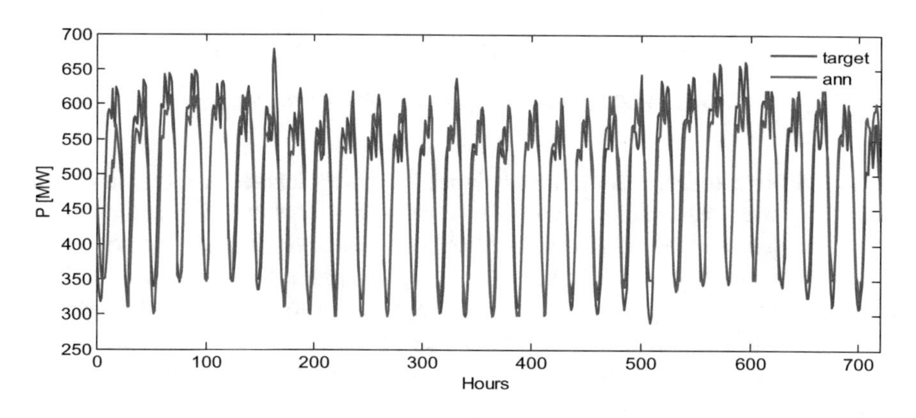

Fig. 3. Results obtained by ANN model

After applying ANN model for mid-term load forecasting, two stage adaptive model based on joint usage of the ANN and EMD approach is applied. First of all on the input data of 720 h of active power consumption is applied EMD approach in order to obtain IMFs and residual. The same methodology as described before for ANN model is applied. Only input variables are not actual values of 720 h of active power consumption, weekday and hours then IMFs and residual obtained by EMD approach. On Fig. 4 are plotted time series of 720 h of active power consumption decomposed into IMFs and residual.

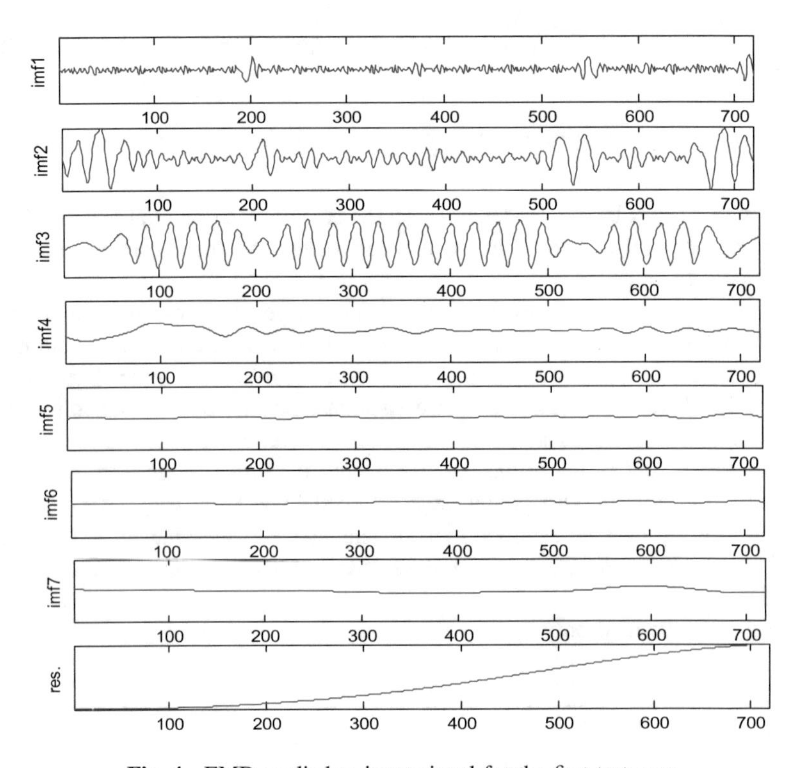

Fig. 4. EMD applied to input signal for the first test case

So ANNs are trained with following input data: IMFs and residual of 720 h values of active power consumption from 2011. Algorithm repeats until minimum error between outputs and targets are achieved. Also herein target data are second 720 h of active power consumption of Bosnia and Herzegovina in 2011, representing mid-term load forecasting. On Fig. 5 is plotted ANN topology of forecasting modelling of the two stage adaptive model based on joint usage of the ANN and EMD approach.

The results obtained with two stage adaptive model based on joint usage of the ANN and EMD approach are plotted on Fig. 6.

In order to validate proposed ANN model and ANN-EMD model on training and testing data set, calculation of the coefficient of correlation r between actual and forecasted values is performed. Equation (12) is coefficient of correlation r.

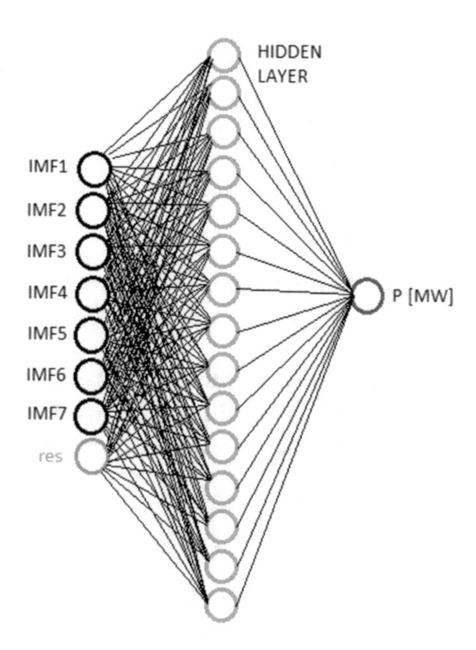

Fig. 5. ANN topology of forecasting modelling

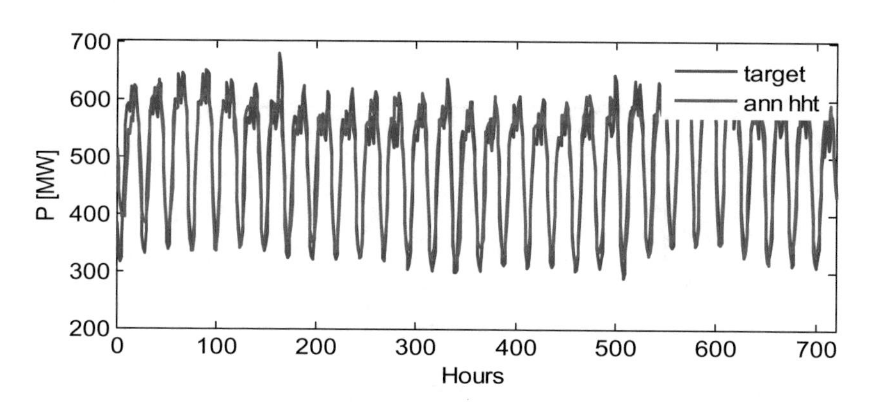

Fig. 6. Results obtained by ANN-EMD model

$$r = \frac{\sum_{i=1}^{n} (X_i - \bar{X}) \cdot (Y_i - \bar{Y})}{\sqrt{\sum_{i=1}^{n} \left(X_i - \overline{X}\right)^2 (Y_i - \bar{Y})^2}}, \quad (12)$$

where: X_i and Y_i are forecasted values and observed values, $\bar{X}$ and $\bar{Y}$ are mean values of entire dataset respectively.

Table 1. Results of a mid-term load forecasting on the basis of ANN model and ANN-EMD model

Model	r
ANN	0.941
ANN-EMD	0.952

Values of the coefficient of correlation r between actual and forecasted values obtained after application of ANN model and ANN-EMD model are presented in Table 1.

From obtained results it can be seen that two stage adaptive based on joint usage of the ANN and EMD approach has no significant improvement in forecasting results over ANN model according to calculate coefficient of correlation from Table 1 and from Figs. 3 and 6.

3.2 Second Test Case - Short-Term Load Forecasting

The input data for the second test case are plotted on Fig. 7. The second test case denotes the short-term load forecasting and is also separated in two parts. In first, ANN model is introduced and the second part of the first test case is application of the two stage adaptive model based on joint usage of the ANN and EMD approach.

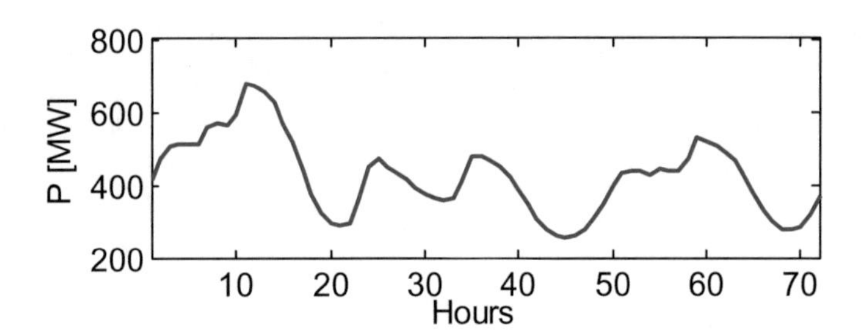

Fig. 7. 72 h of active power consumption, input signal for the second test case

Also in the second test case ANN model for mid-term load forecasting is applied firstly then the two stage adaptive model based on joint usage of the ANN and EMD approach is applied. The same methodology described in the first test case is obtained. But in the second test case input data are 72 h of active power consumption and target data are second 72 h of active power consumption.

The results obtained only applied ANN model are plotted on Fig. 8.

On this input time series is applied EMD approach in order to obtain IMFs and residual. On Fig. 9. are plotted time series of 72 h of active power consumption decomposed into IMFs and residual.

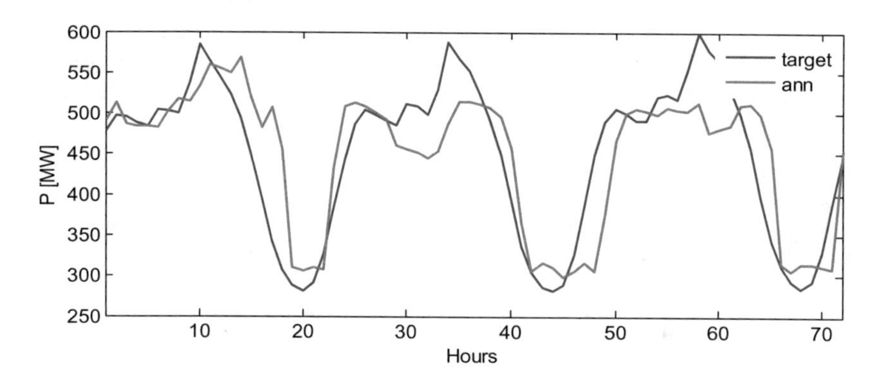

Fig. 8. Results obtained by ANN model

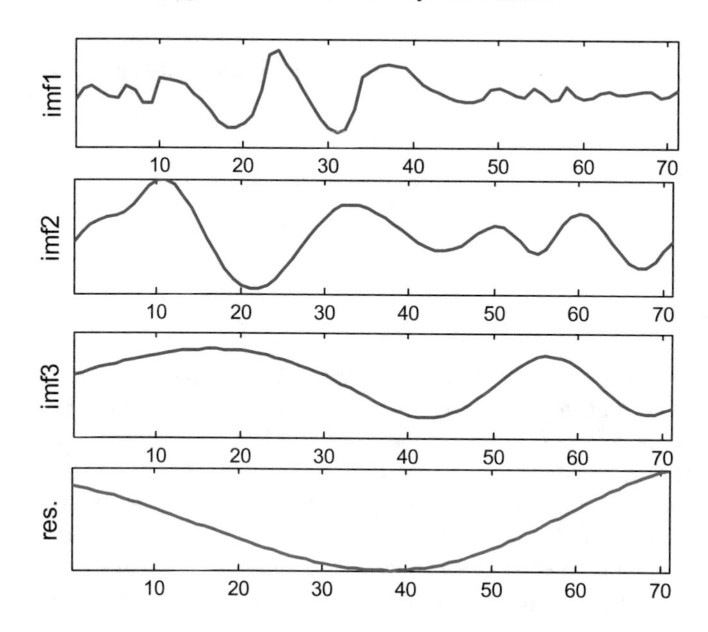

Fig. 9. EMD applied to input signal for the second test case

Input data for the two stage adaptive model are IMFs and residual of 72 h values of active power consumption from 2011. Also herein target data are second 72 h of active power consumption of Bosnia and Herzegovina in 2011, representing short-term load forecasting. The results obtained with two stage adaptive model based on joint usage of the ANN and EMD approach are plotted on Fig. 10.

According to (12) is calculated coefficient of correlation r between actual and forecasted values obtained after application of ANN model and ANN-EMD model and results are presented in Table 2.

From obtained results it can be seen that two stage adaptive based on joint usage of the ANN and EMD approach has significant improvement in forecasting results over ANN model according to calculate coefficient of correlation from Table 2. and from Figs. 8. and 10.

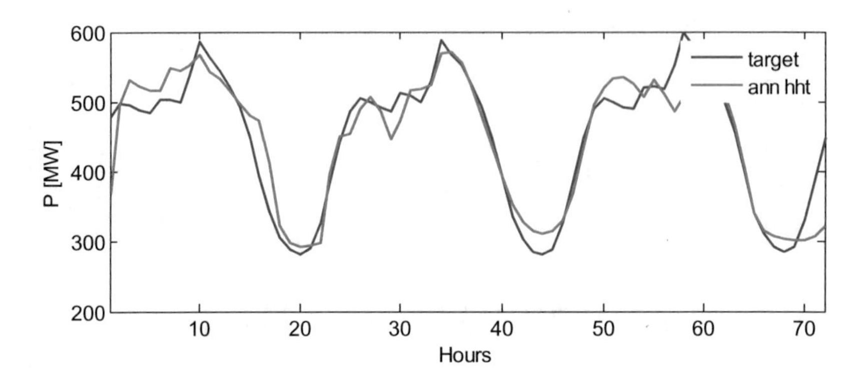

Fig. 10. Results obtained by ANN-EMD model

Table 2. Results of a mid-term load forecasting on the basis of ANN model and ANN-EMD model

Model	r
ANN	0.82
ANN-EMD	0.961

4 Conclusion

Load forecasting has aim to increase energy efficiency, reliable operation of power systems and better quality for power distribution systems planning and management. In this paper methodology that use combination of ANN technique and EMD technique for short-term and mid-term load forecasting is discussed. Observing two test cases it is concluded the accuracy of the forecasting result can be improved by using combined model instead only ANN technique for short-term load forecasting. Also it can be seen that no significant difference in accuracy of the mid-term load forecasting is obtained bay using two stage adaptive model based on joint usage of the ANN and EMD approach.

References

1. Abu-Shikhah, N., Elkarmi, F., Aloquili, O.: Medium-term electric load forecasting using multivariable linear and non-linear regression. Smart Grid Renew. Energy **2**, 126–135 (2011). https://doi.org/10.4236/sgre.2011.22015. http://www.SciRP.org/journal/sgre
2. Tuaimah, F.M., Abass, H.M.A.: Short-term electrical load forecasting for iraqi power system based on multiple linear regression method. Int. J. Comput. Appl. **100**(1) (2014). ISSN 0975-8887
3. Islam, B.U.: Comparison of conventional and modern load forecasting techniques based on artificial intelligence and expert systems. IJCSI Int. J. Comput. Sci. Issues **8**(5), no. 3 (2011)

4. Haykin, S.: Neural Networks: A Comprehensive Foundation, 2nd edn, p. 1104. Williams Publishing House, Jerusalem (2006)
5. Caciotta, M., Giarnetti, S., Leccese, F.: Hybrid neural network system for electric load forecasting of telecommunication station. In: Proceedings of XIX IMEKO World Congress Fundamental and Applied Metrology, Lisbon, Porugal, 6–11 September 2009, pp. 657–661 (2009)
6. Weili, B., Zhigang, L., Quanwei, P., Jian, X.: Research of the load forecasting model based on HHT and combination of ANN. Power Syst. Prot. Control **37**(19), 31–35 (2009)
7. Weili, B., Zhigang, L., Qi, W., Dengdeng, Z.: Load forecasting of power system based on HHT. Sichuan Electr. Power Technol. **32**(3), 9–13 (2009)
8. Liu, Z.G., Bai, W.L., Chen, G.: A new short-term load forecasting model of power system based on HHT and ANN. In: Lecture Notes in Computer Science, vol. 6064, pp. 448–454 (2010)
9. Kutbatsky, V., Sidorov, D., Spiryaev, V., Tomin, N.: On the neural network approach for forecasting of nonstationary time series on the basis of the Hilbert-Huang transform. Autom. Remote Control **72**(7), 1405–1414 (2011)
10. Kutbatsky, V., Sidorov, D., et al.: Hybrid model for short-term forecasting in electric power system. Int. J. Mach. Learn. Comput. **1**(2), 138–147 (2011)
11. Dedovic, M.M., Avdakovic, S., Turkovic, I., Dautbasic, N., Konjic, T.: Forecasting PM10 concentrations using neural networks and system for improving air quality. In: 2016 XI International Symposium on Telecommunications (BIHTEL), pp. 1–6 (2016)
12. Lee, K.Y., Park, J.H.: Short-term load forecasting using an artificial neural network. Trans. Power Syst. **7**(1), 124–132 (1992)
13. Huang, H.E., Shen, Z., Long, S.R., Wu, M.C., Shih, H.H., Zheng, Q., Yen, N.-C., Tung, C. C., Liu, H.H.: The empirical mode decomposition and the Hilbert spectrum for nonlinear and non-stationary time series analysis. Proc. R. Soc. Lond. A **454**, 903–995 (1998)
14. Dedovic, M.M., Avdakovic, S., Dautbasic, N.: Impact of air temperature on active and reactive power consumption - Sarajevo case study. Bosanskohercegovačka elektrotehnika (under revision)

The Small Signal Stability Analysis of a Power System with Wind Farms - Bosnia and Herzegovina Case Study

Semir Nurković[1(✉)] and Samir Avdaković[1,2]

[1] Elektroprivreda BiH d.d. - Sarajevo, Sarajevo, Bosnia and Herzegovina
nurkovic@gmail.com

[2] Faculty of Electrical Engineering, University of Sarajevo, Sarajevo, Bosnia and Herzegovina

Abstract. In the recent years, the issue of the impact of wind farms integration to power system small-signal stability (SSS) has become a topic of increased interest and discussion. Results of SSS analysis of a power system of Bosnia and Herzegovina (BiH) for the base case (system without wind farms connected) and the case with 4 wind farms connected were summarized in this paper. Doubly-fed induction generator (DFIG) and direct drive synchronous generator (DDSG) wind turbine models are used. Analyses are done by computing the eigenvalues of state matrix of the power system around a stable operating point in order to determine the small-signal angular stability of the system with and without WFs connected. Also, three phase short circuits are simulated in two points in the system, with duration of 0.04 [s], in both cases. The power system models in this paper were all built using SIMULINK and the specialized application PSAT (Power System Analysis Toolbox) ver 2.1.10, a software application performed for MATLAB, which performs both, the numerical simulations and linearised eigenstructure analysis. The system responce to small disturbances for the base case (Case A) is compared to the case with WF models (Case B), and conclusions have been made in terms of the effect of addition of the wind farms (WF) into power system on the small-signal angular stability.

1 Introduction

Wind power is the world's fastest growing renewable source. In the 2017 year, the total installed power in the wind farms at global level passed the threshold of 500 GW [2]. Due to the wind energy's randomness, a lot of researching works have been done on simulation for power system SSS with wind power integration [3–12]. SSS is defined as the ability of the power system to maintain synchronism when subjected to small disturbances. SSS analyses are usually called the analyses of static stability, since system response in that case is analysed by linearised differential-algebraic equations of multimachine systems in steady space. Eigenvalue analysis method, based on the linear system theory, has been recognised as one of the broadest methods to analyse SSS [1].

In [3] SSS of wind power systems is analyzed, using squirrel cage induction generator (SCIG) and DFIG. Comparative eigenvalue analyses of those wind turbines on a particular power system is done. Obtained oscillation frequencies determine that

S. Avdaković (Ed.): IAT 2018, LNNS 59, pp. 98–113, 2019.
https://doi.org/10.1007/978-3-030-02574-8_9

two possible oscillation modes are represented, inter-area and local. The use of electronic converter for the DFIG has showed an improved behavior for power system stability.

In [4] it is shown that the network is slightly less damped and that critical fault clearing time is reduced when connecting DFIG or DDSG wind turbine type.

In [5] simulation results show that SCIG based system is marginally stable where as DFIG based system is completely stable and rotor angle stability of DFIG is more predominant than SCIG. DFIG consists of power electronics based controlling system by which it can supply reactive power to the system if necessary and thereby it can enhance SSS of the system.

Reference [6] shows that the type of transmission line between permanent magnet synchronous generator (PMSG) based wind farm and the power grid can affect system's stability. It is more helpful using the cable rather than overhead line in order to improve SSS of the integrated system.

In [7] the approach to analyze impact of increased penetration of DFIG based wind turbines on SSS of a large power system is developed by converting DFIG machines into equivalent conventional round rotor synchronous machine and avaluating the sensitivity of the eigenvalues with respect to inercia.

Reference [8] presents analysis which assess the impact of DDSG based wind turbines on power system SSS. The analysis is repeated for various wind power penetration levels with and without a wind power plant voltage controller. The study found that the inter-area modes were largely unaffected by the increased capacity of the wind power plant. Generally, a very small participation from the wind turbines was found in the system oscillatory modes.

Reference [9] proposes an approach for the separate examination of the impact of affecting factors, i.e. change of load flow/configuration and dynamic interactions brought about by the grid connection of the wind farm, on power system small signal angular stability. Thus, a clearer picture and better understanding of the power system small-signal angular stability as affected by grid connection of the large-scale wind farm can be achieved.

Reference [10] investigates the impacts of large amount of wind power on SSS and the control strategies to mitigate negative effects. The control strategies on WTGs to enhance power system damping characteristic are presented in the paper.

In this paper it is presented how 4 wind farms (WF Mesihovina, WF Jelovača, WF Trusina and WF Podveležje) affect SSS of the existing synchronous machines in 110 kV network of power system of Bosnia and Herzegovina (BiH).

Eigenvalue analysis and simulations of small disturbances in PSAT are performed for the following operating scenarios of BiH power system:

(A) maximum load without connected WFs (Case A or the base case),
(B) maximum load with connected 4 WFs (Case B).

2 Model of a Power System with Wind Farms - Bosnia and Herzegovina Case Study

The real power system of BiH (110 kV, 220 kV and 400 kV network), for maximum system load based on data in [16–18] and [19], was modelled in order to analyze the impact of wind farms on the angular stability of existing synchronous machines in the system, if small disturbance is applied. All simulations were performed in the software application PSAT [13, 14] as a power system toolbox developed for MATLAB [15].

The model of BiH power system includes existing production capacities and future capacities planned to be in operation by 2027 and for which appropriate technical data were available during preparation of the model. Model of BiH power system is based on the following assumption:

- Power transformers are with fixed tap ratio (without regulation);
- Power transformers are without iron losses and shunt admittances are neglected;
- HV transmission lines are presented via ¶ model;
- The limit values of the voltage ranges, line currents and the apparent power of the power transformers are set as found in [16–18, 20];
- In the static model of BiH power system, all synchronous machines are modelled as PV generators, except one "slack" generator. If specified limits for the maximum and minimum reactive power of the synchronous machines on PV bus are reached, PSAT translates PV bus into PQ bus [13];
- The bus loads are modelled as constant PQ loads as long as the voltage of the associated busbar is within the given values. If the voltage limits are violated, the PQ load is converted to a constant impedance [13]:
- In a dynamic model, synchronous machines are modelled as type 6 (saturation is neglected), automatic voltage regulators of synchronous machines are modelled as standard IEEE type I, and turbine regulators of synchronous machines are modelled as type 2 (turbine guvernor type 2) [13, 21, 22];
- Each wind turbine within the wind farm (WF) is modelled separately, i.e. no aggregated models of wind turbines have assumed;
- For the parameters of wind turbines, available data from the literature or data on the web pages of the manufacturer's brochures are used, as well as the values specified in the PSAT application as "default" values [13, 21];
- "Weilbull" wind speed distribution is used in wind model [13, 21];
- Models of a wind turbines with a DFIG and DDSG are explained in [13];
- Power system model for Case A and Case B consists of 192 loads whose total amount, including export to neighboring power systems, is 3344.3 MW.
- The model in Case A is based on the 287 nodes, and 47 synchronous machines (5 termo power plants (TPP) and 16 hydro power plants (HPP)). The total production of all generators (active power), including the slack generator, is 3400.5 MW;
- The model in Case B is based on the 365 nodes, 47 synchronous machines and 70 wind generators (4 WFs). The total production of all generators, including the slack generator, is 3405.6 MW. Introduction of wind power in Case B doesn't displace any conventional unit, but only the active power of the slack generator is reduced to accomodate the power produced by WFs;
- Data on 4 modelled WFs are shown in Appendix (Tables A1 and A2) [13, 23];

3 Eigenvalue Analysis for State Matrix of BiH Power System

For Case A and Case B, calculations are performed for the steady state of the system, for power flows, voltages [kV] and bus angles [°] as well as active and reactive power of the slack generator. Also, calculations of the initial values of the state variables and algebraic variables for the dynamic state of the system are performed (variables of synchronous machines, automatic voltage regulators and turbine regulators, all for Case A and Case B, and variables of wind turbines with DFIG, wind turbines with DDSG, and wind speeds for Case B only).

Starting the eigenvalue analysis routine in the PSAT application, the detailed eigenvalue analysis is conducted for Case A and B. As the real part of eigenvalues ($\lambda_i = \sigma_i + j\omega_i$) indicates damping of the eigenvalue (the more negative real part, the better damped eigenvalue), and the imaginary part gives the frequency of oscillation, the damping ratio ξ is calculated for each of the eigenvalues of the state matrix, and determines a rate of decay of the amplitude of the oscillation as:

$$\xi_i = -\sigma_i / \sqrt{\left(\sigma_i^2 + \omega_i^2\right)} \qquad i = 1, 2, \ldots N \tag{1}$$

where N is order of the system (number of system three-phase buses).

Eigenvalues of the state matrix, for Case A and Case B, are shown in the S-domain in Figs. 1 (3) and 2 (4), respectively.

The results obtained for Case A show that all real parts of a total of 517 eigenvalues (system order 517 × 517) are less than zero. For eigenvalues with an imaginary part (133 complex pairs), the minimum damping ratio is ζ = 5.1845% which is greater than

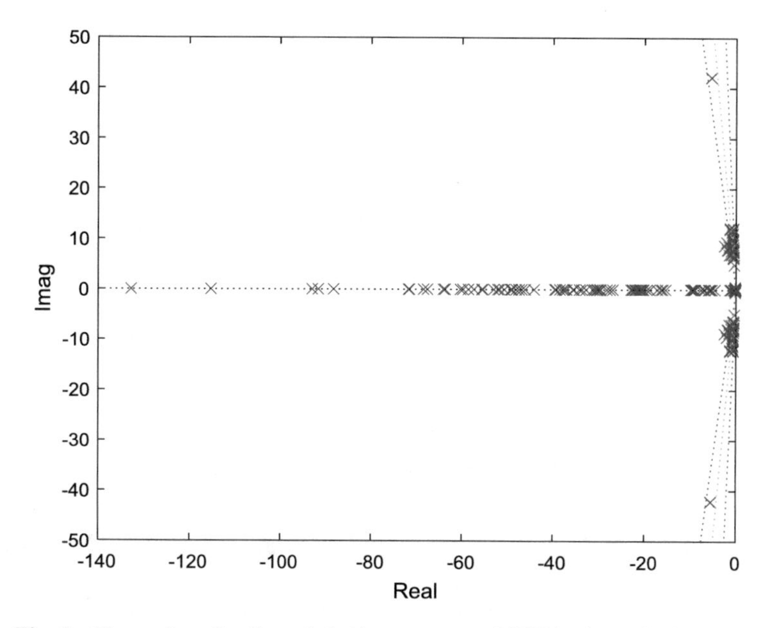

Fig. 1. Eigenvalues for Case A (without connected WFs), shown in S-domain

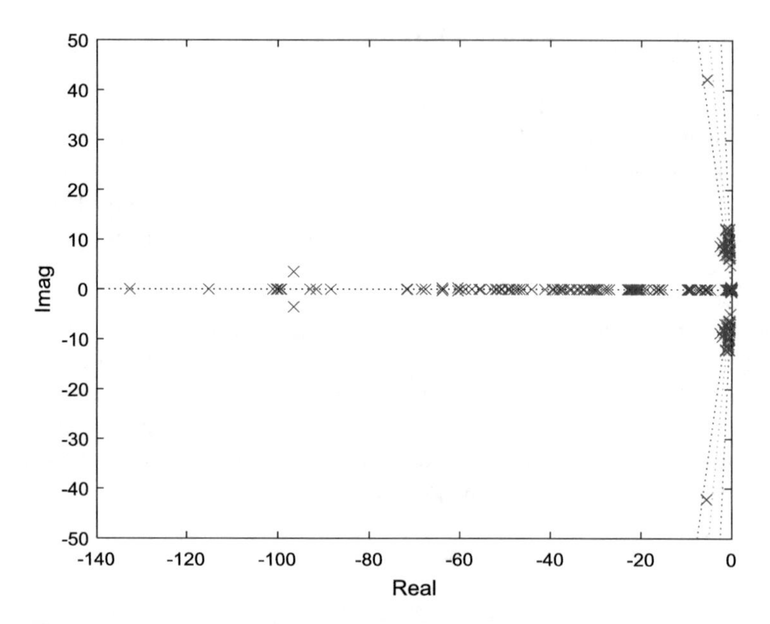

Fig. 2. Eigenvalues for Case B (with connected WFs), shown in S-domain

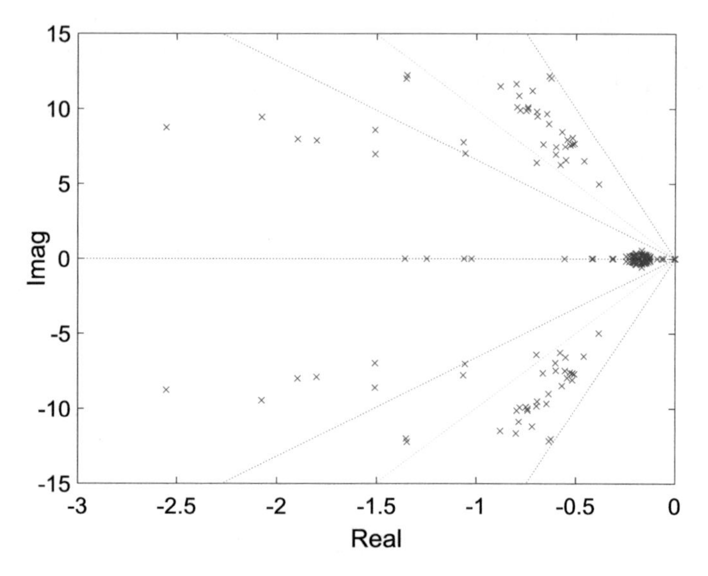

Fig. 3. Enlarged part of Fig. 1 (Case A)

the value $\zeta = 5\%$ considered as the limit of poorly damped oscillations (red dotted line in Fig. 3).

Also, the results obtained for Case B (4 WFs connected) show that all real parts of a total of 867 eigenvalues (order of the system 867×867) are less than zero. For

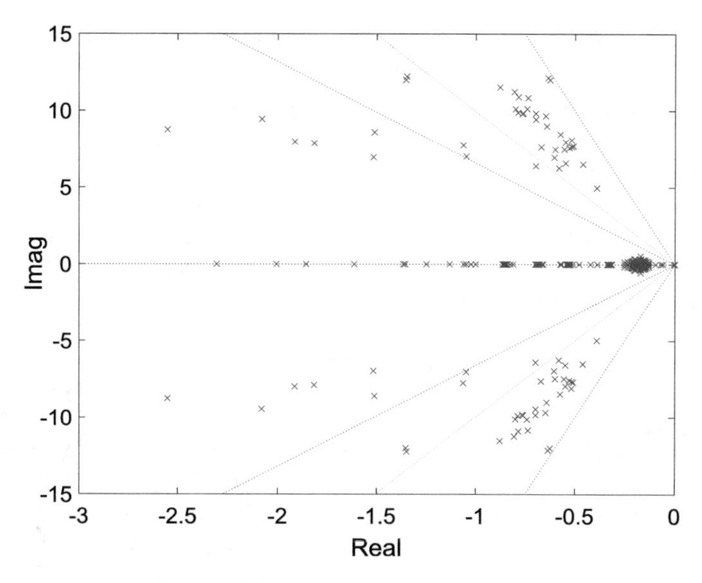

Fig. 4. Enlarged part of Fig. 2 (Case B)

eigenvalues with an imaginary part (136 complex pairs), the minimum damping ratio is identical as in the case without WFs (Case A), i.e. ζ = 5.1845%. The total number of additional eigenvalues due to connection of 70 wind turbines is 350.

280 of mentioned 350 eigenvalues are related to the state variables of the wind turbines (ω_m, θ_p, i_{dr}, i_{qr}, i_{qs} and i_{dc}), and 70 to the wind speed variable v_w where is

ω_m - rotor speed for both types of wind turbines
θ_p - pitch angle for both types of wind turbines
i_{dr}, i_{qr} - rotor currents (frequency converter currents) per d-axes and q-axes, respectively, for wind turbines with doubly-fed asynchronous generator
i_{qs}, i_{dc} - currents on generator and network side, respectively, for wind turbines with direct drive synchronous generator (full AC/DC/AC conversion)

344 of additional 350 eigenvalues, as a result of adding 4 WFs in the model, are negative real numbers which make a stable behavior. A part of them can be seen on the real axis in Fig. 4 (with 4 WFs connected) in comparison with real axis in Fig. 3 (without WFs connected). Remaining 6 of additional 350 eigenvalues are 3 pairs of complex numbers (Table 1).

The real parts of these complex numbers are negatives, while their damping ratio is nearly equal to 1 which is maximum damping rate. Eigenvalues Eig As # 46 and Eig As # 47 can be clearly seen on the Fig. 2. The most associated states (state variables) for these 3 pairs of complex eigevalues are also shown.

On the basis of previously conducted analyses of eigenvalues for Case A and Case B, it can be concluded that adding of 4 WFs in the model don't deteriorate the small-signal stability of the existing synchronous machines in the system. A full PSAT

Table 1. Six complex eigenvalues extracted from PSAT report (order of the system 867 × 867) as a result of adding 4 WFs to the model (Case B)

Eigenvalues	Real part	Imag. part	Damping ratio ζ	Most associated states (state variables)	
Eig As # 46	−96.53031	3.55961	0.999321	iqr_Dfig_51, iqr_Dfig_46	Rotor currents in q-axis of wind turbines in the WF Jelovača
Eig As # 47	−96.53031	−3.55961	0.999321	iqr_Dfig_51, iqr_Dfig_46	
Eig As # 59	−63.86969	0.27881	0.999990	e2d_Syn_10, e2d_Syn_11	Subtranzient voltage in d-axis of syn. generators in the HPP Mlini
Eig As # 60	−63.86969	−0.27881	0.999990	e2d_Syn_10, e2d_Syn_11	
Eig As # 63	−60.40721	0.30972	0.999987	e2d_Syn_15, idr_Dfig_21	Subtranzient voltage in d-axis of syn. generator in the HPP Dubrovnik, and rotor current in d-axis of wind turbine in the WF Trusina
Eig As # 64	−60.40721	−0.30972	0.999987	e2d_Syn_15, idr_Dfig_21	

reports for Case A and Case B with eigenvalues and participation factors are not presented due to lack of space in the paper.

4 Simulation of Small Disturbances in the Model of Power System of Bosnia and Herzegovina

Small disturbances are simulated for Case A and Case B of the BiH power system to confirm results from the previous analysis of the eigenvalues. Variations in WF production, due to variable wind speed, are continuous small disturbances which lead to imbalance of a total production and consumption in power system and rotor angles of synchronous machines are constantly adapted to such disturbances.

Figure 5 shows model of the wind speeds ("Weilbull" distribution) for WF Podveležje as example. Figure 6 shows, due to variations in the production of 70 modelled wind turbines, constant changes of the rotor angle of the synchronous generators in HPP Mostar, as electrically the nearest HPP to the connection point of WFs.

For Case A and Case B, the small disturbance is simulated as a 3-phase short circuit applied at the moment t = 2 [s], with duration of 0.04 [s], in the middle of two overhead lines:

(1) 110 kV OHL SS Podveležje – HPP Jablanica,
(2) 110 kV OHL SS Mesihovina – SS Posušje,

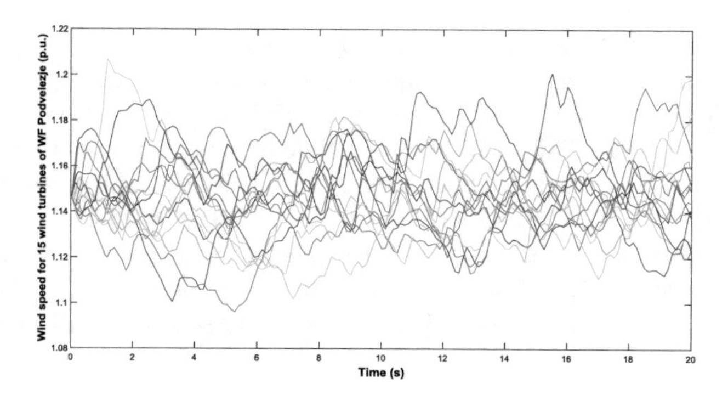

Fig. 5. Model of wind speeds for 15 wind turbines of WF Podveležje

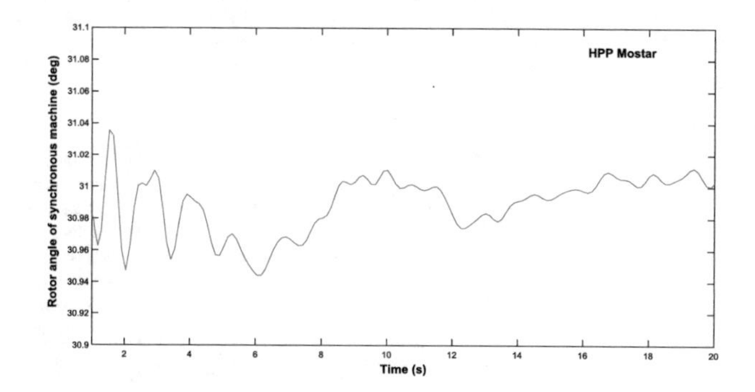

Fig. 6. Rotor angle changes for synchronous machines in HPP Mostar due to wind speed variations for 4 modelled WFs

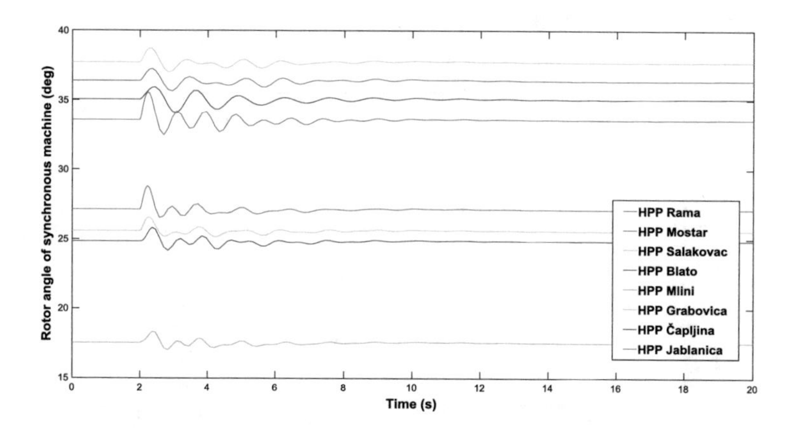

Fig. 7. Case A (without WFs) - Rotor angles of synchonous machines in neighbouring HPPs, if 3-phase short circuit in the middle of 110 kV OHL SS Podveležje – HPP Jablanica is applied at t = 2 [s], with duration of 0.04 [s]

As can be seen from Figs. 7, 8, 9, 10, 11 and 12, the existing power plants (electrically close to the small disturbance) are stable for simulated small disturbances in the middle of 110 kV OHL SS Podveležje - HPP Jablanica for both, Case A and Case B. The largest oscillations of the rotor angle in both cases are in HPP Jablanica, which is electrically closest to the location of the small disturbance, while the largest voltage drops are in HPP Mostar. The amplitudes of the rotor angle oscillations of the synchronous generators for all observed power plants are lower in Case B (Figs. 8 and 10) then in Case A (Figs. 7 and 9). Also, the voltage drop for HPP Mostar, HPP Jablanica and other power plants is slightly larger in Case A (Fig. 11) then in Case B (Fig. 12).

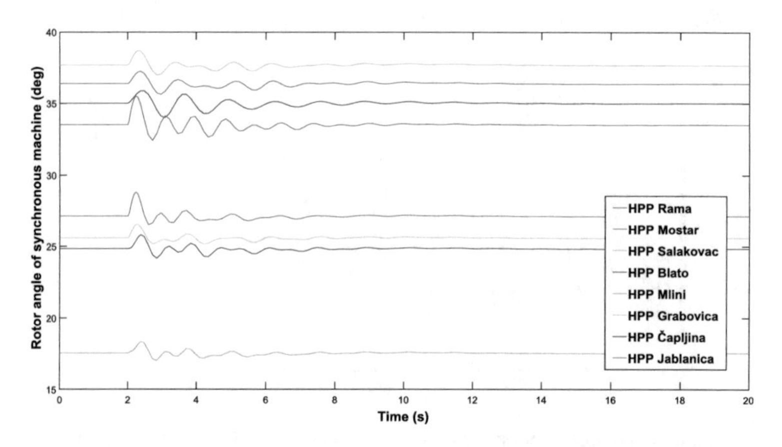

Fig. 8. Case B (with WFs) - Rotor angles of synchonous machines in neighbouring HPPs, if 3-phase short circuit in the middle of 110 kV OHL SS Podveležje – HPP Jablanica is applied at t = 2 [s], with duration of 0.04 [s]

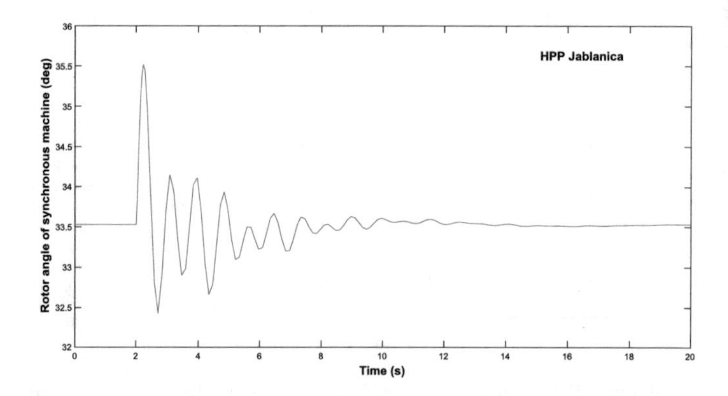

Fig. 9. Case A - Rotor angle of synchonous machine in HPP Jablanica taken from Fig. 5

As can be seen from Figs. 13, 14, 15, 16, 17 and 18, the existing power plants (electrically close to the small disturbance location) are also stable for simulated small

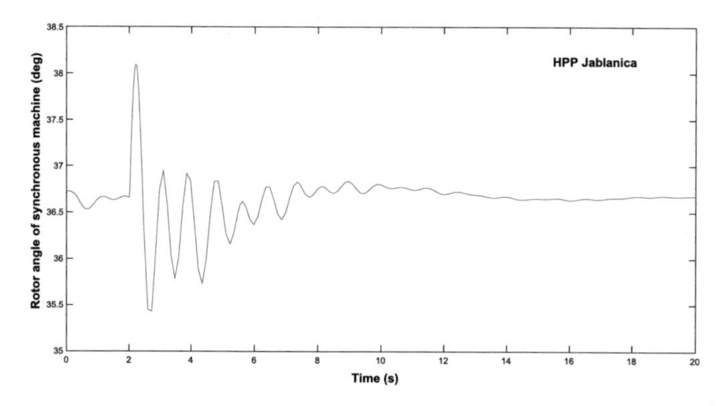

Fig. 10. Case B - Rotor angle of synchonous machine in HPP Jablanica taken from Fig. 6

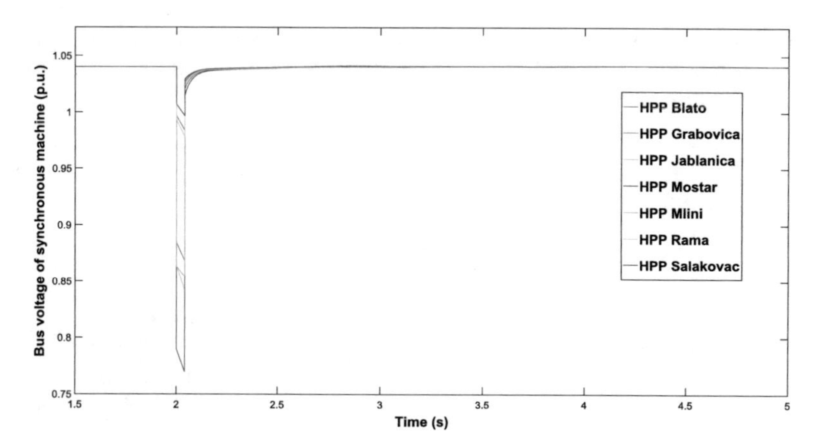

Fig. 11. Case A (without WFs) - Bus voltages of synchronous machines [p.u.] in neighbouring HPPs, if 3-phase short circuit in the middle of 110 kV OHL SS Podveležje – HPP Jablanica is applied at t = 2 [s], with duration of 0.04 [s]

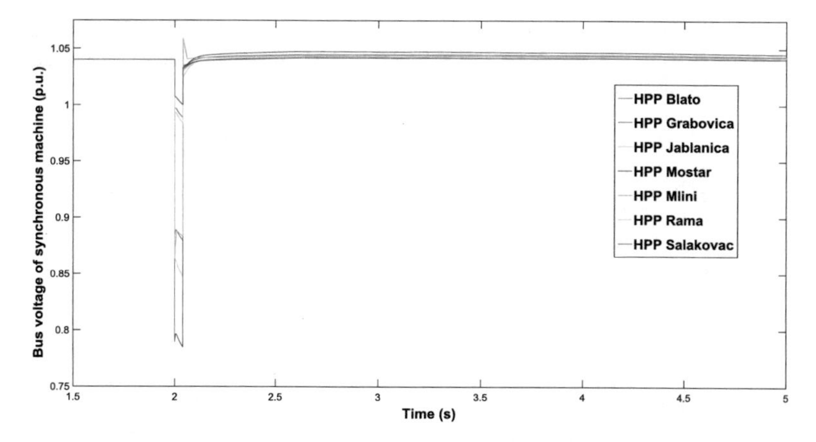

Fig. 12. Case B (with WFs) - Bus voltages of synchronous machines [p.u.] in neighbouring HPPs, if 3-phase short circuit in the middle of 110 kV OHL SS Podveležje – HPP Jablanica is applied at t = 2 [s], with duration of 0.04 [s]

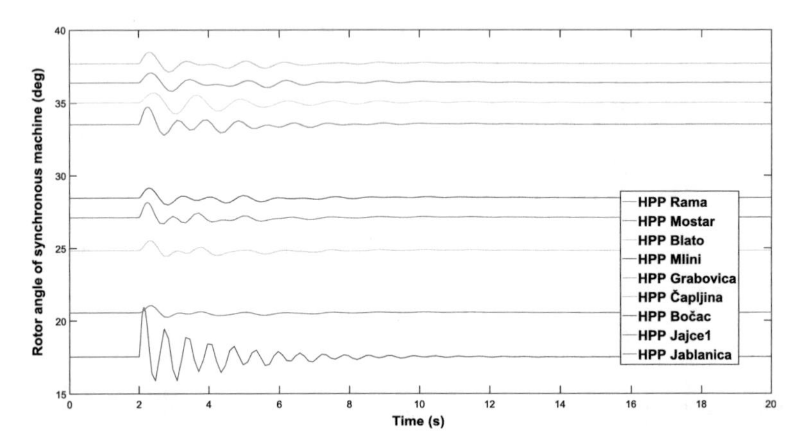

Fig. 13. Case A (without WFs) - Rotor angles of synchonous generators in neighbouring HPPs, if 3-phase short circuit in the middle of 110 kV OHL SS Mesihovina – SS Posušje is applied at t = 2 [s], with duration of 0.04 [s]

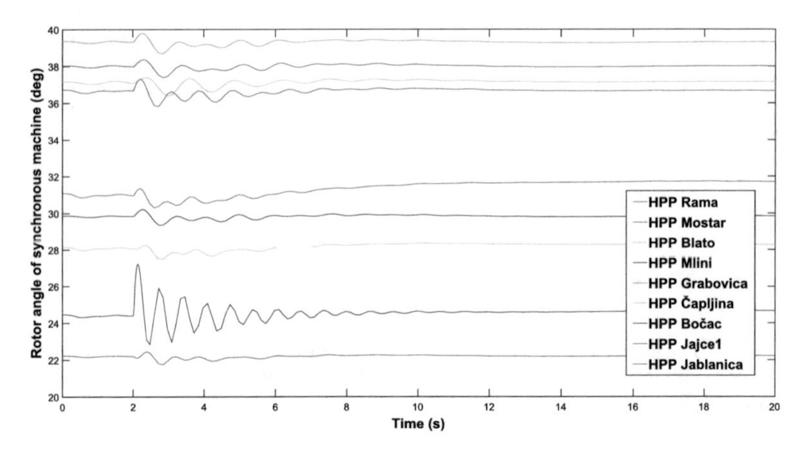

Fig. 14. Case B (with WFs) - Rotor angles of synchonous generators in neighbouring HPPs, if 3-phase short circuit in the middle of 110 kV OHL SS Mesihovina – SS Posušje is applied at t = 2 [s], with duration of 0.04 [s]

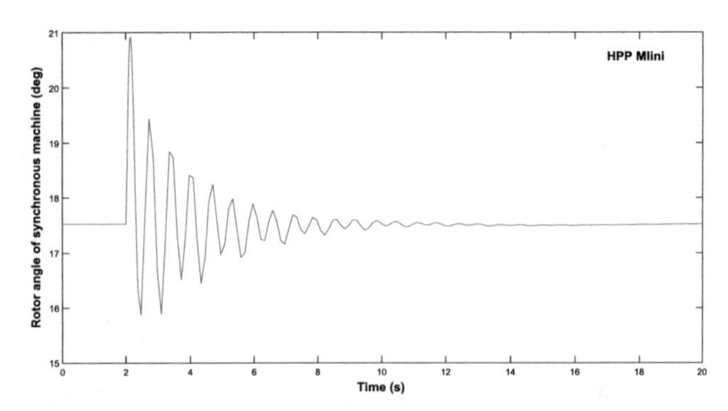

Fig. 15. Case A - Rotor angle of synchonous machines in HPP Mlini taken from Fig. 11

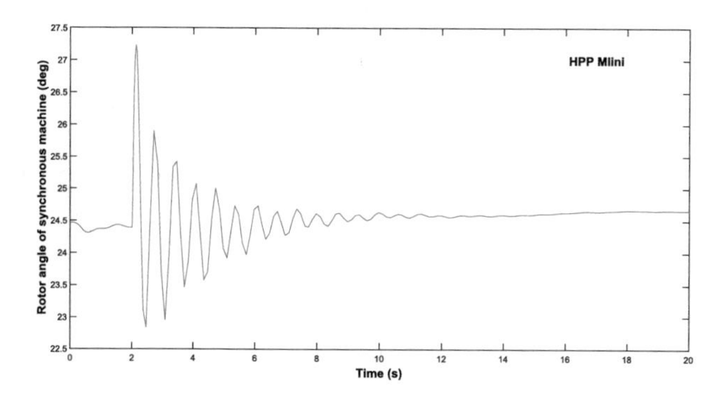

Fig. 16. Case B - Rotor angle of synchonous machines in HPP Mlini taken from Fig. 12

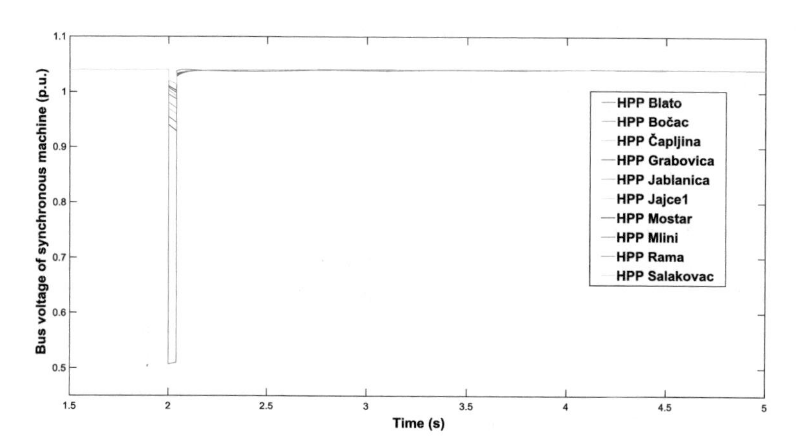

Fig. 17. Case A (without WFs) - Bus voltages of synchronous generators [p.u.] in neighbouring HPPs, if 3-phase short circuit in the middle of 110 kV OHL SS Mesihovina – SS Posušje is applied at t = 2 [s], with duration of 0.04 [s]

disturbances in the middle of 110 kV OHL SS Mesihovina - SS Posušje for both, Case A and Case B. The largest oscillations in the rotor angles in both cases are in HPP Mlini, which is electrically closest to the location of the small disturbance (Figs. 13 and 14). The largest voltage drops occur in the HPP Mlini (Figs. 17 and 18). Here also, it can be seen that the amplitudes of the rotor angle oscillations of the synchronous generators for all observed power plants are lower in Case B (Fig. 14) then in Case A (Fig. 13). Voltage drops are nearly the same in both cases for the observed power plants, including HPP Mlini and HE Mostar (Figs. 17 and 18).

Generally, for both locations of the simulated small disturbance, no differences are noticed in the oscillation frequency and damping time for the rotor angle's oscillations between Case A and Case B.

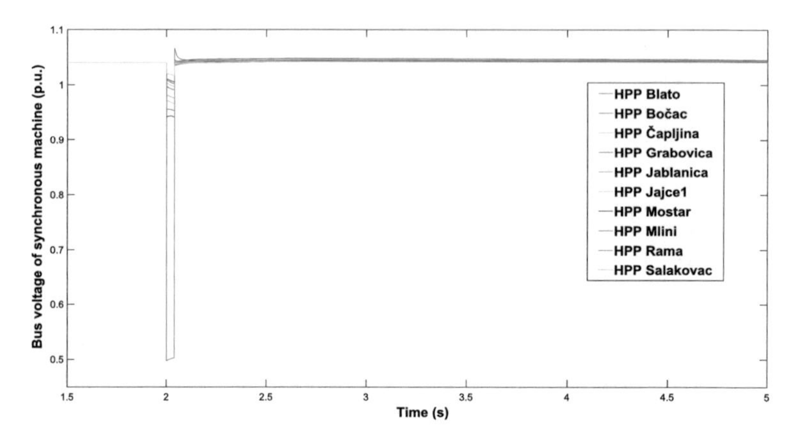

Fig. 18. Case B (with WFs) - Bus voltages of synchronous generators [p.u.] in neighbouring HPPs, if 3-phase short circuit in the middle of 110 kV OHL SS Mesihovina – SS Posušje is applied at t = 2 [s], with duration of 0.04 [s]

5 Conclusion

Based on the selected operating modes of the BiH power system (Case A and Case B), the way the elements of the BiH power system are modelled in PSAT application, available data from literature as input parameters of synchronous machines (including automatic voltage regulators and turbine regulators), and input parameters of the wind turbines, it can be concluded that 4 modelled WFs, generally do not cause a negative impact to the angular stability of the existing synchronous machines in the power system if small disturbance is applied.

The eigenvalue analysis for Case A and B is shown that 344 of additional 350 eigenvalues, as a result of adding 4 WFs in the model of BiH power system, are negative real numbers which make a stable behavior. Remaining 3 pairs of complex eigenvalues have damping ratio nearly equal to 1 (max. damping rate).

The amplitudes of the rotor angle's oscillations and voltage drops of the synchronous generators for almost all observed power plants are lower in Case B (with WFs connected). Generally, for both locations of the simulated small disturbance, no differences are noticed in the oscillation frequency and damping time for the rotor angle's oscillations between Case A and Case B.

The above conclusion can not be considered as general conclusion, but only apply to the chosen model of the BiH power system and simulations. It should be emphasized that the participation of modelled 4 WFs in total production is approximately 4.8% for the selected model of BiH power system, i.e. the wind power penetration is law. In order to check the results and conclusions reached in this paper, it is necessary, on the basis of long-term development plans of the transmission network of BiH, to analyze the influence of a large-scale wind turbines on the angular system stability (with higher WF participation in the total system production), taking into account all future potential locations for the construction of WFs, and the necessity of obtaining the most accurate input parameters for the modelling of the wind turbines, which is not easy task considering the strict confidentiality required by equipment manufacturers. It is also important to note that the PSAT application was used for the modelling of a real EES BiH as free and "open source" software which comes with no warranty conditions. By using highly developed commercial software, one can compare methods of modelling the transmission system, including WFs, with the models presented in this paper in the PSAT application, as well as the numerical methods used to simulate small disturbance in systems, and compare the results obtained.

Appendix

Data on WFs and the values of the parameters for wind turbine models in Case B are shown in Tables A1 and A2, respectively.

Table A1. Wind farms (WF) modelled in power system of BiH (Case B)

Wind farm	Installed power (MW)	Number and unit power of wind generators		Type of wind generator
WF Mesihovina	50.6	22	2.3 MW	Doubly-fed induction generator
WF Jelovača	36	18	2.0 MW	Doubly-fed induction generator
WF Trusina	49.5	15	3.3 MW	Doubly-fed induction generator
WF Podveležje	48	15	3.2 MW	Direct drive synchronous generator with permanent magnet
Total	184.1	70		

Table A2. Values of the parameters for wind turbine models (Case B)

Variable speed wind turbine with doubly fed induction generator	Variable speed wind turbine with direct drive synchronous generator	WF Mesihovina	WF Jelovača	WF Trusina	WF Podveležje
Power, voltage and frequency ratings [MVA, kV, Hz]	Power, voltage and frequency ratings [MVA, kV, Hz]	[2.45 0.69 50]	[2.3 0.69 50]	[3.6 0.69 50]	[3.33 0.69 50]
Stator resistance Rs and reactance Xs [p.u. p.u.]		[0.01 0.10]	[0.01 0.10]	[0.0333 0.10]	
	Stator resistance Rs [p.u.]				0.01

(*continued*)

Table A2. (*continued*)

Variable speed wind turbine with doubly fed induction generator	Variable speed wind turbine with direct drive synchronous generator	WF Mesihovina	WF Jelovača	WF Trusina	WF Podveležje
Rotor resistance Rr and reactance Xr [p.u. p.u.]		[0.01 0.08]	[0.01 0.08]	[0.01 0.08]	
	Direct and inverse reactances Xd and Xq [p.u. p.u.]				[1 0.8]
Magnetization reactance Xm [p.u.]		3.0	3.0	3.0	
	Constant field flux Psi_p [p.u.]				1.00
Inertia constants Hm [kWs/kVA]	Inertia constants Hm [kWs/kVA]	2	2	3	4
Pitch control gain and time constant Kp, Tp [p.u. s]	Pitch control gain and time constant Kp, Tp [p.u. s]	[10 3]	[10 3]	[10 3]	[10 3]
Voltage control gain Kv [p.u.]		10	10	10	
	Voltage control gain and time constant Kv Tv [p.u. s]				[10 1]
Power control time constant Te [s]		0.01	0.01	0.01	
	Active and reactive power control time constants Tep Teq [s s]				[0.01 0.01]
Number of poles p and gear box ratio [int -]	Number of poles p and gear box ratio [int -]	[4 1/91]	[4 1/106.8]	[4 1/89]	[120 1]
Blade length and number [m int]	Blade length and number [m int]	[53.00 3]	[47.50 3]	[54.70 3]	[55 3]
Pmax and Pmin [p.u. p.u.]	Pmax and Pmin [p.u. p. u.]	[1 0.000]	[1 0.000]	[1 0.000]	[0.915 0.0]
Qmax and Qmin [p.u. p.u.]	Qmax and Qmin [p.u. p.u.]	[0.6 −0.6]	[0.6 −0.6]	[0.5 −0.5]	[0.6 −0.6]

References

1. Kundur, P.: Power System Stability and Control. McGraw-Hill, New York (1994)
2. GWEC (Global Wind Energy Council), Annual Market Update 2017. http://gwec.net/publications/global-wind-report-2/
3. López, Y.U., Domínguez, J.A.: Small Signal Stability Analysis of Wind Turbines (2015)
4. Engström, S., Persson, J., Olsson, P., Lundin, U.: Wind Farms Influence on Stability in an area with High Concentration of Hydropower Plants, September 2011
5. Chandra, D.R., Kumari, M.S., Sydulu, M., Grimaccia, F., Mussetta, M., Leva, S., Duong, M. Q.: Impact of SCIG, DFIG Wind Power Plant on IEEE 14 Bus System with Small Signal Stability Assessment (2014)
6. Wei, Z., Shaojian, S.: The Small Signal Stability Analysis of A Power System Integrated with PMSG-based Wind Farm (2014)
7. Gautam, D., Vittal, V., Harbour, T.: Impact of increased penetration of DFIG-based wind turbine generators on transient and small signal stability of power systems. IEEE Trans. power Syst. **24**(3), 1426–1434 (2009)
8. Knüppel, T., Nielsen, J.N., Jensen, K.H., Dixon, A., Østergaard, J.: Small-signal stability of wind power system with full-load converter interfaced wind turbines. IET Renew. Power Gener. **6**(2), 79–91 (2012)
9. Du, W., Bi, J., Wang, T., Wang, H.: Impact of grid connection of large-scale wind farms on power system small-signal angular stability. CSEE J. Power Energy Syst. **1**(2), 83–89 (2015)
10. He, P., Wen, F., Ledwich, G., Xue, Y.: Small signal stability analysis of power systems with high penetration of wind power. J. Modern Power Syst. Clean Energy **1**(3), 241–248 (2013)
11. Liu, W., Ge, R., Li, H., Ge, J.: Impact of large-scale wind power ıntegration on small signal stability based on stability region boundary. Sustainability **6**(11), 7921–7944 (2014)
12. Sun, B., He, Z., Jia, Y., Liao, K.: Small-signal stability analysis of wind power system based on DFIG. Energy Power Eng. **5**(04), 418 (2013)
13. Power System Analysis Toolbox, Documentation for PSAT version 2.0.0, Federico Milano, 14 February 2008
14. Power System Analysis Toolbox ver 2.1.10 (software); Federico Milano, Juni 2016. http://faraday1.ucd.ie/psat.html
15. MATLAB R2016b - student use (software); The MathWorks Inc. 1984–2016. www.mathworks.com/products/matlab.html
16. Indikativni plan razvoja proizvodnje za period 2018–2027; NOSBiH, Mart 2017
17. Dugoročni plan razvoja prenosne mreze 2016–2025; Elektroprijenos BiH, Maj 2016
18. Dugoročni plan razvoja prenosne mreze 2017–2026; Elektroprijenos BiH, Novembar 2016
19. Procjena potrebne snage za integraciju VE u prenosni sistem BiH; NOSBiH, Juli 2017
20. Grid Code, NOSBiH, 15 December 2016. www.nosbih.ba
21. Power system modelling II, F. Milano (knjiga), University of Castilla – La Mancha (2010)
22. Sauer, P.W., Pai, M.A.: Power system dynamics and stability (1998)
23. Registar prethodnih saglasnosti za priključak vjetroelektrana na EES BiH (ažuriran 14 February 2018). www.fmeri.gov.ba/registar-prethodnih-saglasnosti-za-prikljucak-ve-na-ees-bih.aspx

Classification of Distribution Network Faults Using Hilbert-Huang Transform and Artificial Neural Network

Tarik Hubana[1(✉)], Mirza Šarić[1], and Samir Avdaković[2]

[1] Public Enterprise Elektroprivreda of Bosnia and Herzegovina Mostar, Mostar, Bosnia and Herzegovina
{t.hubana,m.saric}@epbih.ba

[2] Public Enterprise Elektroprivreda of Bosnia and Herzegovina Sarajevo, Sarajevo, Bosnia and Herzegovina
s.avdakovic@epbih.ba

Abstract. Identification and classification of faults in the electrical power system remain one of the most important tasks for the system operators and managers. In particular, high impedance faults (HIF) identification and classification is an especially challenging task due to the physical properties of the waveforms and low neutral voltage. In recent years, there have been significant technology driven advances in this field of research, owned to the introduction and progress of smart grid technologies. However, this topic still remains an open area of research. This paper presents a method for classification of HIF in a medium voltage distribution network, based on the Hilbert-Huang Transform and Artificial Neural Networks. This method was tested on generated signals based on the model of a realistic distribution system. The results indicated that the proposed algorithm is capable to accurately classify HIF in the distribution system. This paper contributes to the existing research by developing and testing, on a model of realistic distribution system, a HIF classification method which offers very efficient and accurate performance.

1 Introduction

The Electrical Power System (EPS) is undoubtedly one of the most complex systems created by humans. The main purpose of the EPS is to generate, transmit and distribute electrical energy to industrial, commercial and household customers. The EPS complexity is owned to the fact that it consists of numerous complex dynamic and interacting elements, which are prone to failures, disturbances and faults. Recent years mark the period of remarkable advancement in the field of planning, operation and management of the EPS. These advancements are mainly owned to the proliferation of advanced information and telecommunication technologies in EPS, but also to the changes of regulatory framework and commitment to a new energy paradigm which comprises of energy transition from conventional to renewable generation technologies and electricity market liberalization. In particular, the introduction of smart grid technologies is regarded to be the main postulate for technology driven advancements of EPS operation and obliteration of outdated technologies and planning principles.

S. Avdaković (Ed.): IAT 2018, LNNS 59, pp. 114–131, 2019.
https://doi.org/10.1007/978-3-030-02574-8_10

However, numerous challenges still lay ahead of the visionary EPS which is smart, renewable, resilient and self-healing. One such challenge is the identification and classification of EPS faults. Even in the very early stages of EPS development, the use of high capacity electrical generating power plants and concept of the grid, i.e. synchronized electrical power plants and geographical displaced grids, required fault detection and operation of protection equipment in minimum possible time so that the power system can remain in a stable operating conditions. In modern EPS, fault identification, classification and clearance gains in importance due to the increasing regulatory and customer pressure for improvement of reliability indices, but also due to the system interconnection related issues. The faults on electrical power system lines are required to be detected, classified correctly and cleared by protective systems as soon as possible. The protection system used for line protection can also be used to initiate the other relays to protect the power system from unwelcome events which can not only cause the outages, but also numerous power quality problems such as voltage sag propagation throughout the network [1].

There are various types of power system faults. One possible classification of power system faults groups the faults according to the value of fault resistances. According to this classification scheme, there are two major categories of earth faults. The first group consists of the faults with resistances mostly below a few hundred ohms. In order to clear this type of faults, the circuit breaker tripping is usually required. These faults are most often flash-overs to the grounded parts of the network and it is possible to perform the distance computation for them. The second group of faults are the high impedance faults (HIF) which have the fault resistances in the order of thousands of ohms. In this case, the neutral potentials are very low and difficult to detect so it is not unusual that continued network operation with a sustained fault is possible [2]. Generally, HIFs are regarded as faults with current values in the range from 0 to 75 A in an effectively grounded system [3]. Their detection and classification is one of the major challenges faced by the system operators. Even if numerous contributions have been provided [4], this topic still remains an important and entirely open area of research [5].

Generally, a suitable fault detection system provides an effective, reliable, fast and secure input signal for relaying operations. The application of a pattern recognition technique could be useful in discriminating the faulty and healthy electrical power system. It also enables the user to differentiate among the phases and determine which phase of a three phase power system is experiencing a fault. This paper presents the results of the research conducted in the area of HIF classification. In particular, this paper presents a HIF in a distribution network classifier, based on Hilbert-Huang Transform (HHT) and Artificial Neural Network (ANN). The ANNs are very powerful in identifying the faulty pattern and classification of fault by pattern recognition [6]. Also, ANN have been applied to various engineering applications and can be generally considered as a fast and accurate method for classification. It is proposed that the combination of ANN with HHT, which is applied to the signal at the pre-processing stage, could be a promising approach for the classification of the HIF in the distribution network.

The rest of the paper is organized as follows. First the basic theoretical concepts relevant for the development of the proposed algorithm are presented, followed by the outline of the algorithm structure and computational procedure. Then, then the results of the proposed algorithm application on a model of realistic distribution network are presented and discussed, together with the proposition of future research directions. Finally, the main conclusions resulting from this research are presented.

2 Theoretical Background

The following section will present the basic theoretical concepts that are relevant for the development of the proposed method.

2.1 Hilbert-Huang Transform

HHT is made up of two parts: one part is to get IMF by handling signals with EMD, and another one is to get the time-frequency spectrum by handling IMF with Hilbert transform, called Hilbert Spectral Analysis. Any signal can be broken down into some IMFs and a residue function through handling it with EMD [7].

2.1.1 Hilbert Spectral Analysis

The purpose of HHT is to demonstrate an alternative method to present spectral analysis tools for providing the time-frequency-energy description of time series data. Also, the method attempts to describe nonstationary data locally. Rather than a Fourier or wavelet based transform, the Hilbert transform was used, in order to compute instantaneous frequencies and amplitudes and describe the signal more locally [8].

The PV denotes Cauchy's principle value integral [8]:

$$H[x(t)] \equiv \hat{y}(t) = \frac{1}{\pi} PV \int_{-\propto}^{\propto} \frac{x(t)}{t-\tau} d\tau \tag{1}$$

An analytic function can be formed with the Hilbert transform pair as shown in Eq. (2) [8]:

$$z(t) = x(t) + i\hat{y}(t) = A(t)e^{i\theta(t)} \tag{2}$$

Where:

$$A(t) = \sqrt{x^2 + \hat{y}^2} \tag{3}$$

$$\theta(t) = \operatorname{arctg}\left(\frac{\hat{y}}{x}\right), \quad \text{and} \tag{4}$$

$$i = \sqrt{-1} \tag{5}$$

A(t) and θ(t) are the instantaneous amplitudes and phase functions, respectively. The instantaneous frequency can then be written as the time derivative of the phase, as shown in Eq. (6) [8]:

$$\omega = \frac{d\theta(t)}{dt} \tag{6}$$

After transforming all IMFs with Hilbert transform, a series of analytic functions and instantaneous frequencies can be obtained.

2.1.2 Empirical Mode Decomposition

The EMD algorithm is one component of the HHT method. The algorithm attempts to decompose nearly any signal into a finite set of functions, whose Hilbert transforms give physical instantaneous frequency values [8]. These functions are called intrinsic mode functions (IMFs). The algorithm utilizes an iterative sifting process which successively subtracts the local mean from a signal. The sifting process is repeated until the signal meets the definition of an IMF, which will be explained shortly. Then, the IMF is subtracted from the original signal, and the sifting process is repeated on the rest. This is repeated until the final residue is a monotonic function. The last extracted IMF is the lowest frequency component of the signal, better known as the trend. Previously, the sifting process was said to stop when the signal met the criteria of an IMF [8].

The definition of an IMF, is a signal which has a zero-mean, and whose number of extrema and zero-crossings differ by at most one [9]. IMFs are considered monocomponent functions which do not contain riding waves [9]. Once a signal has been fully decomposed, the signal D(t) can be written as the finite sum of the IMFs and a final residue as shown in Eq. (7).

$$D(t) = R_n(t) + \sum_{j=1}^{n} IMF_j(t) \tag{7}$$

Using previous equations, the analytic function can be formed as shown in Eq. (8) [8].

$$D(t) - R_n(t) = Re\left[\sum_{j=1}^{n} A_j(t)e^{i\int \omega_j(t)dt}\right] \tag{8}$$

Also, for reference, Eq. (9) shows the Fourier decomposition of a signal, x(t) [8].

$$D(t) = Re\left[\sum\nolimits_{j=1}^{n} A_j e^{i\omega j t}\right] \tag{9}$$

Notice that the EMD decomposition can be considered a generalized Fourier decomposition, because it describes a signal in terms of amplitude and basic functions whose amplitudes and frequencies may fluctuate with time [9].

2.2 Artificial Neural Network

An Artificial Neural Network (ANN) can be described as a set of elementary neurons that are usually connected in biologically inspired architectures and organized in several layers. Simply put, an elementary neuron is like a processor that produces an output by performing a simple non-linear operation on its inputs. A weight is attached to each and every neuron and training an ANN is the process of adjusting different weights tailored to the training set. An Artificial Neural Network learns to produce a response based on the inputs given by adjusting the node weights. Hence we need a set of data referred to as the training data set, which is used to train the neural Network [10].

Artificial neural network (ANN) can be applied to fault detection and classification effectively because it is a programming technique, capable to solve the nonlinear problems easily. The problems in which the information available is and in massive form can be dealt with. The fault classification method requires a neural network that allows it to determine the type of fault from the patterns of IMF signals, which are generated from the values measured from a three phase distribution underground cable at one terminal.

2.2.1 Back Propagation Neural Network

In the Back propagation neural network (BPNN) the output is feedback to the input to calculate the change in the values of the weights. One of the major reasons for taking the back propagation algorithm is to eliminate the one of the constraints on two layers ANNs, i.e. similar inputs lead to the similar output. The error for each iteration and for each point is calculated by initiating from the last step and by sending calculated the error backwards. The weights of the back-error-propagation algorithm for the neural network are chosen randomly, feeds back in an input pair and then obtain the result. After each step, the weights are updated with the new ones and the process is repeated for the entire set of inputs-outputs combinations available in the training data set provided by the developer. This process is repeated until the network converges for the given values of the targets for a predefined value of error tolerance. The entire process of back propagation can be understood by Fig. 1. This entire process is adopted by each and every layer in the entire the network in the backward direction [11]. The proposed algorithm uses the Mean Square Error (MSE) technique for calculating the error in each iteration. The algorithm of BPNN is as follows [11]:

Forward propagation:

$$a_j = \sum_{i}^{m} w_{ji}^{(1)} x_i \tag{10}$$

$$z_j = f(a_j) \tag{11}$$

$$y_j = \sum_{i}^{M} w_{kj}^{(2)} z_j \tag{12}$$

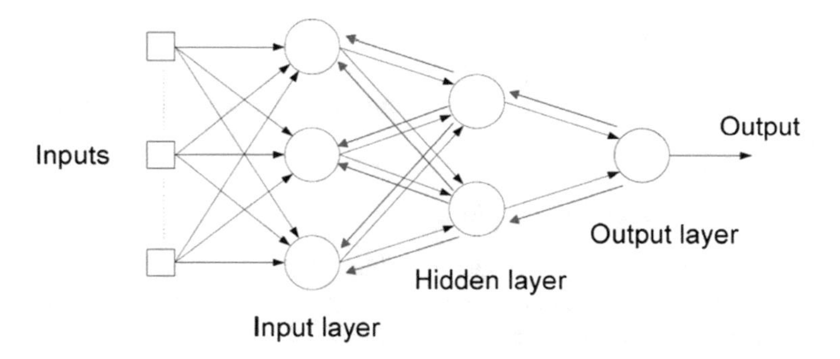

Fig. 1. Structure of back propagation ANN

Output difference:

$$\delta_k = y_k - t_k \tag{13}$$

Back propagation for hidden layers:

$$\delta_j = (1 - z_j^2) \sum_{k=1}^{K} w_{kj} \delta_k \tag{14}$$

The gradient of error with respect to first layer weights and second layer weights are calculated. In this step the previous weights are updated. In the previous equations a_j presents weighted sum of inputs, w_{ji} weight associated with the connection, x_i inputs, z_i activation unit of (input) that sends a connection to unit j, δ_k derivative of error at kth neuron, $\underline{y_i}$ ith output, y_k activation output of unit k, t_k corresponding target of input and δ_j derivative of error w_{rt} to a_j. The MSE for each output in each iteration is calculated by [11]:

$$\mathrm{MSE} = \frac{1}{N} \sum_{1}^{N} (E_i - E_o)^2 \tag{15}$$

where *N* is the number of iterations, *Ei* is actual output and *Eo* is out of the model [11].

Several different training algorithms for ANN are available. All these algorithms use the gradient of the performance function to determine how to adjust the weights to minimize performance. Levenberg-Marquardt optimization technique is used in the implemented ANN structure. In the proposed algorithm, IMF signals are combined and grouped and represent a unique 'signature' for each fault. After that, the ANN is trained with a large set of this data, and becomes capable to detect and identify PDN faults. It is important to note that the proposed classifier is fully customizable, very flexible and can be extended to include additional parameters and criteria.

3 Algorithm Development

This section presents an outline of the algorithm structure and computational procedure of the proposed method.

3.1 Basic Principles of the Algorithm

The proposed method algorithm is shown in Figs. 2 and 3, where Fig. 2 presents the algorithm during the ANN training process, and Fig. 3 the algorithm of the classifier in the power distribution system. The proposed classifier is completely adjustable, and by simple upgrades it is possible to implement additional parameters and criteria. The voltage in one end of the medium voltage underground cable is measured in various simulated scenarios and by applying the HHT EMD process the IMF's are obtained. This IMF's are used to create a unique system's signature during each fault scenario, and thus present a great input for ANN classifier. Finally, the ANN is trained with this data and previously known outputs.

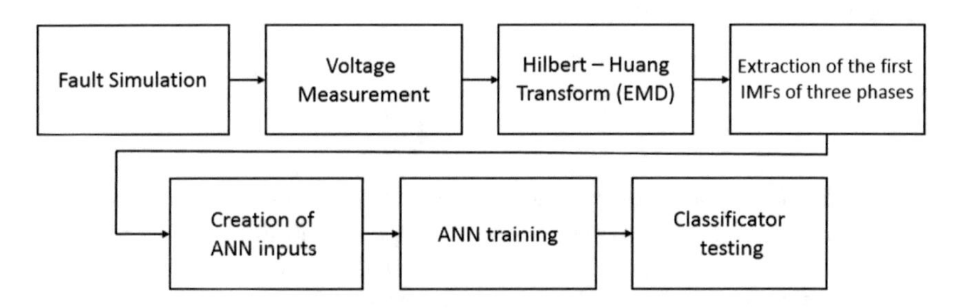

Fig. 2. Proposed method algorithm during the data preprocessing

Once created classifier is easy to use and implement in the system. One present industrial computer, paired with voltage measurements is enough for the system implementation. Measured voltages during the faulty conditions are sent to the HHT-ANN classifier, and after successful classification process, the trip signal can be sent or appropriate controls can be undertaken.

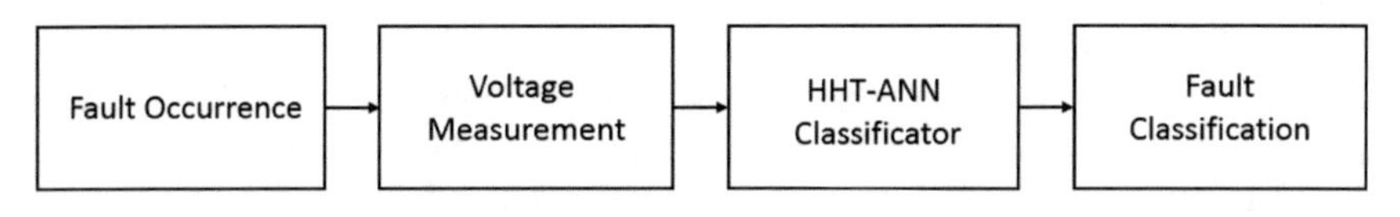

Fig. 3. Proposed classifier algorithm

3.2 Measurement Data Preprocessing

To get the characteristic quantity of the voltage signal it should be processed with Hilbert-Huang transform. High-frequency signals will be generated when a short-circuit fault occurs which reside in IMF1. So the instantaneous frequency of IMF1 will mutate immediately the short-circuit fault takes place [7]. Through EMD decomposition, the original current signal can be changed into some IMFs. For instantaneous amplitudes of the voltage of the same line, their range and changing rate will increase sharply after short circuit happens. High-frequency signals included in short-circuit will shunt through stray capacitance between phases or earth and phases [7].

In this paper, high frequency signal IMF components are chosen to be the characteristic quantity for the further analysis. The HHT is applied to the measured voltage signals, and the first three IMF components during the single line fault with various fault resistance values are shown in Figs. 4, 5 and 6.

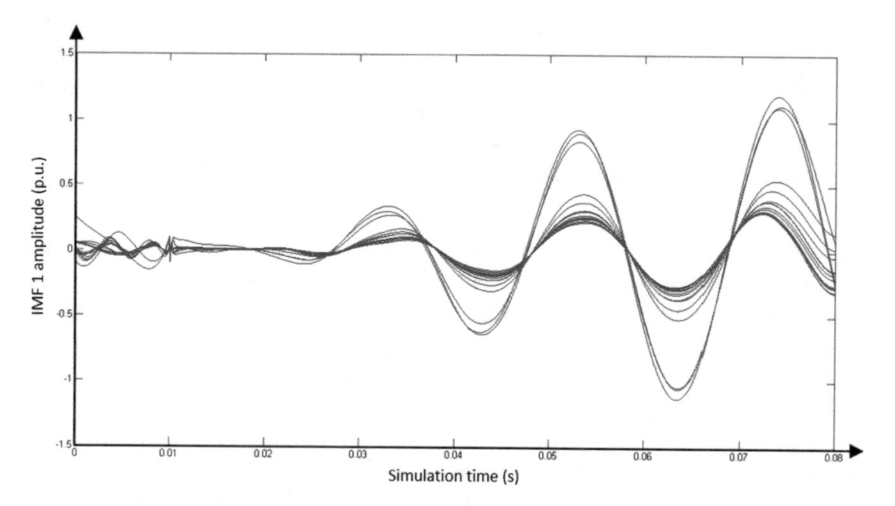

Fig. 4. IMF 1 of phase A, during the LG fault with various fault resistances

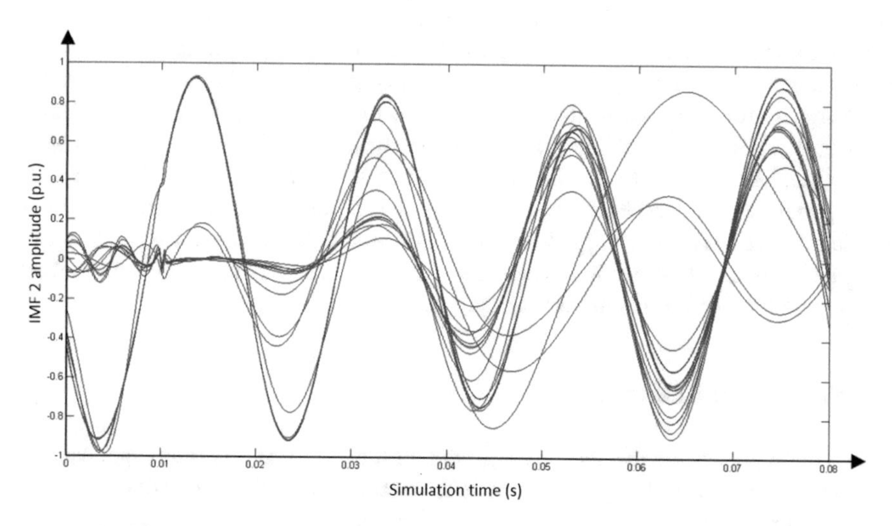

Fig. 5. IMF 2 of phase A, during the LG fault with various fault resistances

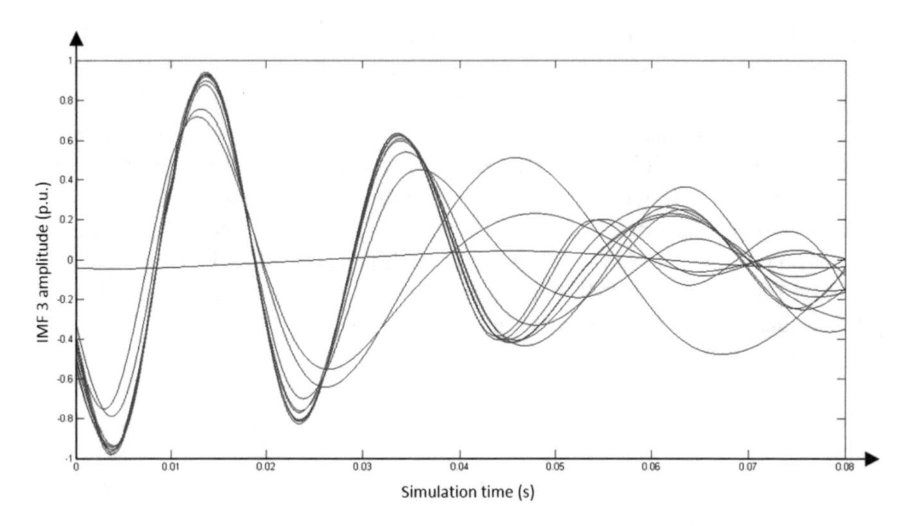

Fig. 6. IMF 3 of phase A, during the LG fault with various fault resistances

However, the number of IMF components depends of the signal complexity, and in various fault scenarios, different number of IMF components occur. Because of that, the proposed algorithm works with first IMF components, since the first IMF component is always present in every signal.

3.3 Training and Testing

Simulations are carried out in different fault scenarios for getting various fault patterns. The system model as well as ANN training is carried out using MATLAB and MATALB Simulink simulation software. Three different fault types (LG, LLG, LLLG faults) are simulated. The fault resistances ranged from 20 to 600 Ω. The number of neurons in hidden layers, is determined empirically by experimenting with various network configurations, and 11 hidden neurons are chosen since it presents a compromise between satisfying accuracy and training speed. The ANNs are fed with the sixty signals for every particular fault scenario, resulting in 180 total scenarios. The classification ANN output layer has three neurons, indicating which fault is taking place (LG, LLG or LLLG) are involved in the fault event. A total of 180 simulations were generated, and 80% of them were used for training, and the other 10% for validating and 10% for testing. Once trained, the networks performance is tested using a validation data set. The suitable network, which showed satisfactory results is finally created, and ready for use in various fault scenarios.

4 Results and Discussion

This section presets the results and discussion of the proposed algorithm application on a model of realistic distribution network, together with the proposition of future research directions.

4.1 Test System

The developed test system represents a part of a real power distribution system in the area of the City of Mostar. Consumers are fed over 35/10 kV transformer, via 10 kV underground cables, 10/0.4 kV distribution transformers and finally over 0.4 kV low voltage network. The test system is developed in MATLAB Simulink simulation software as shown in Fig. 7.

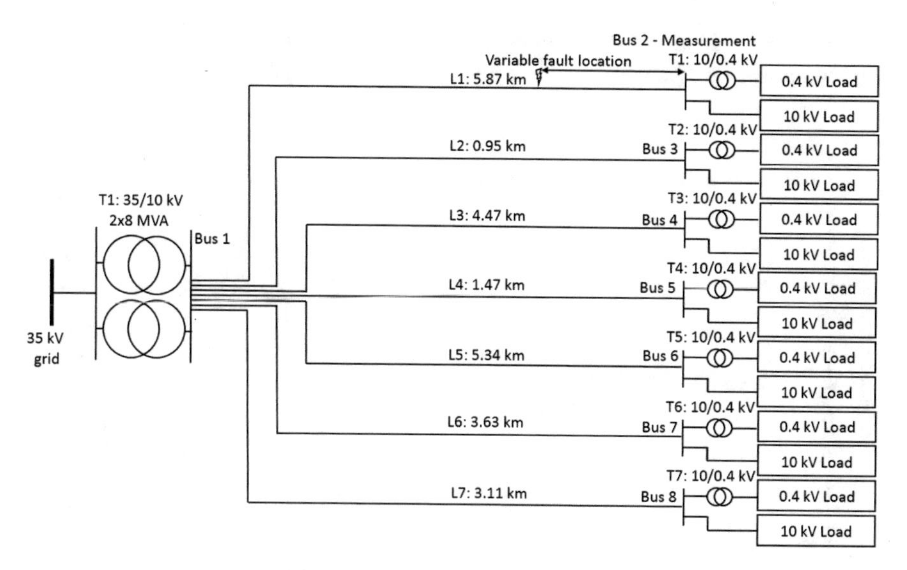

Fig. 7. The developed test system

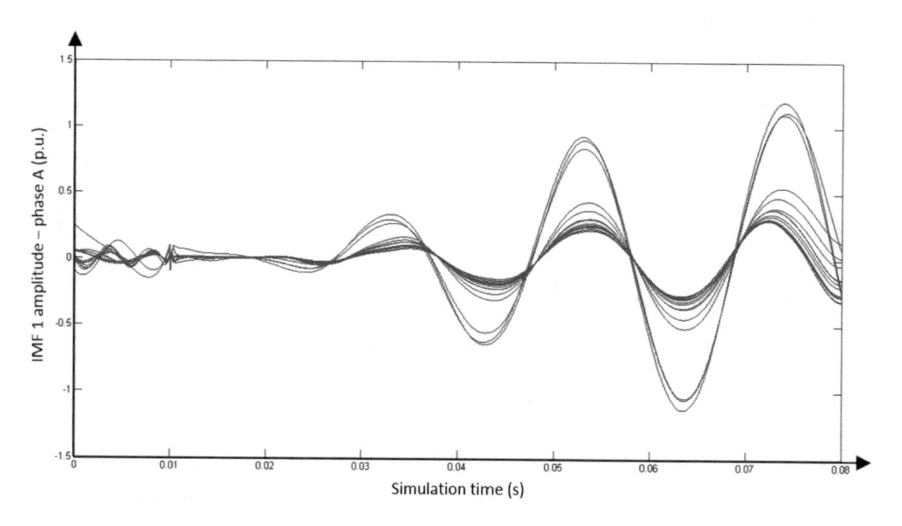

Fig. 8. IMF 1 of phase A, during the LG fault with various fault resistances

Faults are simulated on the 10 kV busbar of the 10/0.4 kV transformer, over various fault locations (in the range from 20 to 600 Ω). The LG, LLG and LLLG faults are simulated. Voltage measurements devices are connected to the 10 kV busbar of the 10/0.4 kV transformer, with the sampling frequency of 3.2 kHz, which is the sampling frequency of the present relays in the power distribution system.

4.2 Simulation Results

During the classification process in distribution network, it is necessary to monitor the IMF component's behavior in all three phases, since every voltage phase has a

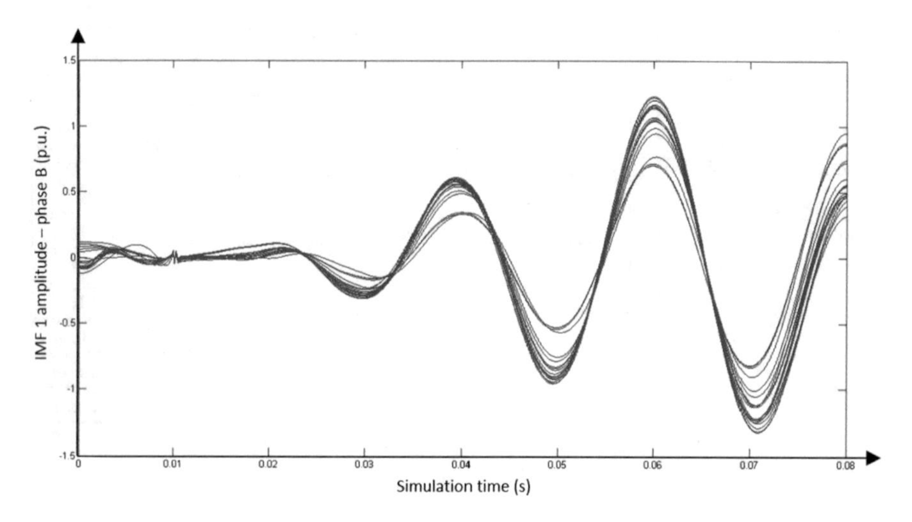

Fig. 9. IMF 1 of phase B, during the LG fault with various fault resistances

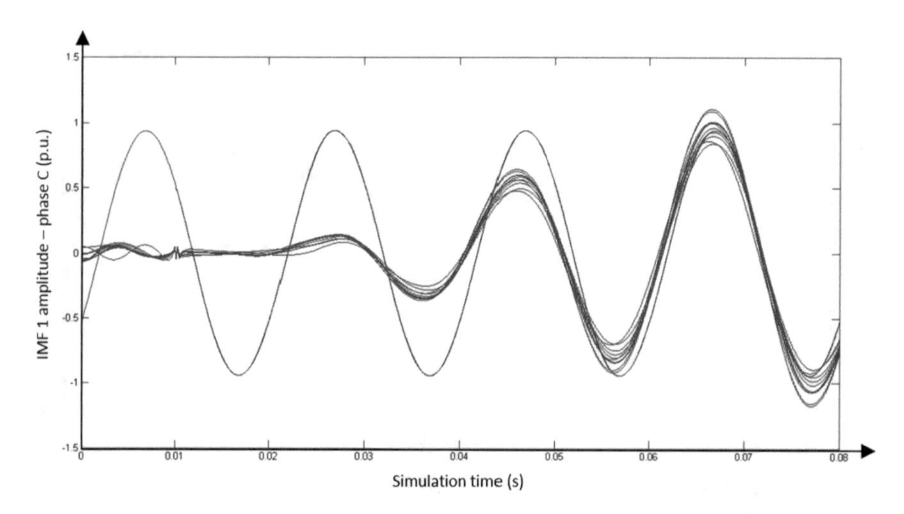

Fig. 10. IMF 1 of phase C, during the LG fault with various fault resistances

characteristic behavior during the fault. As previously mentioned, only the first IMF components will be used by the classifier. The first IMF components of every voltage phase during the LG fault are shown in Figs. 8, 9 and 10.

Figures 11, 12 and 13 show the first IMF components during the LLG fault.

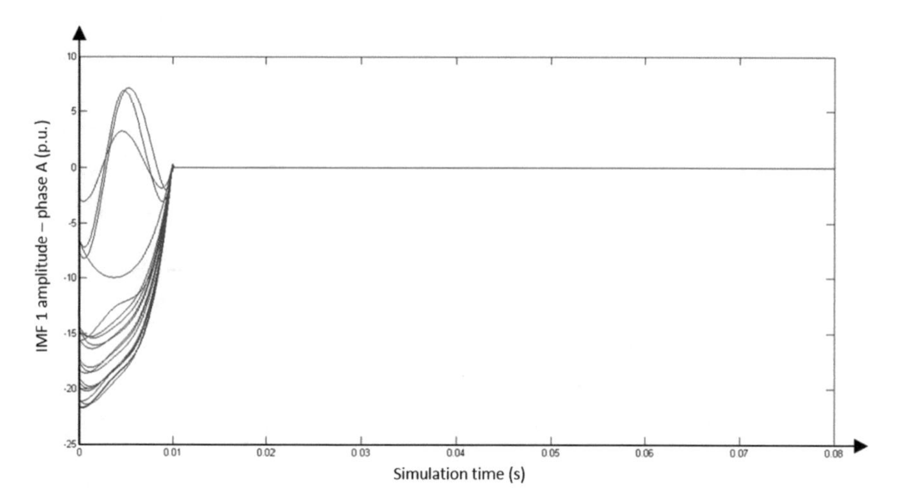

Fig. 11. IMF 1 of phase A, during the LLG fault with various fault resistances

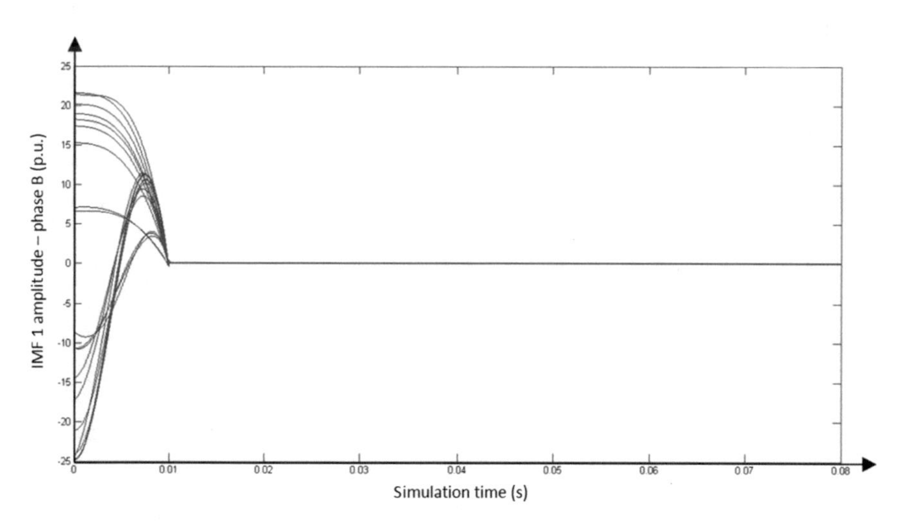

Fig. 12. IMF 1 of phase B, during the LLG fault with various fault resistances

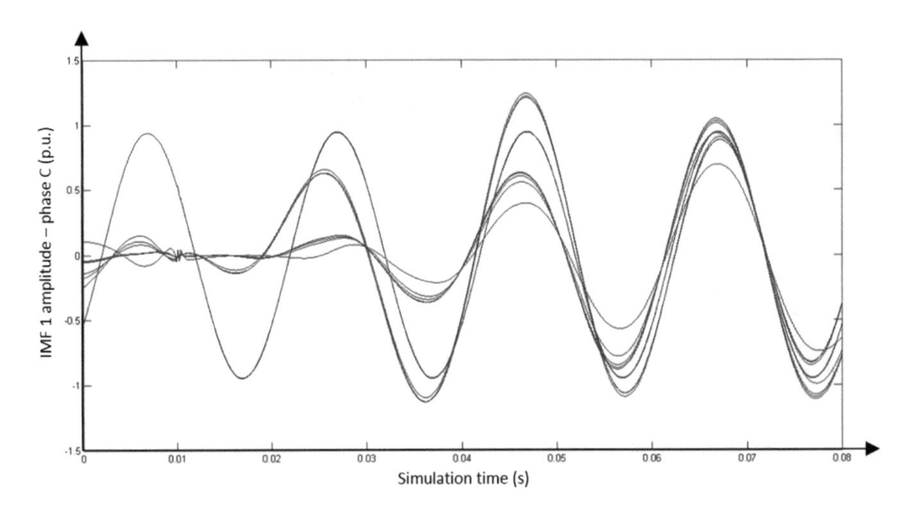

Fig. 13. IMF 1 of phase C, during the LLG fault with various fault resistances

Figures 14, 15 and 16 show the first IMF components during the LLLG fault.

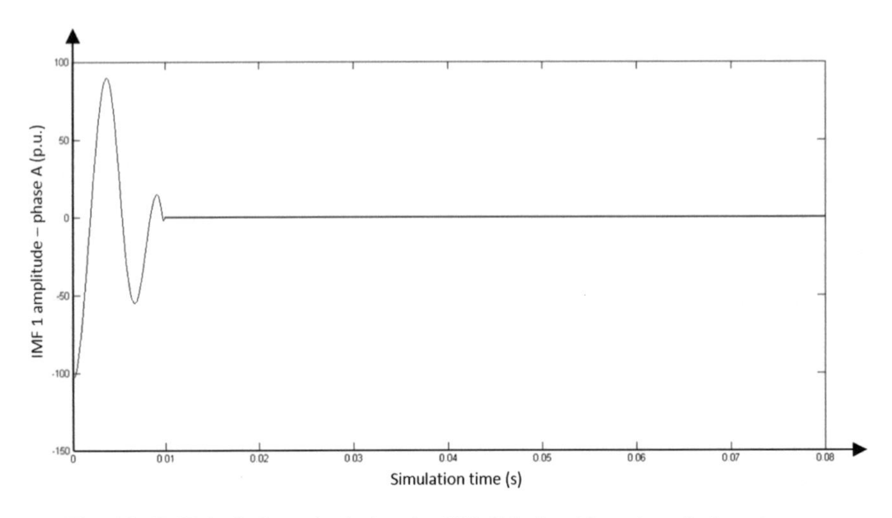

Fig. 14. IMF 1 of phase A, during the LLLG fault with various fault resistances

Figures 11, 12 and 13 show that the value of the first IMF component in the fault conditions is close to zero, but the IMF of the healthy phase has different shape for various fault resistances. Figures 14, 15 and 16 show that the fault resistance does not affect the IMF amplitude during the LLLG fault. Since the classifier is based on ANN, it is necessary to create appropriate ANN inputs. In order to reflect the system behavior in the best manner, a unique signal that contains IMF's of each phase is created. By grouping the signals into one signal, a unique system's signature that responds to

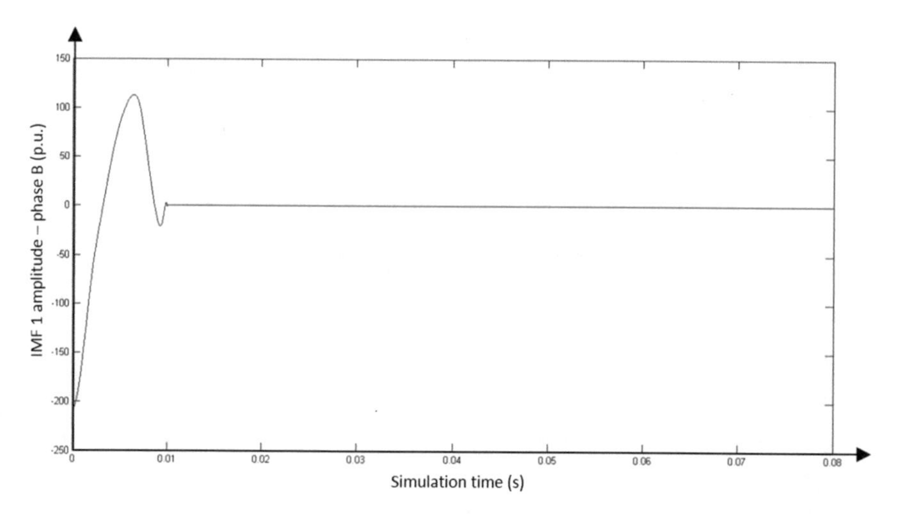

Fig. 15. IMF 1 of phase B, during the LLLG fault with various fault resistances

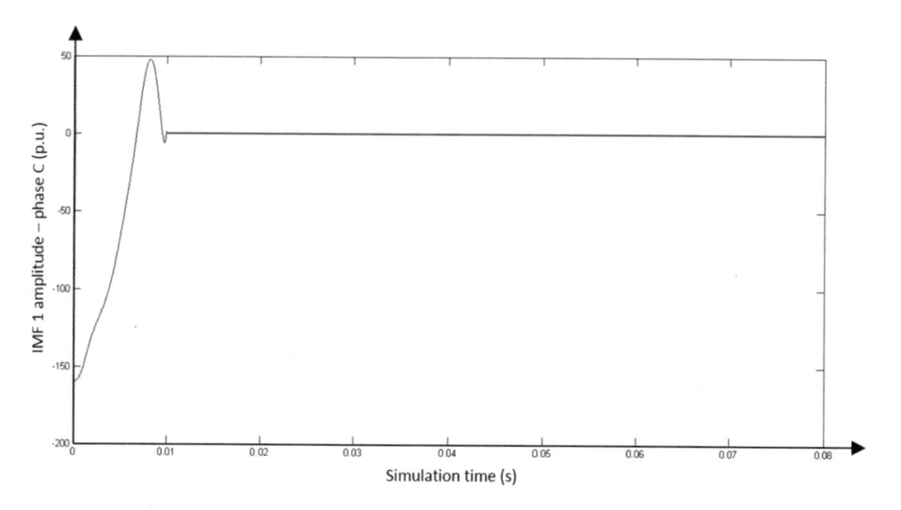

Fig. 16. IMF 1 of phase C, during the LLLG fault with various fault resistances

specific fault scenario is created. Figures 17, 18 and 19 show the grouped IMF signals for each fault type with 60 different fault resistances.

After the grouped signals are created, the ANN inputs are ready. The designed network takes in the set of total 180 inputs. The neural network has three outputs, each of them corresponding to the three fault. Hence the outputs are either 0 or 1 denoting the absence or presence of a fault, and the various possible permutations can represent each of the various faults accordingly.

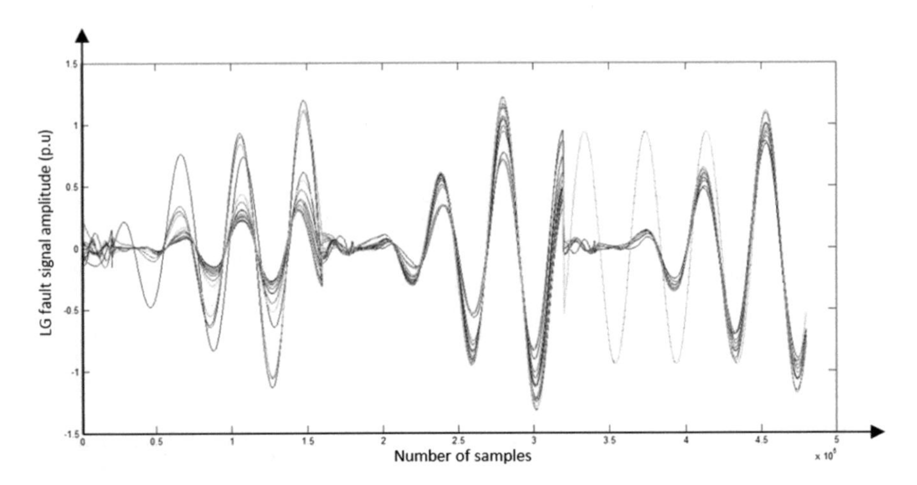

Fig. 17. Grouped IMF signal in case of LG fault for various fault resistances

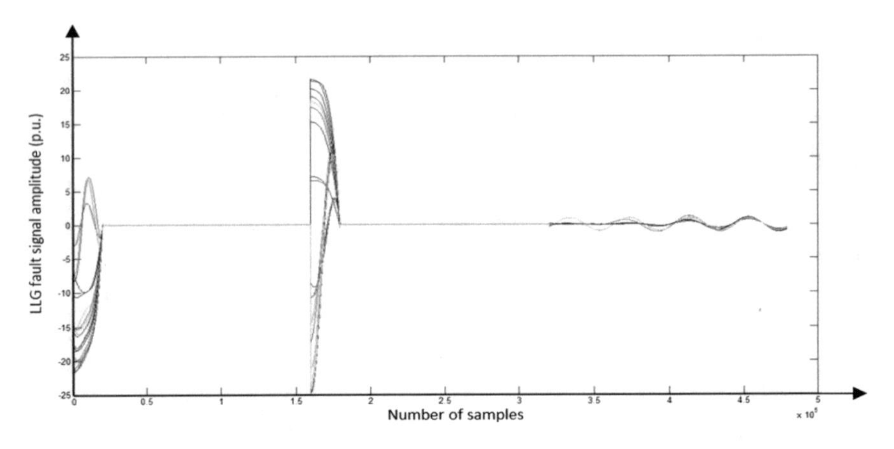

Fig. 18. Grouped IMF signal in case of LLG fault for various fault resistances

4.3 Test of the HHT-ANN Algorithm

After the training process, the ANN is created, with the MSE of 3.469e-4. However, the ANN outputs are not exactly 0 or 1, due to the classification error. This means that he outputs are not unambiguous, but one additional step where the highest value of each output is set to 1, and all other to 0, fixes this problem (Fig. 20).

In order to test the created ANN, the ANN will be tested with completely new fault scenarios, with fault resistance values that ANN is not trained to. New 15 random fault scenarios are simulated, and the proposed classifier is tested. Results of 8 tests are shown in Table 1.

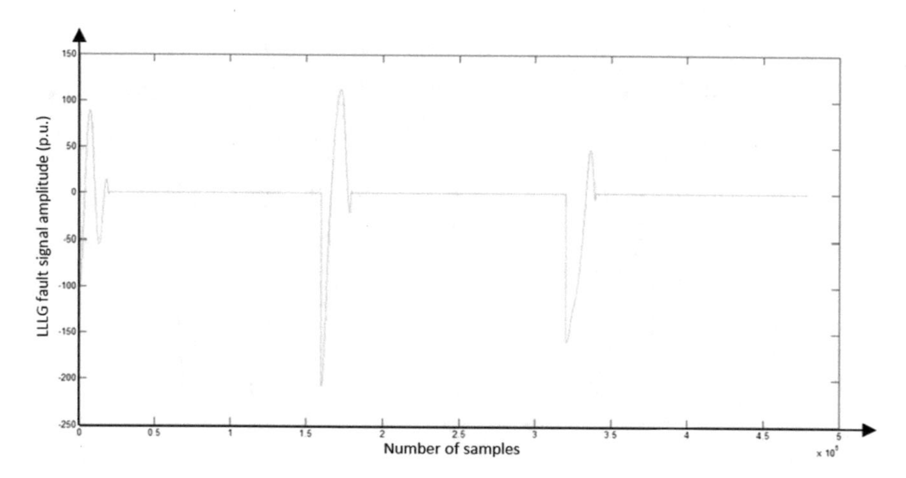

Fig. 19. Grouped IMF signal in case of LLLG fault for various fault resistances

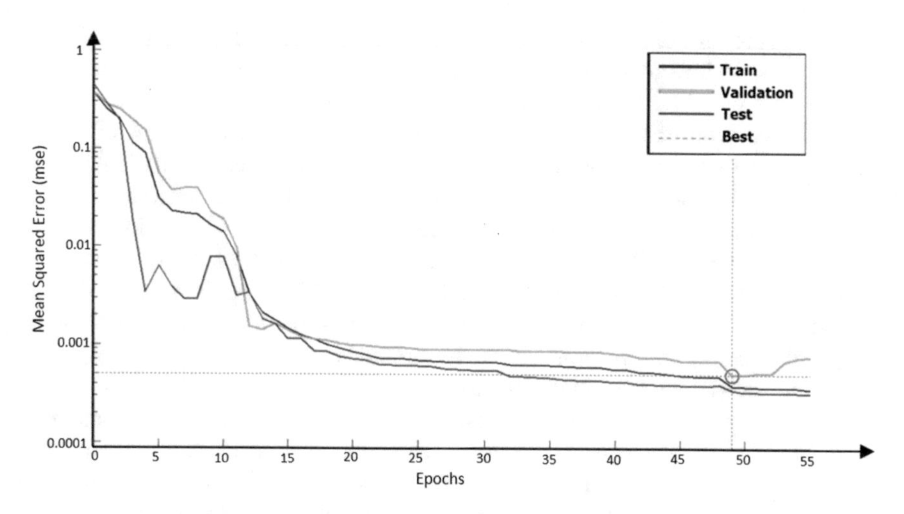

Fig. 20. The ANN training performance

Table 1 shows the efficiency of the proposed HHT-ANN classifier. The accuracy in the resistance range from 20 to 600 Ω is 100%. Future research directions will be to take the fault location and other possible fault types into consideration. This research is part of a large project, whose goal is to develop a new more accurate fault detector and classifier that overcomes the problems of the traditional protection devices.

Table 1. Output of the HHT-ANN classifier

Resistance (Ω)	Desired output (fault type)				Actual output (fault type)			
	LG	LLG	LLLG	No fault	LG	LLG	LLLG	No fault
20	1	0	0	0	1	0	0	0
110	0	0	0	1	0	0	0	1
200	0	1	0	0	0	1	0	0
290	0	0	1	0	0	0	1	0
380	1	0	0	0	1	0	0	0
470	0	1	0	0	0	1	0	0
470	0	1	0	0	0	1	0	0
560	0	0	0	1	0	0	0	1

5 Conclusion

Classification of faults, especially the HIFs in distribution networks is a serious problem, because the existing protection devices are unable to detect nor classify such failures. Thus, an unconventional approach to the problem is required, and therefore a fault classifier based on HHT and ANN is proposed as a possible solution.

For the purpose of algorithm testing, a test system based on a part of the real distribution network from Bosnia and Herzegovina is developed in MATLAB Simulink. Results obtained by applying the proposed algorithm demonstrated that the algorithm presents a good solution for classification of faults in distribution networks. Accuracy of the HHT-ANN algorithm is 100% in the 20–600 Ω range of fault resistances. It is important to point out that the developed ANN is not immune to the large load and switching state changes, and in order to overcome this, further ANN training is required. The algorithm turns out to be promising when it comes to the future heuristic power system protection devices.

This study is a part of an ongoing research whose objective is to contribute to the development of an advanced protection system that has satisfactory accuracy for detection and classification of the power distribution network faults. And important future research should extend the number of system elements and scenarios that can lead to the false fault classification.

References

1. Hubana, T., Begić, E., Šarić, M.: Voltage sag propagation caused by faults in medium voltage distribution network. In: Hadžikadić, M., Avdaković, S. (eds.) Advanced Technologies, Systems, and Applications II, IAT 2017. Lecture Notes in Networks and Systems, vol. 28. Springer, Cham (2018)
2. Hanninen, S.: Single phase earth faults in high impedance grounded networks. Technical Research Centre of Finland, VTT Publications, Espoo (2001)
3. Wang, B., Geng, J., Dong, X.: High-impedance fault detection based on non-linear voltage-current characteristic profile identification. IEEE Trans. Smart Grid **9**(4), 3783–3791 (2017)

4. Hubana, T., Šarić, M., Avdaković, S.: Approach for identification and classification of HIFs in medium voltage distribution networks. IET Gener. Transm. Distrib. **12**(5), 1145–1152 (2018)
5. Šarić, M., Hubana, T., Begić, E.: Fuzzy logic based approach for faults identification and classification in medium voltage isolated distribution network. In: Hadžikadić, M., Avdaković, S. (eds.) Advanced Technologies, Systems, and Applications II, IAT 2017. Lecture Notes in Networks and Systems, vol 28. Springer, Cham (2018)
6. Jamil, M., Sharma, S., Singh, R.: Fault detection and classification in electrical power transmission system using artificial neural network. Springer Plus **4**, 334 (2015). https://doi.org/10.1186/s40064-015-1080-x
7. Guo, Y., Li, C., Li, Y., Gao, S.: Research on the power system fault classification based on HHT and SVM using wide-area information. Energy Power Eng. **5**(4B), 138–142 (2013). https://doi.org/10.4236/epe.2013.54B026
8. Bradley, B.L.: The Hilbert-Huang Transform: theory, applications, development - PhD thesis. University of Iowa, Iowa City (2011)
9. Huang, N.E., Wu, Z.: A review on Hilbert-Huang transform: method and its applications to geophysical studies. Rev. Geophys. **46**(2), 1944:9208 (2008)
10. Ayyagari, S.B.: Artificial Neural Network Based Fault Location for Transmission Lines - Master's theses. University of Kentucky, Lexington (2011)
11. Haykin, S.: Neural Networks: A Comprehensive Foundation. Macmillan College Publishing Company, London (1994)

Distributed Generation Allocation: Objectives, Constraints and Methods

Mirza Šarić[1(✉)], Jasna Hivziefendić[2], and Nejdet Dogru[2]

[1] Public Enterprise Elektroprivreda of Bosnia and Herzegovina Mostar, Mostar, Bosnia and Herzegovina
m.saric@elektroprivreda.ba

[2] International Burch University Sarajevo, Sarajevo, Bosnia and Herzegovina
{jasna.hivziefendic,nejdet.dogru}@ibu.edu.ba

Abstract. This paper introduces the distributed generation allocation problem (DGAP) from the point of view of objectives, constraints and methods used to formulate and solve the problem and presents the results of the theoretical and empirical research review conducted in this field of engineering. The first part of the paper presents the fundamental concepts of the DGAP and proposes a classification of this problem based on objectives, constraints and methods. The second part presents a detailed discussion of some of the most important and frequently used objectives and constraints and methods used to solve DGAP. The main objective of this paper is to present the current state of this field of research, unveil possible conflicting results of previous investigations and finally, to identify the gaps that exist in order to justify the future research of this topic.

1 Introduction

Distributed generation (DG) promises to offer numerous benefits to the power system. However, the integration of renewable energy sources, transforms the distribution networks into an active, bidirectional system which is by far more complex system in terms of operation and management. Therefore, DG integration can potentially cause numerous challenges, ranging from technology based [1] and regulatory issues [2], to economic [3], environmental [4] and social [5] concerns. DG offers an alternative approach to solving the problems of distribution capacity expansion, network security and management issues [6] and it can be used to improve various network parameters [7]. In order to achieve the positive impact of DG on network parameters, an appropriate location and size for DG installation must be determined. Even if the DGAP has been investigated for past few decades, it continues to be a relevant topic and attracts new research efforts [8] and is becoming an increasingly important area of research [9]. DG integration studies have become a standard utility procedure which proceeds any connection approval. However, studies regarding DG sizing and allocation issues are not as that numerous [10]. DG projects will be beneficial only if DG planning is optimized using single or multi-objective function under certain operating constraints [11]. DGAP fundamentally exists because of the questions:

S. Avdaković (Ed.): IAT 2018, LNNS 59, pp. 132–149, 2019.
https://doi.org/10.1007/978-3-030-02574-8_11

(1) Are the decisions of Distribution System Operators (DSO) and private investor regarding the allocation and timing of DG installation the same [12]?
(2) Will DG owners schedule their units in order to maximise the benefits of the DG integration [12]
(3) How is the final result of the optimisation procedure influenced by the input parameter uncertainty [13]?

The answer to the first two questions is almost surely no, because the DSO have limited control over DG allocation and design [14]. Some of the recent developments in this area of research are presented in a comprehensive reviews [8, 15] and [16] which describe optimization methods such as cuckoo search, ant colony, bacterial foraging, genetic algorithm and quasi teaching learning based optimization, classical approaches, basic search methods, society and physic inspired algorithms, nature inspired algorithms and hybrid intelligent methods. The recent comprehensive reviews of DGAP from the point of view of various power system parameters is presented in [17]. The state of art of DGAP techniques and their influence on the ongoing research efforts in this field is presented in [14]. In addition to optimization approach to DG planning, there exist integrated, comprehensive evaluation frameworks based on various evaluation indices and analytic hierarchy process for evaluation of the influence of DG on the distribution system [18]. Reference [19] defines a framework which includes seven most important categories of potential sources of DG (PV) benefits and costs. Finally, there is also an evidence of DG allocation planning using multi criteria decision making [20]. However, the formulations of DGAP as an optimization problem remain the most efficient and most frequently used approach for finding the best solution.

2 Formulation of the DGAP

The objective of the typical optimal DGAP is to find the best locations and sizes of DG in order to improve network parameters, taking into account various DG and network constraints [9]. The selection of objectives is an important consideration which influences the optimizations results. DGAP can be addressed as a single or a multi objective optimization problem, with two or more objective functions. However, formulations with more than three objective functions are rarely attempted due to computational complexity and Pareto front representation. Formally, an optimization problem can be defined as follows [21]:

Given a function f: S $\rightarrow$ R from some set S to the set of real numbers, the goal of the optimization is to determine an element x in S such that:

$$f(\mathrm{x}_1) \leq f(\mathrm{x}), \forall \mathrm{x} \in \mathrm{S} \tag{1}$$

or such that:

$$f(\mathrm{x}_1) \geq f(\mathrm{x}), \forall \mathrm{x} \in \mathrm{S} \tag{2}$$

where S is a search space and is a subset of the Euclidean space R_n, f is an objective function, x and x_0 are the elements of S and called candidate or feasible solutions. The advantages of the single objective formulation of the DGAP include simplicity and faster convergence. Examples of single objective optimization applied to DGAP are presented in [10] and more recently in [22] and [23]. However, this type of DGAP formulation is less common because it does not provide a wider insight into the system behavior. For this reason the multi objective formulation of the problem is more frequently used. The multi objective algorithms are more complex, but are capable to create robust models which more realistically represent the system and offer decision makers with the opportunity to select the best option based on the trade-off between conflicting objectives. Defining the $S \subseteq \mathbb{R}^{n_x}$ as n_x- dimensional search space and $\mathcal{O} \subseteq \mathbb{R}^{n_x}$ as the objective space, then the aim of multi objective optimization problem (MOOP) is to find the vector [24, 25]:

$$\mathrm{x} = (\mathrm{x}_1, \mathrm{x}_2, \ldots, \mathrm{x}_{\mathrm{nk}})^{\mathrm{T}} \in \mathrm{S} \tag{3}$$

which will satisfy the m inequality constraints:

$$\mathrm{g}_{\mathrm{i}}(\mathrm{x}) \geq 0 \;\; \mathrm{i} = 1, 2, \ldots, \mathrm{m} \tag{4}$$

the p equality constraints:

$$\mathrm{h}_{\mathrm{i}}(\mathrm{x}) = 0 \;\; \mathrm{i} = 1, 2, \ldots, \mathrm{p} \tag{5}$$

and will optimize the vector function:

$$\mathrm{f}(\mathrm{x}) = (\mathrm{f}_1(\mathrm{x}), \mathrm{f}_2(\mathrm{x}), \ldots, \mathrm{f}_{\mathrm{n}}(\mathrm{x})) \in \quad \mathcal{O} \subseteq \mathbb{R}^{\mathrm{n}_{\mathrm{x}}} \tag{6}$$

Now, assuming that the aim of the optimization problem is to minimize the objective function, the DGAP can be formulated as the problem of finding the optimum of the function. For a single objective problem formulation, the function is defined as:

$$OF = \min(f_i) \tag{7}$$

For a multi objective problem formulation, where n in the number of objective as, the function is defined as:

$$\mathrm{MOF} = \min(\mathrm{f}_1 + \ldots + \mathrm{f}_{\mathrm{n}}) \tag{8}$$

3 DGAP Objectives

The most frequently considered planning objective include minimization of power and energy losses, voltage and reliability improvements, DG power output maximization, cost minimization [8], maximization of DG capacity, maximization of profit and maximization of a benefit/cost ratio [9]. Technical advancement and regulatory changes encourage the proposition of additional objectives such as the total emission reduction [7] and operational cost minimization [26]. The list of possible objectives is long and depends on regulatory and decision making requirements. The following section list some of the most important DGAP objectives.

3.1 Power Losses

The active power loss is one of the most important and frequently used objective function. Most algorithms which deal with DGAP, address the issue of power loss reduction. This objective function is defined as the active power loss minimization and can be represented as:

$$\min(f_l) = \sum\nolimits_i 3|I_i|^2 R_i \tag{9}$$

where f_l is the power losses function, I_i and R_i are the current and resistance in the ith branch, respectively. There have also been several attempts to quantify the energy losses and allocate them to particular individual DG or large load [27]. Power losses are largely dependent on the location of DG [28]. Methods for power loss calculation for a complex distribution network are usually based on load flow studies and analysis of real and reactive power flows which is then used to calculate the total power loss of the system as the sum of losses of individual components [29]. The power loss reduction is the most commonly used objective function in the DG allocation problem [11] in both single and multi-objective optimization problems [22]. Most of these studies take into consideration load flows with specific loading conditions and do not take into consideration stochastic volatility of distributed generation which is addressed in [30]. DG is often used together with network reconfiguration in order to reduce loses, as shown in [29, 31] and [7]. In addition to DG placement, shunt capacitor units can be installed in order to additionally reduce losses and enhance the voltage profile by providing reactive power to the system, as shown by [32].

3.2 Voltage Profile

In DGAP, the voltage drop minimization is defined as an important objective function. The aim is to find the solution which minimizes the voltage drop at a location with the worst voltage profile. This objective can be written as:

$$\min(f_v) = \sqrt{\frac{P_{ij}\left(R_{ij}^2 + X_{ij}^2\right)}{R_{ij}}} \tag{10}$$

where f_v is the voltage function, P_{ij}, R_{ij} and X_{ij} are the active power, resistance and reactance of branch between nodes i and j, respectively. Voltage profile is one of the most important and frequently considered quantities in the DGAP domain. Voltage levels can be used both objective function, but also as inequality constraint [20]. DG allocation for voltage stability enhancement is a well-documented topic and it continues to be a vibrant area of research [33–35]. The same applies to optimal voltage control strategies considering DG [36]. Reference [37] discusses a simplified voltage stability method in the distribution system, considering DG. Further, [34] presents a study on combined capacitor and DG allocation for voltage stability improvement while [35] considers voltage improvements, losses and load variability in the presence of DG.

3.3 Reliability of the Distribution System

Reliability improvement is one of the possible benefits of DG installation in distribution network [38]. However, the reliability improvement comes at the cost of a need for islanding operation of DG, which requires careful technical and safety considerations. The reliability assessment of the power system with DG is different from classical reliability calculations which do not capture the DG influence. In order to reflect the DG influence on reliability, the distributed generation supply during interruption (DGSI) and distributed generation supply duration during interruption (DGSDI) are defined as [39]:

$$DGSI = \frac{\text{DG Power Output}}{\text{Energy Not Supplied}} = \frac{\sum P_{DG,i}}{ENS} \tag{11}$$

$$DGSI = \frac{\text{DG Supply Duration}}{\text{System Average Interruption Duration}} = \frac{\sum P_{DG,}AID_i}{SAIDI} \tag{12}$$

DGAP for reliability improvement has been a vibrant field of study [40]. The review of DGAP in the distribution power system networks from the point of view of power system reliability enhancement is presented in [17]. Reference [41] present a study of DG planning with the amalgamation of economic and reliability considerations. A multi-objective function used to determine optimize DG allocation in terms of power loss minimization, reliability and voltage profile improvement is presented in [42]. Reference [43] presents an efficient probabilistic-chronological matching modelling for DG planning and reliability assessment in power distribution systems. In practical systems, utilities are becoming increasingly interested in adopting DG allocation to enhance the reliability [40]. Based on the substantial body of evidence, it can be concluded that DG influence on reliability continues to be important and relevant research topic.

3.4 Value of Deferred Investment

DG have potential to defer investment in system capacity extension. DG installation can extend the transformer lifetime span and provide economic benefits which follow an exponential-shaped trajectory, as a function of DG penetration level [44]. The Value of Deferred Investment (VDI), which can be determined by categorizing the reduction of variable costs or deferring capacity investments is an important DGAP consideration. The cost reduction associated with deferring capacity investments can be accurately estimated [19]. Defining P_{limit} as substation maximum capacity power, P_{max} as the maximum power demand at the substation, ΔP as load increase rate and, finally, n as the number of years after which capacity extension will be required, the required number of years n can be easily mathematically determined as follows [19]:

$$\mathrm{P_{limit}} = \mathrm{P_{max}} \times (1 + \Delta\mathrm{P})^{\mathrm{n}} \tag{13}$$

$$(1 + \Delta\mathrm{P})^{\mathrm{n}} = \frac{\mathrm{P_{limit}}}{\mathrm{P_{max}}} \tag{14}$$

$$\mathrm{n} \times \log(1 + \Delta\mathrm{P}) = \log\mathrm{P_{limit}} - \log\mathrm{P_{max}} \tag{15}$$

$$\mathrm{n} = \frac{\log\mathrm{P_{limit}} - \log\mathrm{P_{max}}}{\log(1 + \Delta\mathrm{P})} \tag{16}$$

Using the number of years after which the investment in capacity expansion is necessary and defining IC as the investment cost, the present value of investment (PV) can be calculated as:

$$PV = \frac{IC}{(1 + r)^{n}} \tag{17}$$

These steps are repeated for the case of DG installation, which will result in the reduction of peak power delivered at the substation. Defining $\mathrm{P_{dg}}$ as the power of the installed DG and n as the number of years, after which the investment is required, it can be stated that:

$$P_{limit} = \left(P_{max} - P_{dg}\right) \times (1 + \Delta P)^{n'} \tag{18}$$

$$(1 + \Delta P)^{n'} = \frac{P_{limit}}{P_{max} - P_{dg}} \tag{19}$$

$$n' \times log(1 + \Delta P) = logP_{limit} - \log\left(P_{max} - P_{dg}\right) \tag{20}$$

$$n' = \frac{logP_{limit} - \log\left(P_{max} - P_{dg}\right)}{log(1 + \Delta P)} \tag{21}$$

$$PV' = \frac{IC}{(1 + r)^{n'}} \tag{22}$$

The value of deferred investments can be obtained by subtracting PV′ form PV, taking into account the appropriate correction factor which considers the effect of inflation:

$$VDI = (PV - PV') \tag{23}$$

The deferred value can be evaluated in terms of n, so the function to be maximized can be written as:

$$maxf_n = \frac{logP_{limit} - \log(P_{max} - P_{dg})}{log(1 + \Delta P)} \quad (24)$$

3.5 DG Dynamic Stability Analysis

Generator dynamic stability studies have been traditionally reserved for transmission systems [45]. DG are usually not scrutinized in this sense mainly because they are connected to radial networks and because the influence of eventual tripping is usually localized. However, DG sizing and sitting can be used as a measure for power system transient stability enhancement [46]. The review of DGAP in the distribution power system networks from the point of view of power system oscillations and system security enhancement is presented in [17]. The broader context is mainly concerned with the increase in the popularity of micro grids and the capability of network to operate in both standalone and grid connected mode. In this sense, transient stability and small signal analysis are becoming important and are required in order to ensure safe and effective operation of the micro grid. This is mainly because of an interaction of control systems of DGs, switching operations required for transition from standalone to grid connected mode (and vice versa) and low inertia inherent to small DG. For these reasons, the DG stability analysis represents an important future research direction which will try to ensure safe and effective operation [47].

4 DGAP Constraints

In the absence of constraints, the feasible space is the same as the search space or $\mathcal{F} = \mathcal{S}$ and there would be no need to perform planning or the optimization of the power system. The constraints can be defined as equality constraints such as active (or reactive) power balance and inequality constraints such as voltage limits, line and transformer current limits. The comprehensive review of DG planning constraints is presented in [9, 11] and more recently in [16]. The most commonly considered constraints include voltage limits, line and power transformer thermal rating, phase angles, active and reactive power flows, short circuit ratings, total line loss, tap position, reliability, total harmonic voltage distortion, power factor, budget and limited buses for DG installation [9]. This list does not exhaust the list of all the constraints considered in DG allocation problems. The summary of objectives and constraints considered in this work is summarized in Fig. 1. Additional examples of constraints include the number of DG units, power factor limit, total loss limit, tap position limit and cost, as discusses in [11]. There are also some other constraints that might be included in standard connection approval studies, such as total harmonic distortion (THD) and flickers. However, at the planning stage, it is not always possible to simulate these coefficients. THD and flickers are most often determined by measurements during the DG testing and commissioning stage. Mainly for this reason, many papers omit investigation of additional power quality indicators.

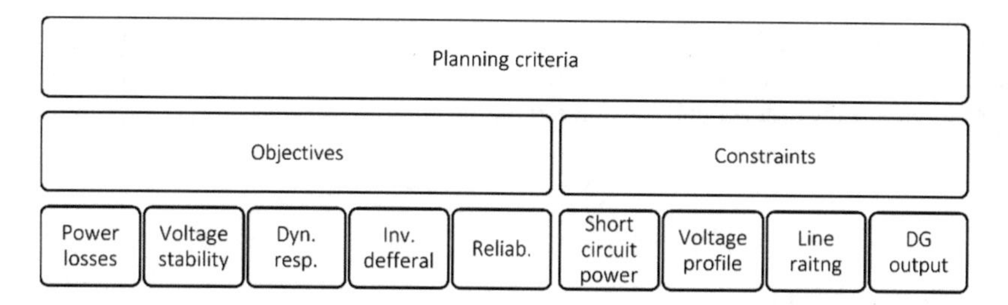

Fig. 1. Simple illustration of DG planning objectives and constraints

4.1 Short Circuit Power Criteria

The short circuit power (sometimes called fast voltage variation) criteria is not frequently used in research papers. However, it is frequently used by utility as extremely important, eliminatory criteria [48]. It is introduced in order to ensure that the maximum voltage change during the DG switching operations will not exceed the prescribed values. According to [48], the maximum permitted fast voltage variation at the DG connection point is 4%. Defining S_{ng} as the rated power of DG, S_{ks} as the three phase short circuit power at the point of the DG connection and k as the ratio between the DG start and rated current, the maximum rated power of DG allowed to be installed in a node, must ensure that [48]:

$$\Delta V = k \times \frac{S_{ng}}{S_{ks}} \tag{25}$$

Synchronous generators and inverters are currently the predominant DG technologies in Bosnia and Herzegovina. The coefficient k = 1 for synchronous generators and inverters and k = 2 for asynchronous generators [48]. The conditions that need to be satisfied in the case of synchronous generators (and inverter) and asynchronous generators, respectively are defined as:

$$\sum S_{ng} \leq \frac{S_{ks}}{25} \tag{26}$$

$$\sum S_{ng} \leq \frac{S_{ks}}{50} \tag{27}$$

4.2 Short Circuit Limit

The previous should not be confused with the short limit criteria, which is calculated in order to ensure that the fault current contribution of DG will not cause the increase of total fault current beyond the permitted values. This criteria is important from the point of view of power system protection and equipment rating. Defining the SCC_{dg} as the short circuit current with DG and SCC_{rated} as the rated short circuit current, then the short circuit limit criteria can be defined as:

$$SCC_{dg} \leq SCC_{rated} \tag{28}$$

4.3 Voltage Profile Limits

The voltage profile limit is one of the most frequently considered constraints. This is not surprising, considering the importance of voltage quality for safe and efficient power system operations. This constraint states that the resulting voltage V_i at all nodes in the network must be within prescribed values. Defining N as the set of power system nodes, which contains the number of elements equal to the number of nodes under consideration, then voltage profile constrained is defined as:

$$V_{i,min} \leq V_i \leq V_{i,max} \quad \forall i \in N \tag{29}$$

4.4 Line Current Limit

Similarly to the voltage profile limit, the line current limit is one of the most frequently considered constraints of the DGAP. In order to satisfy this constraint, currents of each line must be smaller or equal to the rated current. Defining B as the set of power branches under consideration, the maximum line rating constraint is defined as:

$$I_i \leq I_{i,max} \quad \forall i \in B \tag{30}$$

4.5 DG Active Power Output Limit

It is desirable that all power delivered by DG is consumed by local load. The DG active power output limit constraint is introduced in order to prevent transformer back feed, which might have a negative influence on the network. The DG active power output should in all cases be limited to the maximum system demand. Defining $P_{i,gen}$ as generator installed at node i and $P_{i,load}$ as load at node i, DG active power output constraint is defined as:

$$P_{i,gen} \leq P_{i,load} \quad \forall i \in N \tag{31}$$

The more general form of the constraint, which defines both the upper and lower limits of DG output power is described by the following equation:

$$P_{igen}^{min} \leq P_{i,gen} \leq P_{igen}^{max} \tag{32}$$

where P_{igen}^{min} and P_{igen}^{max} represent the lower and the upper DG power output limit, respectively, while $P_{i,gen}$ represent the DG output for the ith configuration. The limits are set according to various criteria such as the maximum power transformer rating (upper limit) or minimum load (lower limit).

4.6 DG Reactive Power Limits

This constraint is introduced in order to prevent voltage variations and excessive power losses by keeping the power factor within prescribed values. Defining Q_{igen}^{min} and Q_{igen}^{max} as the lower and the upper DG reactive power output limit, respectively, and $Q_{i,gen}$ as the DG reactive power output for the ith configuration, the reactive power limit criteria can be defined as:

$$Q_{igen}^{min} \leq Q_{i,gen} \leq Q_{igen}^{max} \tag{33}$$

Similarly, defining Q_{iabs}^{min} and Q_{iabs}^{max} as the lower and the upper DG reactive power absorbed limit respectively and $Q_{i,abs}$ as the DG reactive power absorbed for the ith configuration, the reactive power limit criteria can be defined as:

$$Q_{iabs}^{min} \leq Q_{i,abs} \leq Q_{iabs}^{max} \tag{34}$$

4.7 Voltage Phase Angle Limit

The phase angle limit criteria is introduced in order to ensure that the voltage phase at each bus remains within prescribed values. Defining δ_i^{min} and δ_i^{max} as the upper and lower bounds of the bus voltage phase variation respectively and δ_i as the voltage angle at the ith bus, then the voltage phase angle limit criteria can be defined as:

$$\delta_i^{min} \leq \delta i \leq \delta_i^{max} \quad \forall i \in N \tag{35}$$

4.8 Power Transformer Capacity Limit

Similarly to the branch capacity rating, the substation power transformer capacity rating represents an important limitation that must be satisfied. In fact, substation capacity extension is likely to be more complex than single line replacement. Defining the S_i^{load} and S_i^{rated} as the total apparent power supplied by the transformer and the rated apparent power of the transformer respectively, and T as the set which defines the number of power transformer under consideration, then the power transformer capacity limit criteria is defined as:

$$S_i^{load} \leq S_i^{rated} \; \forall i \in T \tag{36}$$

5 Methods for Solution of the DGAP

The models used in DG planning can be classified in three groups as conventional methods (Linear Programming, Non Linear Programming like AC optimal power flow (OPF) and continuous power flow, Mixed Integer Non-Linear Programming, and Analytical intelligent search-based methods), intelligent methods (Simulated Annealing, Evolutionary Algorithms, Tabu Search, Particle Swarm Optimization) and fuzzy methods introduced in order to account for fuzziness of input variables [11]. Reference [15] classifies DG allocation problem into four main groups: classic approaches, sensitivity analysis based approaches, metaheuristic based approaches and hybrid approaches. The comprehensive review of optimization techniques applied to the integration of distributed generation from renewable energy sources is presented in [49]. Figure 2 shows a graphical presentation of DGAP classification based on objectives, constraints and method as used in this paper. The latest studies are concerned with identifying new possible hybrid optimization methods which might lead to better solutions and improvements of DGAP [49].

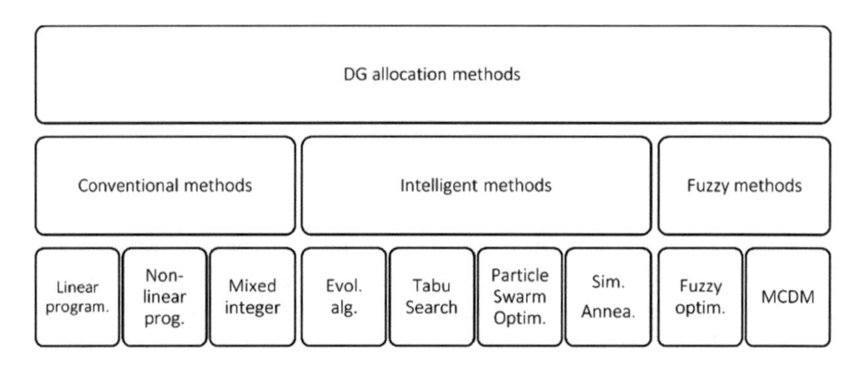

Fig. 2. DG planning classification based on methods

5.1 Classical Approaches

Classical approaches to DG allocation are somewhat less frequently used when compared with intelligent methods. Their advantages include relative simplicity, fast convergence and stability. However, they require a number of conditions, including linearity, continuity and convexity of the objective functions which are not met by numerous engineering applications [15]. Reference [50] use sequential quadratic programming to determine the capacity optimization of DG and storage in a grid-connected and standalone operation, for both dispatchable and non dispatchable units. Results indicate that using this method, the power losses and energy costs are reduced. Reference [26] formulates DGAP as a stochastic mixed-integer linear problem and proposes an allocation method which minimizes the operation and maintenance cost. Reference [51] formulates DGAP as a mixed-integer program and finds the optimal locations, sizes and mix of dispatchable and intermittent DG units. Reference [52] use mixed integer non-linear programming for optimal DG allocation in electricity markets. Additional example of classical method applied to DGAP is presented in [53], which uses dynamic programming search methods to find a global optimization solution for voltage stability maximization and power loss reduction.

5.2 Intelligent Methods

Advantages of artificial intelligence based optimization methods include efficient performance, the need for fewer iteration, the ability to analyze complex systems and substantial accumulation of existing knowledge, while their disadvantages are complexity, premature convergence, instability, setting parameters, unstable results and uncertain convergence [16]. Intelligent optimization methods can be successfully applied in problems that are not linear and do not require continuity and convexity of the objective functions. These are very important factors which promoted intelligent optimization methods use in numerous engineering applications. In addition, there are hybrid intelligent methods for DGAP solution and their advantages include efficient performance, capability to solve complex problems and fast convergence, while disadvantages are larger setting requirements, implementation complexity and smaller accumulated body of knowledge [16]. The most frequently used DG allocation methods are based on genetic algorithm [54]. Reference [34] uses genetic algorithm for optimal DG and capacitor allocation for voltage stability improvements. Optimal reconfiguration and DG allocation in balanced and unbalanced distribution system using GA is presented in [55]. [33] uses a combination of GA and PSO for optimisation of DG allocation. Reference [56] uses a GA-based Tabu search method for optimal placement of multi types DG. An invasive weed optimization algorithm is used for optimal DG allocation to improve voltage and reduce power losses by [57]. A hybrid heuristic search optimization technique for optimal placement and sizing of DG and shunt capacitors for power loss minimization in radial distribution networks is shown by [58]. A cuckoo search algorithm for determination of optimal location and sizing of DG units considering voltage stability losses and load variations is shown in [35]. A novel method for DG placement in the radial distribution network by symbiotic organism search algorithm for real loss minimization is presented in [22]. An optimal siting and sizing of DG including reconfiguration using a uniform voltage distribution algorithm (UVDA) based heuristic method with aim of minimizing power losses is presented in [31]. An integrated approach of network reconfiguration with distributed generation and shunt capacitor placement for power loss minimization in radial distribution networks using the hybrid heuristic search algorithm based on harmony search algorithm and particle artificial bee colony algorithm is presented by [59]. Also, an incorporation of distributed generation and shunt capacitor in radial distribution system for techno-economic benefits is presented in [60]. Reference [23] presents a stud krill herd algorithm for multiple DG placement and sizing in a radial distribution system. A novel whale optimization algorithms for optimal renewable resources placement in distribution networks is presented by [38].

5.3 Fuzzy Methods

Traditional deterministic models need to include a large number of scenarios in order to provide some significant information, while fuzzy approach can provide significant information in a single fuzzy model [61]. Fuzzy logic is used to account for a varying degree of uncertainty, imprecisions and contradictions in a semi structured problem domain which requires a comprehensive model for knowledge analysis [62]. Methods based on fuzzy set and fuzzy logic have become very popular in DGAP because of their capability to model uncertainty. Also, the qualitative information, based on expert opinion or survey, can be readily included in fuzzy models. Reference [63] presents a multi-objective approach for the improvement of efficiency by considering DG. A fuzzy C-means clustering algorithm is used in [64] for distribution system reconfiguration for annual energy loss reduction considering variable DG profiles. Reference [65] uses a fuzzy based on multi objective optimization base PSO for optimal placement of multiple DG and capacitors considering load uncertainty in order to reduce power losses and improve voltage and balancing index. The optimal multi-objective reconfiguration and capacitor placement of distribution systems with the hybrid Big Bang–Big Crunch algorithm in the fuzzy framework for loss reduction and voltage improvement is proposed by [66]. Fuzzy models for decision making are implemented as control algorithms and are particularly useful for addressing the uncertainty of the input information [67]. They behave more like expert systems than fuzzy control algorithms because they are modelled by human expert knowledge and can only be confirmed by testing their outcomes [68]. For this reason, fuzzy decision models can be used in DGAP. It was demonstrated that fuzzy sets and fuzzy multi criteria decision making methods can provide an alternative to traditional optimization techniques in the task of the DGAP [20]. This heuristic approach is interesting because it removes weaker configurations and reduces the search space. Another advantage is flexibility and extensibility, since it can include numerous goals and constraints at little or no increase in the computational cost. [69] describes a method for the electricity distribution system planning which considers load growth, distributed generation, asset management, quality of supply and environmental issues and utilize a number of discrete evaluation criteria within a an MCDM environment to examine and assess the trade-offs between alternative solutions. Finally hybrid fuzzy algorithms, such as fuzzy-genetic algorithm are suitable for application to DGAP [70].

6 Conclusion

This chapter presented an introduction to DGAP in terms of objectives, constraints and methods used to solve it. The results of the theoretical and empirical research review conducted in this field of engineering research are also presented. The first part of the chapter presented the fundamental concepts of the DG allocation problem and proposed a classification of this topic based on objectives, constraints and methods, together with the review of the past recent research. The second part of this chapter presents details of some of the most important and frequently used objective and constraints used in DGAP. It was demonstrated that the DGAP research is well-documented and that it continues to be a vibrant research area with many research opportunities. It can also be concluded that there is no significant conflict of the results in the accumulated body of knowledge. However, results depend on the method and input parameters used in the problem. The accumulated body of knowledge suggests that it is not possible to offer a general solution to the optimal DG allocation and sizing problem. It is necessary to consider every power system separately and include its relevant technical attributes, as well as criteria proposed by decision makers and existing regulatory environment. The body evidence further suggests that DG allocation problem continues to attract some of the most recent research contributions and remains a vibrant research topic with numerous existing and future research opportunities.

References

1. Adefarati, T., Bansal, R.: Integration of renewable distributed generators into the distribution system: a review. Renew. Power Gener. **10**(7), 873–884 (2016)
2. Costello, K.: Major challenges of distributed generation for state utility regulators. Electr. J. **28**, 8–25 (2015)
3. Bizuayehu, A., de la Nieta, A., Contreras, J., Catalao, J.P.S.: Impacts of stochastic wind power and storage participation on economic dispatch in distribution systems. IEEE Trans. Sustain. Energy **7**(3), 1336–1345 (2016)
4. Ameli, A., Farrokhifard, M., Davari, E., Oraee, H., Haghifam, M.: Profit-based DG planning considering environmental and operational issues: a multiobjective approach. IEEE Syst. J. **11**(4), 1–12 (2015)
5. Liang, H., Yang, L., You, Y.: Research on the planning of grid-connected capacity of distributed generations based on the maximum social benefits. In: International Conference on Renewable Power Generation (RPG 2015), Beijing (2015)
6. Piccolo, A., Siano, P.: Evaluating the impact of network investment deferral on distributed generation expansion. IEEE Trans. Power Syst. **24**(3), 1559–1567 (2009)
7. Esmaeili, M., Sedighizadeh, M., Esmaili, M.: Multi-objective optimal reconfiguration and DG (Distributed Generation) power allocation in distribution networks using Big Bang-Big Crunch algorithm considering load uncertainty. Energy **103**, 86–99 (2016)
8. Theo, V., Lim, J., Ho, W., Hashim, H., Lee, C.: Review of distributed generation (DG) system planning and optimisation techniques: comparison of numerical and mathematical modelling methods. Renew. Sustain. Energy Rev. **67**, 531–573 (2017)
9. Georgilakis, S., Hatziargyriou, N.: Optimal distributed generation placement in power distribution networks: models, methods, and future research. IEEE Trans. Power Syst. **28**(3), 3420–3428 (2013)
10. Anwar, A., Pota, H.: Loss reduction of power distribution network using optimum size and location of distributed generation. In: 2011 21st Australasian Universities Power Engineering Conference (AUPEC), Perth (2011)
11. Payasi, R., Singh, A., Singh, D.: Review of distributed generation planning: objectives, constraints and algorithms. Int. J. Eng. Sci. Technol. **3**, 133–153 (2011)
12. Soroudi, A., Ehsan, M., Caire, R., Hadjsaid, N.: Possibilistic evaluation of distributed generations impacts on distribution networks. IEEE Trans. Power Syst. **26**(4), 2293–2301 (2011)

13. Ganguly, S., Samajpati, D.: Distributed generation allocation on radial distribution networks under uncertainties of load and generation using genetic algorithm. IEEE Trans. Sustain. Energy **6**(3), 679–688 (2015)
14. Jain, S., Kalambe, S., Agnihotri, G., Mishra, A.: Distributed generation deployment: state-of-the-art of distribution system planning in sustainable era. Renew. Sustain. Energy Rev. **77**, 363–385 (2017)
15. Jordehi, A.: Allocation of distributed generation units in electric power systems: a review. Renew. Sustain. Energy Rev. **56**, 893–905 (2016)
16. Pesaran, M., Huy, P., Ramachandaramurthy, V.: A review of the optimal allocation of distributed generation: objectives, constraints, methods, and algorithms. Renew. Sustain. Energy Rev. **75**, 293–312 (2017)
17. Singh, B., Sharma, J.: A review on distributed generation planning. Renew. Sustain. Energy Rev. **76**, 529–554 (2017)
18. Jin, W., Shi, X., Ge, F., Zhang, W.W.H., Zhong, C.: Comprehensive evaluation of impacts of connecting distributed generation to the distribution network. J. Electr. Eng. Technol. **12** (2), 621–631 (2017)
19. Hoff, T., Wegner, H., Farmer, B.: Distributed generation: an alternative to utility investment in system capacity. Energy Policy **24**(2), 137–147 (1996)
20. Šarić, M., Hivziefendić, J., Konjić, T.: Distributed generation allocation using fuzzy multi criteria decision making algorithm. In: International Conference on Smart Systems and Technologies (SST), Osijek (2017)
21. Bandyopadhyay, S., Saha, S.: Some single- and Multiobjective Optimization Techniques, in Unsupervised Classification, pp. 17–58. Springer, Heidelberg (2013)
22. Das, B., Mukherjee, V., Das, D.: DG placement in radial distribution network by symbiotic organisms search algorithm for real power loss minimization. Appl. Soft Comput. **49**, 920–936 (2016)
23. ChithraDevi, S., Lakshminarasimman, L., Balamurugan, R.: Stud Krill herd Algorithm for multiple DG placement and sizing in a radial distribution system. Int. J. Eng. Sci. Technol. **20**(2), 748–759 (2017)
24. Engelbrecht, A.: Computational Intelligence: An Introduction. Wiley, New Jersey (2007)
25. Coello Coello, C., Lamont, G., van Veldhuizen, D.: Evolutionary Algorithms for Solving Multi-Objective Problems. Kluwer Academic Publishers, New York (2002)
26. Ikeda, S., Ooka, R.: A new optimization strategy for the operating schedule of energy systems under uncertainty of renewable energy sources and demand changes. Energy Build. **125**, 75–85 (2016)
27. Gonzalez-Longatt, F.: Impact of distributed generation over power losses on distribution system, in Barcelona, 9–11 October 2007 (2007)
28. Fidalgo, J., Fonte, D., Silva, S.: Decision support system to analyse the influence of distributed generation in energy distribution networks. In: Optimization in the Energy Industry. Energy Systems, pp. 59–77. Springer, Heidelberg (2009)
29. Rajaram, R., Kumar, K., Rajasekar, N.: Power system reconfiguration in a radial distribution network for reducing losses and to improve voltage profile using modified plant growth simulation algorithm with Distributed Generation (DG). Energy Repors **1**, 116–122 (2015)
30. Ren, H., Han, C., Guo, T., Pei, W.: Energy losses and voltage stability study in distribution network with distributed generation. J. Appl. Math. **2014**, 1–7 (2014)
31. Bayat, A., Bagheri, A., Noroozian, A.: Optimal siting and sizing of distributed generation accompanied by reconfiguration of distribution networks for maximum loss reduction by using a new UVDA-based heuristic method. J. Electr. Power Energy Syst. **77**, 360–371 (2016)

32. Ramadan, H., Bendary, A.S.: Particle swarm optimization algorithm for capacitor allocation problem in distribution systems with wind turbine generators. Electr. Power Energy Syst. **84**, 143–152 (2017)
33. Moradi, M., Abedini, M.: A combination of genetic algorithm and particle swarm optimization for optimal DG location and sizing in distribution systems. Int. J. Electr. Power Energy Syst. **34**(1), 66–74 (2012)
34. Pradeepa, H., Ananthapadmanabha, T., Sandhya Rani, D., Bandhavya, C.: Optimal allocation of combined DG and capacitor units for voltage stability enhancement. Procedia Technol. **21**, 216–223 (2015)
35. Poornazaryan, B., Karimyan, P., Gharehpetian, G., Abedi, M.: Optimal allocation and sizing of DG units considering voltage stability, losses and load variations. J. Electr. Power & Energy Syst. **79**, 42–52 (2016)
36. Castro, J., Saad, M., Lefebvre, S., Asber, D., Lenoir, L.: Optimal voltage control in distribution network in the presence of DGs. Int. J. Electr. Power Energy Syst. **78**, 239–274 (2016)
37. Liu, K., Sheng, W., Lijuan, H., Liu, Y., Meng, X., Jia, D.: Simplified probabilistic voltage stability evaluation considering variable renewable distributed generation in distribution systems. IET Gener. Transm. Distrib. **9**(12), 1464–1473 (2015)
38. Reddy, P., Reddy, V., Gowri Manohar, T.: Optimal renewable resources placement in distribution networks by combined power loss index and whale optimization algorithms. J. Electr. Syst. Inf. Technol. **28**, 669–678 (2017)
39. AlMuhaini, M.: Impact of distributed generation integration on the reliability of power distribution systems. In: Distributed Generation Systems, pp. 453–508. Elsevier (2017)
40. Adefarati, T., Bansal, R.: Reliability assessment of distribution system with the integration of renewable distributed generation. Appl. Energy Part 1 **185**, 158–171 (2017)
41. Battu, N., Abhyankar, A., Senroy, N.: DG planning with amalgamation of economic and reliability considerations. Int. J. Electr. Power Energy Syst. **73**, 273–282 (2015)
42. Khalesi, N., Rezaei, N., Haghifam, M.: DG allocation with application of dynamic programming for loss reduction and reliability improvement. Int. J. Electr. Power Energy Syst. **33**(2), 288–295 (2011)
43. Alotaibi, M., Salama, M.: An efficient probabilistic-chronological matching modeling for DG planning and reliability assessment in power distribution systems. Renew. Energy **99**, 158–169 (2016)
44. Agah, S., Abyaneh, H.: Quantification of the distribution transformer life extension value of distributed generation. IEEE Trans. Power Deliv. **26**(3), 1820–1828 (2011)
45. Saric, M., Penava, I.: Transient stability of induction generators in wind farm applications. In: 14th International Conference on Environment and Electrical Engineering, Krakow (2014)
46. Razazadeh, A., Sedighizadeh, M., Alavian, A.: Optimal sizing and sitting of distributed generation for power system transient stability enhancement using generic algorithm. Int. J. Comput. Theory Eng. **1**(5), 387–390 (2009)
47. Mumtaz, F., Bayram, I.: Planning, operation, and protection of microgrids: an overview. Energy Procedia **107**, 94–100 (2017)
48. JP EP BiH-b, Tehnička preporuka JP EP BiH d.d. Sarajevo, TP 17, JP EP BiH (2016)
49. Abdmouleh, Z., Gastli, A., Ben-Brahim, L., Haouari, M., Al-Emadi, N.A.: Review of optimization techniques applied for the integration of distributed generation from renewable energy sources. Renew. Energy **113**, 266–280 (2017)
50. Sfikas, E., Katsigiannis, Y., Georgilakis, P.: Simultaneous capacity optimization of distributed generation and storage in medium voltage microgrids. Int. J. Electr. Power Energy Syst. **67**, 101–113 (2015)

51. Wang, Z., Chen, B., Wang, J., Kim, J., Begovic, M.: Robust optimization based optimal DG placement in microgrids. IEEE Trans. Smart Grid **5**(5), 2173–2182 (2014)
52. Kumar, A., Gao, W.: Optimal distributed generation location using mixed integer non-linear programming in electricity markets. IET Gen. Trans. Distrib. **4**(2), 281–298 (2010)
53. Esmaili, M., Firozjaee, E.C., Shayanfar, H.A.: Optimal placement of distributed generations considering voltage stability and power losses with observing voltage-related constraints. Appl. Energy **113**, 1252–1260 (2014)
54. Georgilakis, P., Hatziargyriou, N.: A review of power distribution planning in the modern power systems era: models, methods and future research. Electr. Power Syst. Res. **121**, 89–100 (2015)
55. Taher, S., Karimi, M.: Optimal reconfiguration and DG allocation in balanced and unbalanced distribution systems. Ain Shams Eng. J. **5**(3), 735–749 (2014)
56. Mohammadi, M., Nafar, M.: Optimal placement of multitypes DG as independent private sector under pool/hybrid power market using GA-based Tabu Search method. Int. J. Electr. Power Energy Syst. **51**, 45–53 (2013)
57. Rama Prabha, D., Jayabarathi, T.: Optimal placement and sizing of multiple distributed generating units in distribution networks by invasive weed optimization algorithm. Ain Shams Eng. J. **7**(2), 683–694 (2016)
58. Muthukumar, K., Jayalalitha, S.: Optimal placement and sizing of distributed generators and shunt capacitors for power loss minimization in radial distribution networks using hybrid heuristic search optimization technique. Int. J. Electr. Power Energy Syst. **78**, 299–319 (2016)
59. Muthukumar, K., Jayalalitha, S.: Integrated approach of network reconfiguration with distributed generation and shunt capacitors placement for power loss minimization in radial distribution networks. Appl. Soft Comput. **52**, 1262–1284 (2017)
60. Dixit, M., Kundu, P., Jariwala, H.: Incorporation of distributed generation and shunt capacitor in radial distribution system for techno-economic benefits. Int. J. Eng. Sci. Technol. **20**(2), 482–493 (2017)
61. Miranda, V., Matos, M.A.C.C.: Distribution system planning with fuzzy models and techniques. In: 10th International Conference on Electricity Distribution, CIRED 1989 (1989)
62. Egwaikhide, I.: Fuzzy modeling of uncertainty in a decision support system for electric power system planning. In: Hampel, R., Wagenknecht, M., Chaker, N. (eds.) Fuzzy Control. Advances in Soft Computing. Fuzzy Control, Advances in Soft Computing, vol. 6, pp. 387–396. Physica, Heidelberg (2000)
63. Syahputra, R.: Fuzzy multi-objective approach for the improvement of distribution network efficiency by considering DG. Int. J. Comput. Sci. Inf. Technol. **4**(2), 2012 (2012)
64. Tahboub, A., Pandi, V., Zeineldin, H.: Distribution system reconfiguration for annual energy loss reduction considering variable distributed generation profiles. IEEE Trans. Power Deliv. **30**(4), 1677–1685 (2015)
65. Zeinalzadeh, A., Mohammadi, Y., Moradi, M.: Optimal multi objective placement and sizing of multiple DGs and shunt capacitor banks simultaneously considering load uncertainty via MOPSO approach. Int. J. Electr. Power Energy Syst. **67**, 336–349 (2015)
66. Sedighizadeh, M., Bakhtiary, R.: Optimal multi-objective reconfiguration and capacitor placement of distribution systems with the Hybrid Big Bang-Big Crunch algorithm in the fuzzy framework. Ain Shams Eng. J. **7**(1), 113–129 (2016)
67. Saric, M., Hivziefendic, J.: Management of the power distribution network reconstruction process using fuzzy logic. In: Advanced Technologies, Systems, and Applications, pp. 155–172. Springer, Heidelberg (2017)

68. Saric, M.: Fuzzy approach for evaluating risk of service interruption used as criteria in electricity distribution network planning. In: 12th Symposium on Neural Network Applications in Electrical Engineering (NEUREL), Belgrade (2014)
69. Espie, P., Ault, G., Burt, G., McDonald, J.: Multiple criteria decision making techniques applied to electricity distribution system planning. IEE Proc. Gener. Transm. Distrib. **150**(5), 527–535 (2003)
70. Nayanatara, C., Baskaran, J., Kothari, D.: Hybrid optimization implemented for distributed generation parameters in a power system network. Int. J. Electr. Power Energy Syst. **78**, 690–699 (2016)

The Effect of Summer Months and the Profitability Assessment of the PV Systems in Bosnia and Herzegovina

Faruk Bešlija[1] and Ajla Merzić[2(✉)]

[1] Department of Electrical and Electronics Engineering, Faculty of Engineering and Natural Sciences, International University of Sarajevo, Hrasnicka Cesta 15, 71 210 Ilidza, Sarajevo, Bosnia and Herzegovina

[2] Department for Strategic Development, Public Enterprise Elektroprivreda of Bosnia and Herzegovina, Vilsonovo Setaliste 15, 71 000 Sarajevo, Bosnia and Herzegovina

a.merzic@epbih.ba

Abstract. The renewable energy sources share in a total electricity production and consumption has been growing at the high rate in the previous years. The reasons behind it are the long-terms goals and strategies that the countries agreed to comply with, in order to make improvements about the energy efficiency, pollution reduction and the RES share in the electricity production, which would ultimately lead to a cleaner environment, more stable electricity supply and better control over the electricity demands.

Bosnia and Herzegovina is one of the countries that obliged to implement these long-term goals and strategies. Adopting a set of regulations and strategies, the country is currently taking on a big challenge of increasing the RES share in the electricity production, that should eventually lead to the massive changes in the dominant types of the generation units. In this paper, we will observe the photovoltaic power systems in Bosnia and Herzegovina in terms of their seasonal performance and their profitability at this point.

The paper will be given in five parts. The introductory part will briefly describe the concept behind the national long-term goals and strategies. The second part will analyze the seasonal impact of the summer months on the PV systems at two different representative locations. The third part will be the profitability assessment of the equally specified PV power plants at two abovementioned locations. In the fourth part, the results of the previous two parts will be presented and discussed. Finally, a conclusion will be given.

1 Introduction

In 2010, the Government of the Federation of Bosnia and Herzegovina brought a Regulation on Usage of the Renewable Energy and Cogeneration Sources [1]. The specified aims of this Regulation were to encourage the increase in production of the electricity from these sources, as well as their consumption. The concrete long-term goals include the administrative simplification of the usage of the RES, lowering the usage of the fossil fuels, coming closer to the Kyoto goals, long-term supply and

S. Avdaković (Ed.): IAT 2018, LNNS 59, pp. 150–163, 2019.
https://doi.org/10.1007/978-3-030-02574-8_12

efficiency in usage of the energy and opening of the new working places and enterprise development in energetics.

In 2017, the Government presented the second working version of the 2035 Framework Energy Strategy of the Federation of Bosnia and Herzegovina, which addresses the long-term goals of the aforementioned Regulation, presents the dynamics of the implementation of these goals and makes future considerations based on the data obtained from the previous years [2]. According to the data presented in this Strategy, the country has experienced a 11.6% growth in production of the electricity from the PV power plants between 2012 and 2015, resulting in around 7 MW of installed capacity. It is estimated that, by 2035, the total installed capacity from this source will reach 97.6 MW.

One of the key reasons for this estimation is the market price drop of the technology, which is estimated to cause the unit investment cost drop of 57% in the following 10 years and the average production cost drop of 59% in the following 5 years (Figs. 1 and 2).

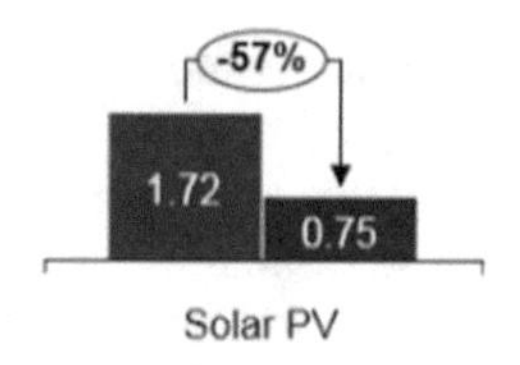

Fig. 1. Investment costs comparison (2015–2025) in 1000 €/kW
Source: IRENA – The power to change: solar and wind cost reduction potential to 2025

The aims of this paper are to further explore the energy production characteristics of the PV systems by observing the impact that the summer months conditions have on the performances of the PV systems at two selected locations in Bosnia and Herzegovina, and to estimate the profitability of these systems, given currently active regulations. For these purposes, two different locations have been chosen as the representative locations of the inland (Tešanj) and the coastal (Čapljina) areas. Summer month effects have been considered through the temperature influence on PV efficiency, as well as solar irradiation values.

The solar irradiance data and the characteristics of the PV systems were calculated using the PVGIS and NREL PVWatts calculators. The temperature averages were obtained from Meteoblue, a service developed at the University of Basel (Switzerland). Other relevant data were obtained from the official documents by either the Federal Government or Regulatory Energy Commission of Federation of Bosnia and Herzegovina (FERK), unless stated otherwise.

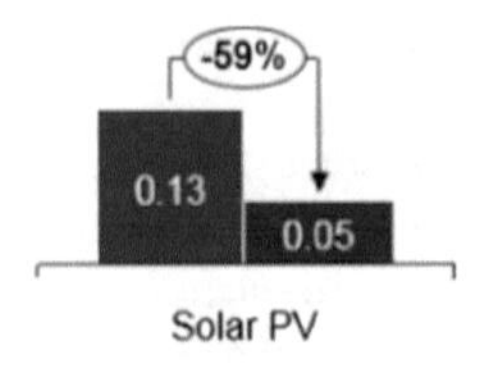

Fig. 2. Average production cost of electricity (2015–2020) in 1 €/kWh
Source: IRENA – The power to change: solar and wind cost reduction potential to 2025

2 The Effect of the Summer Months on the Performance of the PV Systems

2.1 Input Parameters

The resource used in PV technologies is the solar irradiance, measured in power per unit area. It consists of the three main components [3]:

- DNI (Direct Normal Irradiance), a portion of irradiance received without being attenuated by the atmosphere
- DHI (Diffuse Horizontal Irradiance), a portion of irradiance received after being reflected by the particles in the atmosphere
- GHI (Global Horizontal Irradiance), total irradiance received by a surface placed horizontally with respect to the ground

Normally, DNI is considerably greater than DHI because it comes directly from the Sun. However, DHI is constantly present, whereas DNI highly depends on the condition of the atmosphere, because it is highly attenuated by the clouds.

Disregarding the solar irradiance, which is the main component in the power output calculations, the power output itself depends on several other factors, such as temperature, snow, dust, shading, installation losses and degradation with age.

In the next section, we will observe the effect of the seasons on the power output of the PV systems. Assuming all other losses can be addressed equally at both inland and coastal locations, we will focus on the temperature impact along with the average monthly irradiance specific for the location.

2.2 Seasonal Performances of the PV Systems

Considering the STC reference temperature of 25 °C, we would like to know how much is the operation of the PV panels affected by the increase of the temperature for every degree over this reference temperature. Since our proposed systems will use the standard crystalline silicon modules, the efficiency drop will be the value between 0.45% per °C [4] and 0.47% per °C [5]. For the calculation, we will use the average of these two values. Below are the 30-year monthly averages of the temperatures at our specified locations.

We notice that, in July and August, the temperature at our inland location exceeds the reference temperature by 3 degrees, which results in the efficiency drop of 1.38%.

Computing the average annual efficiency drop due to these conditions results in 0.23% (Table 1).

Table 1. Tešanj (inland) temperature averages (source: meteoblue)

Jan	Feb	Mar	Apr	May	Jun	Jul	Aug	Sep	Oct	Nov	Dec
6 °C	7 °C	12 °C	17 °C	22 °C	25 °C	28 °C	28 °C	24 °C	19 °C	13 °C	7 °C

For our coastal location, the reference temperature is exceeded from June to September. That results in the efficiency drops of 0.92%, 2.3%, 2.76% and 0.46% for months of June, July, August and September, respectively. Computing the average annual efficiency drop due to these conditions results in 0.537% (Table 2).

Table 2. Čapljina (coastal) temperature averages (source: meteoblue)

Jan	Feb	Mar	Apr	May	Jun	Jul	Aug	Sep	Oct	Nov	Dec
10 °C	11 °C	14 °C	18 °C	23 °C	27 °C	30 °C	31 °C	26 °C	21 °C	15 °C	11 °C

While these numbers may seem insignificant at first, one should note that the average annual efficiency change as low as 0.5% may cause a profit change of 5% over the time span of 12 years. Now, we will consider the irradiance at the given locations. The average monthly irradiance is given below (Table 3).

Table 3. Average monthly irradiance in kW/m^2

	Jan	Feb	Mar	Apr	May	Jun	Jul	Aug	Sep	Oct	Nov	Dec
Čapljina	83,2	94,6	143	173	192	202	225	223	176	139	88	84,6
Tešanj	54	63,5	114	144	154	164	191	185	133	101	70,4	52,7
Percentage difference	54,07	48,98	25,44	20,14	24,68	23,17	17,8	20,54	32,33	37,62	25	60,53

We conclude that the highest difference in irradiance is measured in December (60.53%), and the lowest in July (17.8%). Let us consider the same data on the seasonal level.

From the Table 4, we can read that the highest percentage difference in irradiance is during the winter months (December, January, February), whereas the lowest is during the summer months (June, July, August). Since we mentioned above the effect of the cloudiness on the irradiance, one can assume that this huge percentage difference during the winter months is the result of the precipitation differences between the two locations.

These informations will be used later while discussing the financial impact of the summer months.

Table 4. Average seasonal irradiance in kW/m^2

	Winter	Spring	Summer	Autumn
Čapljina	262,4	508	650	403
Tešanj	170,2	412	540	304,4
Percentage difference	54,17	23,3	20,37	32,39

3 Profitability Assessment

3.1 Background

In order to understand the economical background and observe the profitability of the PV systems at two given locations, we need to analyze the current regulations regarding these systems applicable to the electricity market of Bosnia and Herzegovina.

Currently, the model in use is the model of the feed-in tariffs with partial responsibility. This non-market principle means that the purchase prices for the electricity produced from the production units based on renewable energy sources are determined administratively, excluding the market risks [6]. The partial responsibility refers to the balancing of the system, and only some of the electricity producers (according to their installed capacities) are required to pay this compensation.

The model of the feed-in tariffs in Bosnia and Herzegovina is defined by the Law on Using the Renewable Energy and the Efficient Cogeneration Sources. The legal entity that is financed within the feed-in tariff system (given fulfilment of all the requirements of this Law) is the privileged electricity producer. This producer is entitled to a 12-year contract by which all the electricity produced at the unit using RES will be bought according to the guaranteed purchase price, determined every 18 months by FERK [7].

The incentive funds for the production units based on RES are collected from the end buyers and distributed among the privileged electricity producers. According to the currently active Decision on Reference Prices by FERK, these funds are taken from the referent price of the electricity, which is a product of the weighted realized price and the incentive coefficient of 1.2 for the electricity produced from RES [8].

3.2 Financial and System Parameters

In this section, we will list the key financial and system parameters for our proposed PV power plants. For further understanding, a few clarifications will be made.

Own-to-loan funds ratio represent the law requirement by which at least 20% of the funds invested in the PV power plant must be provided by the investor. This parameter will not be considered in detail in the cash flow computation, but it will be commented on using notional approach.

Unit investment costs and *operation and maintenance costs* represent the cost of installing the PV module and the annual cost to keep it operating optimally, normalized to 1 kWp installed. The computation of these costs takes in consideration the regional (such as costs of technology) and the national (such as work and documentation costs) parameters [9]. The parameters comprised by these costs are:

- Panel and inverter price, medium voltage grid access connection fee
- Hardware cost, mounting construction, electrical and mechanical installation work (per panel)
- Yearly operating cost, yearly insurance cost, documentation cost (given as percentage of the main investment)
- Connection cable (per km)
- Yearly maintenance (per square meter)
- Equipment technical lifetime (years)

However, one should note that these costs do not include the land and the eventual road infrastructure costs. Due to high variability of these two parameters, they will be excluded from the further estimations.

All data listed in Table 5 are sourced either from the Decision by the FERK (valid from September 2017 to March 2019, marked with star below) or the aforementioned Law. It should be mentioned that the estimations of these costs by FERK are supported by more than 15 acknowledged academic papers (stated in the Decision), thus, can be taken as very reliable.

Table 5. Financial parameters

Own-to-loan funds ratio	20%–80%	Balancing compensation	Only above 150 kW
Installed capacity*	1 MW (mini)	Annual working hours	1500*
Contract length	12 years	Reference price of electricity	0,105858 KM/kWh*
Tariff coefficient*	2,4534	Guaranteed purchase price	0,25971 KM/kWh*
Unit investment costs*	2304 KM/kW	Annual operation and maintenance costs	70 KM/kW*

Regarding the Table 6, system efficiency for the crystalline silicon PV panels is proposed by NREL PVWatts calculator, system losses are proposed by both NREL PVWatts and PVGIS and the degradation with age is given according to the study by Jordan and Kurtz [10].

Table 6. System parameters

Technology in use	c-Si PV panels	System losses	14%
System efficiency	15% (standard)	Degradation with age (yearly)	0.5%

Given the inputs above, we may proceed with the cash flow calculations for both inland and coastal proposed PV power plants.

3.3 Estimation of the Cash Flow

To perform the cash flow calculations, we will observe the period of 12 years for which the producer has the privileged status.

Table 7. Cash flow calculation (Tešanj, inland)

Year	Investment	Production (kWh per kW installed)	Cash Inflow (KM)	Cash Expenditures (KM)	Cash Flow (KM)	Cumulative Net Cash Flow (KM)
0	2304000				-2304000	-2304000
1		1113,6	289213,06	70000	219213,06	-2084786,94
2		1108,03	287766,99	70000	217766,99	-1867019,95
3		1102,49	286328,16	70000	216328,16	-1650691,8
4		1096,98	284896,51	70000	214896,51	-1435795,28
5		1091,49	283472,03	70000	213472,03	-1222323,25
6		1086,04	282054,67	70000	212054,67	-1010268,58
7		1080,61	280644,4	70000	210644,4	-799624,18
8		1075,2	279241,18	70000	209241,18	-590383
9		1069,83	277844,97	70000	207844,97	-382538,03
10		1064,48	276455,75	70000	206455,75	-176082,28
11		1059,16	275073,47	70000	205073,47	28991,18
12		1053,86	273698,1	70000	203698,1	232689,28

The total investment cost is given as the product of the unit investment cost and the installed capacity in kW. The total operation and maintenance cost is given as the product of the annual operation and maintenance cost and the installed capacity in kW.

The total cash inflow is given as the product of the guaranteed purchase price for the period of 12 years, the average annual amount of the electricity produced in kWh (here, taken from PVGIS) and the time of 12 years. Here, the degradation with the age is included in the calculations (Figs. 3 and 4).

Table 8. Cash flow calculation (Čapljina, coastal)

Year	Investment	Production (kWh per kW installed)	Cash Inflow (KM)	Cash Expenditures (KM)	Cash Flow (KM)	Cumulative Net Cash Flow (KM)
0	2304000				-2304000	-2304000
1		1418,4	368372,66	70000	298372,66	-2005627,34
2		1411,31	366530,8	70000	296530,8	-1709096,54
3		1404,25	364698,15	70000	294698,15	-1414398,39
4		1397,23	362874,66	70000	292874,66	-1121523,73
5		1390,24	361060,28	70000	291060,28	-830463,45
6		1383,29	359254,98	70000	289254,98	-541208,47
7		1376,38	357458,71	70000	287458,71	-253749,76
8		1369,49	355671,41	70000	285671,41	31921,65
9		1362,65	353893,06	70000	283893,06	315814,71
10		1355,83	352123,59	70000	282123,59	597938,3
11		1349,05	350362,97	70000	280362,97	878301,27
12		1342,31	348611,16	70000	278611,16	1156912,43

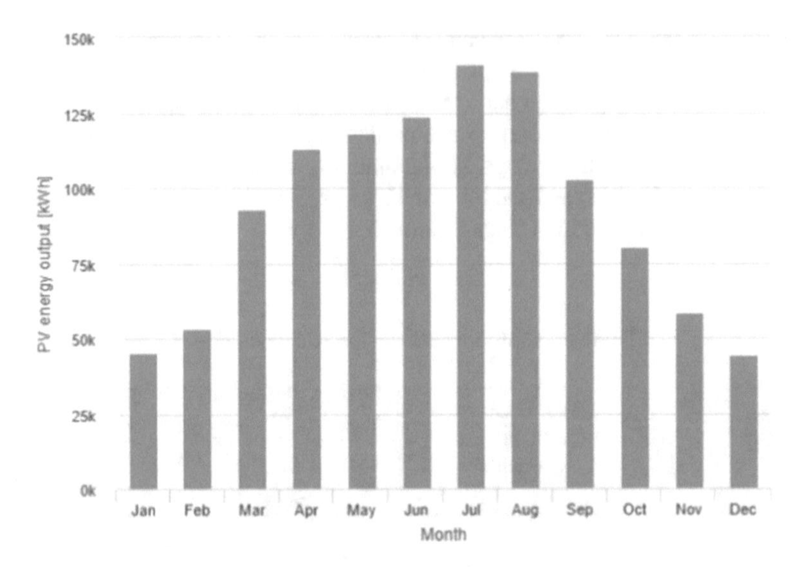

Fig. 3. Monthly energy output (Tešanj, inland)
Source: PVGIS

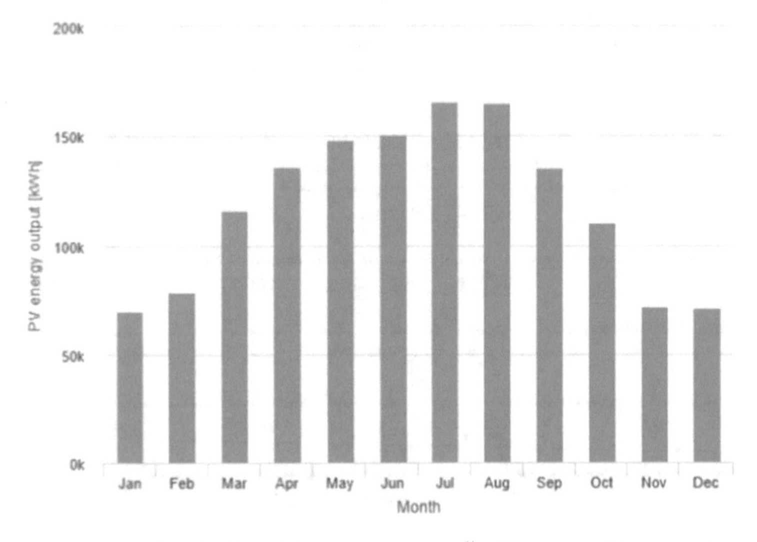

Fig. 4. Monthly energy output (Čapljina, coastal)
Source: PVGIS

4 Results and Discussion

In this part, we will review and discuss the results obtained. Firstly, let us list the results of the cash flow calculations.

- For the inland location:
  - The expected payoff time is during the 11th year
  - The expected profit after the 12 years of the privileged contract is 232689.28 km, which is 10.1% of the total investment
- For the coastal location:
  - The expected payoff time is during the 8th year
  - The expected profit after the 12 years of the privileged contract is 1156912.43 km, which is 50.21% of the total investment

These values are expected for the 1 MW systems, given annual average difference in irradiation of around 300 kWh/m2 or roughly 20%. We may conclude that both these systems are cost-effective, with the coastal one resulting in a considerably higher (nearly 5 times) profit. However, to assess the profitability of the system after the 12th year, one should take into account the following factors:

- The average lifetime (and the usual warranty) of the PV panels is 20 to 25 years
- The average lifetime of an inverter is between 5 and 15 years and it is safe to assume that one inverter change will be required within the 12 years [10]
- The purchase price after the expiration of the contract is the reference price of the electricity (for our particular system, that is nearly 2.5 times less than the guaranteed purchase price)
- The reference price will still be updated every 18 months, so the reference prices after 12 years should be estimated

However, this does not include the own-to-loan funds ratio. In order to precisely address this issue, a separate analysis regarding the best available loans for this purpose would be required. Nevertheless, this parameter is too important to be excluded, hence, a few bold assumptions will be made for the purpose of the analysis.

Since it is already specified that the cash flow calculations will be observed for 12 years of the privileged status, a 12 years loan repayment time will also be assumed. The annual loan installment, computed as the product of the total loan amount and the interest rate, will be added to the annual cash expenditures within the cash flow (Tables 9, 10 and 11).

Table 9. Loan specifications

Own funds 20%	Loan funds 80%	Repayment time	Interest rate type	Interest rate	Annual loan installment
460800	1843200	12 years	Fixed	5%	161280

Now, let us review the financial impact of the summer months.

In the table above, there are several ratios observed as an indicator of the financial impact that the summer months have on the presumed PV systems.

The first is summer-to-winter power output. This ratio represents how many times is the amount of power generated during the summer months (the greatest production

Table 10. Adjusted cash flow calculations (inland)

Year	Investment	Production (kWh per kW installed)	Cash Inflow (KM)	Cash Expenditures (KM)	Cash Flow (KM)	Cumulative Net Cash Flow (KM)
0	460800				-460800	-460800
Pay-off (9th)		1075,2	277844,97	231280	46654,97	9141,97
12		1053,86	273698,1	231280	42418,1	140529,28

Table 11. Adjusted cash flow calculations (coastal)

Year	Investment	Production (kWh per kW installed)	Cash Inflow (KM)	Cash Expenditures (KM)	Cash Flow (KM)	Cumulative Net Cash Flow (KM)
0	460800				-460800	-460800
Pay-off (4th)		1397,23	362874,66	231280	131594,66	76556,27
12		1342,31	348611,16	231280	117331,16	1064752,43

period) greater than the amount of power generated during the winter months (the lowest production period). Observing the table above, one can deduce that the summer months play reasonably more significant role for the inland location.

The second parameter to be taken is how much of the total power output at the given location is produced during the summer months. This straightforward indicator shows that the total power output is composed of the summer months share almost equally for the both locations.

Regarding the third parameter, it represents how much of the total yearly income comes from the production during the summer months. Although the annual cash inflow is over 27% larger for the coastal location (refer to Tables 7 and 8, column Cash Flow), we see from the Table 12 that the cash inflow is merely 15% larger for the coastal location when only the summer months are considered.

How much money should be deduced because of the summer months temperature impact? Let us consider the annual averages given in Tables 13 and 14.

Table 12. Summer months financial impact

	Summer-to-winter power output	Summer months share in total power output	First year summer months cash inflow
Tešanj	3.17	37.85%	109467.14
Čapljina	2.48	34.29%	126314.99

Table 13. Summer months financial impact (inland)

	Jun	Jul	Aug
Average temperature	25 °C	28 °C	28 °C
Efficiency drop due to temperature	0%	1.38%	1.38%
Energy output (kWh/kWp)	124	141	139
Non-reduced energy output (kWh/kWp)	124	142.9	140.9
Money deduced due to efficiency drop	0	493.45 km	493.45 km

Table 14. Summer months financial impact (inland)

	Jun	Jul	Aug
Average temperature	27 °C	30 °C	31 °C
Efficiency drop due to temperature	0.92%	2.3%	2.96%
Energy output (kWh/kWp)	151	166	165
Non-reduced energy output (kWh/kWp)	152.4	169.8	169.9
Money deduced due to efficiency drop	363.59 km	986.9 km	1168.7 km

The total money deduced due to temperature impact during the summer months is 986.9 km for the inland and 2519.19 km for the coastal location. Over the span of 12 years, including the efficiency drop due to aging of the equipment, this would make 11547.91 km for the inland and 29477.55 km for the coastal location. This makes 2.51% of the total own funds invested for the inland and 6.4% of the total own funds invested for the coastal location.

From all of the supplied data, we may conclude that the summer months in Bosnia and Herzegovina act as a balancing period, bringing closer the energy output (within 20% difference) and the cash inflow (within 15% difference) of the inland and the coastal areas. This is partially supported by the climate conditions, because there are no major losses due to the increased temperature and the cloudy skies.

5 Conclusion

The PV technologies have been around at the market of Bosnia and Herzegovina for some time already, and the prices and the costs have balanced out during that time. Compared to the beginnings, the tariff coefficients for the PV power plants have significantly decreased as a result of the price drop of the PV technologies.

It is not only the magnitude of the numbers that is affected by these changes. In the past, the drop of the investment and the production costs was usually followed by a prolonged payoff time, and it is natural to expect the same would happen again. Also, the payoff time is affected by the nature of the investment funds. It has been shown in this paper that the share of the loan funds in the investment reduces the final profit, but also the payoff time and makes the investment more plausible.

As mentioned in the introductory part, it is expected that the average production cost of electricity from PV technologies undergoes a drop of 59% by 2020, resulting in a 0.05 € per kWh produced. Taking in consideration that the guaranteed purchase prices are formed relative to the production cost of the electricity from a specific type of unit, we may say that the year 2020 would bring us a step closer to the cease of the feed-in tariff system.

The best indicator on how close we are to this will be presented to us in March 2019, when FERK brings a new Decision on the Guaranteed Purchase Prices. It is reasonable to expect that the new tariff coefficients will be closer to the unit factor than ever. This would also be interesting from the investment point of view, and how the loans for this purpose would be affected by these deduced investment costs.

Regarding the temperature effect on the PV panel performance, it can be concluded that, while the extreme temperature averages in Bosnia and Herzegovina are uncommon, a huge seasonal difference emerges between the inland and the coastal locations, due to increased cloudiness and precipitation in months other than summer. The best indicator of how much more the summer months are valuable to the PV systems in the inland than at the coastal locations is the summer-to-winter power output given in Table 12.

Furthermore, it should be remarked that all data provided in this research are indicative, based on the assumptions and the estimations of the used irradiance calculators. Prior to making an investment, it is strongly recommended to obtain the measured data for the given location.

This is especially significant for the winter months irradiance, because the data obtained from the satellite may be biased due to snow, which can sometimes be interpreted as clouds by the software. Also, even the partial snow coverage of the panels reduces their performance significantly. Therefore, one should take into account the frequency of the snowfall, but, more importantly, how long the snow stays on the panels before it is cleaned, slides off (due to panel inclination) or melts (due to temperature).

Other relevant data (that are either roughly approximated or not considered at all by the used irradiance calculators) that may affect the system and should be measured are the wind speed and direction (for systems where there is a space between the PV panels and the surface they are attached to, a wind flow under the panels will normally cause the cooling of the panels), the dust and dirt concentration in the air (depending on weather and terrain, dust particles covering the panels reduce the power output, the panels have to be cleaned at times) and the exact position and the installation method of the panels (angle at which panels are placed, proximity of other objects, terrain configuration and system position relative to it, self-shading for fixed-array systems, all these may result in shading, which notably affects the system power output).

References

1. Vlada FBiH. Uredba O Korištenju Obnovljivih Izvora Energije I Kogeneracije, 1 June 2010. http://www.fbihvlada.gov.ba/bosanski/zakoni/2010/uredbe/13hrv.html. Accessed 17 Mar 2018
2. Vlada FBiH. Okvirna Energetska Strategija FBiH Do 2035, 6 June 2017. Accessed 17 Mar 2018
3. Vashishtha, S.: Differentiate between the DNI, DHI and GHI? 03 May 2012. https://firstgreenconsulting.wordpress.com/2012/04/26/differentiate-between-the-dni-dhi-and-ghi/. Accessed 17 Mar 2018
4. Dubey, S., Sarvaiya, J.N., Seshadri, B.: Temperature dependent photovoltaic (PV) efficiency and its effect on PV production in the world – a review. Energy Procedia **33**, 311–321 (2013)
5. Dobbs, A.P.: PVWatts Version 5 Manual. NREL Publications, 4 September 2014. Accessed 17 Mar 2018
6. Nižić Krstinić, M., Hustić, A.: Modeli Odgovornosti Povlaštenih Proizvođača Obnovljivih Izvora Energije Za Odstupanje Od Planova Proizvodnje. Zbornik Veleučilišta U Rijeci3, pp. 93–110 (2015). Accessed 17 Mar 2018
7. Parlament FBiH. Zakon O Korištenju OiEiK, 30 August 2013. Accessed 17 Mar 2018
8. FERK. Odluka O Referentnoj Cijeni, 3 July 2017. Accessed 17 Mar 2018
9. Ścigan, M., Gonul, G., Türk, A., Frieden, D., Prislan, B., Gubina, A.F.: Cost-Competitive Renewable Power Generation: Potential across South East Europe (2017)
10. Jordan, D.C., Kurtz, S.R.: Photovoltaic Degradation Rates — An Analytical Review. NREL Publications, June 2012. Accessed 17 Mar 2018
11. Formica, T.J., Khan, H.A., Pecht, M.G.: The Effect of Inverter Failures on the Return on Investment of Solar Photovoltaic Systems. http://ieeexplore.ieee.org/document/8039151/. Accessed 17 Mar 2018

Near Zero-Energy Home Prediction of Appliances Energy Consumption Using the Reduced Set of Features and Random Decision Tree Algorithms

Lejla Bandić(✉) and Jasmin Kevrić

International Burch University, Sarajevo, Bosnia and Herzegovina
{lejla.bandic, jasmin.kevric}@ibu.edu.ba

Abstract. This paper presents methods for prediction of energy usage of different appliances in homes. Dataset comprising 14804 samples include measurements of weather from a nearby airport station, temperature and humidity sensors from a wireless network and recorded energy use of lighting fixtures. These measurements are sorted into 32 features, from which 17 were filtered and showed to be sufficient for energy usage prediction. Two methods for prediction were trained and tested: Random forest and Random tree. The performance of the methods was studied and it has been showed that the random forest gives better results than random tree method and that it has good performance in prediction of energy use of appliances.

Keywords: Random forest (RF) · Random tree · Appliances · Energy Prediction

1 Introduction

Considering that appliances represent a significant part of the electrical energy demand (20–30%), the understanding of the energy usage of the appliances in the buildings has been the subject of numerous research studies [1, 2]. The majority of papers showed that several parameters and appliances have a positive impact on the electrical energy consumption. Those include number of appliances, computers, TV, electric oven, freezer, washing machine, dishwasher among others. In addition, weather and temperature parameters showed to be relevant to be used for prediction of energy use [3].

In order to understand and quantify relationship between different parameters for energy use, regression models are used. In the past years, studies have used models such as neural networks, multiple regression, support vector machines, random forest, etc. [4–6]. The methods have considered parameters such as weather variables, time of the day, weekdays, months, yesterday's consumption. In the case when there is increased number of appliances or parameters, it is important to clarify which ones represent the main contributors to the consumption.

In [6], authors presented the study on classification of on-off states of appliances consumption signatures. Random Forest, Hoeffding Tree and Bayes Net were the

S. Avdaković (Ed.): IAT 2018, LNNS 59, pp. 164–171, 2019.
https://doi.org/10.1007/978-3-030-02574-8_13

methods used for classification and pattern recognition. Best results were obtained when random forest was applied to dataset used.

In [7], data driven prediction models of electrical energy consumption of appliances are described. The study was done on the same dataset which is used in this paper. Boruta algorithm was used for feature selection, and recursive feature elimination (RFE) for feature ranking. Only 5 features were omitted in the final analysis. Four regression methods were trained and tested: multiple linear regression, support vector machine with radial kernel (SVM-radial), random forest (RF) and gradient boosting machines (GBM). Results showed that RF and GBM improve prediction accuracy compared to the SVM-radial and multiple linear regression. The best accuracy was obtained using GBM.

This paper represents the prediction of energy use of the appliances using Random Forest and Random Tree models, using different feature selection algorithms than the ones in [7], as it yields only 15% dataset reduction. The aim is to investigate whether greater dataset reduction could yield comparable results. Prediction was carried out using data that include measurements of weather from a nearby airport station, temperature and humidity sensors from a wireless network and recorded energy use of lighting fixtures. Results are compared with results given by [7].

2 Dataset

Dataset comprising 14804 samples include measurements of weather from a nearby airport station, temperature and humidity sensors from a wireless network and recorded energy use of lighting fixtures of a low-energy house. House description is given in [7]. Measurements are sorted into 32 features that are showed in Table 1.

Since dataset contains numerous features, it is desirable to find out which parameters are the most important and which ones do not improve the prediction of the energy consumption. CfsSubsetEvaluator, feature selection algorithm from Weka, together with Best first were used to address this task.

There are three main benefits of feature selection. Feature selection increases the classifier's capability for prediction, it is fast and cost effective and it improves comprehensibility of the process through which data is generated.

It is used to get subsets which are not redundant and more relevant in order that overall accuracy is either improved or retained while the size of dataset is decreased [8].

As a result of feature selection, 17 features showed to improve the prediction of the energy consumption. They include features with numbers 3, 5, 6, 7, 8, 9, 14, 15, 17, 19, 22, 23, 24, 25, 28, 29 and 30.

Table 1. Data features

Number of features	Data features	Units
1	Appliances energy consumption	Wh
2	Light energy consumption	Wh
3	T1, Temperature in kitchen area	C
4	RH1, Humidity in kitchen area	%
5	T2, Temperature in living room area	C
6	RH2, Humidity in living room area	%
7	T3, Temperature in laundry room area	C
8	RH3, Humidity in laundry room area	%
9	T4, Temperature in office room	C
10	RH4, Humidity in office room	%
11	T5, Temperature in bathroom	C
12	RH5, Humidity in bathroom	%
13	T6, Temperature outside the building (north side)	C
14	RH6, Humidity outside the building (north side)	%
15	T7, Temperature in ironing room	C
16	RH7, Humidity in ironing room	%
17	T8, Temperature in teenager room 2	C
18	RH8, Humidity in teenager room 2	%
19	T9, Temperature in parents room	C
20	RH9, Humidity in parents room	%
21	To, Temperature outside (from weather station)	C
22	Pressure (from weather station)	mm Hg
23	RHo, Humidity outside (from weather station)	%
24	Windspeed (from weather station)	m/s
25	Visibility (from weather station)	km
26	Tdewpoint (from weather station)	C
27	Random variable 1 (RV_1)	Non dimensional
28	Random variable 2 (RV_2)	Non dimensional
29	Number of seconds from midnight (NSM)	s
30	Week status (weekday (1) or weekend (0))	Factor/categorical
31	Day of week (Monday-Sunday)	Factor/categorical
32	Date time stamp	year-month-day hour:min:s

3 Methods

3.1 Random Forest

Random forest is a powerful machine-learning method that was proposed by Breiman [9]. It is used for classification, regression and unsupervised learning and it can be applied in many fields. Random forest is decision tree algorithm.

The basic idea of a decision tree is to iteratively partition data into boxes using simple rules that minimize the error at each split (node). Each node is split using the best split among all variables, in standard trees. Instead of searching for the best feature while splitting a node, random forest searches for the best feature among a random subset of features. This process creates a wide diversity, which generally results in a better model [10].

It has only two parameters: the number of variables in the random subset and the number of trees. Performance of RF is not sensitive to their values.

3.2 Random Tree

Random tree is a decision tree method that uses a tree-like graph or model of decisions. It is an ensemble learning algorithm that generates many individual learners. The algorithm can deal with both classification and regression problems.

Random trees represent the combination of two existing algorithms: single model trees are combined with random forest ideas. Model trees are decision trees where every single leaf holds a linear model which is optimized for the local subspace described by this leaf. Random trees employ random forest procedure for split selection and thus induce reasonably balanced trees where one global setting for the ridge value works across all leaves, thus simplifying the optimization procedure [11].

3.3 Feature Selection

Feature subset selection represents the process of identifying and removing as much irrelevant and redundant data as possible. Therefore, it reduces the dimensionality of the data and it allows learning algorithms to operate faster and more effectively.

For feature selection, in this paper, Attribute Evaluator – CfsSubsetEval in Weka was used. It is the evaluator that calculates the value of a feature subset by taking into account the predictive capability of every feature together with the amount of redundancy between features [12].

Search method used is Best first. This method is an artificial intelligence algorithm which backtracking facility of the search path. Best first moves through the search space by making local changes to the current feature subset. However, if the path being explored start to appear not so good, the best first search can back-track to a better previous subset and continue the search from there [12].

4 Results

In order to compare the performance of the regression models, two performance measurement indices are used: the root mean square error (RMSE) and the mean absolute error (MAE).

$$\mathrm{RMSE} = \sqrt{\frac{\sum_{n}^{i=1} (X_i - \bar{X}_i)^2}{n}} \tag{1}$$

$$\text{MAE} = \frac{\sum_{n}^{i=1} |X_i - \bar{X}_\imath|}{n} \tag{2}$$

Where X_i is the actual measurement, $\bar{X}_\imath$ is the predicted value and n is the number of measurement.

In order to obtain results and see the performance of the proposed regression models, four different cases were performed. First case represents usage of 50% of instances of the dataset for training and 50% for testing. For the second case 66% of instances were used for training and 34% for testing. Third case represents usage of 90% of instances for training and 10% for testing. For the fourth case 10-CV (cross validation) was used.

In Table 2 results using random forest are shown. The best performance was obtained using 10-CV, with RMSE of 69.039 and MAE of 33.3789.

Table 2. Random forest results

Random forest		
	RMSE	MAE
50% train, 50% test	75.3417	38.1093
66% train, 34% test	71.4505	34.9544
90% train, 10% test	70.4462	34.0677
10-CV	69.2039	33.3789

Table 3 represents the results obtained using random tree. The fourth case, showed the best performance once more. RMSE for random tree is 96.0062 and MAE is 41.0405.

Table 3. Random tree results

Random tree		
	RMSE	MAE
50% train, 50% test	104.0531	46.0443
66% train, 34% test	99.6701	43.6195
90% train, 10% test	97.8488	41.0865
10-CV	96.0062	41.0405

Comparing the results obtained, it can be concluded that random forest has better performance than random tree algorithm.

Results obtained from the [7] are shown in Table 4. According to [7] gradient boosting machines (GBM) gave the best results, when compared with random forest, multiple linear regression and support vector machine (radial). For the results shown in Table 4, GBM and RF was performed using 27 features, which is 10 features more than used in this paper. That can be the reason why [7] obtained better results, but not by a huge margin.

Table 4. Results from [7]

	RMSE	MAE
GBM – Gradient Boosting Machines	66.65	35.22
RF – Random Forest	68.48	31.85

Table 5. Comparison of features used

Number of features	Data features	[7] paper	This paper
2	Light energy consumption	+	
3	T1, Temperature in kitchen area	+	+
4	RH1, Humidity in kitchen area	+	
5	T2, Temperature in living room area	+	+
6	RH2, Humidity in living room area	+	
7	T3, Temperature in laundry room area	+	+
8	RH3, Humidity in laundry room area	+	+
9	T4, Temperature in office room	+	+
10	RH4, Humidity in office room	+	
11	T5, Temperature in bathroom	+	
12	RH5, Humidity in bathroom	+	
13	T6, Temperature outside the building	+	
14	RH6, Humidity outside the building	+	+
15	T7, Temperature in ironing room	+	+
16	RH7, Humidity in ironing room	+	
17	T8, Temperature in teenager room 2	+	+
18	RH8, Humidity in teenager room 2	+	
19	T9, Temperature in parents room	+	+
20	RH9, Humidity in parents room	+	
21	To, Temperature outside (from weather station)	+	
22	Pressure (from weather station)	+	+
23	RHo, Humidity outside (from weather station)	+	+
24	Windspeed (from weather station)	+	+
25	Visibility (from weather station)		+
26	Tdewpoint (from weather station)	+	
27	Random variable 1 (RV_1)		
28	Random variable 2 (RV_2)		+
29	Number of seconds from midnight (NSM)	+	+
30	Week status (weekday (1) or weekend (0))	+	+
31	Day of week (Monday-Sunday)	+	
32	Date time stamp		
Total features		27	17

In Table 5 comparison of features used by both papers is represented. It can be concluded that the most important parameters used for prediction of energy consumption are NSM, weather and temperature parameters.

RF using 17 parameters has RMSE of 69.039 and MAE of 33.3789 (Table 2). RF using 27 parameters has RMSE 68.48 and MAE 31.85 (Table 4). From these results, together with Table 5, it can be concluded that in order to get slightly improved results, it is preferable to include inside humidity parameters into consideration.

Table 6. GBM with different sets of features

GBM		
	RMSE	MAE
No lights - 26 features	66.21	35.24
No lights and no weather data – 21 features	68.59	36.21
No temperature and humidity inside house – 9 features	72.64	40.32
Only weather and time – 8 features	72.45	40.73

Table 6 represents the results of GBM model with different set of features [7]. When compared with RF with 17 features it can be seen that proposed algorithm has satisfying performance. Parameters that are of the most importance include: NSM (Number of seconds from midnight), weather parameters (pressure, humidity, windspeed, visibility), and temperature parameters inside of the house.

5 Conclusion

In this paper, random forest and random tree models were developed to predict the electrical energy consumption of the appliances. The algorithms and solutions are implemented in the Weka environment.

The presented methodology can be efficiently used for prediction of energy use of appliances. RF model improve RMSE and MAE of predictions when compared to random tree. Parameters that increase the accuracy of the prediction for RF include: NSM (Number of seconds from midnight), weather parameters (pressure, humidity, windspeed, visibility), and temperature inside of the house. When compared to previous studies and results obtained, it is preferable to include humidity parameters into consideration as well.

Future work and research can be linked to prediction of energy use of appliances using improved regression methods and more detailed analysis of the impact of different parameters on the electrical energy consumption of the appliances.

References

1. Kavousian, A., Rajagopal, R., Fischer, M.: Ranking appliance energy efficiency in households: utilizing smart meter data and energy efficiency frontiers to estimate and identify the determinants of appliance energy efficiency in residential buildings. Energy Build. **99**, 220–230 (2015)
2. Cetin, K.S., Tabares-Velasco, P.C., Novoselac, A.: Appliance daily energy use in new residential buildings: use profiles and variation in time-of-use. Energy Build. **84**, 716–726 (2014)
3. Jones, R.V., Fuertes, A., Lomas, K.J.: The socio-economic, dwelling and appliance related factors affecting electricity consumption in domestic buildings. Renew. Sustain. Energy Rev. **43**, 901–917 (2015)
4. Arghira, N., Hawarah, L., Ploix, S., Jacomino, M.: Prediction of appliances energy use in smart homes. Energy **48**(1), 128–134 (2012)
5. Ling, S.-H., Leung, F.H., Lam, H., Tam, P.K.: Short-term electric load forecasting based on a neural fuzzy network. IEEE Trans. Ind. Electron. **50**(6), 1305–1316 (2003)
6. Salihagic, E., Kevric, J., Dogru, N.: Classification of ON-OFF states of appliance consumption signatures. In: 2016 XI International Symposium on Telecommunications (BIHTEL) (2016)
7. Candanedo, L.M., Feldheim, V., Deramaix, D.: Data driven prediction models of energy use of appliances in a low-energy house. Energy Build. **140**, 81–97 (2017)
8. Rizwan, M., Waseem, S., Ejaz, A.: Maximum relevancy minimum redundancy based feature subset selection using ant colony optimization. J. Appl. Environ. Biol. Sci. **7**(4), 118–130 (2017)
9. Breiman, L.: Random forests. Mach. Learn. **45**, 5–32 (2001)
10. Liaw, A., Wiener, M.: Classification and regression by random forest. R News **2**, 18–22 (2002)
11. Kalmegh, S.: Analysis of WEKA data mining algorithm REPTree, simple cart and randomtree for classification of Indian news. IJISET - Int. J. Innov. Sci. Eng. Technol. **2**(2), 438–446 (2015)
12. Hall, M.A.: Correlation-based feature subset selection for machine learning. Hamilton, New Zealand (1998)

Experience in Work of Automatic Meter Management System in JP Elektroprivreda B&H d.d. Sarajevo, Subsidiary "Elektrodistribucija", Zenica

Ahmed Mutapcic(✉) and Adnan Memic

Public Enterprise Elektroprivreda B&H, d.d. Sarajevo, Sarajevo, Bosnia and Herzegovina
a.mutapcic@epbih.ba

Abstract. The electricity distribution company Subsidiary "Elektrodistribucija" Zenica started with the introduction of the system of remote reading and management of electric power meters (Automatic Meter Management/Automatic Meter Reading System) in 2009. The multiple projects have been started and completed and currently 17.989 automatic electric meters are managed by AMM system. These meters are implemented for different types of customers such as, metering points at the sites of power exchange with neighbouring power companies, metering points for the medium voltage customers (10 kV, 20 kV and 35 kV), metering points for primary customers (low voltage and power above 23 kW) and 199 different secondary customers which are supplied with electricity by medium - low voltage transformer stations.

By the implementation of the last project, 31 transformation areas into the AMM system were introduced. This paper will give a short description of the stages in the implementation of this last project through which new automatic meters have been introduced as well as present relevant experience gained so far through this project.

1 Introduction

The electricity distribution company Subsidiary "Elektrodistribucija" Zenica has begun with introduction of remote reading and control of electricity meters (system AMM/AMR) in 2009. The first such project is included metering points in the customer category households, 503 measurement points in two (2) transformer areas in parts of distribution, PJD Zenica and PJD Travnik (Table 1).

Electric meters and concentrators of the LandisGyr manufacturer were included in the above-mentioned transformer areas, and the remote reading software of the SEP2W manufacturer Iskraememco was used, by which we have tested and validated the interoperability of this system using two different manufacturers' devices.

After that we started with the activities on introduction metering points of the final customers at the medium voltage, the place of supply/delivery of electricity with neighbouring electro-distribution subsidiaries of JP Elektroprivreda BiH and other power companies, other consumption of 23 kW and below 23 kW with metering points

S. Avdaković (Ed.): IAT 2018, LNNS 59, pp. 172–186, 2019.
https://doi.org/10.1007/978-3-030-02574-8_14

Table 1. The first two transformer areas are registered in the system AMM

Town	Transformer area TA	Number of meters in the AMM system
Zenica	Jalija Mostar	286
Travnik	Željeznička stanica	217
Total:		503

priority on remote repetition stations of GSM and RTV operators into the AMM system. These measuring points are a total of 1.584 and are connected at the voltage levels 35 kV, 10 (20) kV, 6 kV and 0.4 kV (Table 2).

Table 2. Measuring points MV, OC and exchange sites which are introduced into the AMM system

Consumption category	Number of electric meters in the AMM system
Medium voltage (MV)	238
Energy supply/delivery sites	75
Other consumption over 23 kW	966
Other consumption bellow 23 kW repeater stations	305
Total	1.584

Following the implementation of the installation of the AMM system at the mentioned metering points, activities on installation of the counters for the AMM system have been started on a certain number of transformer areas selected according to the defined criteria so that in the AMM system there are currently 199 transformer areas, i.e. a total of 16,405 customers at low voltage.

Also, the introduction of metering points of end-customers of irregular planners for the consumed electrical energy, where remote disconnection gave excellent results in increasing the level of charging of the consumed electrical energy into the AMM system has started (Table 3).

Table 3. Total number of customers in the AMM system

Subsidiary	Medium voltage	Reception points	Other consumption	Households	Total
ED Zenica	238	75	1.271	16.405	17.989

This paper will give an overview of past experiences of introducing metering points into the AMM system that have been collected so far.

By the implementation of the last project, 31 transformation areas into the AMM system were introduced. The aim of the project was extending the existing AMM infrastructure by increasing the number of remote measuring points and creating a favourable environment from the aspect of the opening of the electricity power market in BiH. In each transformer area, measuring cabinets with data concentrators, control meters, electricity metering transformers of appropriate transmission ratio, surge arrester and communication equipment with the AMM centre were installed into the transformer units.

A GPRS network is used as a communication path from the concentrator and the meter to the AMM centre. For optimization purposes and more efficient resource utilization, control meters in substations are connected to the concentrator via the RS485 loop.

Activities to further expand the AMM system are currently taking place.

2 Technical Solution of the System

The AMM system implemented in the Subsidiary "Elektrodistribucija" Zenica is based on the software solution of the SEP2W System by the Iskraemeco manufacturer. Electrical energy meters and concentrators in TS 10(20)/0.4 kV are by the Iskraemeco, LandisGyr and Itron manufacturers, and other communication and related equipment are from various renowned manufacturers.

Depending on the available resources, the available communication connections were used, mostly GPRS/GSM, optical link and digital radio link.

The equipment installed in the system as well as the functioning and basic characteristics of the SEP2W system will be described further on.

2.1 SEP2W System

The AMM service platform provides services for automatic metering management. The primary task is to read the data from measuring devices using different protocols and different communication channels, to save data to the SEP2W database and exchange the measuring data with other systems. The modular structure of this platform enables expansion for future needs.

The SEP2W system consists of metering devices, data concentrators and communication modules at the lowest level, and the service platform for automatic reading of measuring devices at the control centre level. The SEP2W system can be installed on Microsoft server platforms.

The architecture and solutions of the SEP2W AMM system are based on the fact that a large part of data collection is done according to a defined schedule in the AMM centre. The AMM centre is an integrated scheduler that allows data collection based on the prepared transactions that are stored in the database. The second part of the data collection, i.e. so-called collection on demand, is performed through the meter reading service.

The following basic system services are available to the user:

- automatic reading (at the communication level),
- access to the measuring devices,
- readings of the measuring devices,

- control of the measuring device switch (on, off, power limitation),
- Web services,
- SEP2W reports.

2.2 Equipment (Meters, Concentrators)

The implemented AMM system in the distribution network of the Subsidiary "Elektrodistribucija", Zenica consists of single-phase and three-phase electric meters of Iskraemeco manufacturer, Landis Gyr and Itron electric meters (active and reactive energy, for direct or half-indirect measurements), where the meters have a DLC/GPRS modem data transfer installed. In the low-voltage side, MV/LV transformer stations have are built-in data concentrators of the Iskraemeco and LandisGyr manufacturers, where the concentrate communicates with the counters over the low-voltage DLC communication network and with the centre via GPRS communication, optics or via digital radio modems.

Each electric meter can automatically switch to repeat mode and transfer data from the meter with which a data concentrator cannot establish a direct link due to an increase in effective distance between the electric meters point and data concentrator, as well as to increase the efficiency of data collection. Since a few hundred metering points can be connected to a single data concentrator, a single data concentrator is installed into the MV/LV transformer station by one power transformer. The DLC infrastructure is used in urban, suburban and rural areas, where it is possible to connect points with meters to a distance of one kilometre or more.

The electric meters, which are in the AMM system, are constructed for multi-tariff measurements of active and reactive energy and power. The electric meters can measure energy in one direction flow (taken energy), two direction energy streams (retrieved and delivered energy), or measuring absolute energy values per phase. The meters have an optical port for local communication and an integrated DLC/GPRS modem for remote communication.

The data concentrator is an automatic data collection device from a meter with an embedded DLC modem. It uses DLMS/COSEM protocol for DLC communication. The software application in the data concentrator is responsible for finding meters in the AMM system, managing communication with meters and reading and storing their data as well as transferring these data to the data collection centre. For communication with the data collection centre, the concentrator uses various communication media such as GSM, GPRS and Ethernet.

In addition to concentrators, single phase and three-phase meters, within this AMM system, meters for industrial customers and metering points for the supply/delivery of electrical energy from Iskraemeco manufacturers are included.

These meters communicate with the AMM centre through various communication channels (GSM/GPRS, radio connection, PSTN, ISDN, Ethernet …).

Electronic multifunctional three-phase meters are designed to measure and register active and apparent energy for two directions of electrical energy and reactive energy in four quadrants, maximum power for the above-mentioned energy. The meters, in addition, have the ability to register time profiles and energy quality parameters. They can be connected to the network directly, semi-indirectly or indirectly. The meters are made as universal for the nominal phase voltage of the network in the range of 57.7 V to 230 V. The meters and technical characteristics of the meters correspond to international standards for Active Class 1 and 2 active IEC 62053-21 and IEC 62052-11 (IEC 61036-11) or class IEC 62053-22 (IEC 60687) as well as class 3 and 2 of reactive energy meters IEC 62053-23 (IEC 61268).

These electric meters are also used as sum of meters for measuring the total energy taken in the transformer area which are connected to the data concentrator.

3 Overview of Individual Transformation Areas Included into the AMM System

Below are a table of TPs that are included in the AMM system with the date on the number and type of meters (Table 4).

3.1 Exploitation of the AMM System

The frequency of collecting data that enable the implemented AMM system with the appropriate representation from the SEP2W2012 application is shown in Table 5.

3.1.1 Meter Reading and Use of Data from the AMM System for Calculation Purposes

Below you will find the statistics of the successfulness of the DLC meter reading for TP Podvinci 1 in the Breza Business Unit (Fig. 1).

The data that is collected through the AMM system is primarily used for the calculation of electrical energy. The read data is transmitted via the Billing exporter application to the SOEE electrical energy billing application (Fig. 2).

3.1.2 An Overview of the Consumption Logic on Transformer Stations 10(20)/0.4 kV

On all 31 TS 10(20)/0.4 kV covered by this project as well as the existing AMM system, control meters have been installed, whose function is to control consumption in transformer areas and to locate losses in the power distribution network more accurately.

A special application module for SEP2W is used to control consumption. The control is performed by first calculating the sum of the daily tariff data of all the built meters of the customers in the subject transformer area, for the monthly period to which control is applied. Then this sum is compared with the monthly consumption result registered on the control meter. The periods for which the control or calculation of losses can be different from one day to one year, with the obligation to use the same period used for the control meter and for all other customer meters.

The results of this comparison are shown in the aforementioned additional SEP2W module in the form of a report on the absolute and percentage value of the difference between the two measurements.

Table 4. An overview of transformer area per PJD which are introduced into the AMM system

No.	PJD	Name of the TA	Single-phased meters (pcs)	Three-phased meters (pcs)	Total meters (pcs)	Communication towards the AMM centre
1	Breza	Podvinci 1	72	13	85	GPRS
2	Breza	Podvinci 2	108	16	124	GPRS
3	Doboj	Šije G. polje-2	43	24	67	GPRS
4	Kakanj	Ponijeri	206	5	211	GPRS
5	Kakanj	Vrheblje	158	4	162	GPRS
6	Maglaj	Donji ulišnjak	76	37	113	GPRS
7	Olovo	Musići	38	3	41	GPRS
8	Tešanj	Kološević	65	27	92	GPRS
9	Tešanj	Koprivci	61	13	74	GPRS
10	Vareš	Mijakovići	59	5	64	GPRS
11	Visoko	Goduša	182	25	207	GPRS
12	Visoko	Tokmići	53	1	54	GPRS
13	Visoko	Kula Banjer 1	59	13	72	GPRS
14	Olovo	Musići II	26	4	30	GPRS
15	Zavidovići	Borovnica	54	86	140	GPRS
16	Zavidovići	Mednik	63	30	93	GPRS
17	Zavidovići	Maoča	78	69	147	GPRS
18	Zenica	Jastebac	73	9	82	GPRS
19	Zenica	Gladovići	118	7	125	GPRS
20	Zenica	Jastrebac 2	7	0	7	GPRS
21	Zenica	Orahovica	65	5	70	GPRS
22	N. Travnik	Bistro	74	2	76	GPRS
23	Vitez	Brize-Bobaši	42	21	63	GPRS
24	Bugojno	Humac	35	41	76	GPRS
25	Bugojno	Hum	17	5	22	GPRS
26	Bugojno	Rovna II	50	25	75	GPRS
27	G. Vakuf	Voljevac I	41	37	78	GPRS
28	G. Vakuf	Boljkovac	59	40	99	GPRS
29	D. Vakuf	Oborci 2	34	32	66	GPRS
30	D. Vakuf	Biokovine	69	4	73	GPRS
31	Fojnica	Ščitovo 2	93	56	149	GPRS
32	Total		2.178	659	2.837	

Table 5. Data transmitted to the AMM system

Data name	Data type	Frequency of the data collection
Active energy, in higher and lower tariff	Billing	Daily + monthly + when needed
Reactive energy, in higher and lower tariff	Billing	Daily + monthly + when needed
Power in higher and lower tariff	Billing	Daily + monthly + when needed
Load profile	Other	Monthly + when needed
		When needed
Events book	Other	When needed
Values of voltage and electricity	Other	When needed
Time on the meter	Other	When needed
Quality parameters of the electric energy	Other	When needed

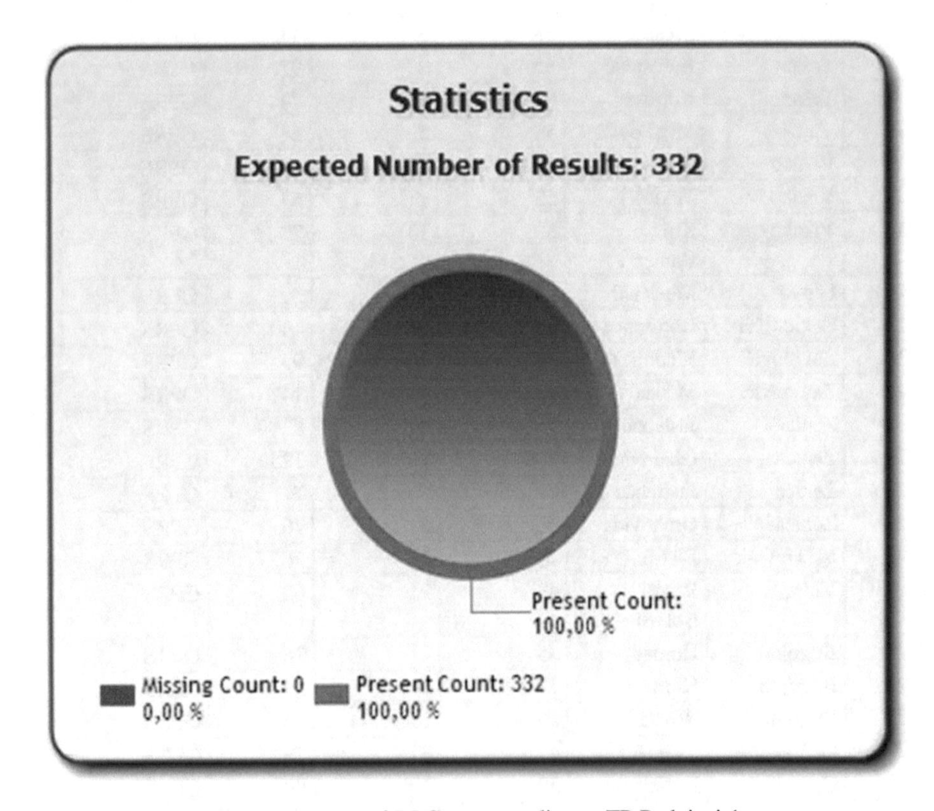

Fig. 1. Success of DLC meter reading at TP Podvinci 1

The value of calculated losses on the observed 31 transformer areas ranges from:

- TP Biokovine 4.97% to
- TP Maoča 9.75%

Figure 3 shows the report on losses in the results of TP Boljkovac-PJD Gornji Vakuf.

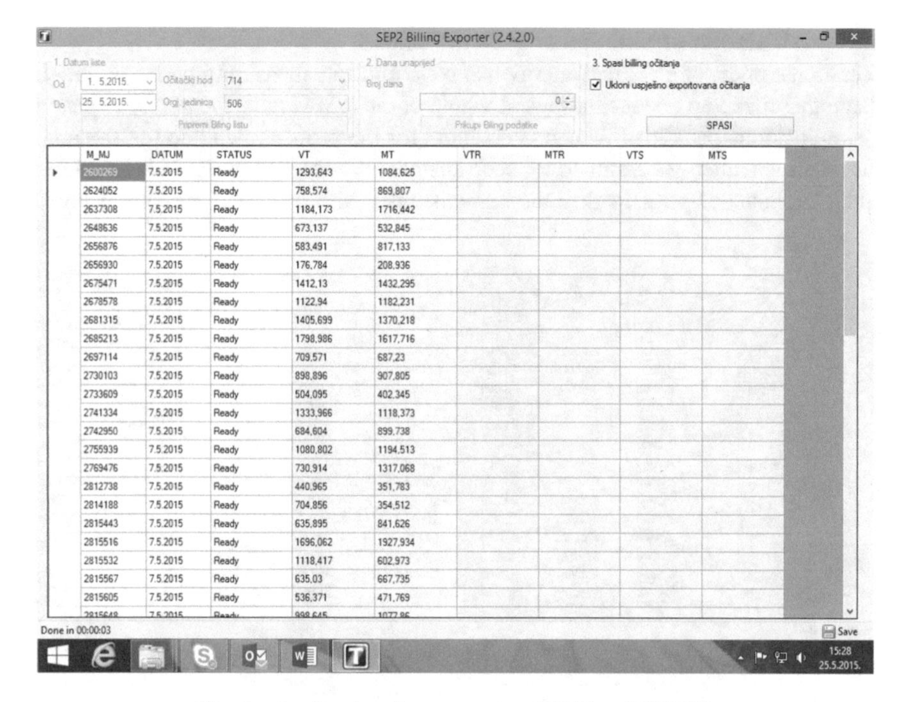

Fig. 2. Application for connecting SOEE and SEP2W

MT381_51757251_A	356,423 kWh
MT381_51757252_A	199,590 kWh
MT381_51757253_A	805,750 kWh
MT381_51757254_A	734,040 kWh
MT381_51757255_A	13,949 kWh
MT381_51757256_A	1.108,197 kWh
MT381_51757441_A	851,122 kWh
ME381_51803604_A	0,000 kWh
ME381_52023829_A	287,912 kWh
MT880_35792027	14.240,380 kWh
ME381_51921783_A	0,000***
Cumulative meters Mx371 at P2LPC:	27.340,719 kWh
Control meter MT8x1:	29.240,380 kWh
Difference (kWh):	1.899,661 kWh
Difference-losses at TP (%):	6,496 %

Fig. 3. Report from SEP2W – logical control

3.1.3 Calculation of Simultaneous Peak Load

For end-customers of electrical energy that have multiple measuring points on the same voltage that make up one technological whole of the AMM have the ability to calculate the simultaneous peak load, which is essential for the proper calculation of the invoice of the electrical energy consumed for such customers. Attached in Fig. 4 is an overview of the simultaneous peak load to the end-customer who has three measuring points.

Time	Rudnik_AbidLolic_Jednovremeni_Aktivna_T1 [kW]	Rudnik_AbidLolic1_T1 [kW]	Rudnik_AbidLolic2_T1 [kW]	Rudnik_AbidLolic_T1 [kW]	Rudnik_AbidLolic_Jednovremeni_Aktivna_T2 [kW]	Rudnik_AbidLolic1_T2 [kW]	Rudnik_AbidLolic2_T2 [kW]	Rudnik_AbidLolic [kW]
1.4. 0:15	0,0000	0,0000	0,0000	0,0000	222,9600	104,4000	0,0000	118,5600
0:30	0,0000	0,0000	0,0000	0,0000	250,9600	130,8000	0,0000	120,1600
0:45	0,0000	0,0000	0,0000	0,0000	378,9600	248,4000	0,0000	130,5600
8:15	421,2800	307,2000	0,0000	114,0800	0,0000	0,0000	0,0000	0,0000
8:30	377,9200	259,2000	0,0000	118,7200	0,0000	0,0000	0,0000	0,0000
23:15	0,0000	0,0000	0,0000	0,0000	175,6800	64,8000	0,0000	110,8800
23:30	0,0000	0,0000	0,0000	0,0000	213,7600	98,4000	0,0000	115,3600
23:45	0,0000	0,0000	0,0000	0,0000	245,3600	130,8000	0,0000	114,5600
30.4. 0:00	0,0000	0,0000	0,0000	0,0000	240,0800	126,0000	0,0000	114,0800
0:15	0,0000	0,0000	0,0000	0,0000	286,9600	169,2000	0,0000	117,7600
0:30	0,0000	0,0000	0,0000	0,0000	400,8800	265,2000	0,0000	135,6800
23:30	0,0000	0,0000	0,0000	0,0000	211,6000	102,0000	0,0000	109,6000
23:45	0,0000	0,0000	0,0000	0,0000	246,4000	136,8000	0,0000	109,6000
1.5. 0:00	0,0000!	0,0000!	0,0000!	0,0000!	284,5600!	174,0000!	0,0000!	110,5600!

Summary

	Rudnik_AbidLolic_Jednovremeni_Aktivna_T1 [kW]	Rudnik_AbidLolic1_T1 [kW]	Rudnik_AbidLolic2_T1 [kW]	Rudnik_AbidLolic_T1 [kW]	Rudnik_AbidLolic_Jednovremeni_Aktivna_T2 [kW]	Rudnik_AbidLolic1_T2 [kW]	Rudnik_AbidLolic2_T2 [kW]	Rudnik_AbidLolic [kW]
Maximum	868,2400 kW 29.4. 11:30				808,8000 kW 4.4. 1:45			

Fig. 4. Report from SEP2W-simultaneous peak load

3.1.4 Analysis of the Curve Loads

In accordance with the requirements of the open market for electrical energy for all customer categories, the AMM system has the ability to collect data for the analysis of curve loads. For this purpose, for the customers of the medium voltage, direct data collecting from the meters on the realized load is made, while for the other customers, an alternative load curve is made (Figs. 5 and 6).

3.1.5 The Reading Analysis of Electrical Energy and Voltage Values

Analyzing the readings of voltage and electrical energy by the phases of the measuring points of the end customers of electrical energy we are currently controlling the measuring site. Any deviations of the readings (voltage 0 V, electricity 0 A at the operating time) provide an alarm for checking and control of the measuring point. Also every malfunction on the power plants is recorded in the SEP2W base directly over the voltage values shown in Figs. 7, 8 and 9.

3.1.6 Power On/Off, Power Limitation and Other AMM System Functionalities

The electrical energy meters used in the AMM system also enable the remote switch on/off function of the end-customers from the network and the limitation of the paid approved power supply.

Time	ActivePower-Plus_T1 [kW]	ActivePower-Plus_T2 [kW]	ReactivePower-Plus_T1 [malfunction]	ReactivePower-Plus_T2 [malfunction]
1.4. 0:15	0,0000	104,4000	0,0000	100,8000
0:30	0,0000	130,8000	0,0000	111,6000
0:45	0,0000	248,4000	0,0000	208,8000
1:00	0,0000	271,2000	0,0000	226,8000
8:15	307,2000	0,0000	231,6000	0,0000
8:30	259,2000	0,0000	205,2000	0,0000
8.4. 0:00	0,0000	242,4000	0,0000	162,0000
0:15	0,0000	302,4000	0,0000	224,4000
11:30	591,6000	0,0000	452,4000	0,0000
11:45	610,8000	0,0000	457,2000	0,0000
12:00	572,4000	0,0000	426,0000	0,0000
30.4. 0:00	0,0000	126,0000	0,0000	109,2000
0:15	0,0000	169,2000	0,0000	135,6000
0:30	0,0000	265,2000	0,0000	193,2000
0:45	0,0000	387,6000	0,0000	308,4000
8:15	314,4000	0,0000	280,8000	0,0000
8:30	244,8000	0,0000	248,4000	0,0000
8:45	222,0000	0,0000	160,8000	0,0000
9:00	124,8000	0,0000	76,8000	0,0000

Summary	Active Power Plus_T1 [kW]	Active Power Plus_T2 [kW]	ReactivePower-Plus_T1 [malfunction]	ReactivePower-Plus_T2 [malfunction]
Maximum	**610,8000 kW 8.4. 11:45**	**579,6000 kW 7.4. 1:00**	**457,2000 malfunction 8.4. 11:45**	**381,6000 malfunction 18.4. 10:45**

Fig. 5. Analysis of the curve load for end customers at the high voltage

The end-customer switch-on/off function allows remote switch-on/off of customers according to the supplier's request due to the unpaid invoice for the consumed electrical energy, which results in a reduction in the cost of hiring the workforce.

If the end customer does not allow the shutdown due to the debit for consumed electrical energy, the metering point is replaced to a different available location and by installing the AMM meters, the shutdown function is enabled. This activity has brought great results in reducing the debts for consumed electrical energy and has animated the buyers unwilling to settle debts.

Date	Result[kWh]	Status [hex]
8.5.2015 11:00	1637,487	0
8.5.2015 11:00	1637,519	0
8.5.2015 12:00	1638,657	0
8.5.2015 13:00	1638,705	0
8.5.2015 14:00	1639,758	0
8.5.2015 15:00	1639,758	0
8.5.2015 16:00	1639,758	0

Fig. 6. Analysis of the curve load for end customers of the household category

Time	Voltage L1 [V]	Voltage L2 [V]	Voltage L3 [V]
12/25/ 12:15 AM	61.00	61.40	61.50
12:30 AM	60.50	60.90	60.90
12:45 AM	60.40	60.80	60.90
1:00 AM	60.50	60.90	61.00
8:00 PM	60.80	61.10	61.10
8:15 PM	60.80	61.10	61.00
8:30 PM	34.30	79.80	79.70
8:45 PM	0.00	104.20	104.10
9:00 PM	0.00	104.20	104.10
9:15 PM	0.00	104.20	104.10
9:30 PM	0.00	104.40	104.40

Graphs

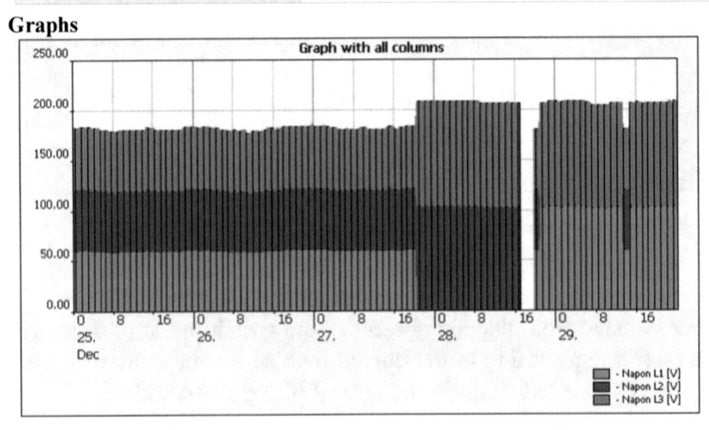

Fig. 7. Readings of the voltage of the customer at medium voltage, grounding phase L1 at 8:45 h

Time	Voltage L1 [V]	Voltage L2 [V]	Voltage L3 [V]
6:45 AM	59.30	59.70	60.60
7:00 AM	59.00	59.40	60.30
7:15 AM	58.70	59.10	60.10
7:30 AM	58.60	59.00	60.00
7:45 AM	58.60	58.90	59.90
8:00 AM	70.80	71.10	72.30
8:15 AM	69.40	69.80	70.90
8:30 AM	59.30	59.70	60.70
8:45 AM	58.80	59.20	60.10
9:00 AM	58.60	59.00	59.90
9:15 AM	58.50	58.90	59.80

Fig. 8. Readings of the voltage of the customer at medium voltage malfunction in TS 110/10 kV, increased voltage by 20% in the time of 8:00 to 8:15 h.

Time	Electricity L1 [A]	Electricity L2 [A]	Electricity L3 [A]
5:15 AM	0.27	0.23	0.32
5:30 AM	0.27	0.23	0.32
5:45 AM	0.27	0.23	0.32
6:00 AM	0.27	0.22	0.31
6:15 AM	0.38	0.34	0.40
6:30 AM	0.90	0.72	0.82
6:45 AM	1.49	1.28	1.37
7:00 AM	2.22	1.96	2.12
7:15 AM	2.90	2.71	2.83
7:30 AM	2.90	2.74	2.85
7:45 AM	2.90	2.72	2.83
Summary			
	Electricity L1 [A]	Electricity L2 [A]	Electricity L3 [A]
Average	1.23 A***	1.12 A***	
Strict Average			1.32 A

Fig. 9. Readings of the currents of the customer at medium voltage, start of the working process at 06:30 h

The power limitation function is important due to the limiting power demand for end-customers whose connection power is less than 23 kW which is prescribed by the relevant regulations to the distributors. The use of the AMM meter that has an integrated element (switch) with a limiter function makes it greatly easier for the distributors to fulfill this requirement.

Other functionalities of the AMM system are:

- Examining events on the electrical energy meter
- Analysis of voltage conditions and electricity size
- Different types of reports
- Defining the schedule of the meter reading.

4 AMM System Advanced Functions

By implementing the AMM system into the distribution network of the Subsidiary of "Elektrodistribucija", Zenica has achieved numerous improvements in the functioning of the system. From the aspect of installation of the AMM meters improvements are the following:

- New electronic metering devices have been installed in the metering points, which has improved the measure-technical characteristics, and the meters of older generations have been dismantled this reducing electricity losses,
- In the Subsidiary of "Elektrodistribucija", Zenica from the average monthly invoiced energy of 85,000,000 kWh via AMM, the system is reads about 45,000,000 or 52.94%, the first-rate and accurate reading which reduces the losses of electrical energy,
- The new meters have an integrated switching clock, which eliminates the effect of end customer's influence on the installed clock,
- Integrated limitation and processing with the same switch-off function, the mentioned switch also serves for remote off/on management of the meter,
- Grading the limitation with a degree of 1 kW and even <1 kW (so far there have been limitators of a specific electricity in A),
- Simple switch-on by pressing the button from the end-user after the limitation process,
- Integrated remote reading module,
- New meters are smaller in outline and possess the listed integrated modules which have greatly influenced the measurement of the meter closets for individual housing units and especially for multi-storey residential buildings.

From the aspect of introducing remote reading improvements are:

- the ability to remote reading of the distant repeaters on the geographic elevations for which it was it is necessary in the previous period to hire special vehicles and additional work force,
- the possibility of remote reading of entire transformer areas (reduction of the number of executives),
- logical control of electricity consumption (locating unauthorized consumption),
- the ability to collect data to analyze the curve load,
- ability to analyze voltage conditions and electricity sizes,
- detection of unauthorized access to the meter (removing the lid of the meter connector from unauthorized persons).

In addition to the above mentioned advantages in the current practice of the AMM system, a number of work difficulties have been identified, which are solved in coordination with the manufacturer of the implemented system. Most often, this refers to the percentage of readings per individual transformer area which is below 95%. One of the reasons for this step is the interruption in the GPRS network, which is solved with the telecom operators.

In our AMM system, the interoperability at the software level, i.e. the software of the manufacturer of the Iskraemeco SEP2W is in function, and it is possible to read the data from the data centre of the manufacturers of Landis Gyr and Itron, which was applied to our first TA of our subsidiaries, which were introduced in AMM system, i.e. TA Jalija Mostar-PJD Zenica and TA Željeznička stanica-PJD Travnik. For the future expansion and operation of the AMM system, a very important characteristic is interoperability that is fully introduced into our subsidiary by the implementation of the upcoming projects.

Over the past period, the equipment manufacturers have improved the characteristics of the equipment related to interoperability and standardization of the protocols, so today there are several manufacturers whose equipment is fully interoperable, i.e. interoperability can be established at concentrator level. This new generation of equipment on this function has the approval of the IDIS (Interoperable Device Interface Specification).

5 Conclusions

In the JP Elektroprivreda BiH, subsidiary "Elektrodistribucija", Zenica, the gained experience in the use of AMM systems point to a number of positive effects that this system has and will have to the organization of the measurement and calculation functions of electrical energy in the future.

The consumption control function and the calculation of distribution losses at the level of the transformer area indicate that this AMM system function will be of crucial importance for different analysis of losses in the distribution network, and for proposing and implementing measures for their reduction in specific areas. The improvement activity on profiling and aggregating energy consumption and load data will enable the implementation of various energy analyzes, improvements in the planning process of distribution network development and exploitation and the implementation of load management measures.

The process of introducing the AMM system allows for a reduction in the number of employees in the direct reading process, further reducing the possibility of errors occurring in the process, while simultaneously demanding a greater number of skilled employees educated to work with the system. The mentioned improvements greatly affect the more efficient operation of the measurement process.

Also, the ability to create load curves and replacement curves is necessary in accordance with the requirements of the open electricity market, the emergence of more suppliers and the development of more efficient and more flexible tariff models of electricity calculation.

References

1. Extension of the AMM system in the Subsidiary “Elektrodistribucija”, Zeica-Project of derived state, Iskraemeco d.o.o. Sarajevo, April 2015
2. Introducing of the MDM system and expanding AMI infrastructure of JP Elektroprivreda BiH d.d. Sarajevo, Subsidiary “Elektrodistribucija”, Zenica, Iskraemeco d.o.o. Sarajevo (2013)
3. Implementation of the AMR/AMM system in the Electricity distribution system JP Elektroprivreda BiH d.d. Sarajevo at the measuring points of end-customers at the voltage of 6 kV, 10 kV, 20 kV i 35 kV and measuring points of supply/delivery of electrical energy with other electric power subjects in the Subsidiary “Elektrodistribucija”, Zenica Project of derived state, Iskraemeco d.o.o. Sarajevo, September 2012
4. Implementation of the remote reading and electric energy meter management systems (AMR/AMM systems) on seven transformation areas in the electricity supply areas of JP Elektroprivreda BiH d.d. Sarajevo – Subsidiary “Elektrodistribucija”, Zenica Project of derived state, Iskraemeco d.o.o. Sarajevo, April 2011
5. General conditions for the delivery of electrical energy, Regulatory committee for electric energy in the Federation of Bosnia and Herzegovina FERK, October 2014
6. Road map for electric energy and gas – Bosnia and Herzegovina; document adopted on the meeting of the Ministry council of the Energy community, Skoplje, 17 November 2006
7. Decision on the scope, conditions and timing of the opening of the electricity market in Bosnia and Herzegovina, State Regulatory Commission for Electricity (SRCE), Official Gazette of BiH No. 48/06 of 26th June 2006
8. Decision on Amendments to the Decision on the Volume, Conditions and Time Schedule for Opening the Electricity Market in Bosnia and Herzegovina, State Regulatory Commission for Electricity (SRCE), Official Gazette of BiH No. 77/09 of 29th September 2009
9. Rulebook on obtaining qualified customer status, Regulatory Commission for Electricity in Federation of Bosnia and Herzegovina - FERK, September 2006

Financial Impacts of Replacing Old Transmission Lines with Aluminum Composite Core Conductors

Semir Hadžimuratović(✉)

Independent System Operator in Bosnia and Herzegovina, Sarajevo, Bosnia and Herzegovina
s.hadzimuratovic@nosbih.ba

Abstract. Driven by the deregulation of power utilities, power lines are nowadays used to send more electric energy through longer distances to end consumers, compared to previous decades. Over the years, there have been various technical proposals for the improvement of old transmission conductors, with a goal of increasing its ampacity, as well as increasing the reliability of the power system as a whole. Aluminum Conductor Composite Core (ACCC) is the result of an interdisciplinary mission to engineer more efficient lines, using state-of-the-art components that result in significant financial savings for the operators who decide to entrust such a new technology. This paper deals with a comparative analysis of the financial impacts that advanced conductor technologies represent versus standard conductors. The primary focus of this paper is the reconductoring of existing overhead transmission lines using old corridors, contrary to erecting new lines. In the end, an extensive cost comparison example is given to demonstrate the various aspects of revitalization with ACCC conductors.

Nomenclature

AAAC	All Aluminum Alloy Conductor
ACSR	Aluminum Conductor Steel Reinforced
ACSS	Aluminum Conductor Steel Supported
GACSR	Gap Type ACSR
INVAR	Special nickel-iron alloy FeNi36
ACCC	Aluminum Conductor Composite Core
ACCR	Aluminum Conductor Composite Reinforced
HTLS	High Temperature Low Sag Conductor
OHL	Overhead Line
TSO	Transmission System Operator
WPP	Wind Power plants
TTC	Total Transfer Capacity
TRM	Transmission Reliability Margin
NTC	Net Transfer Capacity

S. Avdaković (Ed.): IAT 2018, LNNS 59, pp. 187–197, 2019.
https://doi.org/10.1007/978-3-030-02574-8_15

1 Introduction

Aluminum Conductors Steel Reinforced (ACSR) are the most widespread conductors used for overhead lines (OHL). They have been in use for a very long time, have passed extensive testing and finally have proven themselves in electric power systems. With the deregulation of the electricity markets and open market trade between parties crossing national borders, power lines are becoming more and more loaded in order to transmit the energy from producers to consumers. In the same time, with the economic growth and importance of the energy trade, it is more difficult to build new power corridors, because of, [1]:

- lengthy permitting procedures,
- unavailability of new corridors,
- construction time,
- construction costs.

In the same time period, at the beginning of the 2000's, two major electricity blackouts happened, New York in 2003, and Germany in 2006. In its sequence of events, the New York blackout was caused by a 345 kV transmission line that sagged into a tree and tripped.

OHL have a permissible sag, however when increasing current, the temperature increases, causing the conductor to sag below permissible limits. Which can be problematic if the vegetation under the overhead line in the corridor is not regularly trimmed.

Therefore, engineers started working of High Temperature Low Sag (HTLS) conductors that would enable the line to withstand higher temperatures and maintain an allowable sag at the same time.

High-temperature low sag (HTLS) are conductors capable to withstand high operating temperatures and, therefore, to carry a higher amount of power and/or energy when compared to conventional conductors. The CIGRE Task Force B2.11.03 defines them as conductors designed for applications where continuous operation is about 100 °C or as conductors designed to operate in emergency conditions above 150 °C, [2–4].

New composite technology conductors use a core of composite material around which aluminium conductor wires are wrapped. This design results in an increased tensile strength and reduced weight. Moreover, together with their higher operating temperatures, composite conductors have reduced sag under high loads [2, 5].

The conductivity of aluminium alloy reaches 61% IACS (International Annealed Copper Standard for conductivity). According to other sources [6], the conductivity can be of 63% IACS or even better due to the aluminium strands being "dead soft" (fully annealed). This can translate into higher ratings. Moreover, since the aluminium strands are dead soft, the conductor may be operated at temperatures in excess of 250 °C without loss of strength, [2, 6].

HTLS Conductors are most often used for reconductoring of existing transmission and distribution lines, for use in heavy ice regions, ageing structure, long-span crossing, but also for connecting renewable sources, such as wind power plants, [2, 5, 6].

2 Conductor Comparisons

Overhead line conductors can be classified into two major categories:

- The historically older technology, or standard conductors, or ACSR.
- The advanced technology, or HTLS.

All over the years, different HTLS conductors have emerged, some of which are; ACSS, GACSR, Invar, ACCC, ACCR. Table 1 offers the major differences between the two major conductor categories [1, 7].

Table 1. Comparison of standard vs. HTLS conductors

	Standard	HTLS
Core	Steel	Reinforced steel Steel alloys Composite materials
Outer strands	Concentric (round) aluminum wires	Annealed aluminum Aluminum alloys Compact (trapezoid) strand

Unlike standard ACSR conductors that can be produced in any major conductor manufacturing facility, HTLS conductors, as a result of a more refined, diverse and complicated manufacturing processes are usually produced in several factories around the world.

When analyzing ampacity, temperature and sag parameters, Figs. 1 and 2, for different HTLS conductor types, it can be concluded that both ACCC and ACCR, accomplish low sag at high temperatures by using composite core technology.

The author decided to focus on ACCC, due to better comparisons results, as well as the availability of their manufacturer's (CTC Global) comparison software.

3 Aluminum Conductor Composite Core (ACCC)

Aluminum Conductor Composite Core or ACCC is a type of "high-temperature low-sag" (HTLS) overhead power line conductor manufactured by more than 20 international conductor manufacturers.

The ACCC conductor consists of a hybrid carbon and glass fiber composite core which utilizes a high-temperature epoxy resin matrix to bind hundreds of thousands of individual fibers into a unified load-bearing tensile member [8].

The composite core is surrounded by aluminum strands to carry electrical current (Fig. 3). The conductive strands are generally fully annealed aluminum and trapezoidal in shape to provide the greatest conductivity and lowest possible electrical resistance for any given conductor diameter [8].

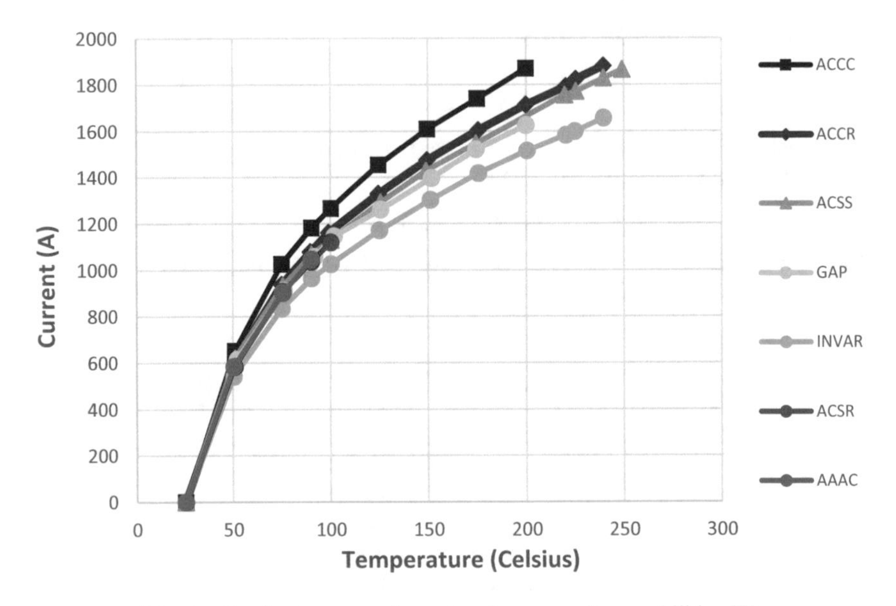

Fig. 1. Conductor comparison showing ampacity capabilities [7]

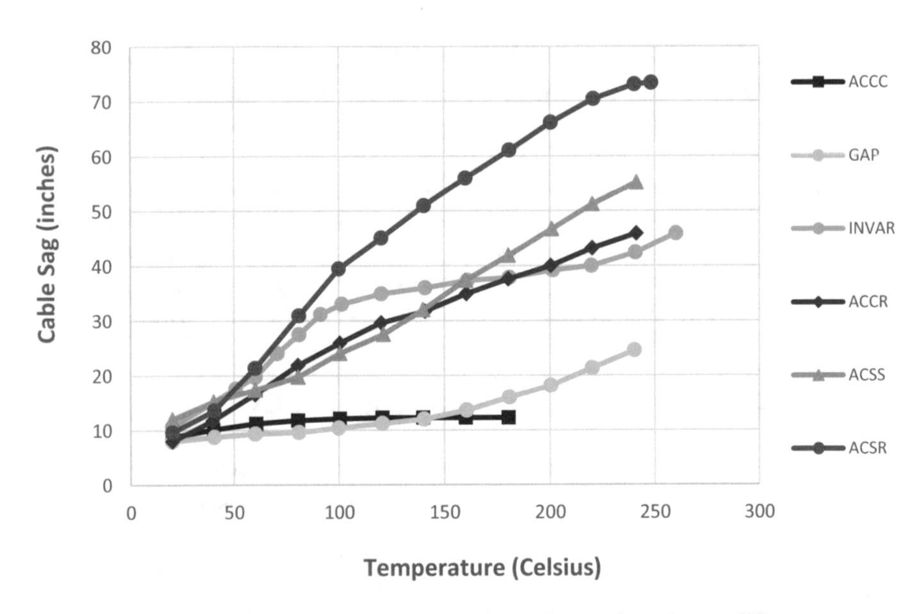

Fig. 2. Sag - temperature comparison of several conductors [7]

The ACCC conductor is rated for continuous operation at up to 180 °C (200 °C short-term emergency), and operates significantly cooler than round wire conductors of similar diameter and weight under equal load conditions due to its increased aluminum content and the higher conductivity [8]. Though the ACCC conductor was initially

Fig. 3. ACSR (left) and ACCC (right) conductors [8]

developed as a "High-Temperature, Low-Sag" (HTLS) conductor to increase the capacity of existing transmission and distribution lines with minimal structural changes, its improved conductivity and reduced electrical resistance makes it ideally suited for reducing line losses on new transmission and distribution lines where improved efficiency and reduced upfront capital costs are primary design objectives [9].

4 Advantages and Disadvantages

This composite strength member provides several advantages, [7–10]:

- It is lighter. The weight saved can be used for more aluminum conductor. ACCC cable uses trapezoidal strands to fit more aluminum into the same cable diameter.
- Softer, fully annealed aluminum can be used for the conductors. ACSR cable uses harder aluminum which contributes to the cable's tensile strength but has about 3% less electrical conductivity.
- It has a much lower coefficient of thermal expansion (CTE) (1.6 ppm/°C) than ACSR (11.6 ppm/°C). This lets the cable be operated at a significantly higher temperature without excessive sag between poles.

While the main disadvantages are:

- The primary disadvantage is its cost; ACCC costs 2.5–3 time as much as ACSR cable.
- Although ACCC has significantly less thermal sag than even other HTLS conductor designs, its core is quite elastic and sags more than other designs under ice load, although a ULS (higher modulus) version is available at a cost premium.
- The conductor has a larger minimum bend radius, requiring extra care during installation.
- The conductor requires special fittings, such as splice and dead-end connections.

5 Financial Impacts

Financial aspects can be divided into the following categories, [7, 11, 12]:

- revitalization costs,
- line losses savings,
- financing the revitalization,
- impact on NTC values.

5.1 Revitalization Costs

The revitalization or reconductoring of ACSR overhead lines is the process of removing old conductors and equipment, in order to replace it with new HTLS equipment.

(A) cost of new conductors,
(B) cost of dismantling existing conductors,
(C) cost of installing new fittings, suspension and insulators,
(D) cost of anti-corrosion protection.

It is important to notice that all of these prices are country-specific and subject to a large number of changes and economic fluctuations. Take for example the price or new HTLS conductors, that are a dependent on the wholesale price of aluminum, as well as the stock exchange rate of US Dollars. The manufacture and delivery can take up to 4–6 months, which in the meantime, affects the final price.

5.2 Line Losses Savings

Besides the technical advantages, including the aspect of increasing throughput capacity and having a lower sag, by replacing a line with composite core conductors, another financial advantages are the decreased line losses.

According to line losses formula, I^2R, each line generates losses, depending on the current and the resistance. By using HTLS conductors, and therefore by decreasing the resistance of the line, a decrease of line losses occurs.

The planned lifetime of an overhead line, depending on different factors and evaluation criteria ranges between 30 to 50 years. Consequently, the cumulative line losses savings over a 30 year period, are an important subject when considering the revitalization.

To that end, the financial feasibility of line revitalization is justified by the savings in line losses over the planned lifetime of the line.

5.3 Financing the Revitalization

The cost of revitalization can be financed in different ways, from the TSO's own resources, credit funds, or as a combination of the above. In any case, a present discounted analysis has to be performed in order to assess the economic aspects of financing the revitalization.

Present Discounted Value, or simply Present Value (PV) is the value of an expected income stream determined as of the date of valuation. The present value is always less than or equal to the future value because money has interest-earning potential [13].

The present value for the assessment of OHL revitalization has to take account of the following financial parameters:

- interest rate,
- repayment period,
- annuities,
- discount rate,
- grace period.

5.4 Impact on NTC Values

Net Transfer Capacity (NTC) is the capacity available for commercial transactions. The NTC value is calculated as follows [14]:

$$\text{TTC} - \text{TRM} = \text{NTC} \tag{1}$$

where,
TTC – Total Transfer Capacity
TRM – Transmission Reliability Margin

In other words, NTC is the maximum allowed transaction possible between two neighboring countries. The capacity varies based on the period of the year, production, and consumption, but also on grid's infrastructure. The maintenance of grid elements, affecting the topology of the grid, can lead to a decrease of NTC values, while the investments in expanding the grid, such as new power lines, or reconductoring with new technologies can increase the NTC value.

However, based on experience, an extensive increase of NTC values can only be expected with the revitalization of significant lines on 220 or 400 kV networks.

The funds collected on the account of allocating capacity present a compelling item for budgeting of TSOs. Therefore, the revitalization of significant lines with HTLS conductors can also be justified and financed by NTC funds.

6 Example

In this chapter, an example of the above financial impacts is given. A cost comparison analysis is given for a 20 km overhead line 110 kV in Bosnia and Herzegovina. In this example, an existing ACSR OSTRICH (Al/Fe 150/25 mm^2), has to be replaced by line with a higher throughput capacity. Two possible options are:

- Building a new line using standard technology ACSR HAWK (Al/Fe 240/40 mm^2) using the existing corridor,
- Revitalizing the existing line, by replacing the conductors and suspension equipment with ACCC ROVINJ.

As it can be observed from Table 2, ACSR HAWK has a higher weight compared to ACSR OSTRICH, therefore existing towers and suspension equipment couldn't bear the weight. For that reason, a new line using ACSR HAWK would require new towers, but it would be possible to use the existing corridor.

Table 2. Comparison of conductors

	ACSR OSTRICH	ACSR HAWK	ACCC ROVINJ
Diameter (mm)	17.27	21.79	17.09
Aluminum area (mm^2)	152	241.7	187.8
Rated strength (kN)	53.8	86.7	71.1
Weight (kg/km)	615	976	576.4
AC resistance at 25 °C (ohm/km)	0.190	0.1198	0.1520
Ampacity (A)	490	640	921

On the other hand, ACCC ROVINJ has a lower weight than ACSR OSTRICH, allowing it to be used on the existing lines. However, because of ACCC technology, new fittings will have to be installed.

According to [15], the cost of building ACSR HAWK is 97.200 €/km (per three phases), while the dismantling of existing towers, foundations and conductors would cost 25.000 €/km. Therefore, the cost of building a new line using ACSR HAWK would sum up to 2.444.000 €.

The cost of conductor ACCC ROVINJ is 10.000 €/km (per one phase). In order to calculate the total length, one has to account for the following factors:

- geographical length: 20 × 3 km,
- sag,
- reserve.

In practice, the sag and reserve take for 10% of the geographical length. So, the total length of conductor ACCC ROVINJ would sum up to 66 km. Accordingly, the cost would be 660.000 €.

Besides the of the new conductor, additional costs are given in Table 3.

Table 3. Revitalization costs

Conductor	Cost of new conductor	Dismantling existing conductors	Cost of installing new fittings, suspension and insulators	Cost of anti-corrosion protection	Total
	€	€	€	€	€
ACCC ROVINJ	660.000	525.000	180.000	230.000	1.595.000

In the following table, the costs for both options are given below:

Table 4. Investment costs

Investment	Investment cost
	€
New ACSR HAWK	2.440.000
Revitalization with ACCC ROVINJ	1.595.000

Therefore, it can be seen that the revitalization with HTLS conductors is about 35% cheaper than building a new line, while at the same time increasing the throughput capacity for more than 34% (according to Table 2).

As for the line loss savings, according to the CTC Global Comparison Software, the following load and energy assumptions are used (Table 5).

Table 5. Comparison assumptions

Assumption	Value
Line length	20 km
Voltage	110 kV
Total peak operating Amps	300 A
Load factor	20%
Active power losses price	48,6 €/MWh

It is important to note that the active power losses prices are is formed as a market price, based on offer and demand. In this case, the price of 48,6 € was taken as reference in this example, base on the average losses price for Bosnia and Herzegovina in 2016, [16].

After performing the calculations, the following results are obtained (Tables 6 and 7).

Table 6. Comparison results - yearly

Conductor	Yearly losses		
	MWh	€ (48,6 €/MWh)	%
ACSR OSTRICH (base conductor)	648	31.492,8	100
ACSR HAWK (option 1)	401	19.488,6	61,9
ACCC ROVINJ (option 2)	514	24.980,4	79,3

In other words, over a 30 year period, compared to the base conductor, ACSR HAWK would earn savings in the amount of 360.126 €, while the savings for ACCC ROVINJ would amount to 195.372 €.

Table 7. Comparison results – 30 years

Conductor	30 years losses	
	MWh	€ (48,6 €/MWh)
ACSR OSTRICH (base conductor)	19.440	944.784
ACSR HAWK (option 1)	12.030	584.658
ACCC ROVINJ (option 2)	15.420	749.412

Table 8. Final comparison results

Option	Conductor	Investment cost	30 year savings	Total cost
		€	€	€
1	ACSR HAWK	2.440.000	360.126	2.079.874
2	ACCC ROVINJ	1.595.000	195.372	1.399.628

By taking into account the investment costs from Table 4, and the calculated 30 year savings, it can be observed that the most profitable option is ACCC ROVINJ (Table 8).

7 Conclusion

The focus of this paper is to present, summarize and categorize the financial aspects related to the reconductoring of standard OHL with technologically advanced composite core conductors. The financial impacts are divided into several categories, including; revitalization costs, line losses savings, financing the revitalization, and finally the impact on NTC values. Within the revitalization costs, the most common expenditures are given for reconductoring projects. An important expense for the TSOs are line losses, and within the Line Losses Savings chapter, it is explained how HTLS conductors can help create savings cause by line losses. The financing aspects of the reconductoring are explained next and at the end, it is explained how revitalization with HTLS conductors can potentially help increase the capacity for commercial transactions.

In the last chapter, a detailed example of a reconductoring project is given, defining all relevant costs and items needed to replace an existing ACSR OSTRICH line with the adequate ACCC conductor. An extensive cost comparison is given, proving and justifying the need to invest in the advanced ACCC technology.

References

1. Hadzimuratovic, S., Fickert, L.: Impact of gradually replacing old transmission lines with advanced composite conductors. In: IEEE ISGT Conference Proceedings, Sarajevo (2018, unpublished)
2. Migliavacca, G.: Advanced Technologies for Future Transmission Grids. Springer, Heidelberg (2013)

3. CIGRE: Results of the questionnaire concerning high temperature conductor fittings Task Force B2.11.032004
4. CIGRE: Considerations relating to the use of high temperature conductors Technical Brochure 331 (2007)
5. Cole, S., Van Hertem, W. Meeus, L.: Technical developments for the future grid. In: Proceedings of International Conference on Future Power, p. 6, 16–18 November 2005
6. Southwire Company: Southwire Overhead Conductor Manual-Introduction to bare overhead conductors OCM (Overhead Conductor Manual) Book CP 1–11. http://www.southwire.com/
7. Techno-economic analysis of the application of the ACCC conductor on the 110 kV OHL Sinj - Dugopolje – Meterize, Energovod (2014)
8. ACCC Engineering Manual, CTC Global Corporation (2011)
9. Kenge, A.V., Dusane, S.V., Sarkar, J.: Statistical analysis & comparison of HTLS conductor with conventional ACSR conductor. In: International Conference on Electrical, Electronics, and Optimization Techniques (ICEEOT) (2016)
10. Alawar, A., Bosze, E.J., Nutt, S.R.: A composite core conductor for low sag at high temperatures. IEEE Trans. Power Deliv. **20**(3), 2193–2199 (2005)
11. Techno-economic analysis of the selection of conductors on OHL 2 × 110 kV Meterize – Dujmovača – Vrboran, Energy Institute Hrvoje Požar & Dalekovod Project Zagreb, January 2016
12. Techno-economic analysis of the selection of conductors on OHL 110 kV Obrovac - Zadar, Energy Institute Hrvoje Požar & Dalekovod Project Zagreb, April 2016
13. Sekhar C.: CAPITAL BUDGETING Decision Methods: Payback Period, Discounted Payback Period, Average Rate of Return, Net Present Value, Profitability Index, IRR and Modified IRR Theory and Data Interpretation Series. Independently Published (2018)
14. Kosarac, M., Carsimamovic, A., Turajlic, H., Vujovic C.: A practical example of increasing net transfer capacity (NTC) by modifying the power system topology. In: Cavtat Conference (2014)
15. Elektroprenos BiH: Prices of goods, works, services and normative standards (2014)
16. Independent System Operator in Bosnia and Herzegovina: Report on auxiliary services and balance market report in Bosnia and Herzegovina for 2016 (2017)

Energy Efficiency Evaluation of an Academic Building – Case Study: Faculty of Electrical Engineering, University of Sarajevo

Amna Šoše[1], Tatjana Konjić[2,3], and Nedis Dautbašić[3](✉)

[1] Systech doo, Sarajevo, Bosnia and Herzegovina
[2] Faculty of Electrical Engineering, University of Tuzla, Tuzla, Bosnia and Herzegovina
[3] Faculty of Electrical Engineering, University of Sarajevo, Sarajevo, Bosnia and Herzegovina
nedis.dautbasic@etf.unsa.ba

Abstract. One of the main sector important for energy efficiency is building. Results of a pilot study related to energy efficiency in an academic building were summarized in this paper. The pilot study was undertaken at the Faculty of Electrical Engineering, University of Sarajevo. Ten necessary steps of developing energy efficiency improvement project were defined. Data of water, heating and electricity consumption and theirs costs were analyzed. Special attention was paid on 15-min load data for the case-study building to identify key characteristics of electricity use. Identification of potential energy savings and improvement of energy efficiency are considered after detailed analyses of available and collected data.

1 Introduction

The energy sector of countries from all around the world is faced with the challenge of sustainable development, i.e. the development which enables steady supply of energy, while at the same time reducing negative influences (e.g. global warming, climate change, etc.). Due to the apparent decrease of current supplies of conventional energy sources and having in mind the increasing prices and energy usage per capita, it can be concluded that the concept of sustainable energy usage, rational planning and the increase of energy efficiency of all elements within the energy sector has a top priority. The concept of sustainable energy development and the impact of energy efficiency in the future creation of economic and political plans play an important role. In this context, the focus is placed on finding clean energy sources and creating the notion of properly energy usage.

Energy efficiency is a process which does not end with implementing measures of improvement; it continues through tracking and confirming planned savings, finding new processes and implementing new measures of further energy efficiency. This is a process which is continuous and cyclical and it leads to systematic and continuous energy management. It is important to state that energy efficiency should not be observed as saving energy. Savings presume giving up certain amounts of energy,

S. Avdaković (Ed.): IAT 2018, LNNS 59, pp. 198–210, 2019.
https://doi.org/10.1007/978-3-030-02574-8_16

whereas efficiency means keeping necessary functionalities while using smaller amounts of it. Efficient energy usage never endangers work and living conditions. Energy efficiency presumes efficient energy usage in all sectors: industries, traffic, services, agriculture, construction and general in objects we use daily [1].

Among all above-mentioned sectors energy efficiency in buildings is complex and very important issue. According to Gul and Patidar energy consumption in non-domestic buildings, causes approximately 19% of the total UK CO_2 emissions [2]. Percentages of energy consumption in buildings are pretty high if compared to other economic sectors and it varies from country to country. According to information presented in [3] and [4] energy use in buildings makes about 30–45% of the global energy demand.

Energy management [5] represents a method to control and decrease energy consumption used by any organization which in the end enables the reduction of costs of energy consumption and the costs of harmful gas emission taxes, by reducing the emission of harmful gases and the impact they cause, and also reducing the risk of problems in supply.

Energy management and tracking the consumption and other energy indicators is one of the main elements in supplying sufficient levels of energy efficiency of a certain building.

Energy characteristics of buildings can be assessed by two models in two different ways:

- Direct model – a traditional approach which is based on theoretical calculations,
- Inverse model – a modern approach which is based on real-time data of the physical characteristics of the building and energy consumption collected by modern equipment for monitoring and measurement [6].

Pilot study of the energy efficiency in the Faculty of Electrical Engineering building is presented in the paper.

The paper has five chapters. Section 2 is related to legal regulations of energy efficiency in Bosnia and Herzegovina (BiH) while also taking into account regulatory basics of the European Union. Basic theoretical background of the process of enhancing energy efficiency in buildings is described in Sect. 3. Available information concerning energy consumption with special review on electricity consumption is analyzed and given in Sect. 4. Conclusions and recommendations for future researches are found in Sect. 5.

2 Regulatory Frameworks for Energy Efficiency in Bosnia and Herzegovina

Energy efficiency in the European Union has become a serious point of interest and research in 2006 with the introduction of the Directive of the European Union on energy and use efficiency and energy services. In 2010, two more directives have been announced: the Directive of the European Union on energy performance of buildings and the Directive of the European Union on labelling of energy-related products. The Directive of the European Union on energy efficiency of the European Union from 2012 has further sparked interest in this field. This Directive sets up rules or a set of bounding measures with a purpose to remove obstacles in the energy sector and improve efficiency in energy supply and consumption. It also sets up a framework of national goals of increasing energy efficiency in the European Union by 20% until 2020 [7].

Since energy efficiency has a pivotal role in the development strategy of the European Union until 2020, the membership of Bosnia and Herzegovina in the Energy Community of South East Europe (since July 1st 2006) [8], and signing the Memorandum of Understanding related to the cooperation with donors active in the field of energy efficiency in Bosnia and Herzegovina (since December 18th 2012) [9], European Union intends to provide support and track the progress of the development of Bosnia and Herzegovina. However, insufficient coordination between entity and state-level bodies, lethargy in making and implementing appropriate and necessary decisions and acts, the lack of adequate institutional framework on the state level, weak capacities and resources for development and investments; represent just a few obstacles in the booming, steady and rapid development and progress of the energy sector in Bosnia and Herzegovina.

Currently, a legal framework for energy efficiency does not exist on the state level, while on the entity level there are two laws on energy efficiency: one in Federation of Bosnia and Herzegovina passed on March 24th 2017 [10] and the other in Republika Srpska on July 15th 2013 [11]. These laws contain all segments of energy efficiency, as well as measures and activities which are required. Passing these laws was based on study documents and relevant strategies of developing energy efficiency [12].

By incorporating energy efficiency and the usage of sustainable energy sources in strategies of energy development and environmental protection, as well as passing these laws, Bosnia and Herzegovina has harmonized its legal framework with the guidelines of the European Union and has committed itself to fulfil these activities. Directives of the European Union thus represent the starting ground for the development and progress of Bosnia and Herzegovina.

3 Project of Energy Efficiency Improvement - Theoretical Background

According to information and the experience of authors presented in [1, 7, 12–16], it is possible to define ten necessary steps in developing a complete energy efficiency improvement project of a certain building. At the beginning, it is very important to define the project assignments with clearly stated goals and expected results. The results will depend on the quality of available and collected data on the physical state of the building, means and the intensity of energy consumption and ultimately the proper analysis of the data. Identification of potential energy savings and improvement of energy efficiency would be able to consider after detailed analyses of available and collected data. Results of the analyses always can lead to applications of some energy efficiency measures. Energy efficiency measures can be divided in three main categories, depending on the costs and the complexity of measures. *Measures of energy efficiency with low investment* (also named as measures without investments or measures of managing energy) are related to unconscionable behavior of building users and eventually improving regular maintenance of the building. It is simple to identify them and the period of payments of invested funds can last from a couple of days to two years. *Measures of energy efficiency with medium investments* mostly tend to revolve around replacing inefficient heating boilers and burners, installing automatic steering systems, energy efficient lighting, systems for heat recuperation, frequency regulators and solar systems, reconstructing heat distribution systems, etc. The payment period usually last from three to five years. *Measures of energy efficiency with high investments* are implemented in projects with high costs and larger investments in increasing the energy efficiency of street lights, installing cogeneration plants, energy remediation of buildings and their energy systems, etc. The payment period is longer than the first two mentioned measures.

Since the project of improving energy efficiency of a building is cyclical and continuous, it does not end by adopting and implementing these measures, instead it continues with further monitoring of energy consumption while also identifying potentially new ways of saving energy. All the steps can be found in Fig. 1. Following detailed explanations of all steps defined in [12], we can divide steps in the general approach (from step 1 to step 5) and detailed approach (from step 6 to step 10).

4 Case Study: Faculty of Electrical Engineering

An energy efficiency pilot study of the Faculty of Electrical Engineering, University of Sarajevo is presented in this chapter. The pilot study includes implementations of the general approach - steps from 1 to 5 presented in Fig. 1. Available and collected energy data of water and heating with special consideration of electricity consumption and load have been analyzed. In accordance to analyzed data potential energy savings and energy efficiency improvements were identified.

4.1 General Approach

According diagram of detailed steps of energy efficiency project development presented in Fig. 1, general approach of the project has been applied in the case of pilot study of energy consumption of the faculty.

Defining the Project Task: The project task is to collect: data of the building's physical state, date of energy consumption (water, heating, electricity) and costs as well as date of daily load diagram. Base on collected and analyzed data the project proposal should lead to improving the building's energy efficiency.

Defining Project Aims and Results: The goal of the project is to propose a set of measures for improving energy efficiency in the building. The desired result of the project is to have more insight in potential measures of improving the energy efficiency of the Faculty's building, mainly with the goal of financial savings.

Energy Overview of the Building and Gathering Necessary Data for Further Analysis: Following data are collected: the physical characteristics of the building,

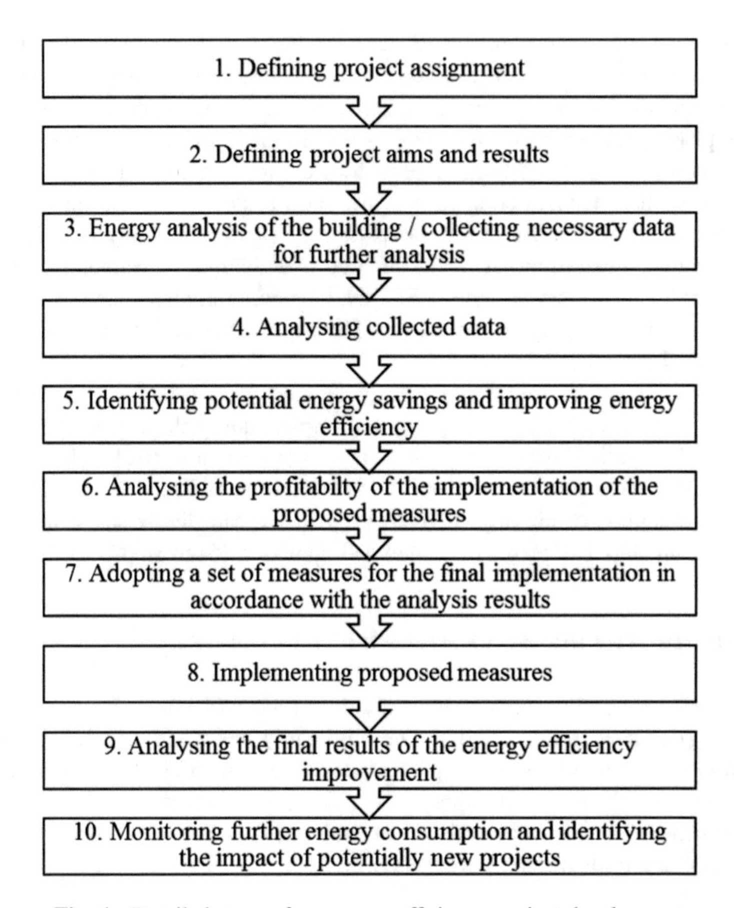

Fig. 1. Detailed steps of an energy efficiency project development

consumption and costs of water, heating and electricity from 2014 to 2016 on monthly level, as well as date related to daily load from January 1st 2016 to August 27th 2016.

Analysis of Collected Data: Collected data in the previous step are the basis for further analysis of consumption and costs related to water, heating and electricity. Detailed overview of collected data can be found in [12].

Identifying Potential Energy Savings and Improving Energy Efficiency: Based on the analyses, possible energy efficiency measures are identified and proposed.

4.2 Collected Data

The basis for collecting all above-mentioned data, given in [12], was a questionnaire "The study on energy efficiency of 300 public buildings in Canton Sarajevo" [14] published in June 2016. In Table 1 main physical characteristics of the Faculty's building can be found.

Table 1. Data on the physical characteristics of the Faculty of Electrical Engineering building

Building characteristics
Build type - classical
Floor numbers (without the ground floor) - 3
No basement, no attic
Stair type – concrete and ceramics
Roof type – hip; Roof material – sheet metal
Outer wall type – brick (52 cm)
Partition wall type – plasterboards + brick
Window glazing - 2
No thermal isolation on outer walls
No thermal isolation on roof or ceiling
No thermal isolation on the ground
Heating
Central
Number of air conditioners (for heating and cooling) - 47
No solar panels
Number of hours of heating per day - 10
Size of the heated space - 5800 m^2
Height of rooms in the heated space – 2.8m
Security
Video surveillance – number of cameras - 28
Fire protection - alarm + fire extinguishers
Lightning protection system
Lighting
Fluorescent lamps 20W - 184
Fluorescent lamps 85W - 10
Reflector 150W - 3
Number of working hours per day - 8 (approximately)

Water Consumption for Period 2014–2016: Based on the available bills for consumed water, monthly water consumption and its costs for 2014–2016, is presented in Figs. 2 and 3, respectively.

Heating Energy Consumption for Period 2014–2016: Based on the available data of heating energy consumption from issued bills, it is possible to showcase the amount of the consumed energy, as well as related costs, as it is shown in Figs. 4 and 5, respectively.

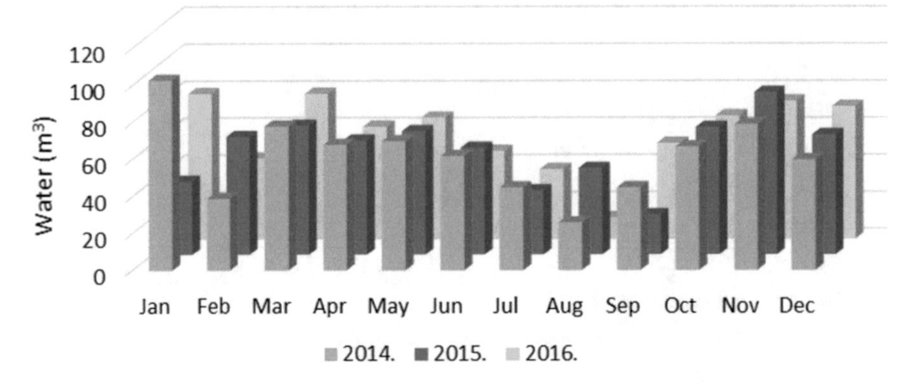

Fig. 2. Total monthly consumption of water for period 2014–2016

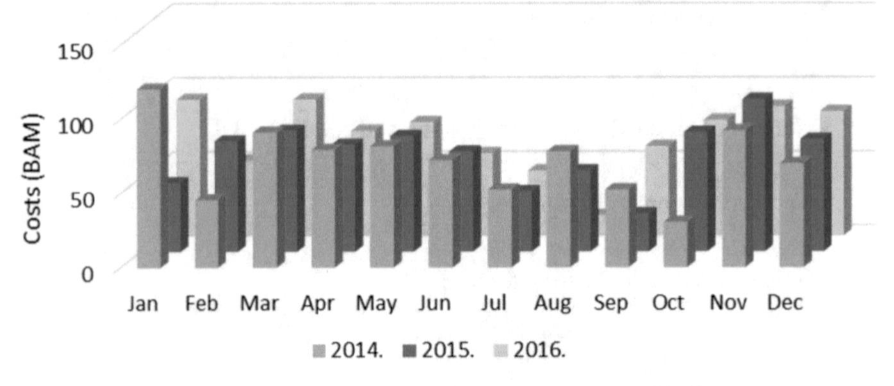

Fig. 3. Total monthly costs of water for period 2014–2016

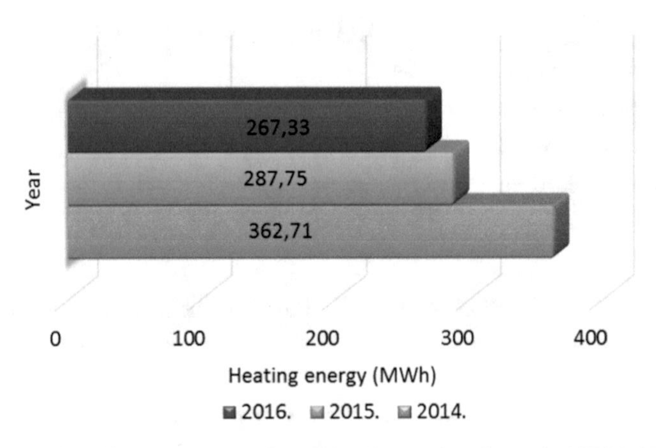

Fig. 4. Total yearly consumption of heating energy for period 2014–2016

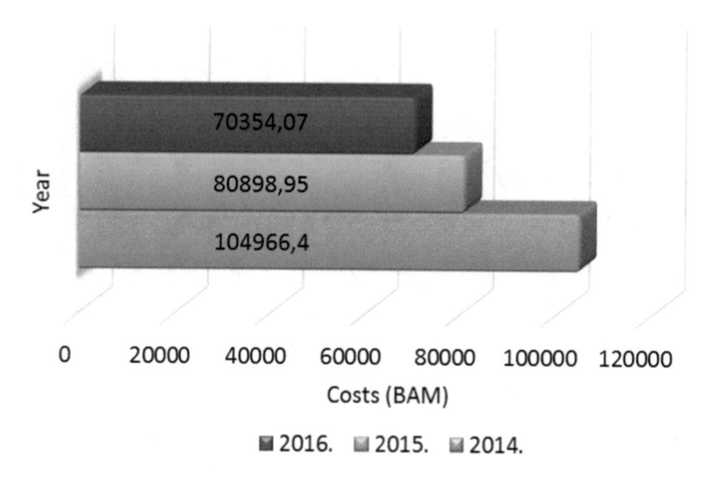

Fig. 5. Total yearly costs of heating energy consumption for period 2014–2016

Electrical Energy Consumption for Period 2014–2016: Based on the available data of electricity consumption from issued bills, it is possible to showcase the electricity consumption, power engaged and costs. Total monthly electricity consumption is presented in Fig. 6 for period 2014–2016.

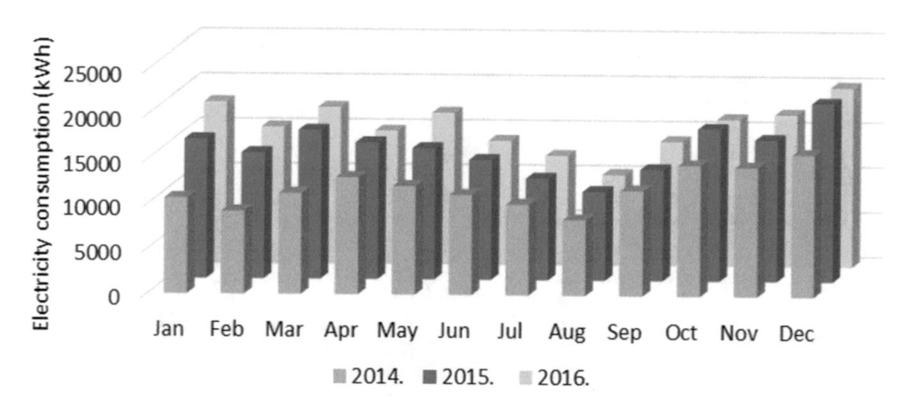

Fig. 6. Total monthly consumption of electrical energy for period 2014–2016

Based on the analyzed data of electricity consumption and the peak power for period 2014–2016, Fig. 7 contains data on monthly costs.

In addition to previously data, 15 min load were also available for the period from January 1st to August 27th, 2016. These data provide us with more detailed knowledge about customer habits of electricity use in the building related to the different activities and their intensity during the school year per months and per day. Figure 8 shows the representative daily load diagrams for January 2016 and August 2016.

Peak power and minimum power for observed period (January–August 2016) on a monthly basis is presented in Fig. 9.

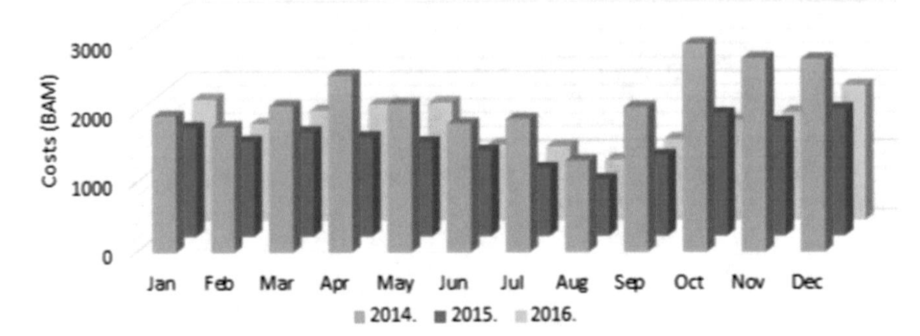

Fig. 7. Total monthly costs of electricity consumption for 2014–2016

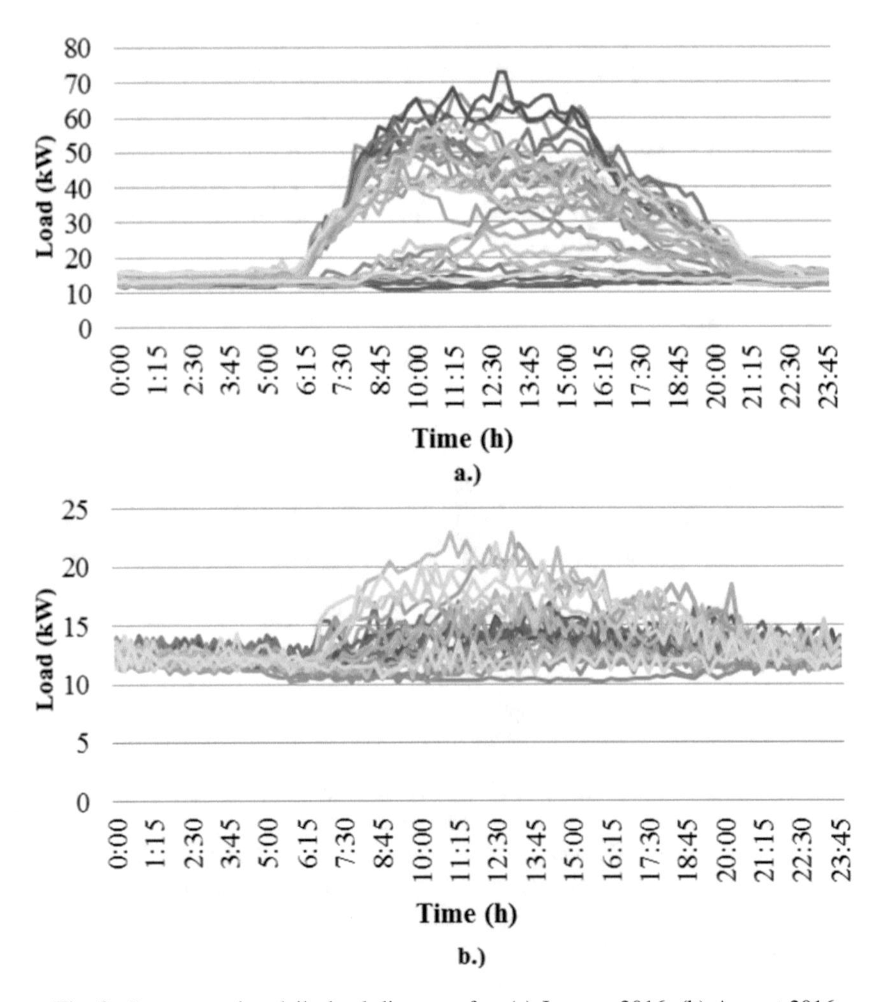

Fig. 8. Representative daily load diagrams for: (a) January 2016, (b) August 2016

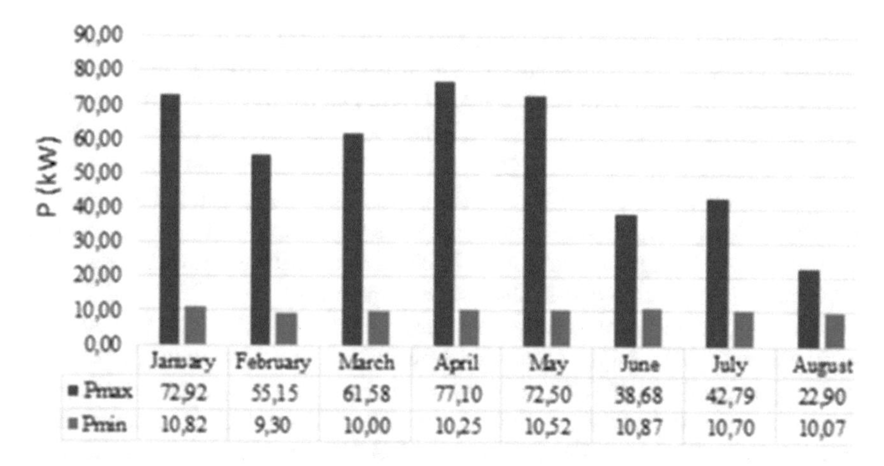

Fig. 9. Pmax and Pmin for the period 01.01.–27.08.2016.

4.3 Analysis of Collected Data

Total annual consumption of water and heating (Figs. 2 and 4, respectively) and their total annual costs (Figs. 3 and 5, respectively) are higher in 2014 compared to 2015 and 2016. The reason for higher consumptions of water and heating is reconstruction of the faculty's building that is finished at the beginning of 2014. It is possible to notice that the consumptions are stable and similar in next two years (2015, 2016).

By analyzing data on electricity consumption, it is possible to notice essential parameters that characterize energy consumption within facility under consideration. Based on daily load diagrams shown in Fig. 8 (shown in details in [12]) it is possible to notice that working time of facility is from 08:00 until 21:00 o'clock, with the maximum activity being expressed in the period from 10:00 until 17: 00 o'clock. During this period, daily load diagrams display higher load value compared to other periods of the day. In the evening and during night, consumption is mostly constant, which corresponds to base consumers (consumers that constantly use electricity, such as servers, necessary indoor and outdoor lights, electronic devices working in standby regime, etc.) [12].

Based on load diagram it is possible to notice that average base load of active power is about 12 kW. If minimum base load of 10 kW is adopted on monthly level, this implies about 7200 kWh of base consumption, which, considering the electricity prices (high tariff - 0.0481 BAM/kWh or low tariff - 0.0240 BAM/kWh) gives sizeable amounts of costs only for base consumption. Monthly amount of base consumption costs is obtained in the way that half of the base consumption is calculated by low tariff, and the other half by high. In total, this amounts to around 260 BAM per month, or 3120 BAM per year.

If additional attention paid to the results of electricity consumption and costs for the observed period 2014–2016 (Figs. 6 and 7, respectively), it is possible to notice that electricity consumption was the lowest in 2014, while largest costs are present in the year. Explanation for this phenomenon lies in the fact that different tariff charging systems were used during the observed period. Unit price for high tariff was 0.1452 BAM/kWh in 2014, and unit price for low tariff was 0.0726 BAM/kWh while these unit prices were for high tariff 0.0481 BAM/kWh, and for low tariff 0.024 BAM/kWh in 2015 and 2016. In addition, unit fee for rated peak power was 14.49 BAM/kWh in 2014, and it was 14 BAM/kWh in 2015 and 2016. Due to the fact that unit fees, for both high and low tariff in 2014, were three times bigger compared to the fees in 2015 and 2016, it is reasonable to see that the final costs are higher, regardless of the fact that smaller amount of energy was consumed in 2014. In addition to the mentioned facts, it is important to emphasize that in the observed period of 2014, 2015 and 2016, the increase of connected power was made twice, for the first time on 21st of July 2014, from 50 kW to 65 kW and the second time on 23rd of June 2016 to the final approved connected power of 77.1 kW. These increases were due to the evident exceeding of the peak power defined by the previous Connection Contract and Electrical Energy Agreement signed between the Faculty of Electrical Engineering as consumer and Public Enterprise Electric Power Company BiH as an authorized distributor of electricity in Bosnia and Herzegovina.

4.4 Identifying Potential Energy Savings and Energy Efficiency Improvements

Based on the analysis of collected data, the first-instance conclusion is made on potential sites of excessive energy consumption and possible measures are enumerated that would lead to the improvement of energy efficiency of the facility in global. First of all, the proposal is to pay attention to the amounts of base consumption and to try to implement certain measures to eliminate this consumption. Specific measures that could be taken are:

1. educating users of the faculty's building (employees and students) and raising awareness about rational use of energy in the facility - low investment measures;
2. place billboards on the importance and instructions on implementing energy efficiency measures at sight locations - low investment measures;
3. detailed analysis of consumption caused by lighting in the building - replacement of existing lighting fixtures with lighting fixtures of greater efficiency and longevity - medium investment measures;
4. utilization of renewable energy sources (e.g. solar energy, i.e. specifically, installation of a solar power plant and connection of its production directly to the faculty building system) - high investment measures;
5. installation of diesel powered aggregates - high investment measures;
6. analysis of the possibility of using new technologies for the purpose of energy production in the facility (e.g. use of piezoelectric materials and methods for energy production) - experimental procedure.

Any of these proposed measures requires a detailed technical-economic analysis, which implies making of feasibility study and cost-effectiveness of the proposed measures implementation. Only after elaborate technical and economic studies, final conclusions can be made for or against the implementation of a certain measure. The proposed measures are based on electricity, its utilization and consumption in the facility, and detailed explanation can be found in [12]. In addition to these measures, it is also possible to consider saving of energy in the building by various other measures, such as reconstruction of the existing physical condition of facility (reconstruction of windows, roof and walls), installation of its own heating system, installation of hot water preparation and consumption system, etc.

5 Conclusion

Some parts of the pilot study related to energy efficiency evaluation of the academic building are presented in the paper. The pilot project undertaken at the Faculty of Electrical Engineering, University of Sarajevo. According to analyzed date of water and heating consumption in the building during three years it is possible to notice that the consumptions are stable and similar with exception of few months during the building's reconstruction. Main attention paid on analyses of electricity consumption and daily load profile that provided information how the building is used, operated and managed on a daily basis. The analysis of consumption and daily load patterns of the building indicated more intensive use in December, January, April and May when teaching and work in laboratories is in full swing in comparison to June, July and August when most of the students are out of the building preparing for exams. High difference between maximum and minimum load were visible especially during above mentioned months. Minimum load is mainly contstant and similar no metter of a day in a week or a month. According to obtained results and other researches in the building a space for application of energy efficiency measures was noted. Low investment measures could be applied immediately and after a detailed technical-economic analysis should be prepared projects for applications of medium and/or high investment measures.

In case of further development of energy efficiency improvement project it is necessary to make detailed energy audit of a building, collect detailed information about physical condition and characteristics of the object, identify electrical devices in use (number of computers, air conditioning devices, light equipment, laboratory equipment, etc.), as well as collecting measurement data about energy consumption for longer period of time. According to gathered data, detailed analyses and adequate conclusions about potential of energy efficiency should be made, with special emphasis on techno-economic analysis of profitability of investments.

Bosnia and Herzegovina has a lot of potential and opportunities for energy efficiency improvements across the entire territory of the country, while not much effort has been given to this subject in past, nor aspects of energy efficiency and environmental pollution were considered as serious as they are. Therefore, energy efficiency presents one of the major development opportunity for Bosnia and Herzegovina, an opportunity which can bring up many job openings and improvements to life standard of its citizens. But, besides opportunities, there is a serious obstacle to development and progress in the field of energy efficiency and energy sector in general caused by political disagreement at the state level. However, more attention and serious consideration will be devoted to energy efficiency and the energy consumption issues, due to Bosnia and Herzegovina's efforts to become a member of the European Union.

References

1. The International Energy Agency: Energy efficiency – market report 2017 (2017). http://www.iea.org/publications/freepublications/publication/Energy_Efficiency_2017.pdf
2. Gul, M.S., Patidar, S.: Understanding the energy consumption and occupancy of amulti-purpose academic building. Energy Build. **87**, 155–165 (2015)
3. Vilnis, V.: Energy Management Principles and Practice. BSI, London (2009)
4. Asimakopoulos, D.A., Santamouris, M., Farrou, I., Laskari, M., Saliari, M., Zanis, G., Giannakidis, G., Tigas, K., Kapsomenakis, J., Douvis, C., Zerefos, S.C., Antonakaki, T., Giannakopoulos, C.: Modelling the energy demand projection of the building sector in Greece in the 21st century. Energy Build. **49**, 488–498 (2012)
5. Pout, C.H., MacKenzie, F., Bettle, R.: Carbon Dioxide Emissions from Non-domestic Buildings: 2000 and Beyond. BRE Energy Technology Centre, Watford (2002)
6. Morvaj, Z., Sučić, B., Zanki, V., Čačić, G.: Priručnik za provođenje energetskih pregleda zgrada, USAID Ekonomija energetske efikasnosti/3E, UNDP BiH, GIZ Konsultacije za energetsku efikasnost, BiH (2011)
7. Directive 2012/27/Eu of The European Parliament and of The Council, of 25 October 2012, on energy efficiency, amending Directives 2009/125/EC and 2010/30/EU and repealing Directives 2004/8/EC and 2006/32/EC
8. Annual Report of the Energy Community Regulatory Board, Energy Community (2007)
9. http://www.capital.ba/eu-usaid-undp-i-giz-potpisali-memorandum-o-razumijevanju/. Accessed 20 Dec 2017
10. Zakon o energetskoj efikasnosti u Federaciji Bosne i Hercegovine, Službene novine Federacije BiH, 24 marta 2017
11. Zakon o energetskoj efikasnosti Republike Srpske, Službeni glasnik Republike Srpske, 15 juli 2013
12. Šoše, A.: Energetska efikasnost - prijedlog projekta poboljšanja energetske efikasnosti na primjeru objekta Elektrotehničkog fakulteta. Master (M.Sc.) thesis, Faculty of Electrical Engineering, Sarajevo (2017)
13. Bukarica, V., Dović, D., Hrs Borković, Ž., Soldo, V., Sučić, B., Švaić, S., Zanki, V.: Priručnik za energetske savjetnike, Program Ujedinjenih naroda za razvoj (UNDP) u Hrvatskoj (2008)
14. Energetska efikasnost u javnom i stambenom sektoru, Razvoj i provedba projekata, Tim projekta USAID Ekonomija energetske efikasnosti, Sarajevo (2014)
15. Priručnik za energetsko certificiranje zgrada, Program Ujedinjenih naroda za razvoj – UNDP, grupa autora, Zagreb (2010)
16. Priručnik za energetsko certificiranje zgrada – Dio 2, Program Ujedinjenih naroda za razvoj – UNDP, grupa autora, Zagreb (2012)

Fault Identification in Electrical Power Distribution System – Case Study of the Middle Bosnia Medium Voltage Grid

Jasmina Čučuković(✉) and Faruk Hidić

JP EP BiH d.d. Sarajevo – Podružnica ED Zenica, Ivana Gundulića bb, 72000 Zenica, Bosnia and Herzegovina
{j.cucukovic, f.hidic}@epbih.ba

Abstract. Power system fault localization is immensely important factor towards faster fault removal and quick comeback to function with minimal disruptions. Thus, it can be achieved less electric power equipment straining and better customer satisfaction. Power lines are one of the most important components in every power system. Overhead lines are exposed to environmental influence and their possibility of failure is much higher compared to other system components. The period of time needed for locating the fault significantly affects the electrical power quality. This study presents some of the practical experiences of fault locating, based on fault information from Disturbance Recorder and the PowerCAD fault calculation, used in lack of sophisticated tools and module for locating the faults, in some parts of electric power system of Bosnia and Herzegovina, real MV distributive systems of Maglaj Municipality (System 1) and Olovo Municipality (System 2). The results of fault analysis for mentioned areas show that this way faults can be located with high accuracy, in order to take quick and effective actions of their removing and re-establish normal power supply.

1 Introduction

Distribution electric power lines are significant component of every electric power system and their protection is needed for establishing system stability and minimizing the damage in case of failure. Overhead distribution network is noticeably exposed to environmental influence and weather conditions so their error and failure possibility is generally higher compared to the other components of the system.

The main task of electric power system is to reliably, safely and economically supply customers with electrical energy. Customers demand continuous and stable electrical energy supply. However, electrical energy customers, mainly because of their position in power supply system, suffer power supply disruption due to faults in any part of electric power system. Research has shown that 94% of faults, which cause power supply disruption, happens in distributive system. Faults can be caused by lightning bolt, wind, ice, snow, birds, insulation breakthrough, fault, equipment ageing, overheating, etc. [1, 2].

S. Avdaković (Ed.): IAT 2018, LNNS 59, pp. 211–223, 2019.
https://doi.org/10.1007/978-3-030-02574-8_17

The most common failure of electric power systems are faults. In three-phase system we can count four types of faults: three-phase, two-phase, two-phase with line to ground fault and one-phase fault. Fault in system is generated when electrical straining of isolation gets higher than isolation dielectric strength. The main cause of faults is usually voltage increase which affects insulation (overvoltage occurrence), decrease of insulation resistance (e.g. lines contact due to their swinging) or increasing voltage and decreasing insulation resistance at the same time. Considering the duration of fault, we separate the causes of it in these groups:

- Transient – overvoltage, insulation weakening, contact due to lines swinging, bridging with foreign combustible body (bird, branch).
- Permanent causes – insulator breakage, insulator pollution, fall one line to another or to grounded part and foreign body bridging that does not burn if it falls on a line.

Momentary cause in a place of fault creates electric arc which maintains it's existence even after the disappearance of it's cause and all the way through until the line is not turned off. After line exclusion, arc extinguishes, and with restart arc would not show again. The appearance of fault in system affects few parameters and has these unwanted results:

- Strong dynamic stringing between lines can be caused due to high initial values of fault current
- Fault current heats up the lines. That is the reason why the line has to be excluded before the appearance of thermal straining and destruction of line
- Fault currents can cause high-level disturbances in telecommunication devices and occurring of life threatening overvoltage
- Near fault, the voltage "breaks" – it decreases to the level that most of the devices (e.g. machines) are turned off by undervoltage protection
- If the fault current runs through soil, at place of fault (line and ground compound) very dangerous potential differences can occur [3].

There are two ways of increasing electrical energy power supply reliability. One of them is to decrease the number of power supply disruption during the year, which is directly connected to new investments and modification of system configuration. The other way is to decrease the duration of power supply disruption. One of the mechanisms for decreasing is malfunction locator implementation. They are supposed to give the information to operator in case the fault current runs through the line which carries the locator or it did not. The duration of fault place disconnection is decreasing by receiving this information, and along with that the duration of power supply disruption of customers who are not affected by this malfunction decreases. Fault locators can be designed the way that they show information about fault current locally or to send it to operator in SCADA (Supervisory Control and Data Acquisition) system [2].

Some of the approaches and methods of fault identification in electrical power system are presented in [4–13].

Authors in [4] analyze application and usefulness of two methods for determining fault locations and their distance to the reference end buses connected by the faulted transmission line. Two used methods are referred to as impedance-based and traveling wave methods. It is found that Impedance-based methods are easier and more widely used than traveling-wave methods. The percentage error due to traveling wave method was zero, but impedance based method had the highest error of 3.91%. The traveling wave method is accurate and can calculate the fault location within a couple of seconds after a fault. But to monitor the wideband transient signal and to process such signal to locate time, requires expensive technological tools. Impedance based method is known to be simple and low cost. It does not require the communication channel to exchange information between relays. Hence, this method is not suitable for fault locating in EHV and UHV where faults are cleared in less than two cycles. This method also causes issue for series compensated line.

An overview to fault location methods in distribution system based on single end measures of voltage and current is given in [5]. The paper presents some of the most relevant methods for fault location in radial power systems. Fault location techniques in distribution systems are classified in four categories: the classical approaches that use fundamental voltages and currents, techniques based in travelling wave theory, approaches based on topological methods and those knowledge-based approaches. This paper presents a review of the classical techniques and knowledge based approaches and also proposes an hybrid approach based on both. Additionally here is presented an hybrid fault location algorithm which takes advantage of both, the algorithmic and the knowledge based methods. The obtained results from fault location methods help utilities in both network operation and network planning.

The comparison of two electric fault location techniques (Ratan Das and Saha) is presented in [6]. Using prefault and fault data (voltages and currents) and a model of the network, the algorithms give an approximation of the actual fault location. Both techniques were implemented on simple lines and tested through simulations. The results obtained, showed that Ratan Das algorithm was giving better approximations to the actual fault. On a second step the algorithms were implemented on a real distribution network. The results also showed that Ratan Das algorithm give better results and is less sensible to fault impedance. An application with a graphical user interface has been created to execute the methods, but also to execute the necessary previous steps: obtain the phasors of the fault, either through a simulation or loading actual registered faults, graphical representation of waveforms and phasors… In conclusion, the Saha method is easy to implement but an evaluation of the fault resistance is needed, and this let the method to be more sensible at uncertainties.

[7] provides a comprehensive review of the conceptual aspects as well as recent algorithmic developments for fault location on distribution system. Several fundamentally different approaches are discussed in the paper together with the factors affecting the assumptions of the underlying concepts and the various criteria used in the different approaches are reviewed. Fault locating strategies using impedance based methods, travelling wave based methods and knowledge-based methods have been reviewed. Most of the fault location techniques discussed have some limitations. Some of them are as follows: The iterative fault location algorithm are generally time consuming and always has the constant risk of running into a diverging solution; Heuristic procedure may take large number of trials and also time in practical distribution systems before identifying the fault location and restoring the power supply to healthy part of system; Almost it requires voltage and current measurements from all the nodes and branches in order to detect the fault location. It is concluded that in comparison

between all methods, knowledge based method seemed to have more accuracy and speed and less cost.

Authors in [8, 9] explore and compare different fault location techniques. Specifically, those studies are focused on impedance-based methods. On the one hand, one-end methods have been analyzed, the main advantage being less deployment needs (only one measuring point). However, they have higher errors than the second set of methods, and have the significant drawback of not being compatible with DG scenarios. On the other hand, multi-end methods have also been studied. This set provides better results than one-end methods. However, this fact mainly occurs because they are supported by a greater measurement infrastructure, which is significantly more expensive than one-end method deployments. Additionally, multi-end methods are compatible with DG scenarios, thus being a perfect solution for SG networks. However, some of the analyzed one-end method are good option for underground networks where a multi-end deployment is not possible.

Results shown in [10, 11] describe the fundamental principles of the travelling wave fault location methods. A comparative analysis is performed between two of these methods, based on changes in the fault records sampling rate. In general, is considered that the methods that use information from both line ends are more robust than methods that use information from only one line end. Although in general, this statement is well founded, the paper demonstrates that is not absolutely correct since the two-end method proved to be more sensitive to the variations that the one-end method. In any case, these results are not alarming since differences between errors are very small, and also the two-end methods have better results at higher sample rates which are the most common used samples for the traveling wave fault location methods. But in some cases this needs to be clarified in order to avoid confusions. The most important thing is to really know the different fault location methods in order to select the one that best suits the available resources.

In [12] the dispersion characteristic of traveling wave is analyzed in time and frequency domain, respectively. The dispersion effect seriously affects the accuracy and reliability of fault location. In this paper, a novel double-ended fault location method has been proposed to overcome the dispersion effect of traveling wave. In the method, a correction algorithm for overcoming the dispersion effect of traveling wave enhances the singularity of the transient traveling wave. The proposed method is tested under various experiment conditions, such as different fault distances, different transition resistances, and different fault inception angles. The simulation experiments demonstrate that, compared with the traditional traveling-wave fault location methods, the proposed method can significantly improve the accuracy of fault location. Moreover, the novel method is suitable for both transposed and untransposed transmission lines. However, there are several other parameters which affect the slope of traveling wave, such as the impedance characteristic of traveling-wave measurement equipment and shunt reactor connected to the line.

Finally, [2] and [13] describe fault location methods based on smart grids, which use digital and other advanced monitoring and management techniques. In [13] fault location is estimated by using artificial neural networks. Neural locator is being trained by using different input data of particular power system grid, and by simulating different fault scenarios. Obtained results show that artificial neural networks can be used successfully in online fault detection and location estimation on power system lines. Utilization of fault locator on 20 kV overhead distribution network IS demonstrated over reliability indicators: SAIFI, SAIDI, CAIDI and ENS. Also, it is demonstrated how much the reliability of the power supply can be increased by installing fault locators. Author in [13] presents the electric power network fault locator based on measurements of electrical energy parameters. It is shown, on the example of the real network, that the ENS can be reduced over 40%, i.e. a few hundreds MWh/year, by installing only one fault locator. Furthermore, it is shown that there is an optimum number of fault locators for a particular network.

2 The Fault Localisation Based on Disturbance Recorder Data

2.1 Two-Phase Fault in System 1

The area of System 1 is powered from three sources, three primary substations: 110/35/10 kV “Maglaj”, 35/10 kV “Maglaj 1-Natron” (installed power 4 MVA in total) and 35/10 kV “Maglaj 2” (installed power 8 MVA in total). 110/35/10 kV “Maglaj” (40 MVA of installed capacity), supplies the overall consumption of 8796 customers, relatively 123 × 10/0.4 kV [14].

2.2 The Model of System 1 MV Distribution Network

For the purposes of this analysis one part of electric power system Bosnia and Herzegovina is modeled, real MV distribution system of Maglaj Municipality. This area is powered from primary substation 110/35/10 kV “Maglaj” (T1 20 MVA, T2 20 MVA). Maglaj MV system consists of 10(20) kV lines and 35 kV lines which connects 110/35/10 kV “Maglaj” to 35/10 kV “Maglaj 1-Natron” and 35/10 kV “Maglaj 2”. Lines, transformers, loads, equivalent nodes (on the 110 kV side of 110/x kV) are modeled on the basis of [15], while all the technical data on the analyzed network are taken from the technical database by JP Elektroprivreda BiH d.d. Sarajevo [14]. Seeing as all the applicable measurements were not available to the authors of this study (peak powers at industrial consumers, measurement at 10(20)/0.4 kV, etc.), peak powers at 10(20)/0.4 kV substations were estimated on the basis of the installed powers of transformers and the total peak load in the observed consumer areas, as well as on historical data on maximum current loads along 10 kV outputs to 35/10 kV “Maglaj 1-Natron” and 35/10 kV “Maglaj 2” [16]. The network was modeled using the Power-CAD software package [15]. Based on available real data, all the required values for the system were calculated and entered (Fig. 1).

2.3 Two-Phase Fault on 10 kV Transmission Line “Jablanica”

The day of 20th, 23rd and 25th of April, and 8th May 2017. Few disruptions of 10 kV transmission line “Jablanica” has occurred, which common characteristic was that malfunction was caused by fault of “R” i “T” phases each time, with almost identical

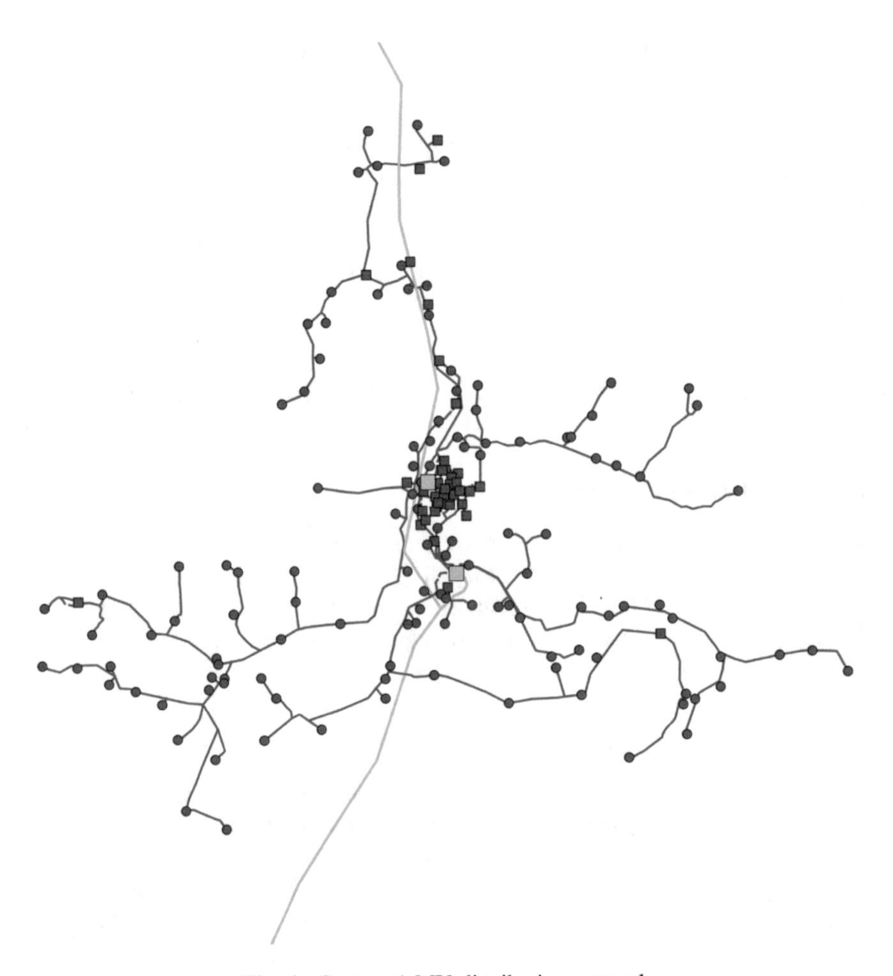

Fig. 1. System 1 MV distribution network

value of fault current $i_m = 3$ kA (Fig. 2), which leads to conclusion that the fault occurred at the same place every time.

The model of MV System 1 is created in Software called "PowerCAD" and based on fault estimate on transmission line "Jablanica" two areas were located at which, during the two-phase fault, mentioned values of current occur, two possible macrolocations were identified (transmission line sections), marked at Fig. 3.

2.4 Two-Phase Fault in System 2

The area of System 2 is supplied from 35/10 kV "Kladanj" over 35 kV transmission line "Kladanj-Olovo" and 35/10 kV "Olovo" (2 × 4 MVA) from one, and from 110/35/10 kV "Vareš" over 35 kV transmission line "Vareš-Nišići" and 35 kV cable line "Nišići-Olovo" from the other side. 35/10 kV "Olovo", total installed power 8 MVA supplies entire consumption of 4642 customers, relatively 95 × TS 10/0.4 kV [14].

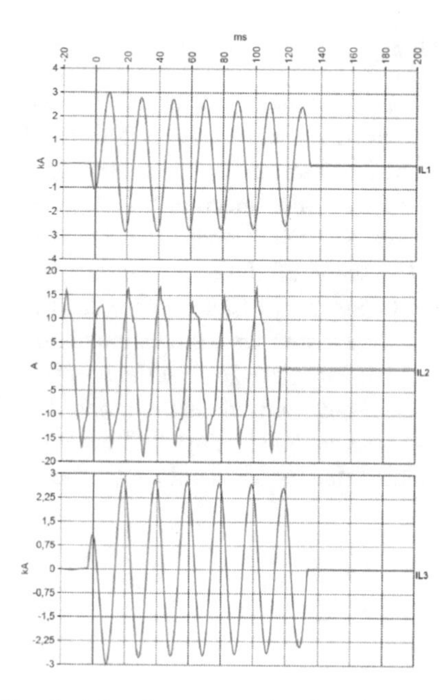

Fig. 2. Disturbance recorder time diagrams

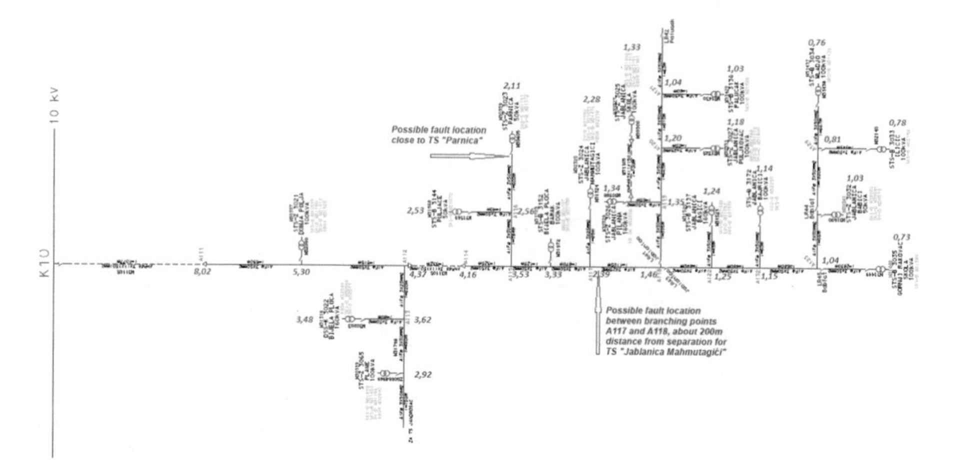

Fig. 3. Possible fault locations estimated using fault current measurements and System 1 PowerCad model results

2.5 The Model of System 2 MV Distribution Network

For the purposes of this analysis one part of electric power system Bosnia and Herzegovina is modeled, real MV distributive system of Olovo Municipality. Total length of MV lines is about 150 km, and total number of power substations 10(20)/0.4 kV from whom the supply of 4642 customers is done, is 95.

Lines, transformers, loads, equivalent nodes (on the 110 kV side of 110/x kV) are modeled on the basis of [15], while all the technical data on the analyzed network are taken from the technical database by JP Elektroprivreda BiH d.d. Sarajevo [14]. Peak power values at 10(20)/0.4 kV substations were estimated on the basis of the installed powers of transformers and the total peak load in the observed consumer areas, as well as on historical data on maximum current loads along 10 kV outputs to 35/10 kV "Olovo" [16]. The network was modeled using the PowerCAD software package [15]. Based on available real data, all the required values for the system were calculated and entered (Fig. 4).

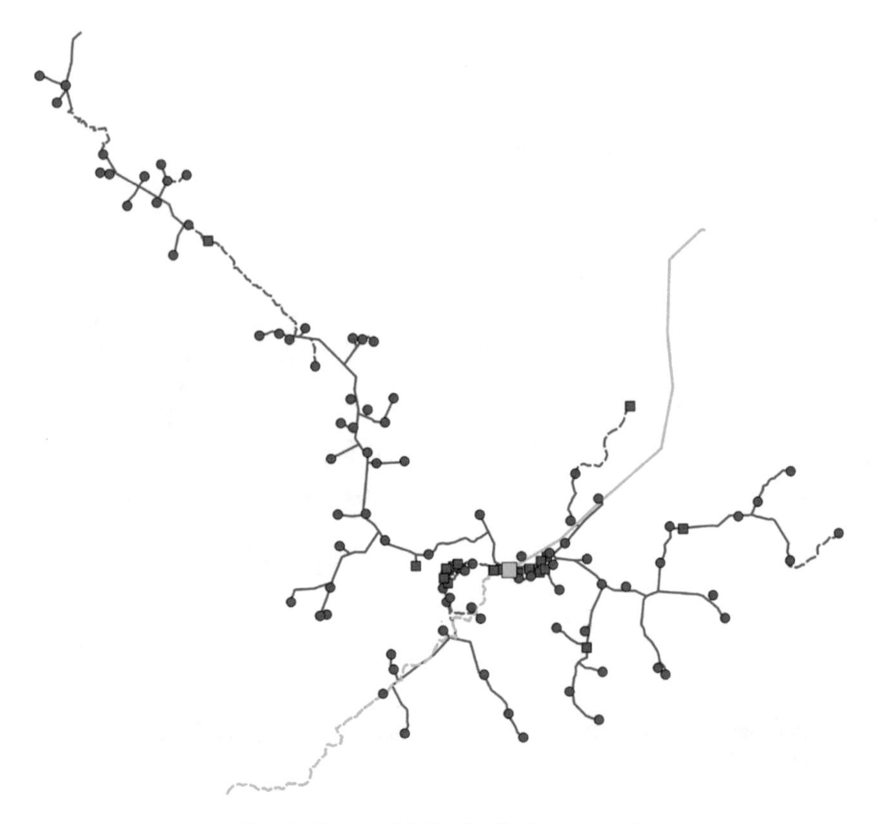

Fig. 4. System 2 MV distribution network

2.6 Two-Phase Fault on 10 kV Transmission Line "Šaševci"

The day of 14th December 2017. Disruption of 10 kV transmission line "Šaševci" occurred due to two-phase fault with RMS value of fault current I_{L1} = 728,012 A and I_{L3} = 743,431 A (Fig. 5).

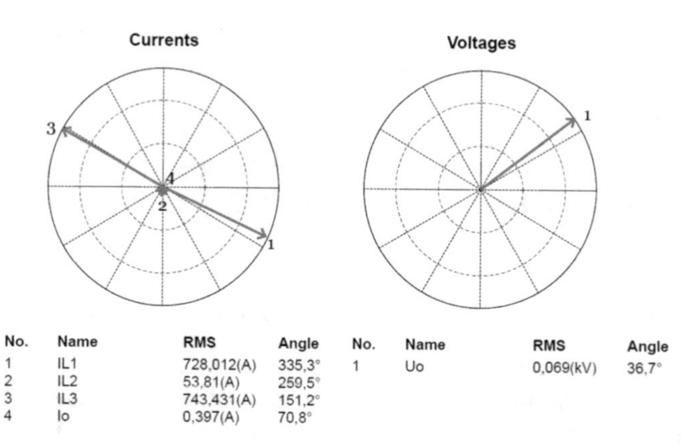

Fig. 5. Disturbance recorder vector diagrams

Based on results of fault analysis in PowerCAD System 2 model, in fact at 10 kV transmission line "Šaševci", three possible fault areas have been detected on which, during two-phase fault, occur approximately mentioned current values. Obtained locations are marked in Fig. 6.

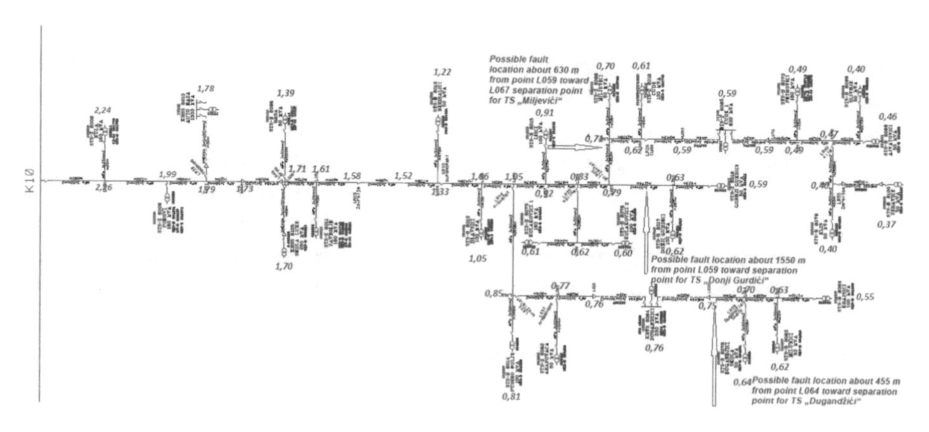

Fig. 6. Possible fault locations estimated using fault current measurements and System 2 PowerCad model results

3 Results and Discussion

3.1 Two-Phase Fault on 10 kV Transmission Line "Jablanica" – System 1

The team for maintenance of system and facilities went to outfield for purpose of visiting 10 kV transmission line "Jablanica" two mentioned targeted sections: first between branching points A117 and A118 and the other on transmission line from point A116 toward power substation "Parnica". It has been found that, near by substation "Parnica", there are few fields of transmission lines on which, during the occurrence of slightly strong wind, branches of nearby trees were touching and connecting AlFe cables of transmission line. The fault location is successfully located, after which the needed actions for removing and preventing the malfunction from happening again were taken.

3.2 Two-Phase Fault on 10 kV Transmission Line "Šaševci" – System 2

The team for maintenance of system and facilities went to outfield for purpose of visiting 10 kV transmission line "Šaševci" three mentioned targeted sections: first on transmission line from point L059 toward L067 separation for substation "Miljevići", second between branching points L059 and separation point for substation "Donji Gurdići", and the third on transmission line from point L064 toward separation for substation "Dugandžići". Just by arriving to the first location it was determined that on the distance of 400 m from separation point for substation "Miljevići" at AlFe cables of transmission line was a tree that caused interphase fault.

3.3 Suggestions for Improving/Future Development – "Semi-automation"

When it comes to future researches, considering all mentioned examples of successful locating the fault areas, it is possible to develop "semi-automation" model of fault locator in a way that the values of fault current Ik2 (which are the result of PowerCAD estimate) input and display graphically, constantly on energetic scheme or for example crossing over with mouse cursor in SCADA scheme for each branching point in system. The possible option is to make excel documents for each system separately with separate sheets for MV transmission lines in which are located graphic displays with current values, so it is possible in very short period of time to read off the section with the fault, by opening appropriate document and sheet. The next Figure shows example of model in which the values of two-phase fault currents are written for every branching point in System 1 (Fig. 7):

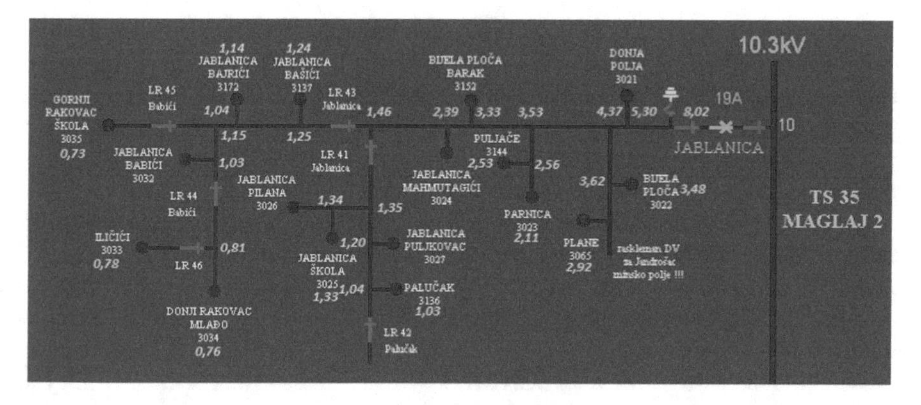

Fig. 7. SCADA sheme with two-phase fault current values written for every branching point in System 1

The other idea is that separate documents show graphical representation of fault current dependence of the distance of primary substation, as it is marked in example of System 1 (Fig. 8):

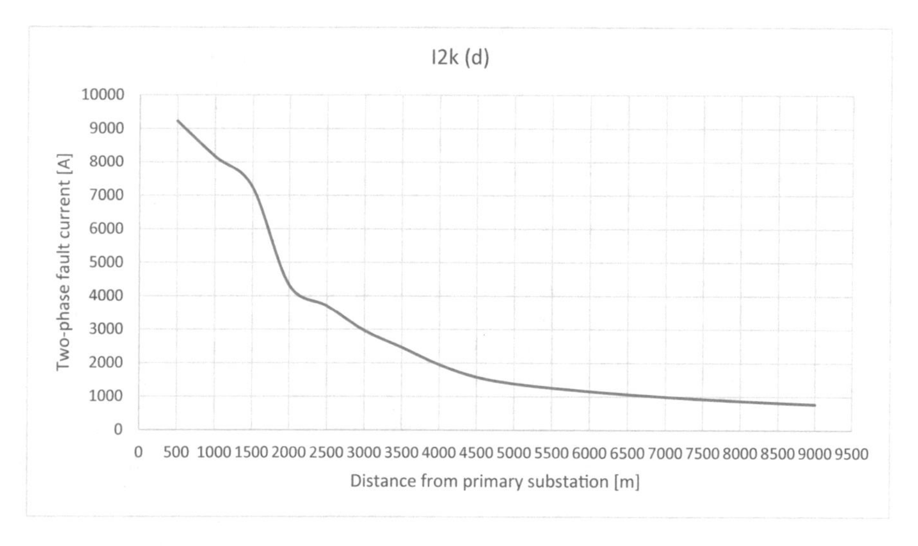

Fig. 8. Graphical representation of fault current dependence of the distance from primary substation

4 Conclusion

Based on analyzed examples, which present real situations in electric power system of Bosnia and Herzegovina, it has shown that fault locating can be done successfully using the information about current values from devices installed in system and estimate of fault current in MV distribution system models.

Suggested model of fault locator, based on information from Disturbance Recorder and estimate faults in systems which were modeled in PowerCAD, can significantly decrease the number of unsuccessful attempts of inclusion lines in failure in order to locate the fault area, and along with that decrease straining of demountable elements as well as possible damage to customers devices which objectively can occur in situations like this.

Based on exact simulations of faults at lines, which were made and analyzed in this study, it can be noticed that suggested method has very high level of accuracy. The estimation of fault area with possible lapse to CCA 100 m shows more than just satisfying results, considering the significantly decrease of time needed to locate and remove the fault on the field.

Input of fault current value I_{k2} SCADA or some other, specific graphical display for every branch point in system gives the possibility of faster and easier fault locating. Based on that, we can conclude that using this method of fault locating opens the possibility of using for semi-automation fault locators on distribution lines.

References

1. Nađ, M.: Identifikatori kvara u distributivnoj mreži, Diplomski rad, Osijek (2016)
2. Mršić, P., Lekić, Đ., Zeljković, Č.: Demonstracija upotrebe lokatora kvarova u distributivnoj mreži. Naučno-stručni simpozijum Energetska efikasnost, Banja Luka (2015)
3. Ivas, M.: Lociranje kvara u razdjelnim mrežama. Fakultet elektrotehnike i računarstva, Sveučilište u Zagrebu (2007)
4. Ghimire, S.: Analysis of fault location methods on transmission lines. Theses and Dissertations, University of New Orleans (2014)
5. Mora, J., Meléndez, J., Vinyoles, M., Sánchez, J., Castro, M.: An overview to fault location methods in distribution system based on single end measures of voltage and current, University of Girona (2014)
6. Vinyoles, M., Meléndez, J., Herraiz, S., Sánchez, J., Castro, M.: Electric fault location methods implemented on an electric distribution network, University of Girona (2005)
7. Mirzaei, M., Ab Kadir, M.Z.A., Moazami, E., Hizam, H.: Review of fault location methods for distribution power system. Aust. J. Basic Appl. Sci. **3**, 2670–2676 (2009)
8. Zimmerman, K., Costello, D.: Impedance-based fault location experience. SEL J. Reliab. Power **1**(1) (2010). Previously presented at the 2006 IEEE Rural Electric Power Conference, April 2006, 8th Annual Georgia Tech Fault and Disturbance Analysis Conference, April 2005, and 58th Annual Conference for Protective Relay Engineers, April 2005. Originally presented at the 31st Annual Western Protective Relay Conference, October 2004
9. Personal, E., García, A., Parejo, A., Larios, D.F., Biscarri, F., León, C.: A comparison of impedance-based fault location methods for power underground distribution systems. Energies J. **9**, 1022 (2016)
10. de Andrade, L., Ponce de Leão, T.: Travelling wave based fault location analysis for transmission lines. In: EPJ Web of Conferences, vol. 33 (2012)
11. Barburas, I.V., Petrovan, T.M., Nasui, I., Bugnar, S., Zah, I.N., Boiciuc, I.: Detecting the fault location using traveling wave. In: 6th International Conference on Modern Power Systems, Rumunija (2015)

12. Jia, H.: An improved traveling-wave-based fault location method with compensating the dispersion effect of traveling wave in wavelet domain. In: Hindawi Mathematical Problems in Engineering Volume (2017)
13. Hubana, T.: Lokator kvara u elektroenergetskom sistemu na osnovu mjerenja parametara kvalitete električne energije. Master rad, Univerzitet u Sarajevu, Elektrotehnički fakultet (2015)
14. JP EP BiH d.d. Sarajevo: DEEO aplikacija (2018). http://deeo/rest/data/tp_sn/986715.html
15. Fractal d.o.o. Split: Obuka za korištenje programskog paketa PowerCAD, Zenica (2009)
16. JP EP BiH d.d. Sarajevo, aplikacija, Historian (2018)

Implementation of Microgrid on Location Rostovo with Installation of Sustainable Hybrid Power System (Case Study of a Real Medium-Voltage Network)

Fatima Mašić[1], Belmin Memišević[1], Adnan Bosović[2](✉), Ajla Merzić[2], and Mustafa Musić[2]

[1] International Burch University, Sarajevo, Bosnia and Herzegovina
[2] Department of Strategic Development, Public Electric Utility Elektroprivreda of Bosnia and Herzegovina, Sarajevo, Bosnia and Herzegovina
a.bosovic@epbih.ba

Abstract. Distributed generation (DG) especially energy acquired from renewable energy sources (RES) plays a significant role in modern power sector due to high carbon emissions around the globe. It is an attempt to reduce these emissions and satisfy electricity demand. Its emerging potential is feasible by implementing microgrids. Higher cost and stochastic nature of intermittent RES are complications for the implementation and operation of such solutions. This paper analyzes economic feasibility and sustainability of implementation of hybrid power system (HPS) consisting of wind generator (WG), photovoltaic system (PVS), diesel generator unit and batteries as storage of energy. Technical analysis of the grid integration and parallel operation of the system and the grid are presented in the paper with an example of a real medium-voltage distribution network operating in Bosnia and Herzegovina. It is shown that implementing such HPS would be beneficial in terms of economy, ecology, as well as in reducing energy losses. Besides, it will reduce power supplying costs and secure better exploitation and utilization of natural renewable energy sources. These technologies positively affect power network by decreasing the risk of network-components overloading, need for network expansions, better exploiting the power-generation facilities based on renewable resources and positively impacting voltage profiles. Moreover, it is shown that the microgrid can operate in island mode with autonomous power supply for consumers. The microgrid could serve as an example to similar remote locations in order to reduce their costs of electricity, acquire more reliable and sustainable power supply, and embrace green future. All analyzes have been done by applying HOMER and DIgSILENT Power Factory professional software tools.

1 Introduction

Microgrid represents a key component of the Smart Grid for increasing system energy efficiency, providing possibility of grid-independence to individual end-user sites and improving power reliability and quality. U.S. Department of Energy Microgrid

S. Avdaković (Ed.): IAT 2018, LNNS 59, pp. 224–242, 2019.
https://doi.org/10.1007/978-3-030-02574-8_18

Exchange Group defined microgrid as "a group of interconnected loads and distributed energy resources within clearly defined electrical boundaries that acts as a single controllable entity with respect to the grid. A microgrid can connect and disconnect from the grid to enable it to operate in both grid-connected or island mode." Both these approaches are carefully modelled and analyzed in this study.

For sustainable development of rural areas electrification has become an effective instrument in both developing and developed countries. Increasing interest has been observed in the deployment of medium to large scale wind-diesel, photovoltaic (PV)-diesel and wind-PV-diesel hybrid power system (HPS) for rural electrification in various countries around the globe [1]. Area of Rostovo, with its own characteristics, RES potential and winter-tourism potential is a good choice for detailed analysis presented in this paper.

Power supply renewable and clean power generation alternative technologies will play an important role in the future due to increased global public awareness of the need for environmental protection and desire for less dependence on fossil fuels for energy production. These technologies include power generation from RES, such as wind, photovoltaic (PV), micro hydro, biomass, geothermal, ocean wave and tides, and clean alternative energy (AE) power generation technologies, such as fuel cells (FCs) and micro turbines (MTs). Diesel generators and reciprocating engines are also still commonly used for a wide range of power applications, particularly in remote areas, despite they are not renewable. Diesel generators are kept in the market because of its high fuel efficiency, diesel engine's mature technology, relatively cheaper price and low fuel cost. They can be considered renewable power sources when fuelled by renewable fuels such as bio-fuel [2].

The key reasons for the deployment of the energy systems mentioned above are reduced carbon emission, improved power quality and reliability, and in some cases, combined heat-and-power (CHP) operation which would significantly increase their overall system efficiency [2], followed by economic benefits since their implementation costs less than conventional grid extensions.

System performance can be improved by hybrid combination of power generation from RE resources (where most of them have intermittent nature), along with storage and/or AE power generation. Proper technology selection and generation unit sizing are essential in the design of hybrid systems for improved operational performance, and dispatch and operation control in order for hybridization to increase reliability [2, 3].

Following the European Union directives and strategy 20-20-20, it is important to promote usage of RESs and reduce current emissions including green house and other harmful gases. Global warming has considered even more seriously than before and power sector is considered to be one of the most responsible reasons for it. This paper promotes sustainable HPS, independently powering micro location of Rostovo tourist center. Making this HPS autonomous power losses will be reduced, because power will be consumed locally with no transmission and distribution losses, which will reduce overall CO_2 emissions.

Electrical energy generation from renewable energy sources (RESs) is of a great interest for Federation of Bosnia and Herzegovina according to Electricity Law in Federation of Bosnia and Herzegovina, where the importance of technologies proposed in this paper is highlighted in state's legislation. RESs became part of a huge change in

Bosnia and Herzegovina's legislation about electrical energy market. Law regarding usage of RES and its efficient cogeneration defines in detail feed-in tariff system that guarantees producer to have certain privileges.

Authors of [4] addressed the issue of supply of the remote consumer areas, at sites with exploitable RES potential. Hence, HPS composed of wind and solar power plants, diesel generator and battery storage at the Rostovo location, is modelled with appropriately chosen powers based on real energy resource data and load profiles of existing customers by using HOMER [25] software.

This paper analyzes the power supply for location Rostovo and microgrid of named location, supplied by a 20 kV feeder Donji Vakuf 2, modelled in DIgSILENT Power Factory software. Different scenarios with different HPS configuration and different network parameters caused by winter or summer season are proposed and analyzed for grid-connected and island mode operation.

In Sect. 2, the related literature is reviewed. Section 3 presents the analyzed case study-network. In Sect. 4, the methodology proposed in this paper is described. Section 5 presents the results. Section 6 draws conclusions.

2 Literature Review

In remote areas where the grid either is not feasible or nonexistent renewable energy based hybrid systems can compete with power from the grid [5]. Hybrid systems such as wind-PV, wind-diesel, wind-PV-diesel and PV-diesel with or without battery backup are now proven cost effective technologies for electricity supply to remote locations [6].

Hybrid Optimization Model for Electric Renewables (HOMER) [7] software is used in some studies [4, 8, 9] to find optimum sizing to minimize the cost of hybrid power system with specific load demand in stand-alone applications. A hybrid power schemes compared to a stand-alone PV system are more sustainable in terms of supplying electricity to a Telecentar due to prolong cloudy and dense haze periods [10].

Need for optimization for unit sizing for reliable and cost effective energy system is caused by high upfront cost of hybrid system. Minimizing excess energy and cost of energy in order to optimize hybrid energy systems is defined by Razak in [11]. In the study [12] is concluded that the optimized wind-PV-battery hybrid system is more cost effective compered to wind-alone, PV-alone system and wind-PV hybrid system for the load with 10% annual capacity of shortage for that hypothetical system in the proposed site whose distance is greater than 10 km from the grid.

In [13], the author analyzed microgrid benefits in improving reliability, energy saving and consumption reduction, environmental protection, investment deferral in transmission and distribution grids from the social perspective. It analyzes its cost and benefits in typical situations by the two typical cases of grid-connected and off-grid microgrids compared with that of distributed generator directly connected to power grid. Although the construction costs of microgrid are high, it is economic to invest in microgrid in view of its social benefits in improving reliability, energy saving and emission reduction, environmental protection and deferral of investment in transmission and distribution grids [13].

It is shown that with the increased load resistance, Rload, the microgrid system in islanding mode stays stable but the critical eigenvalue moves closer to the right-half plane. Changes in current-sharing controller gains have very little effects on the steady-state eigenvalue results. [14]

In [15–17], the stability of large-scale distributed generation systems was analyzed by the state-space model.

Recent years' studies investigated the effects of abnormal conditions, such as extreme weather [18, 19], overloading [20] and mechanical aging [21], on the reliability index in distribution systems. The results reported from these studies show that extreme conditions significantly deteriorate system reliability performance. Microgrids also experience abnormal operating conditions, in the form of overcurrent, overvoltage, under-voltage and abnormal frequency. They potentially threaten microgrid safety or even cause power outages, which are expected to be eliminated or limited by protection systems.

An evaluation strategy is proposed in [22] to quantify the effects of deficient protection scheme on reliability indices in a microgrid. In particular, the evaluation strategy takes into account the trigger probability of protective actions under abnormal operating conditions, such as warranted trips, rejections and malfunctions.

Hence microgrids can operate in grid connected mode or islanded mode in [23], the authors analyzed impact of nonlinear loads in a microgrid in terms of power quality. It is recognized from the study that increase in the number of distributed renewable sources facilitate reduction in total harmonic distortion (THD) in voltage profile of microgrid.

Authors of [24] concluded that the DG impact in a form of small solar systems (photovoltaic) on the power network is significant. When their penetration level is small, the network supports their integration without consequences. When their penetration level is high, the voltage limits can be violated as observed from the model simulation results. Regardless of that, installing PV systems positively affects the power system especially in the period of high consumption, when the power production from PV systems decreases the network load.

3 Analysis of a Real Distribution Network at Location of Rostovo

Presented in this paper is a case-study of a real micro-location. It is situated in Rostovo, central part of Bosnia and Herzegovina, and is considered a location with a potential for winter tourism. It is specific because of its location which is 7 km far from the nearest 20 kV power line supplying this area of total length of 30 km. The aim of this paper is to show that this location can exploit solar and wind resources in order to make HPS and enable sustainable powering of Rostovo and similar places. A comparative analysis is performed in order to present importance of HPSs comparing to conventional power supplying by power distribution network.

A georeferenced scheme and single-line diagram of the analyzed part of the power network are presented in Figs. 1 and 2, respectively.

Fig. 1. Georeferenced scheme of the analyzed network

This location was already analyzed in [4], with different study focus, time sequence and period and methodology. This network can serve as an example of a typical medium-voltage (MV) network for supplying remote areas. Given its wind and solar potential this part of power distribution network is analyzed in order to examine its possibility of being potentially micro grid and enable autonomous power supplying. The current network supplies a total of 51 consumers. Out of a total number 46 of them are residential and 5 of other types including a hotel with capacity of 45 rooms, a ski-lift and three telecommunication repeaters. These consumers are connected to the power grid and are supplied from a 110/20 kV transformer substation of a 20 MVA transformer installed power by 20 kV power lines and cables. Proposed HPS supplies consumers listed above.

Since this location has the winter-tourism potential, power consumption is highest during winter season. The average seasonal and daily load profiles are taken from real measurements from smart metering devices with time step size of 15 min for period from 1 January to 31 December 2016, and are shown in Figs. 3 and 4, respectively.

The maximum load (peak) was in a January weekend. It was expected to be so, since January is when most of tourists spend their winter holidays. During the summer, lower energy consumption was recorded. Maximum load was 154.48 kW and average daily consumption for 2016 was 649.78 kWh/d.

The specific daily load profiles are taken from same data but for two days that are to be analyzed individually: 18 January and 19 July 2016. These two days are chosen for the analysis since their load profiles are the most typical i.e. 18 January for winter and 19 July for summer. They are presented in Figs. 5 and 6.

Data used for the analysis, simulations and calculations considering the solar potential of this area are taken from internet automatically via HOMER Energy

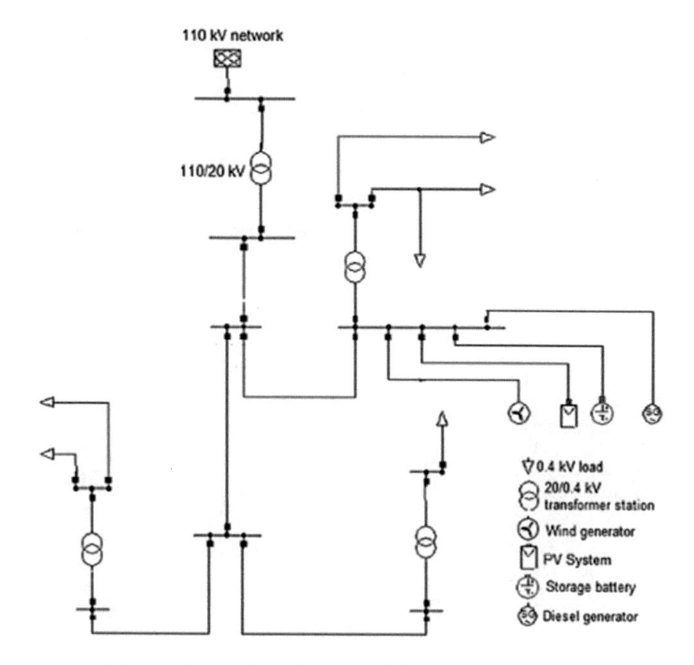

Fig. 2. Single-line diagram of the analyzed network

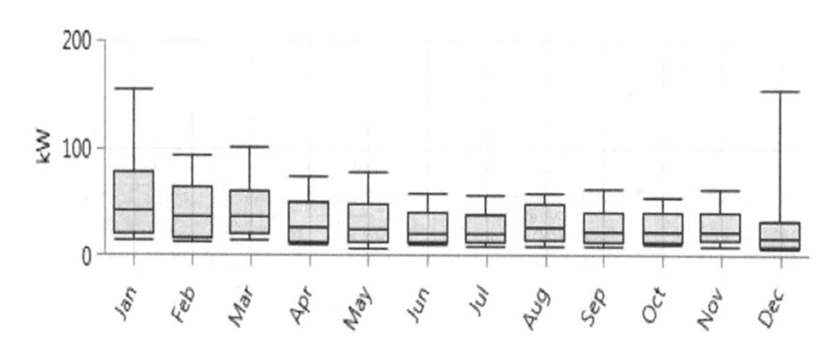

Fig. 3. The seasonal load profile for considered micro location

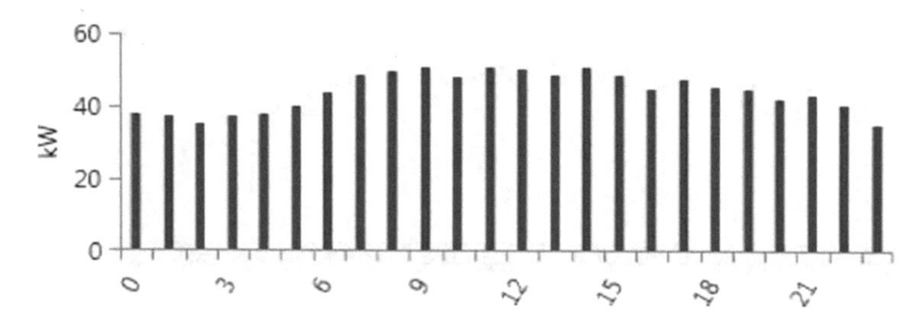

Fig. 4. The daily load profile for considered location

software which is the base software tool for optimizing sizes of the HPS in this paper. These data include NASA's surface meteorology and solar energy data for solar irradiance and temperature. Considering wind potential, wind data are measured in Rostovo on height of 60.5 m.

After the optimal size of the HPS is obtained, PVGIS data for solar irradiance and wind data measured in Rostovo are used to analyze technical aspects of the HPS implementation. According measured data, annual average wind speed is 5.05 m/s [4].

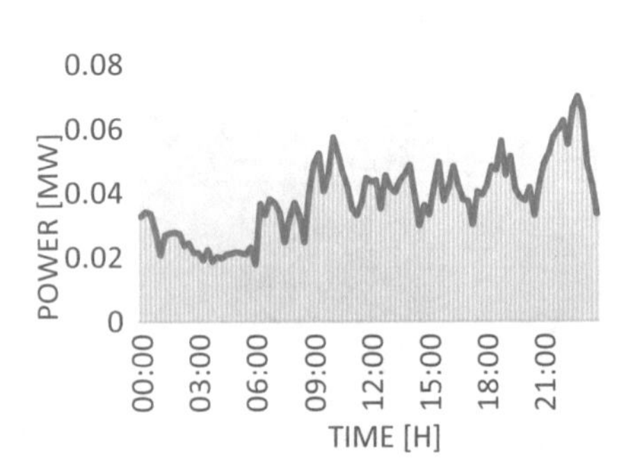

Fig. 5. Active-power load profiles of the analyzed network for 18. January

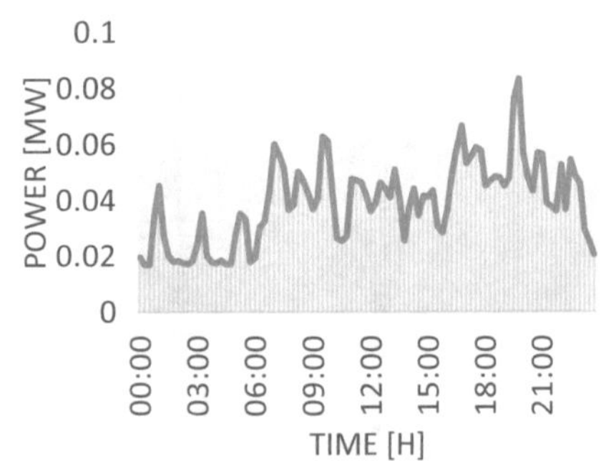

Fig. 6. Active-power load profiles of the analyzed network for 19. July

4 Methodology

The methodology used in this case-study is presented in Tables 1 and 2. Two scenarios in Table 1 are compared in order to present the actual benefits that can be achieved by using HPS which enables sustainable and independent power supply of the consumers differentiating from the conventional supply by power distribution network. The S1 scenario represents the HPS configuration obtained by simulations in the HOMER software tool consisting of PVS, WG, diesel generator and Li-ion battery which is used as the electricity storage. The S2 scenario represents the conventional approach – consumers connected to the grid and centralized generation units.

DIgSILENT Power Factory software which provides the Quasi-Dynamic Simulation toolbox is used. It simulates the power network operation and calculates load flow for different time intervals. This simulation toolbox is used for the analysis with the time interval of a day. HPS from S1 scenario will be analyzed in different scenarios presented in Table 2.

Table 1. Different scenarios for analyzing HPS over conventional supply of consumers

Scenario	PVS	WG	Diesel generator	Battery
S1	100 kW	100 kW	50 kW	100 kWh
S2	n/a	n/a	n/a	n/a

Table 2. Different scenarios for analyzing technical aspect of the implementation of HPS from S1 scenario

Scenario	Season	Grid connection
S1-1	Winter	Yes
S1-2	Summer	Yes
S1-3	Winter	No
S1-4	Summer	No

Figures 7 and 8 show annual monthly-average solar irradiation for analyzed location and annual monthly-average wind speeds.

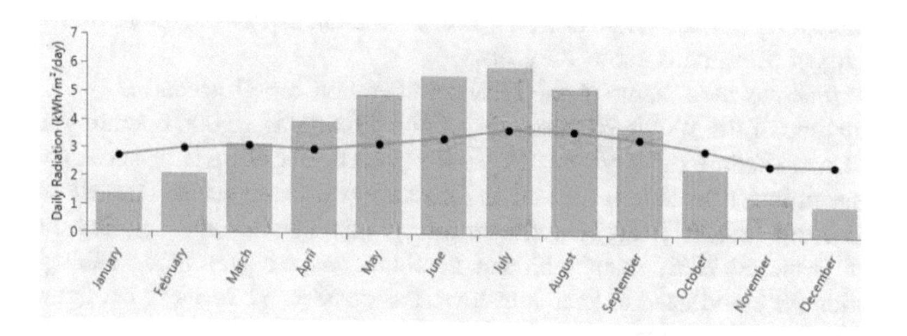

Fig. 7. Annual monthly-average solar irradiation for analyzed location

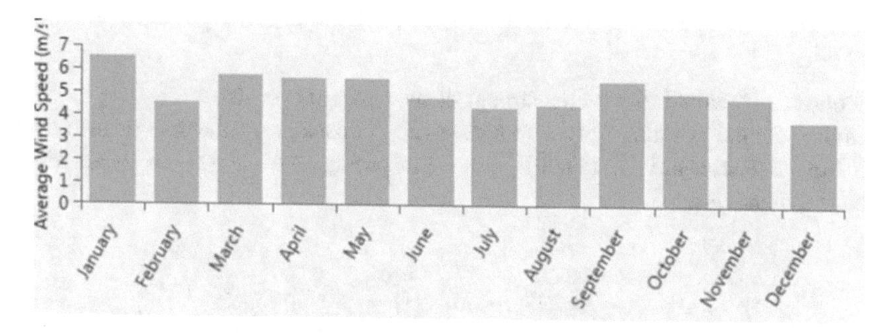

Fig. 8. Annual monthly-average wind speeds for analyzed location

S1 scenario gives the appropriate configuration of the HPS on this area according to the resources available there. This scenario is compared to the S2 scenario which represents the conventional approach – consumers connected to the grid and centralized generation units, as it is the case now. S1 is the most appropriate HPS configuration according to software HOMER Energy which is used to perform optimization functions to give best exploitation of available resources for the lowest net present cost (NPC).

After performing several simulations of the system, software presented configuration described in S1 scenario. It is consisted of flat PVS with 100 kW installed power, Norvento nED 24 WG with 100 kW installed power, diesel generator with 50 kW installed power and two Lithium-ion battery with 100 kWh energy connected via system converter of 62.2 kW. This type of wind turbine is very suitable since it operates well under bad weather conditions which is important because it would be installed in mountain area with low temperatures during winter when the danger of snow and icing is higher.

HOMER software used load following strategy controller mode to dimension elements of HPS. Under the strategy when a generator is needed, it produces only enough power to meet the demand. Load following tends to be optimal in systems with a lot of renewable power that sometimes exceeds the load [25].

Further optimization is performed taking into account additional constraints i.e. discount rate of 8%, inflation rate of 2%, maximum annual capacity shortage allowed to 2% and project life time is set to 25 years.

Figure 9 shows the scheme of the HPS configuration for S1 scenario.

The lifetime of the WG is 20 years, diesel gen. lifetime is 15,000 operating hours, and PVS has a lifetime of 25 years. Required replacements of parts or some units in HPS are taken into calculations as well as operating and maintenance (O&M) costs.

Whilst using HOMER energy software for optimizing sizes and defining configuration of proposed HPS, many different combinations are presented. Taking into consideration NPC and need to keep it as lowest as possible S1 scenario configuration was chosen as an appropriate one for this micro location. Taken the importance of self-sustainability of the area with a tourism potential, which had for a goal to show that similar areas can be self-sustainable and independently powered by generating units at that very place, combining different technologies into one hybrid system acting as a micro grid that can operate even in island mode, several key-sustainability factors were observed:

- Economic sustainability – investment costs, operating costs
- Environmental sustainability – reducing CO_2 emissions, better use of resources
- Social and cultural sustainability – reinforcing the sense of local identity, employment rise.

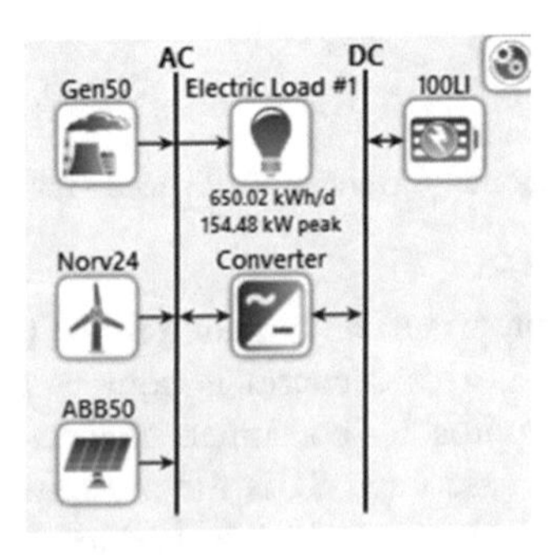

Fig. 9. The scheme of the HPS configuration

Following step by step procedure in choosing the most appropriate solution including all of the references listed above, unique configuration of HPS is elaborated.

S1-1 scenario is used for the case of winter day. Before choosing 18. January for a typical winter day representation, other days were also analyzed and it is concluded that this one is the most representative for the winter daily load profile. This scenario analyzes the microgrid connected to the existing power distribution network.

Also, for the simulation to be exact, solar irradiance and wind speeds are needed in the model. Online free solar photovoltaic energy calculator for PV systems with solar radiations maps PVGIS is used to calculate output power of the PVS in HPS.

These data include series of values with 15-min step size. The output power from the PVS for the 18 January is presented in Fig. 10.

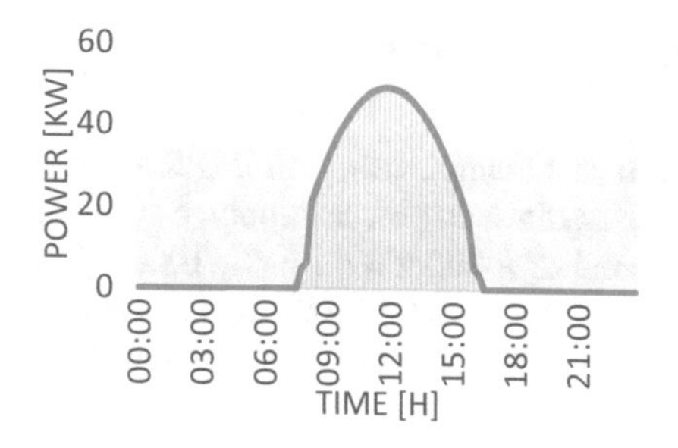

Fig. 10. Active-power output from the PVS in HPS for 18 January

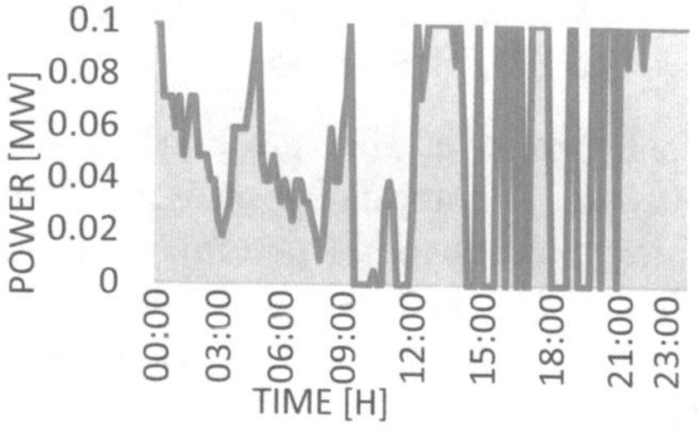

Fig. 11. Active-power output form WG in HPS for 18 January

Wind speeds are taken form the actual measuring station located in Rostovo. Measuring station is equipped with first class anemometer 60.5-m high. Wind speeds are measured in 10-min time step [4]. WG used in this analysis is Norvento nED 24 WG with 100 kW installed power. Output power curve of this WG is available in the product data sheet. The output power from the WG for 18. January is presented in Fig. 11.

Same procedure is repeated for the S1-2 scenario. 19. July is chosen for the analysis and it serves as an example of a typical summer day.

PVS and WG output powers for S1-2 scenario are presented in Figs. 12 and 13, respectively.

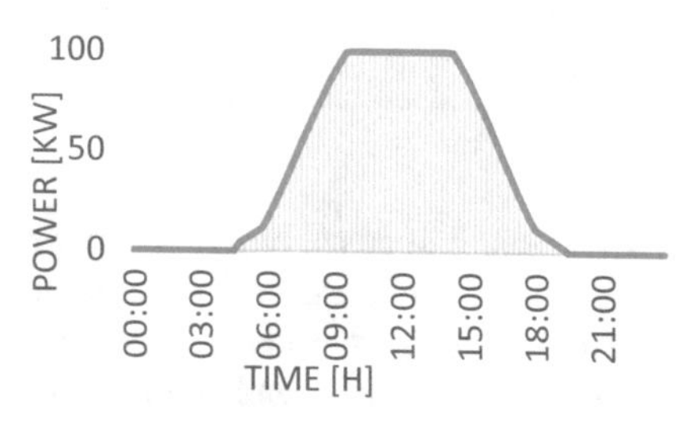

Fig. 12. Active-power output from PVS in HPS for 19 July

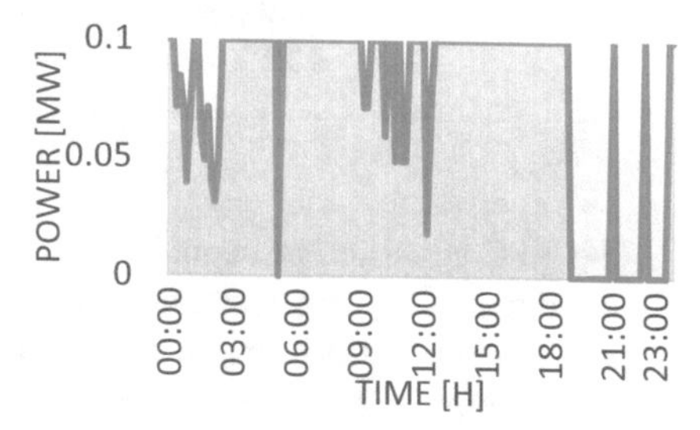

Fig. 13. Active-power output form WG in HPS for 19 July

S1-3 and S1-4 scenarios analyze the implementation of microgrid on location Rostovo, with the installation of wind, solar and diesel power plants in combination with storage as an autonomous HPS with independent power supply of consumers based on technologies used in the HPS. These scenarios present the analysis of the possibility of a modelled microgrid operation in island mode. Installed power of HPS components are same as in the previous scenarios. Only difference between these two scenarios is that S1-3 analyzes operation of microgrid for typical winter day and S1-4 for the summer when connected to the grid.

5 Results and Discussion

5.1 HPS Configuration and Economic Aspects of Implementation (HOMER Results)

Configuration of HPS presented in S1 scenario is obtained using HOMER energy software. A comparative analysis of HPS over traditional power supply by power distribution network is performed. HPS is consisted of a 100 kW in PVS, 100 kW in WG, 50 kW in Diesel Generator, 100 kWh battery and 62.2 kW in converters. Resulting characteristics of S1 scenario and HPS configuration are presented in Table 3.

Table 3. Results for S1 scenario HPS configuration

Scenario	HPS configuration	Storage	Production kWh/year	Renewable percent	Capacity shortage %/year
S1	PVS: 100 kW WG: 100 kW Diesel Gen: 50 kW Converter: 62.2 kW	Li-ion 100 kWh 2 strings	319,003	83.5	1.3

The NPC consisted of capital investment, O&M costs, fuel costs, replacement and salvage obtained by the software, regarding mentioned characteristics of calculation equals $693,560. According to the results, higher costs are for the diesel generator, since it is employed whenever there is no solar irradiance or wind and the price of diesel affects O&M and fuel costs. More detailed costs review is available in Table 4. The prices for each system in HPS are calculated based on available prices on the internet market for simillar products.

Table 4. Net Present Costs (NPC) for HPS

Component	Capital ($)	Replacement ($)	O&M costs ($)	Fuel ($)	Salvage ($)	Total ($)
PVS	111,900	0	12,928	0	0	124,828
WG	110,000	35,069	71,101	0	−19,764	196,407
Diesel gen	12,500	10,317	26,741	142,317	−2,101	189,774
Li-ion battery	72,000	30,548	0	0	−5,749	96,798
System converter	6,222	2,640	0	0	−497	8,365
Total	312,622	78,574	188,158	142,317	−28,111	693,560

In Figs. 14 and 15 graphical representations of NPC and cash flow for the 25 years are presented, respectively. NPC is presented for each component of HPS individually, while cash flow is presented for the overall system.

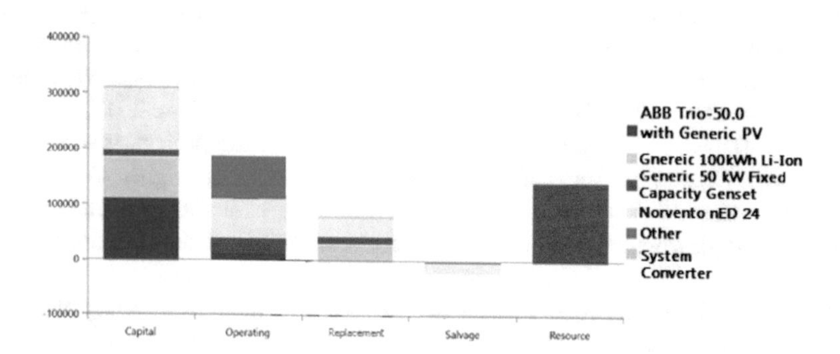

Fig. 14. NPC for the HPS

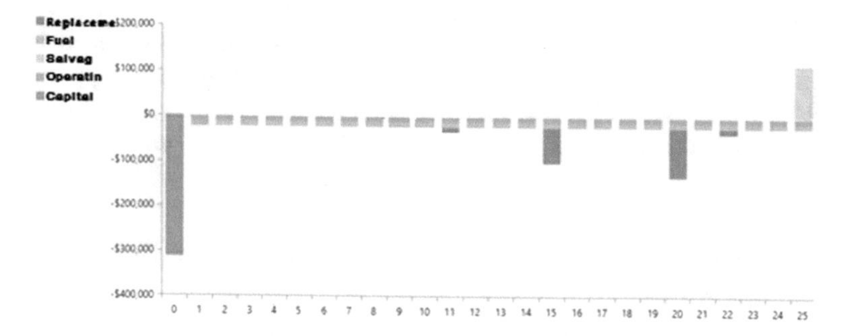

Fig. 15. Cash flow for the HPS

Figure 16 shows monthly average power production by HPS components. The highest production is from WG as expected, since this is a windy location. Power generated from PVS is highest during summer.

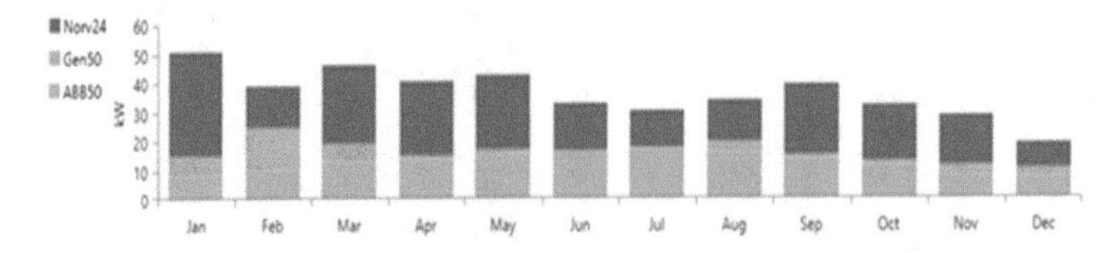

Fig. 16. Monthly average power production by HPS components

Comparing the HPS from S1 with S2 scenario several conclusions can be drawn. Micro location of Rostovo is connected to the power grid, which relies on centralized generation units which dominantly use domestic coal for electricity production, while the rest of production portfolio is based on large hydro power plants, with the average ratio of thermal-hydro energy being 80–20%. The CO_2 emissions are approximately 1000 kg/MWh. The network expansion to this place would cost approximately 1,249,437.00$ without costs for substations or any other equipment [4]. When the cost of electrical energy – 0.12 $/kWh (average price of electricity, June 2017, Agency for Statistics of B&H), for 25 years, which is the life time of the project, is added to this value, 2,014,398.00$ is obtained. This amount of money is higher than the total investment costs and NPC for the proposed HPS. Even excluding the investments in the network infrastructure, total investment costs and NPC of the proposed HPS are lower.

Reliability of the network was taken into consideration by limiting the annual capacity shortage. However, this area during winter experiences extremely low temperatures and snow and ice accumulation are expected, so maybe this 1.3% of yearly shortage will be even higher.

Another important aspect is that software provided emissions report. The conclusions drawn from this report show benefits of the HPS in terms of emissions. It is mentioned that conventional power generation units in Bosnia and Herzegovina have emission rate of approximately 1000 kg/MWh and this HPS produces approximately 319 MWh per year. That will be 319 tons of CO_2 emissions per year, if it is supplied by the power network. But if HPS was autonomous CO_2 emissions would be reduced to 78 tons per year.

5.2 Technical Aspects of HPS Implementation (DIgSILENT Power Factory Results)

In the following Table 5, excess electricity, unmet load and capacity shortage are presented. Since at the beginning maximum capacity shortage percentage has been limited to 2%, optimization has performed taking into account this constraint. Annual capacity shortage percentage obtained after simulation equals to 1.3%.

Table 5. Electrical quantities of HPS

Quantity	kWh/yr	%
Excess electricity	62,640	19.7
Unmet electric Load	1,447	0.61
Capacity shortage	2,993	1.26

Figures 17 and 18 show the powers from the HPS components generated in the microgrid for S1-1 and S1-3 scenarios, respectively.

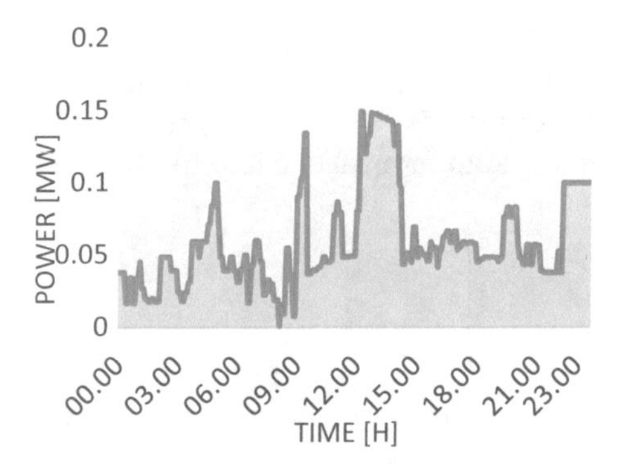

Fig. 17. Active-power generated in the microgrid for S1-1 scenario

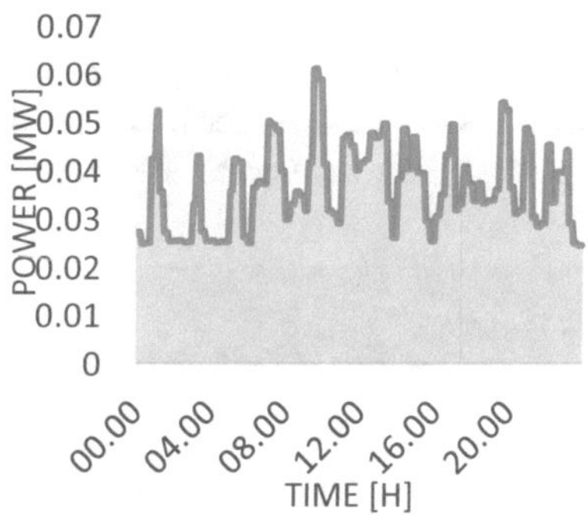

Fig. 18. Active-power generated in the microgrid for S1-3 scenario

Since the Li-ion battery is selected as a reference machine in the model for S1-3 scenario where microgrid operates in island mode, it will balance the active power and one may notice load-followed power generation in the Fig. 18. That implies that surplus energy generated from RESs is sufficient to supply consumers.

Figures 19 and 20 show the powers from the HPS components injected to the microgrid for S1-2 and S1-4 scenarios, respectively.

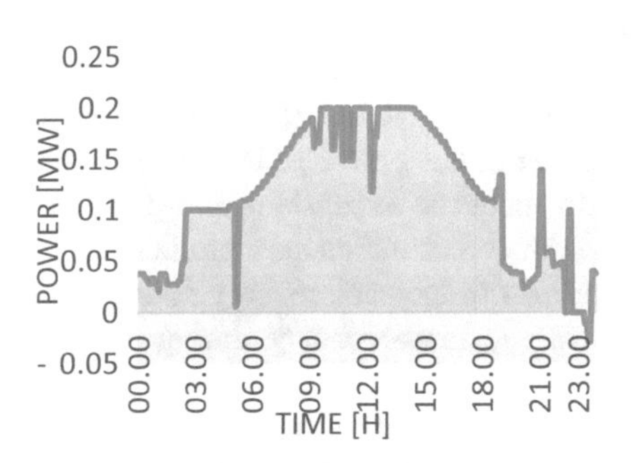

Fig. 19. Active-power generated in the microgrid for S1-2 scenario

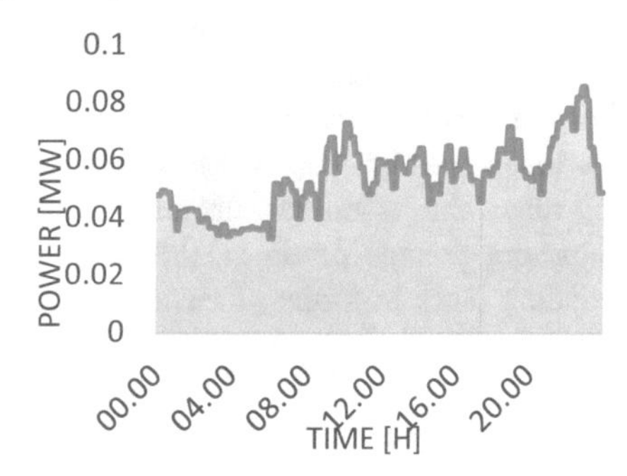

Fig. 20. Active-power generated in the microgrid for S1-4 scenario

Similar conclusions can be drawn for Fig. 19 due to battery operating as slack machine.

For all scenarios, the results obtained from the software show that all of the generated power comes from the single bus i.e. 20 kV side of Rostovo Motel transformer where HPS components are connected. One can conclude that this HPS generates enough power to supply the microgrid of Rostovo. All surplus power can be either stored to storage battery implemented in the HPS or can be delivered elsewhere by the 20 kV medium-voltage power distribution network if the microgrid does not operate in island mode. Either way, there has to be switch between the power network and the microgrid which will enable operating this way.

In the Fig. 21 the voltage drop at Rostovo Village 0.4 kV bus where the voltage values are the lowest comparing to other voltages in the network is shown for all analyzed scenarios. As seen, voltage drop is within ±2.5% Un, which is good in terms of voltage limit defined by EN 50160 European norm for quality of electrical energy.

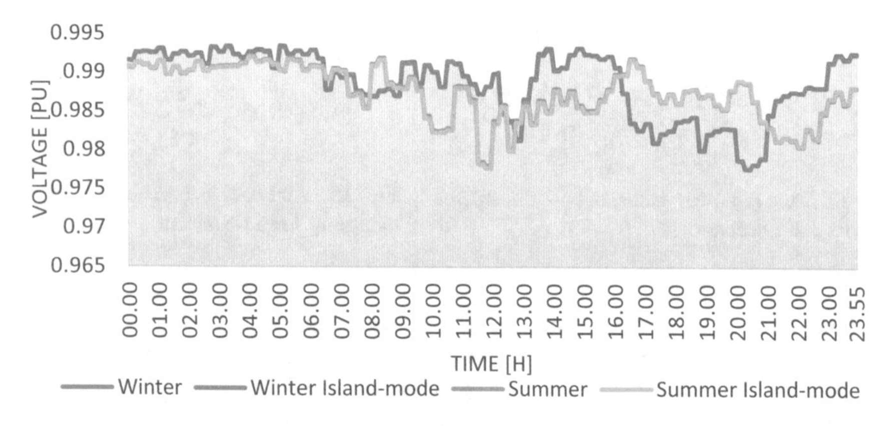

Fig. 21. The voltage drop at Rostovo Village 0.4 kV bus - lowest voltage values in the network

None of the voltage values in the network exceed nominal voltage level.

As seen, implementation of HPS with configuration presented in this paper positively affects the network power system by decreasing the risk of network-components overloading which will be presented, better exploiting the power-generation facilities and renewable resources, decreasing the need for network expansions and positively impacting voltage profiles. HPS implementation does not affect load profiles except reducing them in terms of electricity generation from renewable energy sources and increasing them in terms of storing electrical energy in the battery during off-peak periods.

Figure 22 shows active-power losses for all scenarios during one day. Losses are presented in kWh, where 5-min averages are multiplied by 24 h obtaining the daily losses. Only part of the 20 kV feeder is modelled, including the consumers at the microlocation while other parts of this feeder supply neighboring villages. The values of losses are very close to each other, yet it is easy to understand they are lower when microgrid is operating in island mode for both scenarios analyzing microgrid without connection to the medium-voltage power distribution network. Since power generation is lower during the winter, losses are lower as well.

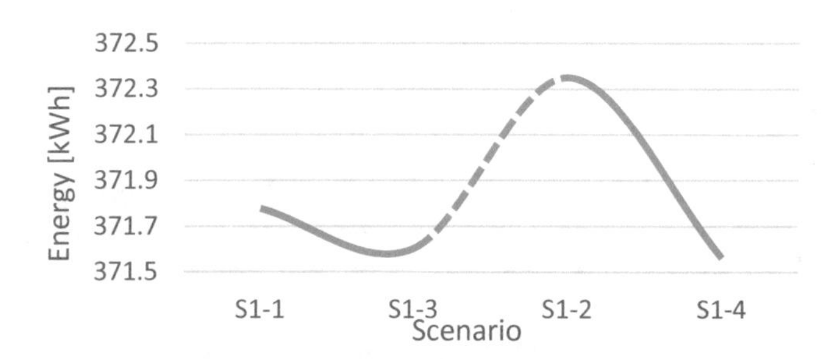

Fig. 22. Active-power losses for all scenarios during one day

When investigating the impact of implementation of microgrid consisting of HPS with different technologies it is important to analyze power grid elements loading i.e. transformers and lines. The transformers and lines loadings for each scenario are going to be presented in a way that highest loading percentage will be shown for two most loaded elements. All other elements that are not presented have lower loading percentage comparing to those presented.

Figures 23 and 24 show the most loaded transformers and lines for the S1-1, S1-2 scenarios, respectively. Loading of elements is S1-3 scenario is the same as in the S1-1, while the loading in S1-4 is the same as in the S1-2 scenario. The conclusion that can be drawn from this is that element loadings in the microgrid are the same no matter if network operates in island mode or it is connected to the power grid.

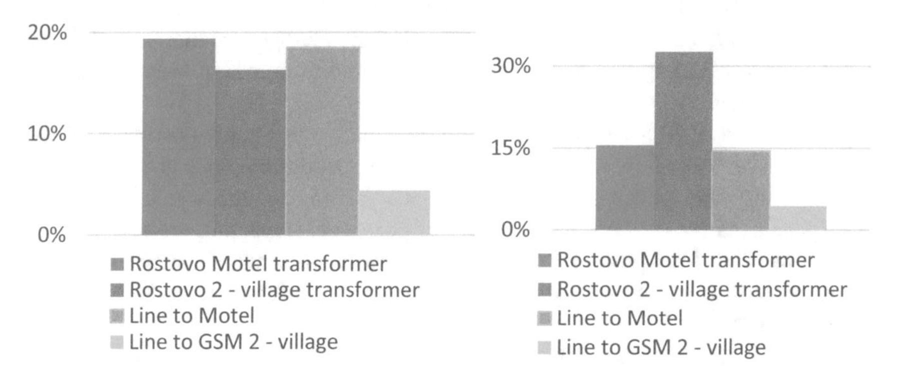

Fig. 23. Element loadings for S1-1 scenario

Fig. 24. Element loadings for S1-2 scenario

The most loaded line – line to GSM 2 transformer station output line is always the same. All other lines and cables are less loaded than this one. All the lines and cables are loaded well below their nominal capacity, and this is the case because all equipment including lines and cables is purchased by type and they are not dimensioned especially for this network. Line to motel is more loaded during winter days because of increase in power consumption comparing to summer. Since ski-lift is connected to the grid via hotel line, its operating has affected this loading.

6 Conclusion

This paper proposes a HPS consisting of PVS, WG, diesel generator and storage battery as an alternative to power distribution grid expansion. Besides economic factors, the analysis showed that other aspects of sustainability are fulfilled including environmental aspects.

The implementation of sustainable HPS for electricity supply of consumers at Rostovo, Bosnia and Herzegovina is analyzed by using different simulation scenarios and investigating different possible solutions and HPS configurations trying to optimize and propose the best solution according to economic aspect.

The conclusion drawn from this study is that proper HPS configuration and implementation of it can positively affect environment, reduce power supplying costs and secure better exploitation and utilization of natural renewable energy sources. Moreover, HPS configuration in this paper is based on actual and real parameters of power consumption and solar and wind potential at the site.

In this way, possible solution for distant places is proposed enabling autonomous power supply for consumers, which will completely rely on local sources in forms of wind and solar power, combined with electricity storage in battery. There is also diesel generator which will serve as a backup of power in cases of outage and other problems affecting reliable power supply.

A proper HPS configuration is proposed in order to satisfy consumers' needs and enable autonomous and independent power supply of consumers. HPS mostly relies on wind and solar resources in combination with electricity storage in batteries and diesel generator as power backup and compensate lack of these resources in order to achieve power flow balance.

Implementation of HPS with configuration presented in this paper positively affects the network power system by decreasing the risk of network-components overloading, better exploiting the power-generation facilities based on renewable resources, decreasing the need for network expansions and positively impacting voltage profiles. HPS does not affect load profiles except reducing them in terms of electricity generation from renewable energy sources and increasing them in terms of storing electrical energy in the batteries during off-peak periods.

To make this microgrid possible to operate in island mode, completely independent of existing power distribution network, ensuring reliable power supply of consumers, as future improvement of this project stability of this network should be analysed and solved in the sense of frequency stability. With the properly developed control and operating mechanisms in order to have stable microgrid, this microgrid will be able to operate in island mode, completely independent of existing power distribution network, ensuring reliable power supply of consumers.

HPS like this one offers the best penetration of renewables, lowest CO_2 emissions and levelized cost of electrical energy is the lowest which will lead to consumers' satisfaction.

References

1. Rehman, S., Al-Hadhrami, L.M.: Study of a solar PV-diesel-battery hybrid power system for a remotely located population near Rafha, Saudi Arabia. Energy **35**, 4986–4995 (2010)
2. Nehrir, M.H., Wang, C., Strunz, K., Aki, H., Ramakumar, R., Bing, J., Miao, Z., Salameh, Z.: A review of hybrid renewable/alternative energy systems for electric power generation: configurations, control and applications. IEEE Trans. Sustain. Energy **2**, 392–403 (2011)
3. Kellogg, W.D., Nehrir, M.H., Venkataramanan, G., Gerez, V.: Generation unit sizing and cost analysis for stand-alone wind, photovoltaic, and hybrid wind/PV systems. IEEE Trans. Energy Convers. **13**(1), 70–75 (1998)
4. Merzic, A., Music, M., Rascic, M., Hadzimejlic, N.: An integrated analysis for sustainable supply of remote winter tourist centers - a future concept case study. TUBITAK **24**, 2821–3837 (2016)
5. Nema, P., Nema, R.K., Rangnekar, S.: A current and future state of art development of hybrid energy system using wind and PV-solar: a review. Renew. Sust. Energy Rev. **13**, 2096–2103 (2009)
6. Manwell, J.F., Mc Gowan, J.G.: Development of wind energy systems for New England islands. Renew. Energy **29**, 1707–1720 (2004)
7. National Renewable Energy Laboratory, HOMER Getting Started Guide Version 2.1 NREL (2005)
8. Ekren, O., Ekren, B.: Size optimization of a wind/PV hybrid energy conversion system with battery storage using simulated annealing. Appl. Energy **87**(2), 592–598 (2010)
9. Khan, M.J., Iqbal, M.T.: Pre-feasibility study of stand-alone hybrid energy systems for applications in Newfoundland. Renew. Energy **30**, 835–854 (2005)
10. Abdullaha, M.O., Yunga, V.C., Anyia, M., Othmana, A.K., Hamida, K.B.A., Taraweb, J.: Review and comparison study of hybrid diesel/solar/hydro/fuel cell energy schemes for a rural ICT Telecenter. Energy **35**, 639–646 (2010)
11. Razak, J.A., Sopian, K., Ali, Y.: Optimization of renewable energy hybrid system by minimizing excess capacity. Int. J. Energy **1**(3), 77–81 (2007)
12. Nandi, S.K., Ghosh, H.R.: Prospect of wind-PV-battery hybrid power system as an alternative to grid extension in Bangladesh. Energy **35**, 3040–3047 (2010)
13. Jin, X.: Analysis of microgrid comprehensive benefits and evaluation of its economy. In: 10th International Conference on Advances in Power System Control, Operation & Management (APSCOM 2015), Hong Kong, pp. 1–4 (2015)
14. Chen, C.L., Lai, J.S., Martin, D., Lee, Y.S.: State-space modeling, analysis, and implementation of paralleled inverters for microgrid applications. In: 2010 Twenty-Fifth Annual IEEE Applied Power Electronics Conference and Exposition (APEC), Palm Springs, CA, pp. 619–626 (2010)
15. Katiraei, F., Iravani, M.R.: Power management strategies for a microgrid with multiple distributed generation units. IEEE Trans. Power Syst. **21**, 1821–1831 (2006)
16. Pogaku, N., Prodanovic, M., Green, T.C.: Modeling, analysis and testing of autonomous operation of an inverter-based microgrid. IEEE Trans. Power Electron. **22**, 613–625 (2007)

17. Mohamed, Y., El-Saadany, E.F.: Adaptive decentralized droop controller to preserve power sharing stability of paralleled inverters in distributed generation microgrids. IEEE Trans. Power Electron. **23**, 2806–2816 (2008)
18. Alvehag, K., Soder, L.: A reliability model for distribution systems incorporating seasonal variations in severe weather. IEEE Trans. Power Del. **26**(2), 910–919 (2011)
19. Rocchetta, R., Li, Y.F., Zio, E.: Risk assessment and risk-cost optimization of distributed power generation systems considering extreme weather conditions. Reliab. Eng. Syst. Saf. **136**, 47–61 (2015)
20. Sun, Y., Wang, P., Cheng, L., Liu, H.: Operational reliability assessment of power systems considering conditional-dependent failure rate. IET Gener. Transm. Distrib. **4**(1), 60–72 (2010)
21. He, J., Sun, Y., Wang, P., Cheng, L.: A hybrid conditions-dependent outage model of a transformer in reliability evaluation. IEEE Trans. Power Del. **24**(4), 2025–2033 (2009)
22. Xu, X., Wang, T., Mu, L., Mitra, J.: Predictive analysis of microgrid reliability using a probabilistic model of protection system operation. IEEE Trans. Power Syst. **32**(4), 3176–3184 (2017)
23. Padayattil, G.M., Thobias, T., Thomas, M., Sebastian, J., Pathirikkat, G.: Harmonic analysis of microgrid operation in islanded mode with nonlinear loads. In: 2016 International Conference on Computer Communication and Informatics (ICCCI), Coimbatore, pp. 1–5 (2016)
24. Memisevic, B., Masic, F., Bosovic, A., Music, M.: Impact of plug-in electric vehicles and photovoltaic technologies on the power distribution network (case-study of a suburban medium-voltage network). Elektrotehniški vestnik, vol. 84, no. 3, Ljubljana (2017)
25. http://www.homerenergy.com

Implementation of Protection and Control Systems in the Transmission SS 110/10(20)/ 10 kV Using IEC 61850 GOOSE Messages

Adnan Cokić[1(✉)] and Admir Čeljo[2]

[1] CET Energy Ltd., Sarajevo, Bosnia and Herzegovina
adnan.cokic@cet-energy.com
[2] Elektroprenos – Elektroprijenos BiH a.d. Banja Luka, Banja Luka, Bosnia and Herzegovina

Abstract. This article describes implementation concept of modern protection and control system in the substation Sarajevo 13 with special reference on using IEC 61850 GOOSE messages for interlocking conditions and protection functions design. Interlocking condition of existing 110 kV gas insulated switchgear has been made by a wire on the GIS local command cabinets' level. For proper control from P&C cabinets and local/remote SCADA level it was necessary to duplicate the interlocking conditions using IEC 61850 GOOSE messages and to protect the system in communication loss case. The 10 kV switchgear is classic air insulated medium voltage switchgear. The reverse interlocking and busbar protection are implemented by blocking high-set current stage I>> on transformer and coupling bays with pickup of high-set current stage I>> on feeder bays. The signal is transmitted via IEC 61850 GOOSE message. This article describes details for correct parameter settings of high-set current stage I>> in entire 10 kV switchgear considering predicable time delay of IEC 61850 GOOSE message because of Ethernet network. Also the IEC 61850 GOOSE messages are used for implementation of arc protection function in 10 kV switchgear.

1 Introduction

Concept of modern protection and control (P&C) system integrates wide spectrum of technological achievements necessary to interconnect all parts of substation (SS) in compact and reliable system.

In this technological era a big steps have been taken to completely change the way that P&C systems works. These changes will affect the complete secondary system in SSs including design, configuration, testing, commissioning and maintenance process of P&C systems. The final goal is standardization and digitalization of whole process.

Development of Ethernet network technologies, in last 15 years, had a major impact on development of P&C systems in power plants and SSs. Many manufacturers have recognized these possibilities even earlier and they developed many communication protocols for communication between P&C relay (IED) and Supervisory control and data acquisition (SCADA). The largest challenge in integrating equipment from different vendors in one functional system was interoperability and solution approach.

S. Avdaković (Ed.): IAT 2018, LNNS 59, pp. 243–254, 2019.
https://doi.org/10.1007/978-3-030-02574-8_19

In order to fulfill all functionality of modern SSs, the integrators had to interconnect all these equipment in one system, using protocol converters (gateways) [1, 7], which had direct impact on costs and reliability of P&C systems.

As result of capabilities offered by Ethernet network technology and aspiration to integrate IEDs from different manufacturers into one compact and functional P&C system, the IEC 61850 standard has been developed. The standard includes and integrates IEC 61850 communication protocol, as many others functionality and benefits which open the possibility to build a digital SSs.

The standard IEC 61850 define complete structure of power plant or SS through logical nodes. In this way, the structure of IED is standardized, i.e. the way these devices are programmed and integrate into the P&C system.

Communication to SCADA systems has also become standardized through recognizable logical nodes, so as possibility of sharing information between IEDs through recognizable IEC 61850 Generic Substation Events mechanism (GSE). GSE mechanism includes IEC 61850 Generic Object Oriented Substation Events (GOOSE) and Generic Substation State Events (GSSE). Also, new extensions of standard has made possible to transmit the sample value (SV) data through the Ethernet network.

Transition between classical and digital SS is step by step process, especially in existing objects which are in reconstruction process.

Implementing of communication protocol IEC 61850 in SS Sarajevo 13 was first step in direction of digital SS. The possibilities that standards IEC 61850 offer are used for the following functionalities in SS;

- Simple communication between IEDs and SCADA system,
- Topology of communication network that provides redundancy (N-1),
- Usage of IEC 61850 GOOSE messages for implementation of interlocking condition on 110 kV and 10 kV switchgear,
- Usage of IEC 61850 GOOSE messages for implementing of reverse busbar protection (reverse blocking) on 10 kV switchgear,
- Usage of IEC 61850 GOOSE messages for implementation of arc protection in 10 kV switchgear.

2 Protection and Control System in the SS 110/x kV Sarajevo 13

Implementation of the project was set up a new modern P&C system in SS Sarajevo 13 which integrates functions of supervision, control, data collection and protection into one functional system. This implies complete hardware and software solutions for implementation of SCADA system and its connection to IEDs as well as to supervisory dispatch centers. New system is based on a distributed architecture with the aim of distributing tasks between processes, Fig. 1. A high level of reliability with maximum availability is achieved by redundancy at the level of IEDs, Remote Terminal Unit (RTU) and communication links.

To achieve communication at station level, IEDs are connected in a ring topology, (3 rings – 1 for 110 kV, and 2 for 10 (20) kV) where one end of ring starts in one Ethernet switch, and ends in another Ethernet switch.

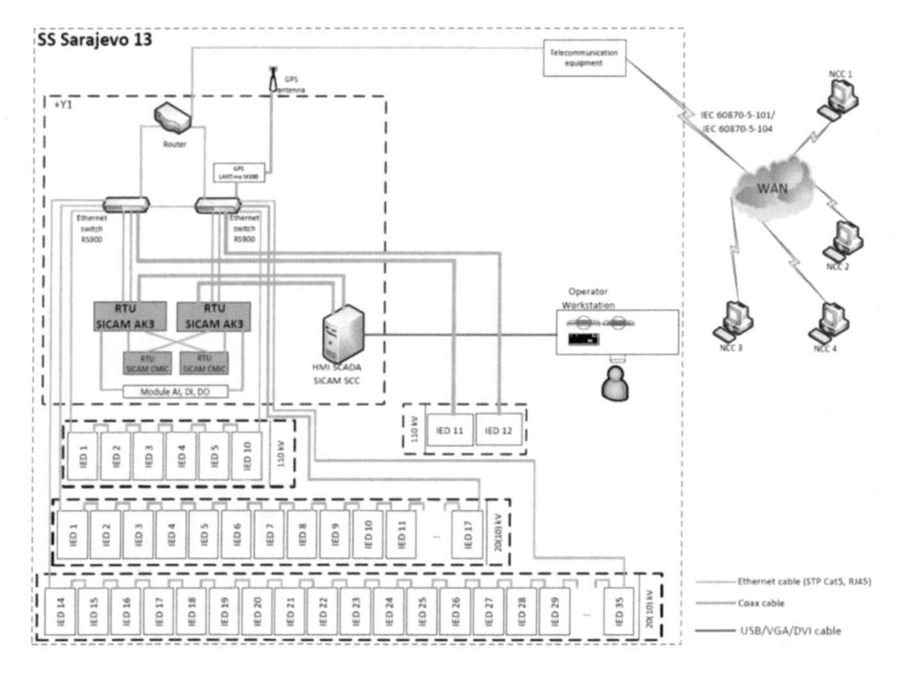

Fig. 1. Block diagram of functional architecture of SCADA system

This topology achieves the redundancy of communication between IEDs and RTUs, as follows:

- At level of physical port of IEDs,
- At level of communication link (in case of ring communication interruption, there is no communication interruption),
- At Ethernet switch level,
- At RTU level.

Network time protocol (NTP)/Simple network time protocol (SNTP) server was used for time synchronization of all devices in SS.

Communication of SCADA server with IEDs is made over two Ethernet switches in a way that SCADA server has two redundant network cards. This achieves the redundancy of network card and redundancy of communication with IEDs.

The 110 kV plant in SS Sarajevo 13 was constructed as a gas SF6 insulated plant (GIS).

The protection system consists of two parts:

- The protection integrated at GIS (pressure sensors for SF6 gas in different gas zones of switchgear),
- The protection integrated in P&C cabinets in command relay room.

The 10 kV switchgear is constructed as a metal-shielded, air-insulated unit for internal assembly and consists of sections A and B, connected with the coupling bay.

The protection system is designed in a way that there is a group of protection that operates locally at the bay level and the group of protection that operates at the level of the entire 10 kV switchgear. The group of protections that operates locally are standard protection function for this type of switchgear (overcurrent protection, sensitive earth fault protection, etc.). The group of protections that operates on the level of entire 10 kV switchgear are:

- Fast reverse busbar protection,
- Protection against arc occurrence.

Both protection functions use IEC 61850 GOOSE message.

3 Communication Standard IEC 61850

Introduction of international standard IEC 61850 sets the foundation for creating a unique communication structure which covers all hierarchical levels in SS. Unlike other communication protocols, IEC 61850 defines strict rules for exchange data between logical nodes, thus allowing interoperability independent of the equipment manufacturer. Interoperability is defined as ability of two or more devices of same or different manufacturers to exchange information on the basis of which internal or external operations are performed [1, 6]. The communication standard defines rules for different parts:

- Client/Server communication (communication rules between IED and SCADA system),
- GOOSE communication (communication between IEDs on process level),
- SV (sampled measured value exchange),
- NTP/PTP (Network time protocol/Precise time protocol),
- High-availability Seamless Redundancy/Parallel Redundancy Protocol (HSR/PRP),
- Teleprotection option.

In this project, three possibilities of standard IEC 61850 have been recognized and used;

- Client/Server communication,
- GOOSE communication,
- NTP protocol.

Logical nodes

Although the IEC 61850 protocol is defined as standard for communication structure, its main contribution is in defining an object model for all equipment in SS.

The greatest shifts and efforts were made in the field of object modeling, that is, the parsing of complex data structures to smaller units [2].

Every IED performs a series of operations grouped into units, called Logical devices (LD). LD further comprises the corresponding Logical nodes (LN), which define functions and associate equipment functions in the SS (e.g. transformers, switches or protective devices). LN includes all relevant mandatory data, properties and extensions of object it represents and also defines a standard for accessing these data.

In principle, there are two basic groups of logical nodes that define the equipment and function:

- Logical nodes that represent equipment at process level (e.g. Circuit breaker, disconnector, etc.),
- Logical nodes that represent functions of SS automation system (e.g. Distance protection, automatic voltage control, etc.).

LN as the smallest object can communicate with another LN built into another IED. In short, LN defines the function in SS in standardized way. A simplified logic model is shown in Fig. 2 [2, 5].

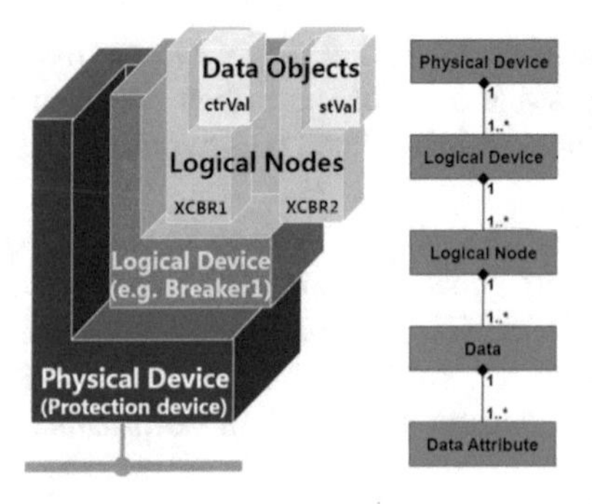

Physical device is accessed via network address

A set of all logical nodes in one device

Function in device (e.g. switch or differential protection)

The value of individual attributes in logical node (e.g. switch position or value of differential protection)

Fig. 2. Simplified logic model scheme

LN contains object data of SS objects. Each of these data consist the corresponding properties that describe the SS object. So, for example, LN XCBR for switch contains data object with one property of stVal - position of the device (on, off, malfunction, intermediate position), another property of ctlVal for switching commands. In this way, the objects and functions in SS are uniquely described by breaking up the complex structure of SS to data details. Therefore, the interoperability of devices from different manufacturers in same SS in accordance with IEC 61850 protocol rules is enabled [2].

IEC 61850 GOOSE messages

The rapid exchange of data directly between IEDs in accordance with IEC 61850 GOOSE mechanism can be used to implement interlocking condition as well as to enhance protection functions. The IEC 61850 GOOSE mechanism is independent of the communication between IEDs (as a server) and RTU/SCADA (as a client) and it is performed as multicast communication.

They are sent to the network with sender identification media access control address (MAC) and message, but without destination address. IED from different manufacturers should be able to "subscribe" to receive GOOSE message from another IED manufacturer on the same communication network, which is one of the biggest benefits of the standard IEC 61850 [2]. The way IEC 61850 GOOSE message is generated and forwarded is showed in Fig. 3. This example shows the exchange of "Picked up I>" signal between two IED. If "Picked up I>" signal is configured for exchange, then sender will generate the signal in cycle form telegram every 0.5 s.

Priority of a signal is determining the duration of cycle. There are a three group of priorities:

- Priority High (Minimum monitoring time: 1 ms, Maximum monitoring time: 500 ms)
- Priority Medium (Minimum monitoring time: 4 ms, Maximum monitoring time: 10000 ms)
- Priority Low (Minimum monitoring time: 10 ms, Maximum monitoring time: 20000 ms).

The times of priorities are freely programmable, and should be in accordance with function used.

Cyclical sending is main security mechanism of IEC 61850 GOOSE message, because in event of IEC 61850 GOOSE message loss for more than specified time limit, the state of interruption of communication can be declared thus avoiding eventual wrong manipulation.

If there is a fault in grid and protection function I> starts, then sender change signal state "Picked up I>", and changes method of generating the IEC 61850 GOOSE message from cyclical to spontaneous. IEC 61850 GOOSE message is generated and repeated after 1 ms, 2 ms, 4 ms, 8 ms … 500 ms. After reaching a time of 500 ms (cyclic time), it proceeds further by cyclic generation method in accordance with selected priority and new signal value "Picked up I>" = 1.

The exchange of SV is based on publisher/subscriber principle. The publisher store SV in output buffer, and makes them available to all logical nodes that are subscribed for. In addition to the SV, a timestamp is set so that subscriber can continuously check and correct the timing of the sampled signals [1].

There are two basic models for sending data of SV:

- SV for protection functions are sent 80 samples per cycle,
- SV for an analysis of quality of electricity is sent to 256 samples per cycle.

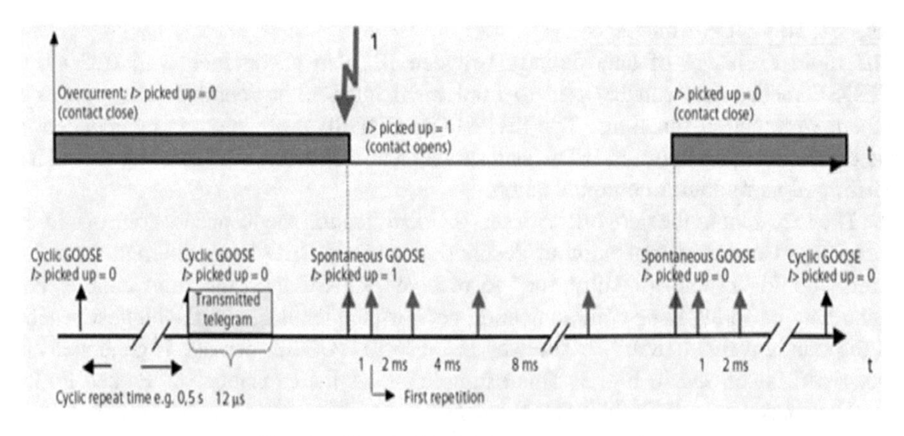

Fig. 3. IEC 61850 GOOSE message [3]

4 Use of the IEC 61850 GOOSE Messages in the SS

Implementation of communication protocol IEC 61850 at station level in SS Sarajevo 13 has made it possible to bind the complete SS into a unique P&C system. IEC 61850 GOOSE messages are used for implementation of following functions:

- Implementation of interlocking condition on 110 kV and 10 kV switchgear,
- Implementing of reverse busbar protection (reverse blocking) on 10 kV switchgear,
- Implementation of arc protection in 10 kV switchgear.

Interlocking system on 110 kV and 10 kV switchgear
Interlocking system for GIS is implemented on two levels:

- Cabinets for local control on GIS,
- Cabinets for P&C at command relay room.

Interlocking conditions of 110 kV switchgear were carried out by the "wire" at level of local control cabinet, as follows:

- Vertical interlocking conditions that defines the control conditions for bay
- Horizontal interlocking conditions that define control condition between various bays.

Horizontal interlocking system is necessary since the GIS consists of two sections, a coupling bay and two busbar grounding disconnector, one at each section.

Control from P&C cabinet level is also enabled by respecting all interlocking conditions. That means that at the level of P&C cabinet, complete interlocking conditions had to be mapped from the local control cabinets, using IEC 61850 GOOSE messages.

Vertical interlocking system, implemented at bay level, is wired into every control bay IED and interlocking system is programmed in logical controller diagrams in a way that monitors the interlocking condition from lower control levels.

Regarding horizontal interlocking conditions, an information exchange solution is implemented through exchange IEC 61850 GOOSE message between 110 kV control IEDs.

The standard IEC 61850 defined use of GOOSE messages for implementation of P&C functions. Typical transmission time of IEC 61850 GOOSE message through communication network is defined by standard and is approximately 10 ms, which is satisfactory speed for implementation of interlocking conditions [4].

Thus, using the possibility of horizontal communication between IEDs in GIS, it is enabled the full functionality of interlocking system. In this way, issuing command from P&C cabinet or higher control level (local or remote SCADA) and not performing its function is not possible, without operator knowledge about interlocking conditions.

A special requirement for this solution is reliability and safety of IEC 61850 GOOSE communication and P&C system itself, since communication network and communication protocol IEC 61850 is used for full functionality.

In event of communication interruption between the IEDs that exchange IEC 61850 GOOSE messages, interlocking conditions and possibility of issuing the command may be disabled. In this case, it is left to duty staff to bypass interlocking conditions implemented at IED control level by switching the interlocking switch to position "control without software blocking". In this case, the command can be issued only from the control IED, and not from SCADA.

IEC 61850 GOOSE messages were also used for implementation of certain interlocking condition in transformer bays at 10 kV switchgear. Since IEDs on this bays does not have a built in switch for control without interlocking conditions (it is impossible to operate without interlocking due to mechanical lock in metal bay itself), in the event of IEC 61850 GOOSE communication loss, a different principle was used in comparison to GIS. In 10 kV switchgear, in event of communication loss and GOOSE messages loss, IEDs are using the replacement value in internal logic of IED, which always allows execution of command.

In this way, the required level of security in case of eventual loss of IEC 61850 GOOSE communication is achieved in 110 kV and 10 kV switchgear.

Reverse busbar protection scheme on 10 kV switchgear

Fast reverse busbar protection on 10 kV switchgear, performed by blocking short-circuit stage (I>>) on transformer bays, is an economical way of protecting the switchgear from short-circuits on busbars and achieving satisfactory selectivity for short-circuit faults on outgoing feeders.

On the 10 kV switchgear reverse busbar protection was performed in such a way that on the incoming bays (transformer bays) on sections A and B, short-circuit stage (I>>) has to be blocked in the case of activating the same stage on some of the outgoing feeder or home transformer. Solution for transmitting blocking signal from the IED on the outgoing feeder to the IED on the transformer bay was performed using IEC 61850 GOOSE message.

The solution for transmitting the blocking signal from IED on feeder to IED on transformer bay was performed using IEC 61850 GOOSE message.

In event of a short circuit on one of outgoing feeder, the IEC 61850 GOOSE blocking message of short-circuit stage (I>>) is routed and forwarded to the IED on transformer bays (sections A and B) and to coupling bay. Interpretation of signal is depending on switching state of coupling bay. For example, if a short circuit happens on one of feeder bays on Section A then a "picked up >>" IEC 61850 GOOSE message is generated from IED on that feeder. If coupling bay is open, this message will block short circuit protection stage I>> for a time interval of 200 ms on IED for transformer bay on section A. The IED for transformer bay on section B will not be affected, and short circuit protection stage I>> will work in base time.

In case that coupling bay is closed, then IEC 61850 GOOSE message will block the short circuit stage I>> on both transformer and coupling bay for a time interval of 200 ms.

In order to achieve sufficient reliability, it was necessary to implement the mechanisms of busbar protection and to satisfy protection selectivity in case of loss of IEC 61850 GOOSE communication. For this reason, reserve short circuit stage I>>> is activated on IEDs for transformer bays. Reserve short circuit stage I>>> is blocked in normal operating condition. In case of IEC 61850 GOOSE communication loss, a reserve short circuit stage I>>> is activated, and a short circuit stage I>> is blocked, thereby generating a signal "GOOSE communication fault". In case any feeder bay loses communication, then short circuit stage I>> is blocked, and reserve short circuit stage I>>> is activated on both transformers.

The standard IEC 61850 specifies maximum time (5 ms) for which IEC 61850 GOOSE message must be transmitted through the communication network from information source to receiver [4].

All IEC 61850 certified manufacturers of IEDs guarantee that equipment meets these requirements.

So, in worst case scenario, the short circuit protection of transformer bay has to be time delayed minimum time that requires the GOOSE signal to be transferred from IED from some of feeder bays to IED on transformer bays. In accordance with test results, a safety margin of 20 ms was set.

Therefore, short circuit protection at 10 (20) kV bays are set in the following manner:

- Feeder bays; I>> activated; tI>> 0 ms (no time delay).
- Transformer bays; I>> activated; tI>> 20 ms
    - If there is no GOOSE blocking signal the protection function operate in base time (20 ms)
    - If there is GOOSE blocking signal the protection function operate in delayed time (200 ms)
    - In case of lose GOOSE communication I>> is blocked, and I>>> is activated with time delayed of 200 ms.

Arc protection in the 10 kV switchgear

Arc protection in 10 kV switchgear is performed in such a way that sensors (in form of a micro switch) are installed on specific sections of the 10 kV bays. These sensors will detect the occurrence of arc, and forward this information to IEDs on bays. The effect of arc protection function depends on location of arc, type of bay, fault current and current switching state of 10 kV switchgear.

Depending on bay type, the sensors are installed as follows:

- Feeder bay
    - Busbar department – BGS1
    - Circuit breaker department – BGP1
    - Cable department – BGK1
- Measurement bay - Section A
    - Busbar department – BGS1
    - Voltage transformer department – BGP1
- Home transformer bay
    - Busbar department – BGS1
    - Circuit breaker department – BGP1
    - Cable department – BGK1
- Coupling bay
    - Busbar department – BGS1
    - Circuit breaker department – BGP1
    - Cable department – BGK1 (to raising bay).

Depending on bay type and the location of the arc in the switchgear, several stage of protection against arc has been parameterized. In certain cases, only signal from the sensor is sufficient to trip the circuit breaker, while in other cases tripping is also conditioned by the occurrence of a fault current (Table 1).

Table 1. Tripping logic of busbar transformer

	Arc protection for measurement bay
Busbar department BGS1 or Voltage transformer department – BGP1	The IED sends a GOOSE signal to IEDs on transformers and coupling bays The IEDs on transformers and coupling bays check amount of fault current and make the tripping decision
	Arc protection for feeder/house transformer bays
Busbar department BGS1 or Circuit breaker department BGP1	The IED trip the local circuit breaker without checking the amount of the fault current The IED sends a GOOSE signal to IEDs on transformers and coupling bays The IEDs on transformers and coupling bays check amount of fault current and make the tripping decision
Cable department BGK1	The IED trip the local circuit breaker with checking the amount of the fault current
	Arc protection for transformer bays
Any Busbar department BGS1 or Circuit breaker department BGP1 on the rest of the switchgear	The IED receives the GOOSE arc signal, and with check of the current conditions and switching state of coupling bay generates the tripping signal for local circuit breaker
	Arc protection for coupling bays
Busbar department BGS1 or Circuit breaker department BGP1	The IED trip the local circuit breaker without checking amount of fault current The IED sends a GOOSE signal to IEDs on transformers bays
Cable department BGK1	The IED trip the local circuit breaker with checking the amount of the fault current
Any Busbar department BGS1 or Circuit breaker department BGP1 on the rest of the switchgear	The IED receives the GOOSE arc signal, and with the check of the current conditions generates the tripping signal for local circuit breaker

5 Conclusion

Implementation of modern P&C system in SS Sarajevo 13 with integration of communication protocol IEC 61850 as one of most important hardware and software platform, significantly reduced investment costs, while fulfilling all technical requirements necessary for exploitation of the electric power SS.

The IEC 61850 standard has been designed in a way that has potential to change way in which P&C systems are designed and tested, and this possibility has been largely used in SS Sarajevo 13.

One of basic advantages of standard is IEC 61850 GOOSE communication mechanism which has been used for realization of several functionalities in SS Sarajevo 13:

- Use of IEC 61850 GOOSE messages for implementation of interlocking condition on 110 kV and 10 kV switchgear,
- Use of IEC 61850 GOOSE messages for implementing of reverse busbar protection on 10 kV switchgear,
- Use of IEC 61850 GOOSE messages for implementation of arc protection in 10 kV switchgear.

Using IEC 61850 GOOSE message for implementation of interlocking system on the GIS had improved diagnostics and tracking the correct interlocking conditions for all control levels. Also, in order to achieve the same functionality by conventional means (wiring), the designer should "duplicate" blocking conditions from local cabinet level in GIS.

Using IEC 61850 GOOSE message for implementation of fast reverse busbar protection and arc protection on 10 kV switchgear had reduced design process and the necessary wiring. If one of IEDs does not send signals via IEC 61850 GOOSE messages, such a state can be detected and alarmed. The classic way of accomplishing this function over wire, cannot alarm the malfunctioning of protection device or breaking the wire, which is certainly the advantage of IEC 61850 protocol and GOOSE mechanism.

Using IEC 61850 GOOSE message in P&C systems is a great advantage in reduced design and wiring requirements, which reduces cost of assembly and testing, and contributes to reliability of the system. Also, exchange of all types of information (binary states, apparatus positions, measurements) opens new possibilities in implementing complex P&C functions, such as, a breaker failure protection or distributed synchro check.

In addition to the many benefits of IEC 61850 GOOSE message, it is also necessary to provide some information regarding modifications in work concept. Changes are reflected in fact that in addition to basic IEDs software, there is an additional software package for configuration of IEC 61850 GOOSE message. The engineering staff must be aware of fact that complete GOOSE "engineering", i.e. information exchange can be "see" and analyze only in additional software package and accompanying SCD (System Configuration Description) file. In addition to this it is necessary to change the way that we make our design documentation.

The disadvantages of using IEC 61850 GOOSE message can be analyzed from standpoint of great dependence of P&C functions on proper operation of communication equipment.

Also, the diagnostics of communication equipment and the design of applications and functions with the help of IEC 61850 GOOSE message require additional engineering knowledge and skills.

References

1. Jurišić, G.: The influence of the IEC 61850 standard on protection and control systems, 11. HRO CIGRE Cavtat, no. 20130004, p. 7, November 2013
2. Vidović, F., Leci, G., Vukić, I., Ožanić, D.: Protocol IEC 61850 – proposal of the technical solution for the secondary system of distribution transformer station 1. CIRED Šibenik, p. 6, 18–21 May 2008
3. Siemens: Efficient Energy Automation with the IEC 61850 Standard Application Examples, Siemens AG Energy Sector, p. 48 (2010)
4. IEC 61850-5: Communication networks and systems in substations – Part 5: Communication requirements for functions and device models, p. 138
5. Brand, K.-P., Janssen, M.: The specification of IEC 61850 based substation automation systems, p. 52. DistribuTech (2005)
6. Higgins, N., Vyatkin, V., Nair, N.K.: Distributed power system automation with IEC 61850, IEC 61499, and intelligent control. IEEE Trans. Syst. Man Cybern. Part C (Appli. Rev.) **41** (1), 81–92 (2011)
7. IEC TR 61850-1: Communication networks and systems in substations – Part 1: Introduction and overview, p. 44 (2013)

Design, Optimization and Feasibility Assessment of Hybrid Power Systems Based on Renewable Energy Resources: A Future Concept Case Study of Remote Ski Centers in Herzegovina Region

Said Ćosić[1(✉)] and Ajla Merzić[2]

[1] Faculty of Engineering and Natural Sciences, Department of Electrical and Electronics Engineering, International University of Sarajevo, Sarajevo, Bosnia and Herzegovina
said_cosic@hotmail.com

[2] Department for Strategic Development, Public Enterprise Elektroprivreda of Bosnia and Herzegovina, Sarajevo, Bosnia and Herzegovina

Abstract. One of the key needs of the modern world, considering the pace and extent of technological advancement and its dependence on electrical energy, is the requirement for secure and reliable electricity supply to all people. Providing constant, reliable and accessible electricity supply has been a major challenge for decades, especially for remotely-located areas whose electrical loads are tremendously difficult to be met because of geographical isolation and sparse population. In a hilly and mountainous country like Bosnia and Herzegovina, which is full of isolated power consumers, this becomes an even greater issue. Moreover, connecting these loads to a grid requires significant infrastructural investments, and vastly increases grid losses in the transmission process. One of the most tangible actions that can help to ensure an affordable, stable and environmentally sensitive energy by overcoming the intermittent nature of renewable energy sources, is the concept of hybrid power systems. This paper offers an analysis of a hybrid power system that serves the load profile of a ski center situated in the northern Herzegovina region. The paper uses quantitative methods of simulation, optimization and sensitivity analysis using HOMER (hybrid optimization model for electric renewable) software, to evaluate the economic feasibility and optimal electrical configuration of the desired off-grid power system, considering the availability of sun and wind resources on the site. The analyses have shown that the proposed model is sustainable, environmentally friendly and economically viable. Furthermore, this paper aims to establish a profound basis for the future infrastructure improvements that can be derived in further researches to optimize power system configuration to suit more load, or to include an estimation of a grid-connected load which serves the excess electricity back to the grid.

S. Avdaković (Ed.): IAT 2018, LNNS 59, pp. 255–268, 2019.
https://doi.org/10.1007/978-3-030-02574-8_20

1 Introduction

In recent years, rapid advancements in technology, automation and industry have created a huge growth of the world energy demand and developed concerns over the future of power systems and their role in energy generation, supply and transportation. For that reason, electric power industry faces challenges far larger than any in its history. These challenges include ensuring security, equity and sustainability of energy supply to the consumers. Utilizing renewable energy sources has been seen as a perfect solution to achieve secure, environmentally friendly, and economically feasible energy as a step towards sustainable development and bright future [1, 2]. In addition, latest studies have shown that the cost for renewable energy devices has been drastically reduced in recent years [3], and an ever-increasing number of countries are aiming to increase the share of renewables in their power systems. Due to those ambitious objectives, the importance of flexible loads and smart energy grids will increase even further [4]. Smart grid is believed to be a future grid that offers increased efficiency, reliability, and environmental friendliness in power generation, transmission, distribution, consumption and management with the advancement of integration in information and communications technology [5]. With the emergence of smart grids, smart meters and novel storage technologies, accelerated employment of the renewable energy technologies has commenced [6]. With the ultimate goal of replacing conventional energy sources to meet the future energy demand for buildings, transportation, and industrial sectors [7], wind and sun energy have been extensively investigated throughout the recent period. An attempt was made by Rehman and Al-Hadhrami [8] to explore the possibility of utilizing power of the sun to reduce the dependence on fossil fuel for power generation to meet the energy requirement of a small village located in Saudi Arabia. Rina et al. continued the constant advancement in the areas of PV systems control, optimization and storage by the development of a microcontroller-based charge controller [9]. Hunter [10] studied in detail the technical and operational characteristics of diesel generation and found various disadvantages of diesel only generation, concluding that introducing wind energy generation to the existing diesel only system is preferable and desirable. Based on a research into the contemporary history of wind power in Denmark, Heymann and Nielsen [11] argue that the hybridization of electric utility regimes by means of innovative adaptation of wind power has proved to be a viable addition to the power system. However, of the key characteristics of the current electricity market is the difficulty of integrating intermittent renewable sources of electricity into the existing grid [12]. Apart from the intermittent nature of RES's, another major issue is the lack of electrical grid in remote areas and the high costs for construction of new transmission lines [13].

A very elegant solution was found in the concept of hybrid power systems. Recently, the hybrid energy systems received most interest because of like these systems can overcome the intermittent nature of renewable energy sources, the over-sizing issue and enhance the reliability of supply [14]. HPSs based on locally available RESs can offer a cost-effective and pollution-free alternative to expensive grid extensions and can decrease fuel transport costs and transmission and distribution losses in remote areas [15, 16]. Hybrid systems can be efficiently used in a variety of applications, ranging from an energy harvester system with energy storage system, energy monitoring system, and maximum power point tracking [17], as a stand-alone solar PV/Fuel Cell/Battery/Generator hybrid power system to serve the electrical load of a commercial building [18], or to serve the electrical load of an educational institution in Indonesia [19]. Aladrois's and Krarti's analysis of a grid connected to a wind farm and a PV system [20] indicates there is a good potential in increasing the renewable fraction and decreasing the carbon dioxide emission while maintaining a relative low cost of energy. Other comparative studies on the use of micro grid power systems for different configurations, such as a hybrid photovoltaic and fuel cell [21], PV systems with energy storage [22], and a triple hybrid system that combines simultaneously generated power from thermoelectric (TE), vibration-based electromagnetic (EM) and piezoelectric (PZT) harvesters for a relatively high power supply capability [23].

This paper, nevertheless, focuses on the issue of supply to isolated consumer areas which have a considerable renewable energy potential, especially in the case study area of northern Herzegovina. According to the findings of [14, 24–26], renewable energy sources may replace the conventional energy sources and be a feasible solution for the generation of electric power at remote locations with a reasonable investment. The purpose of this study was to offer a realistic insight into future possibilities of sustainable power generation using RESs that serves a specific type of load, namely a ski center. The simulations and analyzes were done for a proposed configuration which includes a hybrid power system that utilizes wind and solar energy, as well as sensitivity assessment of other variables. Since this is a future concept, it was assumed that electric vehicles belonging to the staff and visitors of the ski center will be used as storage options. Even though this futuristic concept might not be foreseeable in the near future, considering economical situation in the country, it doesn't make a huge difference in the initial investment due to the fact that the prices of EV's tend to sharply decrease, and better and more affordable storage technologies are being invented. The obvious advantage of using EV's as a storage facility, apart from a lower capital investment and O&M costs, is the fact that EV's compensate for the missing power at the peaks of loading. For HPS configuration evaluations, simulations, and capacity optimization, the world's leading micro-grid modeling software, HOMER (Hybrid Optimization of Multiple Energy Resources), was used.

It is anticipated that the outcome of this investigation may benefit researchers from other developing countries, with similar climatic and economic conditions for designing hybrid photovoltaic-wind-battery systems in isolated areas.

2 Research Methodology

Since the main purpose of this study is to offer a realistic insight into future possibilities of sustainable power generation using renewable energy sources incorporated into hybrid electrical power systems that serve loads at remotely located sites, it will mainly focus on the issue of supply to isolated consumer areas away from developed distribution networks, which have a considerable renewable energy potential, especially in the case study area. For this purpose, comparative analyses between independent power supply through particular HPS configurations were performed. The methodological approaches used in this integrated research are the following:

1. Evaluation of local conditions
2. Renewable resources availability assessment
3. Hybrid system modeling
4. Operational requirements analysis
5. Sensitivity analysis
6. Simulation and optimization

For an easier understanding of the undertaken steps, the above-mentioned steps can be seen in the following flow chart (Fig. 1):

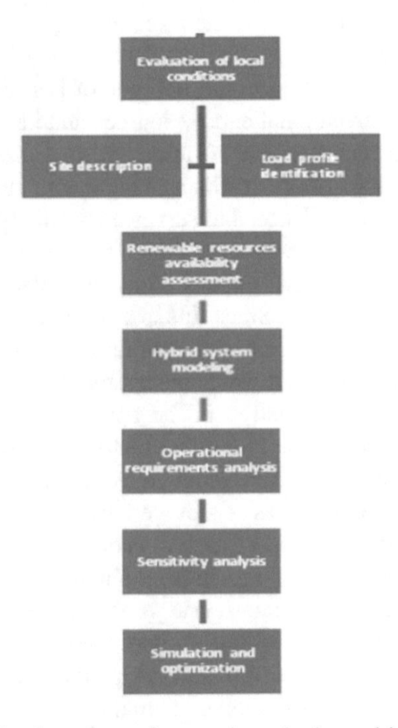

Fig. 1. The flow chart of research methods used in the study

3 Evaluation of Local Conditions

3.1 Site Description

As an example of a consumer whose power supply is to be conceptualized, a winter tourist center which comprises of a 4-floor, medium-size hotel with the area of 4013.4 m^2, a restaurant, a conference saloon, bungalows, and a ski lift equipped with snow makers, was taken into consideration. The location of the ski centre was set to be the northern Herzegovina, mountain Čvrsnica, the area of Risovac Plateau, in the immediate proximity of Blidinje Lake and the Blidinje Nature Park. The analyses were conducted for the micro-location which has the following coordinates: Latitude N 43° 38′ 22″, Longitude E 17° 32′ 44″, and the altitude of 1250 m. The exact location can be easily seen in Fig. 2:

Fig. 2. The case study location

3.2 Load Profile Identification

I order to make a good load profile approximation; a few assumptions have been made. Firstly, to realistically simulate the possibility of developing a completely independent hybrid power system that serves a ski center, the consumption data was taken from a small hotel in Havre, Montana, US, which has a similar base elevation with the case study area (1371 m) and is located at similar geographic latitude (48.16°). Secondly, an analysis of the ski lift as the main load in winter months was conducted by approximating its consumption, considering the actual weather parameters. Furthermore, electric consumption data of the snow makers, restaurant, rooms, and nearby apart-houses were evaluated, added to the ski lift hourly load data, mathematically interpolated and the load profile was made, as can be seen in Figs. 3 and 4. The average daily consumption was 1209.6 kWh, i.e. 50.4 kW per hour. The registered maximal hourly load amounted to 212.48 kW. An uneven peak distribution was present in the graph because the lift operated mostly during the weekends in December, January, February, and the beginning of March, as well as for the Workers' Day, when it was operating for touristic sightseeing. The peak season was in mid-January, where the ski lifts were working almost every day. Lowered consumption is registered during periods without snow, when the ski lift does not work at all.

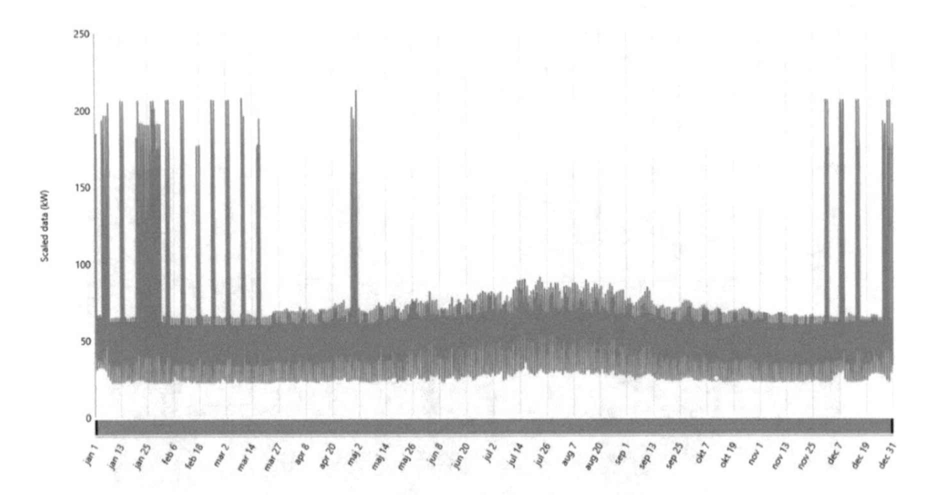

Fig. 3. Annual load profile in [kW]

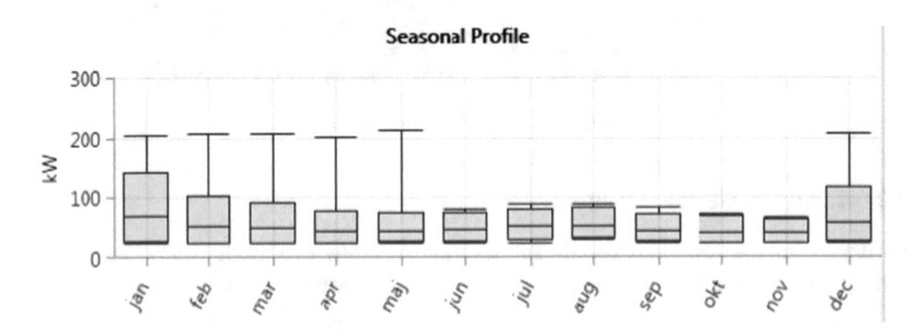

Fig. 4. Monthly load profile in [kW]

4 Renewable Resources Availability Assessment

4.1 Wind Resources

The average hourly wind speed data was taken from the actual weather parameters measured at the height of 50 m above the surface of the earth and stored in the database. The baseline data is a one-year time series representing the average wind speed, expressed in meters per second, for each time step of the year, done by NASA Surface meteorology and Solar Energy database. The average monthly wind speed data for this specific location are given by HOMER as illustrated in Fig. 5. Overall, the annual average of wind speed was 4.16 m/s.

4.2 Solar Resources

The solar radiation data have been calculated from hourly global and diffuse irradiance values provided by Photovoltaic Geographical Information System of the European

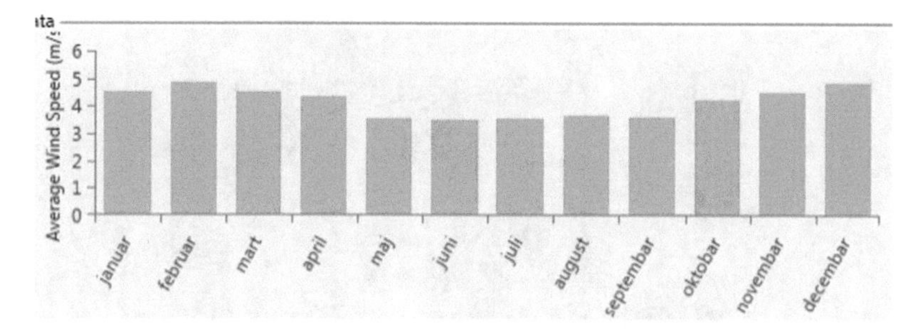

Fig. 5. Monthly wind speed data [m/s]

Commission available inside COSMO database for the given location. The daily average solar global horizontal irradiance at the location is found to be 4.82 kWh/m^2. Figure 6 indicates the monthly average solar GHI data, as well as the corresponding clearness index.

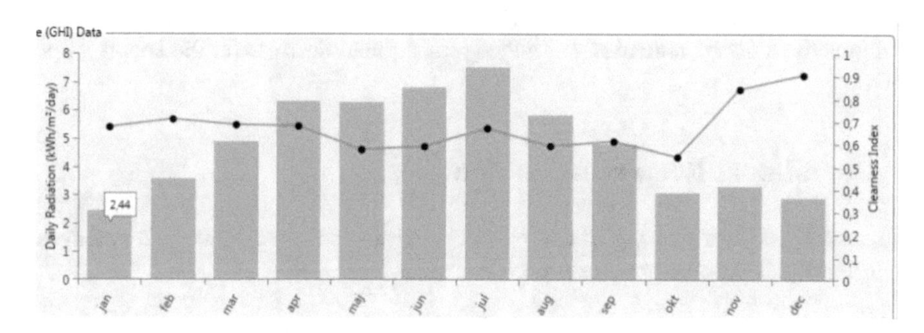

Fig. 6. Monthly average solar global horizontal irradiance and the clearness index at the location

5 Hybrid System Modeling

To satisfy the demand of the previously discussed electrical load by utilizing the above-mentioned renewable energy sources and establish an independent power system, the different system models which consists of a wind power plant, a photovoltaic power plant with storage capabilities and an additional diesel generator. The proposed configuration is shown in Fig. 7.

The configuration includes PV cells of type SPR-X21 for maximum efficiency and the size which depends on the optimization structure, and a system converter of corresponding unit sizes that convert the variable direct current (DC) output of the solar panels into a utility frequency alternating current (AC) that is used by the off-grid network. Due to specificity of the situation, a wind turbine of type L-33 produced by XANT is used. The proposed storage is approximated by generic Li-Ion 100-kWh batteries with different string sizes. It is worth mentioning that the capital cost of the

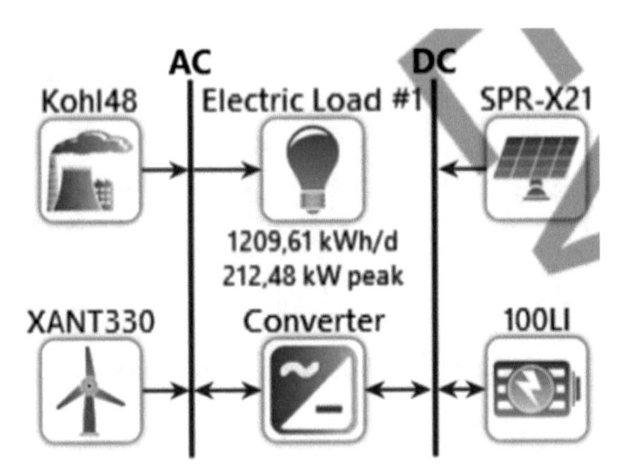

Fig. 7. Hybrid power system configuration

storage was omitted as the storage is foreseen to be in the form of electric vehicles, whose number and nominal capacity agree with the predicted capacity of the ski center. Additionally, a diesel generator was added into simulations to cover the supply of peak loads.

6 Operational Requirements Analysis

This analysis consists of system components analysis, constraints analysis and system control analysis. The summaries are given below.

6.1 System Components Analysis

The system's main components are solar photovoltaic panels, system inverter, wind turbine, storage, and a diesel engine. After a careful estimation of various solar PV cells types and models, it was decided that the panels of flat-panel type X221-335-BLK, manufactured by SunPower, will be used because of their high efficiency (21%) and acceptable nominal power of 335 W, and temperature coefficient of −0.29%/°C. The sizing evaluation of the PV module was done in the search space increments of 10 kW, in the range from 10 to 200 kW. To maintain the flow of energy between the AC and DC components, a system converter with the inverter input efficiency of 95%, rectifier input efficiency of 90% and the expected lifetime of 15 years was chosen, and the different sizes were simulated by HOMER Optimizer tool. The wind turbine was chosen according to the average wind speed data and by comparing a number of wind turbine power curves, and it was concluded that for moderate wind speed that were present at the site, the best solution would be XANT L-33, which has rated capacity of 330 kW and superior performance for the wind speeds at the location. Its lifetime was estimated to be 20 years. As the type of the diesel generator, Kohler 48 kW was chosen due to reduced CO_2 emissions and acceptable price. As already mentioned, it was assumed that majority of the center's guests and staff will be using electric vehicles, and they are modeled as generic 100 kWh batteries with very small capital and no operating costs.

6.2 Constraints Analysis

Constraints are conditions that the system must meet to be feasible. There were three types of constraints used in the simulation: operating reserve, maximum annual capacity shortage and minimum renewable fraction. The software tool calculates the total required operating reserve for each hour by multiplying hourly load percentage, annual peak load percentage, solar and wind power output by the load or output value and adding the result. Maximum annual capacity shortage was set according to the specific configuration, allowing only the shortage maximum of 1% of annual capacity. The minimum renewable fraction was set to be 50% in each case.

6.3 System Control Evaluation

A dispatch strategy is a set of rules used to control generator and the storage bank operation whenever there is insufficient renewable energy to supply the load. Two possible strategies for the system control were introduced and utilized in the simulation process: the load following (LF) and cycle charging (CC). Under the load following strategy, whenever a generator is needed it produces only enough power to meet the demand. Under the cycle charging strategy, whenever a generator has to operate, it operates at full capacity with surplus power going to charge the battery bank.

7 Sensitivity Analysis

Sensitivity analysis is performed by entering multiple values for a particular input variable, in order to determine how important that variable is, and how the solution changes depending on its value. Based on the sensitivity analysis, the demand increasing will affect the operating costs, total NPC and the CoE. As the main indicators that are affected by changes in energy supply and demand, the following cost components are used: initial capital cost, replacement cost, operation and maintenance costs. The initial capital cost of a component is the total installed cost of that component at the beginning of the project. The replacement cost is the cost of replacing a component at the end of its lifetime, as specified by Lifetime parameter in the Component model. The Operation and Maintenance (O&M) cost of a component is the cost associated with operating and maintaining that component.

8 Simulation and Optimization

For simulation and optimization purposes, the software tool called HOMER was used. The HOMER Pro® microgrid software by HOMER Energy is the global standard for optimizing microgrid design in all sectors. HOMER simulates the operation of a system by making energy balance calculations in each time step (interval) of the year. For each time step, HOMER compares the electric demand in that time step to the energy that the system can supply in that time step and calculates the flow of energy to and from each component of the system. HOMER performs the energy balance calculations for each considered system configuration. It then determines whether a configuration is feasible, (i.e., whether it can meet the electric demand under the conditions that you specify), and estimates the cost of installing and operating the system over the lifetime of the project.

9 Results and Discussion

9.1 Energy Yield Analysis

The optimal configuration of the proposed model is shown in Table 1.

Table 1. System architecture for the proposed case

Component	Name	Size
PV	SunPower X21-335-BLK	150 kW
Storage	Generic 100 kWh Li-Ion	80 strings
Wind turbine	XANT L-33	1
System converter	System converter	125 kW
Diesel generator	Kohler 48 kW standby	48 kW
Dispatch strategy	CC	

In order to determine if the proposed design could meet the energy requirement of the ski center, a comparative energy yield analysis was conducted. Its findings are summarized in Table 2. Overall, it can be noticed that the renewable power penetration is quite high and predominant. Furthermore, it can be observed that there is a considerable amount of excess electricity produced which can't be effectively stored or consumed. A good option could be connecting the HPS to the grid and selling electricity at a feed-in tariff. This solution will have both energy and economical benefits for the tourist center. In this case, the consumer wouldn't be a consumer anymore, but rather a prosumer, producing and consuming its own electricity, and at the same time selling the excess electricity to the system.

Table 2. Energy yield summary of the system

Total production (kWh/yr)	556.508
Excess electricity (kWh/yr)	77.139
Unmet electric load (kWh/yr)	1.067
Capacity shortage (kWh/yr)	4.780
Nominal renewable capacity divided by total nominal capacity (%)	90,9
Total renewable production divided by load (%)	114
Total renewable production divided by generation (%)	90,1

To assess the electricity production of different components and compare their efficiencies in satisfying consumer requirements, the graph of monthly output of HPS was given in Fig. 8 with respect to each HPS component. From there, contributions from each single HPS part can be seen and their cumulative output compared. However, in order to make a more complete and realistic picture, economic analyses needed to be conducted.

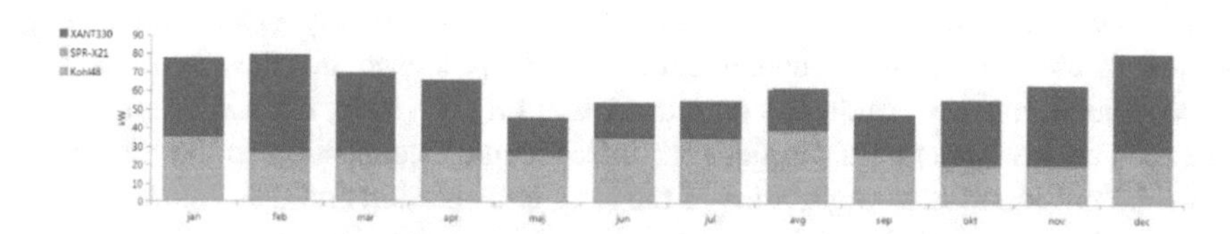

Fig. 8. Monthly average electric production

9.2 Economic Analysis

Some of the most important economic indicators of the proposed HPS design are presented in Table 3, where a breakup of the net present costs (NPC) of the system, capital, operating and replacements were given alongside the annualized costs and the levelized cost of energy. It should be noted that due to the presence of the diesel generator, bulk of the total NPC is accounted for diesel generator resource costs (not shown in the table as a separate category), which might vary according to the fuel price on the market. However, this is not clearly evident as the wind turbine capital and operating cost are considerably smaller due to the choice of wind turbine. This increases the LCoE to almost twice the value of the LCoE in Case A and C. This result doesn't surprise because of the fact that each MWh of electricity produced from renewables-only configurations results in conservation of a considerable amount of fuel, which further decreases marginal generation cost, which was 0.293 €/kWh for the DG.

Table 3. Resulting economic indicators for the considered HPS configurations

Net present system cost [€]	923.344,70
NP capital cost [€]	486.463
NP operating cost [€]	188.875
NP replacement cost [€]	15.894
Annualized system cost [€]	71.425
Annualized capital cost [€]	37.630
Annualized operating cost [€]	14.610
Annualized replacement cost [€]	1.230
Levelized cost of energy [€/kWh]	0.162

However, the choice of the best possible alternative should not be based solely upon the economic indicators. That's why there is a necessity for an integrated analysis of both electrical and economic factors. For more profound indicative consideration on feasibility of the proposed HPS configuration, an average electricity purchase price of 0.11 €/kWh was taken into account. It was found that for the time range of 25 years, the electricity purchase cost would be 1,151,216 €, which exceeds the amounts of initial capital cost for every case. Moreover, that amount of money is even larger than the total simulated NPC. Furthermore, the economic analyses didn't include the possibility of selling the excess electricity to the grid. This would certainly create much bigger because of the significant amounts of excess energy produced. For these reasons, it can be concluded that the simulated configuration is highly viable, especially after taking into account investments in the network infrastructure, connection to the distribution network, and losses that might arise in the long transmission and distribution lines.

Finally, to improve the system's reliability and security, the best solution would be establishing a smart management system within the consumer center that would enable a rational allocation of available energy and consumer needs.

10 Conclusion

This study focuses on the issue of hybrid power systems as one of most tangible actions to overcome the intermittent nature of renewable energy sources. More precisely, it offers techno-economic analysis on a proposed configuration of a remotely located load in the form of a ski center in the northern Herzegovina region. The paper uses quantitative methods of simulation, optimization and sensitivity analysis using HOMER software, to evaluate the economic feasibility and optimal electrical configurations of the desired off-grid power system, considering the availability of sun and wind resources on the site. The analyses have shown that the proposed model is sustainable, environmentally friendly and economically viable. Furthermore, this model is not limited to the considered micro-location. This concept might be used in the broader region of Herzegovina to serve various types of consumers, such as winter resorts, hotels, mountain huts, livestock pastures and other small-to-medium range remotely located loads. The possible locations of the project's applicability are the ski centers at Rujište, Raduša, Vranica, Kupres and similar locations in the larger northern Herzegovina and southern Bosnia region. It is also shown that this concept can be implemented in the future with a broader adoption of electric vehicles, smart grids and renewable energy resources. Most importantly though, this concept will certainly help to ensure the affordable, stable, reliable and environmentally sensitive energy supply to all consumers.

References

1. Diaf, S., Notton, G., Belhamel, M., Haddadi, M., Louche, A.: Design and techno-economical optimization for hybrid PV/wind system under various meteorological conditions. Appl. Energy **85**(10), 968–987 (2008)
2. Markovic, M., Nedic, Z., Nafalski, A.: Comparison of microgrid solutions for remote areas. In: Electrical Power and Energy Conference 2016, EPEC, pp. 1–5 (2016)

3. Islam, M.S., Islam, A.: Economic feasibility analysis of electrical hybrid grid in a city area. In: 2013 Proceedings of IEEE Southeastcon, Jacksonville, FL, pp. 1–6 (2013)
4. Triebke, H., Göhler, G., Wagner, S.: Data analysis of PEV charging events in rural and business environments – a load behaviour comparison. In: IAT University Stuttgart, pp. 1–3 (2015)
5. Farhangi, H.: The path of the smart grid. IEEE Power Energy Mag. **8**(1), 18–28 (2010)
6. Coyle, E.D., Simmons, R.A.: Understanding the Global Energy Crisis. Harnessing Nature: Wind, Hydro, Wave, Tidal, and Geothermal Energy. Purdue University Press, pp. 1–37 (2014)
7. Ghenai, C., Janajreh, I.: Comparison of resource intensities and operational parameters of renewable, fossil fuel, and nuclear power systems. Int. J. Therm. Env. Eng. **5**, 95–104 (2013)
8. Rehman, S., Al-Hadhrami, L.M.: Study of a solar PV-diesel-battery hybrid power system for a remotely located population near Rafha. Saudi Arabia. Energy **35**, 4986–4995 (2010)
9. Rina, Z.S., et al.: Development of a microcontroller-based battery charge controller for an off-grid photovoltaic system. In: IOP Series: Materials Science and Engineering, vol. 226, p. 012138 (2017)
10. Hunter, R., Elliot, G.: Wind-Diesel Systems, Wind-Diesel System Options: A Guide to the Technology. Cambridge University Press, Cambridge (1998)
11. Heymann, M., Nielsen, K.H.: Hybridization of electric utility regimes the case of wind power in Denmark, 1973–1990. RCC Perspect. Energy Transit. Hist.: Glob. Cases Contin. Chang. **2**, 69–74 (2013)
12. Guo, C., Bond, C.A., Narayanan, A.: The Adoption of New Smart-Grid Technologies. Incentives, Outcomes, and Opportunities. RAND Corporation (2015)
13. Jahangiri, M., et al.: Techno-economical assessment of renewable energies integrated with fuel cell for off grid electrification: a case study for developing countries. J. Renew. Sustain. Energy **7**, 023123 (2015)
14. Hassan, Q., et al.: Optimization of PV/WIND/DIESEL hybrid power system in HOMER for rural electrification. J. Phys.: Conf. Ser. **745**, 032006 (2016)
15. Merzic, A., Music, M., Rascic, M., Hadzimejlic, N.: An integrated analysis for sustainable supply of remote winter tourist centers – a future concept case study. Turk. J. Electr. Eng. Comput. Sci. **24**, 3821–3837 (2016)
16. Wang, L., Singh, C.: PSO-based multi-criteria optimum design of a grid-connected hybrid power system with multiple renewable sources of energy. In: IEEE Swarm Intelligence Symposium (SIS), pp. 250–257 (2007)
17. Banjarnahor, D.A., et al.: Design of hybrid solar and wind energy harvester for fishing boat. In: IOP Conference Series: Earth and Environmental Science, vol. 75, p. 012007 (2017)
18. Ghenai, C., Bettayeb, M.: Optimized design and control of an off grid solar PV/hydrogen fuel cell power system for green buildings. In: IOP Conference Series: Earth and Environmental Science, vol. 93, p. 012073 (2017)
19. Nishrina, et al.: Hybrid energy system design of micro hydro-PV-biogas based micro-grid. In: IOP Conference Series: Materials Science and Engineering, vol. 180, p. 012080 (2017)
20. Alaidroos, A., He, L., Krarti, M.: Feasibility of renewable energy based distributed generation in Yanbu, Saudi Arabia. World Renewable Energy Forum, Denver, CO (2012)
21. Mohamed, O.H., Amirat, Y., Benbouzid, M., Elbast, A.: Optimal design of PV/fuel cell hybrid power system for the city of brest in France. In: IEEE ICGE, Sfax, Tunisia, pp. 119–123 (2014)
22. Schmid, A.L., Hoffmann, C.: Replacing diesel by solar in the Amazon: short-term economic feasibility of PVediesel hybrid systems. Energy Policy **32**, 881–898 (2004)
23. Ulusan, H., et al.: Triple hybrid energy harvesting interface electronics. J. Phys.: Conf. Ser. **773**, 012027 (2016)

24. Turner, J., Sverdrup, G., Mann, M.K., Maness, P.C., Kroposki, B., Ghirardi, M., Evans, R.J., Blake, D.: Renewable hydrogen production. Int. J. Energy Res. **32**(5), 379–407 (2008)
25. Munuswamy, S., Nakamura, K., Katta, A.: Comparing the cost of electricity sourced from a fuel cell-based renewable energy system and the national grid to electrify a rural health centre in India: a case study. Renew. Energy **36**(11), 2978–2983 (2001)
26. Syarifah, A.R., et al.: Design of hybrid power system for remote area. In: IOP Conference Series: Materials Science and Engineering, vol. 288, p. 012010 (2018)

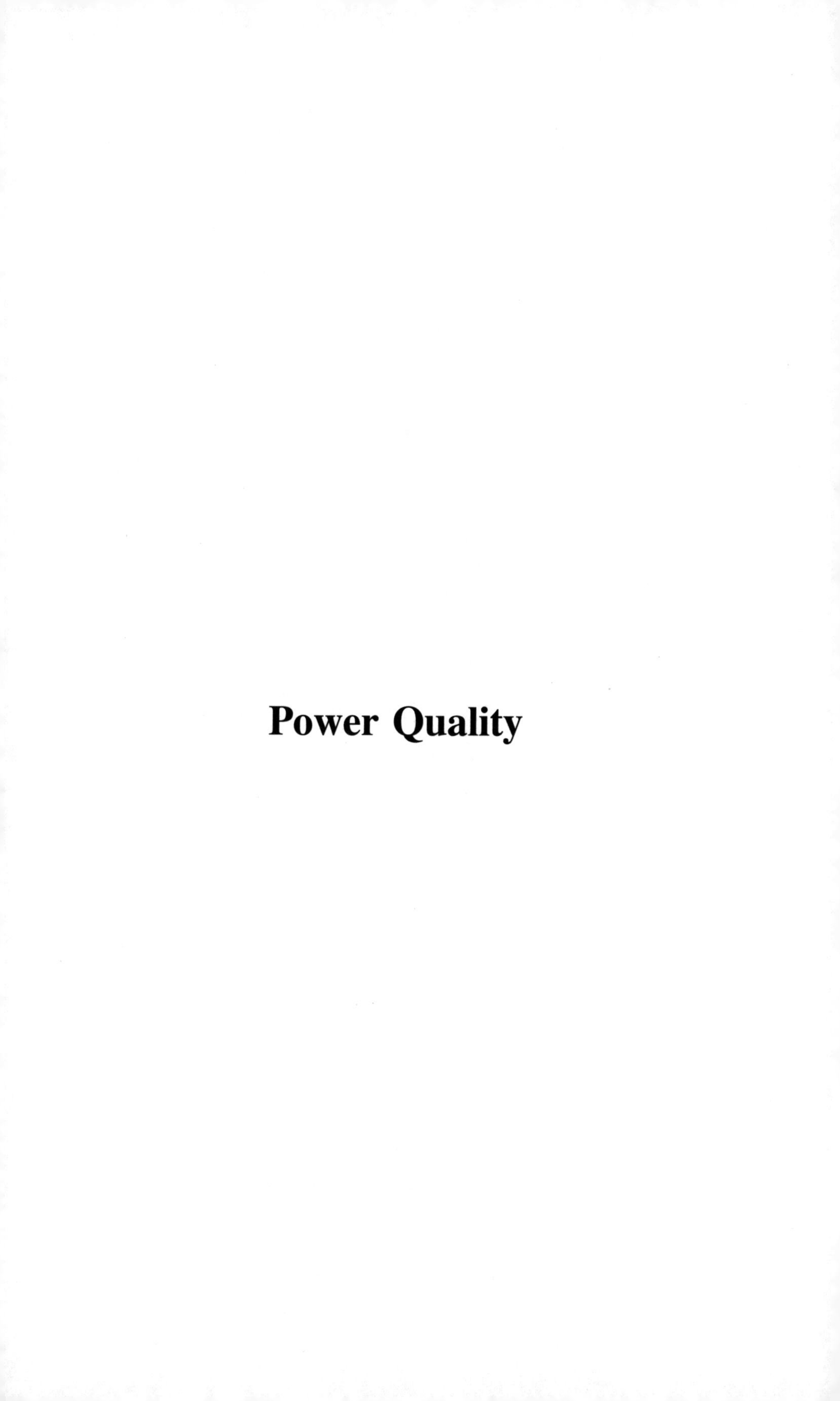

Power Quality

PV Plant Connection in Urban and Rural LV Grid: Comparison of Voltage Quality Results

Ivan Ramljak[1(✉)] and Ivana Ramljak[2]

[1] P.U. "Elektroprivreda HZ HB", Development division Mostar, Mostar, Bosnia and Herzegovina
ivanramljak1985@gmail.com

[2] Elektroprenos-Elektroprijenos BiH a.d. Banja Luka - Company for the Transmission of Electric Power in Bosnia and Herzegovina, Mostar, Bosnia and Herzegovina
ivana.marincic1988@gmail.com

Abstract. Renewable energy sources (RES) today are basic part of power system. Their penetration is present on each voltage level, from high voltage (HV) to low voltage (LV) level. Each analysis of power system regarding power grid analysis must take in account presence of RES or planned penetration of RES. RES have various impact on power grid where they are connected, in terms of load flow analysis, protection schemes, power quality and others. That impact can be positive, but negative as well. Each case of connecting RES on power grid must be analysed separately. Regarding power quality, impact of RES on point of common coupling (PCC) in power grid can be in terms of reliability and voltage quality. When one talks about voltage quality, impact of RES on PCC can be in different aspects. The most influential parameters are: voltage variations, voltage unbalance, flickers and total harmonic distortion (THD). RES impact on PCC depends on distributed generator type and strength of the grid (short circuit current and rated current of generator ratio) in PCC. Evaluation of RES impact on PCC regarding voltage quality must be performed considering some Norms (Standards) and/or Regulations. This paper deals with impact of two photovoltaic (PV) plants with equal characteristics, one connected in strong (urban area) and second one in weak (rural area) distribution grid in PCC. The impact is considered through analysis the most vulnerable voltage quality parameters, mentioned above.

1 Introduction

Main voltage quality parameters describing impact of PV plant on distribution grid are: voltage variations, voltage unbalance, flickers P_{lt} and THD. Those parameters are defined and explained in [1, 2]. Grid connected PV plants cause voltage increasing in PCC. Voltage increasing is due to currents generated by the inverters of PV plants [3]. Voltage variation depends on grid size and PV plant rated power [4]. Voltage unbalance is property of one phase inverters of PV plants and can be great (for one phase inverters) [5]. According to [6], PV plant does not have any significant impact on flicker values and there is no direct relationship between cloud days and flicker values. Voltage distortion occurs due to the currents demanded by nonlinear loads (current

S. Avdaković (Ed.): IAT 2018, LNNS 59, pp. 271–278, 2019.
https://doi.org/10.1007/978-3-030-02574-8_21

distortion). These currents flow through the impedance of the grid and results are voltage harmonics [3]. If the size of PV plant is relatively low respect to short circuit currents in PCC of the grid, there can not be significant impact of PV plant on voltage distortion in PCC.

Impact of RES on PCC regarding voltage quality parameters must be evaluated using some Standards and/or Regulations. The most used Standards are EN 50160:2011, IEEE Standards and IEC 61000 Series. The most influential Standard in region is EN 50160:2011. EN 50160:2011 is Standard for voltage quality supply published by CELENEC (European Committee for Electrotechnical Standardization). Standard gives quantitative characteristics of voltage in normal conditions [1]. This Standard has purpose to describe the characteristics of voltage parameters regarding permissible deviations. It analyses voltage characteristics regarding:

- frequency,
- waveform,
- symmetry and
- magnitude.

Measurement of voltage quality according to EN 50160:2011 requires some instruments specially created for that purpose (defined in IEC 61000-4-30 Standard) and measurement period should be one week. All measurements performed according to EN 50160:2011 are strictly defined with IEC 61000-4-x Standards (for example IEC 61000-4-7 describes harmonic measurements and IEC 61000-4-15 flicker measurements) [1, 2]. Limits of voltage quality parameters which are in scope for this paper are listed in Tables 1 and 2.

Table 1. Supply voltage parameters according to EN 50160:2011

Supply voltage parameter	Statistical evaluation	Compliance limit
Supply voltage variations	95% of the time in 1 week 100% of the time in 1 week	$U_c \pm 10\%$ U_c + 10%/−15%
Rapid voltage changes (and flicker)	95% of the time in 1 week	$P_{lt} \leq 1$
Supply voltage unbalance	95% of the time in 1 week	<2%
Harmonic voltage	95% of the time in 1 week 95% of the time in 1 week	See Table 1 THD < 8%

Table 2 shows harmonic limits according to EN 50160:2011.

Table 2. Harmonic limits according to EN 50160:2011

Individual harmonics											
THD	3	5	7	9	11	13	15	17	19	21	23
8.0%	5.0%	6.0%	5.0%	1.5%	3.5%	3.0%	0.5%	2.0%	1.5%	0.5%	1.5%

Set of Standards regarding voltage quality is released too by IEEE (Institute of Electrical and Electronics Engineers). List of those Standards by area can be found in [2]. Tables 3 and 4 show limits of voltage quality parameters which are in scope for this paper (to which authors had access).

Table 3. Supply voltage parameter according to IEEE standards

Supply voltage parameter	Statistical evaluation	Compliance limit
Rapid voltage changes (and flicker)	95% of the time 95% of the time	$P_{lt} \leq 0.8$ $Ps_t \leq 1$
Harmonic voltage	–	THD < 5%

Table 4 shows harmonic limits according to IEEE Standards.

Table 4. Harmonic limits according to IEEE standards

THD	3-n
5.0%	3.0%

Tables 5 and 6 show limits of voltage quality parameters which are in scope for this paper (to which authors had access), issued by IEC (International Electrotechnical Commission). Those parameters are from IEC 61000 Series Standard.

Table 5. Supply voltage parameter according to IEC 61000 series standard

Supply voltage parameter	Statistical evaluation	Compliance limit
Supply voltage variations	15 min	$U_c \pm 10\%$
Rapid voltage changes (and flicker)	–	$P_{lt} \leq 0.6$ (0.8) $Ps_t \leq 1$
Supply voltage unbalance	–	< 2%
Harmonic voltage	–	THD < 8%

Table 6 shows harmonic limits according to IEC 61000 Series Standard.

Table 6. Harmonic limits according to IEC 61000 series standard

THD	3	5	7	9	11	13	15	17
8.0%	5.0%	6.0%	5.0%	1.5%	3.5%	3.0%	0.3%	2.0%

2 Voltage Quality Measurements in PV Plants

PV plants analysed in this paper are identical PV plants in terms of rated current/power (number and rated current/power of inverters). Both PV plants are connected on LV grid in distribution cabinet of PCC. PCC in both cases is point where PV plant delivers energy to LV grid and where consumer consumes energy from PV plant/LV grid. It is all shown on Fig. 1. Values of short circuit currents are taken from Distribution system operator data.

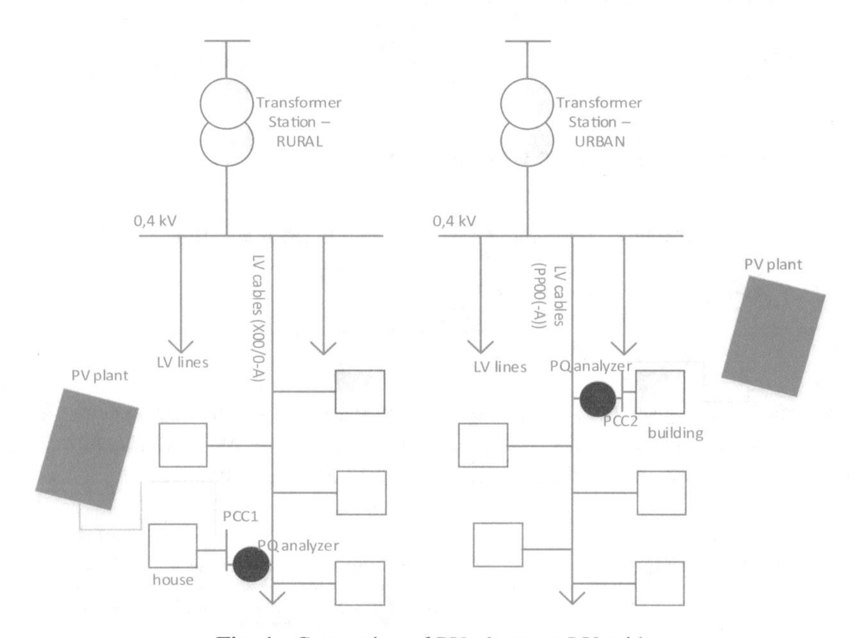

Fig. 1. Connection of PV plants on LV grid

PCC example and measurement method is shown on Fig. 2. Measurements were performed with high quality instrument according to IEC 61000-4-30.

Differences of these two PV plants are:

- one PV plant is connected in rural area (weak grid) one the other one in urban area (strong grid),
- PV plant in rural area is connected in PCC in distribution cabinet of family house and in case of urban area it is a commercial building,
- PV plant in urban area is connected with cables and the one in rural area with overhead lines,
- energy produced by PV plant in urban area is consumed mostly in PCC by commercial building (almost no energy flowing to grid) and
- energy produced by PV plant in rural area mostly flows to grid side, because house as consumer is negligible compared to rated current of PV plant.

Fig. 2. PCC and measurement method

Summary of differences of these two PV plants is presented in Table 7, where I_r (A) is rated current of PV plant, P_r (kW) rated power, and $I_{S.C}$ short circuit power at PCC.

Table 7. Summary of differences of PV plants

PV plant	LV lines	PCC	I_r (A)	P_r(kW)	$I_{S.C}$ (A)	$I_{S.C}/I_r$
Rural	Overhead lines (X00/0-A)	Distribution cabinet in connection point	34.94	23	1,450	41.49
Urban	Cables (PP00(-A))	Distribution cabinet in connection point	34.94	23	14,400	417.51

3 Results of Voltage Quality Measurements

Voltage quality measurement was performed for both PV plants according to EN 50160:2011 with PQ instruments, a week before PV plant connection and a week with parallel work of PV plant and LV distribution grid. Measurements were performed in PCC. For both (rural and urban) PV plants, maximum values of analysed voltage parameters are shown, before and after PV plant connection (Table 8). Difference is presented with Δ, what is actually impact of PV plant on PCC after PV plant connection. Positive Δ is increasing of voltage parameter value, and negative Δ is decreasing of voltage parameter value. Voltage parameter values are given per cent, except flickers P_{lt} which are given in absolute values (in accordance with EN 50160:2011).

Table 8. Results of voltage quality measurements for PV plants in PCC

Voltage parameter	Measurement time	Value before connection	Value after connection	Δ	Value before connection	Value after connection	Δ
			Rural			Urban	
Voltage variations (%)	95% week	4.99	5.11	0.12	1.40	1.04	−0.36
	100% week	5.73	7.11	1.38	1.40	1.10	−0.30
Unbalance (%)	95% week	1.33	1.34	0.01	0.82	0.87	0.05
	100% week	1.98	1.60	−0.38	1.09	1.25	0.16
Flickers P_{lt}	95% week	0.946	0.983	0.037	1.16	1.17	0.01
	100% week	1.391	1.244	−0.147	1.505	1.214	−0.291
THD (%)	95% week	2.48	2.45	−0.03	2.39	2.41	0.02
	100% week	2.67	2.70	0.03	2.46	2.49	0.03

In Table 9 is presented summary of voltage quality parameter results for PV plants before and after connection, for 95% of week. Up arrow line presents increasing of parameter values after PV plant connection, and down arrow line presents decreasing of parameter values after PV plant connection. Table 10 presents the same as Table 9 but for 100% of week.

Table 9. Summary of voltage quality parameter results for 95% of week

PV plant location	Voltage variations (%)	Unbalance (%)	Flickers P_{lt}	THD (%)
Rural	↗	↗	↗	↙
Urban	↙	↗	↗	↗

Table 10. Summary of voltage quality parameter results for 100% of week

PV plant location	Voltage variations (%)	Unbalance (%)	Flickers P_{lt}	THD (%)
Rural	↗	↙	↙	↗
Urban	↙	↗	↙	↗

It is visible from Table 8 that only voltage flickers (P_{lt}) are above permissible value for both cases of PV plants and according to all Standards (EN, IEEE and IEC). But, high level of flickers is presented regardless of PV is, or it is not connected on distribution grid. Exception is PV plant in rural area for EN 50160:2011 (before and after PV plant connection). Voltage flickers (P_{lt}) can be generally voltage quality problem in distribution grids [7]. It is interesting to note that flickers are pronounced in urban area where nonlinear loads are more present (commercial building).

All other parameters are in accordance with all analysed Standards, before and after PV plant connection.

It must be noted that impact of PV plant connected in rural area on voltage variations is not negligible (even it is in this case in accordance with Standards). Increasing of voltage value for 1.38% after PV plant connection is quite a great increase, especially if it is known that absolute voltage increasing is 7.11%. This is due weak grid, and could be greater problem if grid is even weaker or another PV plant at the same LV feeder should be connected. Voltage variations can be potentially a quite problem for PV plants connection, regarding strength of the grid and rated power of PV plant.

There is no apparently any correlation for unbalance and THD before and after PV plant connection, considering PV plants connection place (in strong or weak grid).

4 Conclusion

Voltage quality measurements for PV plants in urban and rural area are performed on 0.4 kV voltage level in PCC. Obtained results imply problems mainly with voltage flickers, and potentially problems with voltage variations. Main reason for voltage flickers existence is variable (dynamic) work mode of electrical loads connected in PCC (especially in urban commercial building). That is why flickers are greater in urban area. Those flickers are quiet independent of presence of PV plants. Voltage increase in weaker grid is related with strength of the grid and can be potentially great problem for weak grids (weaker than this one). Problem becomes more serious if some other PV plant should be connected on the same LV feeder. There was not found any correlation for voltage unbalance and THD, before and after PV plant connection, considering PV plant connection in strong or weak grid.

References

1. Ramljak, I.: Influence of PV plant connection on power quality in point of common coupling in distribution network. Bosanskohercegovačka elektrotehnika **8**, 73–82 (2014). (In Croatian)
2. Tokić, A., Milardić, V.: Power quality. PrintCom, Tuzla (2015). (In Croatian)
3. Gonzalez, P., Romero-Cadaval, E., Gonzalez, E., Guerrero, M.A.: Impact of grid connected photovoltaic system in the power quality of a distribution network. In: Technological Innovation for Sustainability, DoCEIS (2011)
4. Kopicka, M., Ptacek, M., Toman, P.: Analaysis of the power quality and the impact of photovoltaic power plant operation on low-voltage distribution network. In: Electric Power Quality and Supply Reliability Conference, Rakvere, Estonia, June 2014

5. Sikorski, T.: Power quality in low-voltage distribution network with distributed generation. Przgled Elektrochniczny **1**(6), 34–41 (2015)
6. Bletterie, B., Heidenreich, M.: Impact of large scale photovoltaic penetration on the quality of supply a case study at a photovoltaic noise barrier in Austria. In: 19th European Photovoltaic Solar Energy Conference and Exhibition, Paris, France, June 2004
7. Ramljak, I., Perko, J., Žnidarec, M.: Voltage flickers as power quality problem in industrial sector. In: Advanced Technologies, Systems, and Applications II. Springer, Heidelberg (2018)

Monitoring of Non-ionizing Electromagnetic Fields in the Urban Zone of Tuzla City

Vlado Madžarević, Majda Tešanović(✉), and Mevlida Hrustanović-Bajrić

Faculty of Electrical Engineering, University of Tuzla, Tuzla, Bosnia and Herzegovina

Abstract. During the recent years the problem about the monitoring of electromagnetic field pollution attracts the increasing attention of both scientists and national authorities. The topic of the problem lies in the increasing interest towards the local, regional, and global aspect study and control of the electromagnetic pollution. The need of receiving regular and accurate information about the changes in environmental electromagnetic radiation appeared.

This paper analyzed the cause of natural and man-made electromagnetic environment pollution, pointed out the harm of public health and the city normal production and living order and the city sustainable development. This paper suggested that comprehensive treatment measures should be taken to solve the problem of electromagnetic environment pollution in cities including the reasonable planning and layout, the real powerful supervision, the advertisement and education, and the high level of protection technology.

This paper has been written to show analysis of results measured in the urban zone. The measurement is realised by instruments ESM-100 and PCE – EM 29. For this reason, in this study electromagnetic radiation (EMR) measurements were conducted on four different days in Tuzla.

Measured EM field levels are compared with the limits that are determined by the International Commission on Non-Ionizing Radiation Protection (ICNIRP) and World Health Organization WHO.

Keywords: Electric field strength · Magnetic field strength Electromagnetic radiation (EMR) · Electromagnetic field (EM) measurement

1 Introduction

Electromagnetic (EM) waves are radiated from many sources, natural and man-made, that produce electromagnetic pollution. Radio and TV transmitters, base stations, power lines, transformers, electrical household appliances medical equipment and etc. cause EM pollution. The fast growth in the use of mobile communication services also causes an increase in the exposed EMR levels. Increasing demand for communicating from any place, multimedia usage push cellular system operators to install more base stations as each base station works within a limited geographical region and for limited number of users [1, 2]. Each base station is an EMR source, and a large number of base stations result in a long-term, probably lifelong exposure. However, the potential health

S. Avdaković (Ed.): IAT 2018, LNNS 59, pp. 279–288, 2019.
https://doi.org/10.1007/978-3-030-02574-8_22

effects they may cause is an important research topic [3]. There are international standards and limits on effects of EMR on human health. Commonly accepted limits in many countries including European Union and USA are recommended by ICNIRP [4], which is recognized by the World Health Organization (WHO).

Numerous studies have been conducted adverse biological and health effects of low-level non-ionizing radiation [5–12]. Latest study has been published in 2017 indicating that free radical damage explains the increased cancer risks associated with residential exposure to power lines, occupational exposure to extremely low frequency (ELF) and radio frequency (RF) transmitters In 1996, the World Health Organization (WHO) established the International EMF Project to address the health issues associated with exposure to EMF. The EMF Project is currently reviewing research results and conducting risk assessments of exposure to static and ELF electric and magnetic fields. Clinical studies on biological effects of EMF has indicated that being aware of the EMF levels are so critical. Dangerous region of the electromagnetic pollution caused by the electric fields around power line has been determined by Yougang et al. [13].

The goal of this study is to determine the EMR levels in urban zone like is Tuzla city which is one of the most populated (120000) city BiH based on the short term EMR measurements conducted at differently populated parts of Tuzla city using two instruments ESM-100 and PCE-EM 29. The short-term measurements were performed during three days at 62 different locations. Considering that the measurements results may be affected by the factors such as number of users, base station usage density, the distance from the base station and the existence of line of sight path, measurements were repeated on some critical locations.

2 Measurement of Non-ionizing EM Fields

Monitoring of environment regarding electromagnetic field impact put seriously theoretical, technical and organizational tasks and it is connected directly with problems for protection of peoples and environment from unfavorable influence of EMF in wide frequency band from 0 Hz to 300 GHz. Today, there is tendency in technological and technical improvement of telecommunication means radiating electromagnetic energy and enlarging its number. It is mastered new frequency bands, enlarged net of radio connections (including mobile) and radio transmission.

To ensure protection of population European Council approved and issued *Recommendation (1999/519/EC)* for limitation submission of population to EMF radiation with frequencies from 0 Hz to 300 GHz, which have to be received like a minimum requirements from all EU member - countries. Every country – member can accept and realize also more strict requirements when it is necessary. Other basic normative document is *Directive 2004/40/EC* for protection of workers issued by the European Council and European parliament with obligation every country member of EU to realize them like minimum requirements not later than 2008.

There are international standards and limits on effects of EMR on human health. The limits are recommended by the International Commission on Non-Ionizing Radiation Protection (ICNIRP), which is recognized by the World Health Organization (WHO) and based on the assumption of 24-h exposure, Tables 1 and 2.

Table 1. Reference levels of electromagnetic fields [14]

Frequency f	Electric field E(V/m)	Magnetic field H(A/m)	Magnetic flux density B(μT)	Power density Sekv(W/m^2)
9 kHz–100 kHz	87	5	6.25	-
100 kHz–150 kHz	87	5	6.25	-
0,5 MHz–1 MHz	87	0.73/f	0.92/f	-
1 MHz–10 MHz	$87/f^{1/2}$	0.73/f	0.92/f	-
10 MHz–400 MHz	28	0.073	0.092	2
400 MHz–2000 MHz	$1.375/f^{1/2}$	$0.0037/f^{1/2}$	$0.0046/f^{1/2}$	f/200
2 GHz–10 GHz	61	0.16	0.20	10
10 GHz–300 GHz	61	0.16	0.20	10

Table 2. Limited values of electric and magnetic field (public exposure) [14]

Frequency f (Mhz)	Electric field E(V/m) (public exposure)	Magnetic field H(A/m) (public exposure)
0.009–0.15	34.8	2
0.15–1	34.8	0.292/f
1–10	$34.8/f^{1/2}$	0.292/f
10–400	11.2	0.0292
400–2000	$0.55/f^{1/2}$	$0.00148/f^{1/2}$
2000–300000	24.4	0.064

Authorities of Bosnia and Herzegovina established a set of standards related to electromagnetic field radiation and human exposure, Table 3.

It is necessary to emphasize that in Bosnia and Herzegovina there are no regulations in this area (Law on Non-Ionizing Radiation Protection and Regulation about limited values of electric and magnetic field (public exposure), provided by the Law on Electricity "Official Gazette of Federation of B&H" number 41/02 i 38/05, Chapter VIII-Electrical energy production, Article 32, "Facilities for the production of electricity must meet the established criteria of environmental protection, and ensure permanent control of environmental impact".

2.1 Measurement Methodology

The EMF measurements were carried out with a two instruments:

Table 3. B&H standards [14]

	Standard
1.	**EN 50364:2001** **BAS EN 50364:2002** Limitation of human exposure to electromagnetic fields from devices operating in the frequency range 0 Hz to 10 GHz, used in Electronic Article Surveillance (EAS), Radio Frequency Identification (RFID) and similar Applications
2.	**EN 50371:2002** **BAS EN 50371:2007** Generic standard to demonstrate the compliance of low power electronic and electrical apparatus with the basic restrictions related to human exposure to electromagnetic fields (10 MHz–300 GHz) – General public
3.	**EN 50385:2002** **BAS EN 50385:2005** Product standard to demonstrate the compliance of radio base stations and fixed terminal stations for wireless telecommunication systems with the basic restrictions or the reference levels related to human exposure to radio frequency electromagnetic fields (110 MHz–40 GHz) – General public
4.	**EN 50360:2001** **BAS EN 50360:2005** Product standard to demonstrate the compliance of mobile phones with the basic restrictions related to human exposure to electromagnetic fields (300 MHz–3 GHz)

1. The ESM-100 3D Field meter is patented, hand-held instrument which allows easy measuring of alternating electric and magnetic fields at the same time, independant of direction and corresponding to one common point. Special features of this instrument are:

   - Simultaneous isotropic measurement of E and H fields
   - Frequency range from 5 Hz–400 kHz
   - Measuring range 1nT-20mT and 0.1 V/m–100 kV/m
   - 6 channel FFT and osciloscope function
   - High precision ± 5%
   - Standardized measurements according EN V50166, DIN VDE 0848
   - Long time recording over 24 h
   - Long term memory capacity for 65520 readings
   - Up to 45 h of continous operation

2. The PCE – EM 29 instrument for measurements of high frequency EM fields

   - 3D sensor
   - Calculating average value of EM field
   - Long term memory capacity for 65520 readings
   - Measure values in mV/m, V/m, mA/m, mW/m^2
   - LCD display
   - veliki zaslon
   - frequency range 50 MHz–3.5 GHz
   - powered by batteries

Measurements were carried out in the area of Tuzla, on 62 specific locations. Locations were chosen specifically, on the places with daily large groups of people.

Locations were chosen mostly close to public facilities such as schools (primary and secondary), colleges, playgrounds, shopping centers, parks and open spaces in the town center.

EMR measurements were conducted on the route illustrated in Fig. 1 during busy times of daily hours (10.00–15.00).

In Fig. 1 each numbered red dot indicates the measurement location roughly.

Based on the international standards and ordinances released by ICNIRP and ICTA, the average E value at a location was recorded after six minutes long measurement.

The measurement locations are shown in the Fig. 1.

A closer view of measurement location points on the territory of Tuzla, one part is shown of Fig. 1.

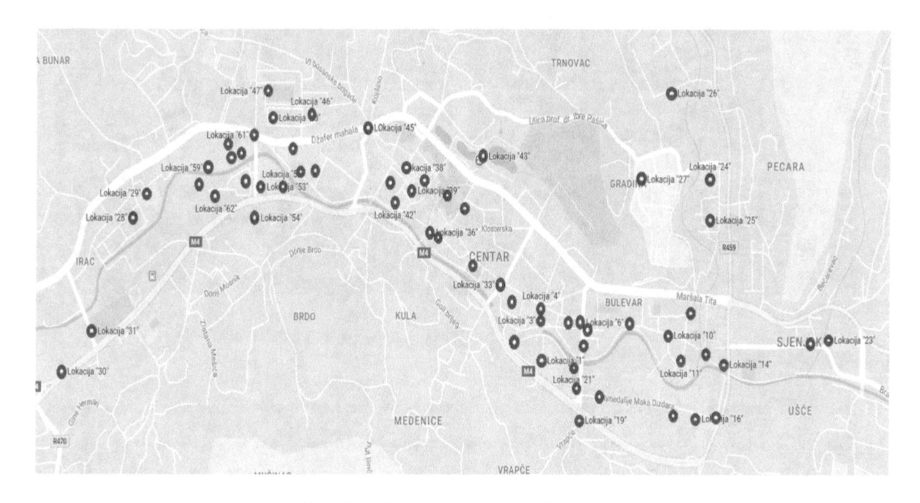

Fig. 1. The measurement locations

The measurements were taken on in period 20–25 November 2017 (Fig. 2).

Fig. 2. A closer view of measurement location points on the territory of Tuzla-part 1

It should be noted that the measurements are realized under favorable weather conditions, no rainfall and very low temperatures.

The meter was held one meter above the ground and placed in the center of the probe. Both field strengths at various distances were measured.

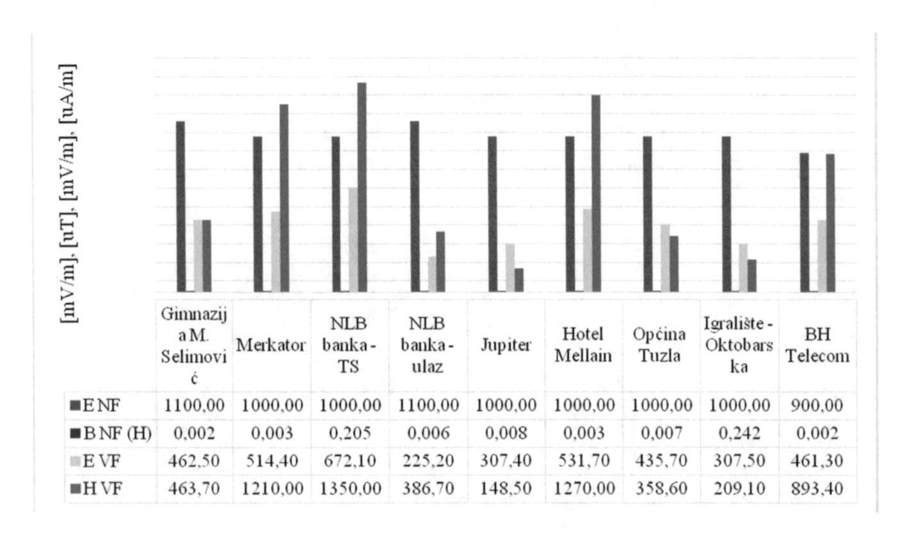

	Gimnazija M. Selimović	Merkator	NLB banka - TS	NLB banka - ulaz	Jupiter	Hotel Mellain	Općina Tuzla	Igralište - Oktobarska	BH Telecom
E NF	1100,00	1000,00	1000,00	1100,00	1000,00	1000,00	1000,00	1000,00	900,00
B NF (H)	0,002	0,003	0,205	0,006	0,008	0,003	0,007	0,242	0,002
E VF	462,50	514,40	672,10	225,20	307,40	531,70	435,70	307,50	461,30
H VF	463,70	1210,00	1350,00	386,70	148,50	1270,00	358,60	209,10	893,40

a)

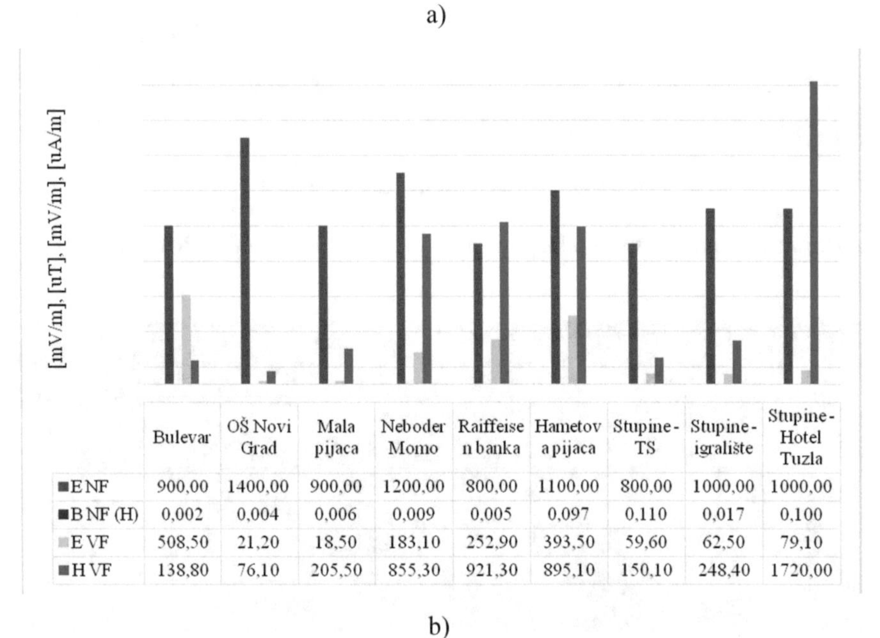

	Bulevar	OŠ Novi Grad	Mala pijaca	Neboder Momo	Raiffeisen banka	Hametova pijaca	Stupine - TS	Stupine - igralište	Stupine - Hotel Tuzla
E NF	900,00	1400,00	900,00	1200,00	800,00	1100,00	800,00	1000,00	1000,00
B NF (H)	0,002	0,004	0,006	0,009	0,005	0,097	0,110	0,017	0,100
E VF	508,50	21,20	18,50	183,10	252,90	393,50	59,60	62,50	79,10
H VF	138,80	76,10	205,50	855,30	921,30	895,10	150,10	248,40	1720,00

b)

Fig. 3. (a)–(c) Comparison of results of measurements of electric and magnetic fields at 27 of 62 locations

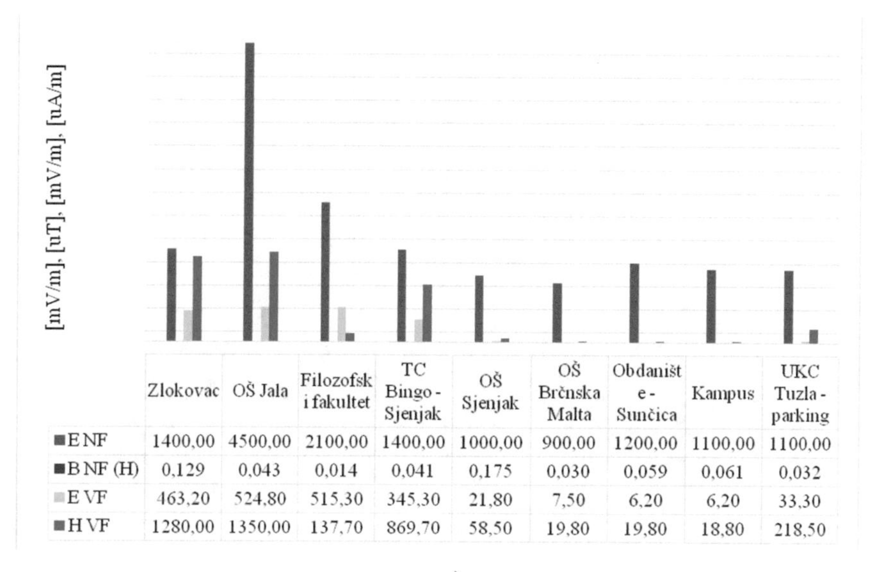

	Zlokovac	OŠ Jala	Filozofsk i fakultet	TC Bingo - Sjenjak	OŠ Sjenjak	OŠ Brčnska Malta	Obdaništ e - Sunčica	Kampus	UKC Tuzla - parking
E NF	1400,00	4500,00	2100,00	1400,00	1000,00	900,00	1200,00	1100,00	1100,00
B NF (H)	0,129	0,043	0,014	0,041	0,175	0,030	0,059	0,061	0,032
E VF	463,20	524,80	515,30	345,30	21,80	7,50	6,20	6,20	33,30
H VF	1280,00	1350,00	137,70	869,70	58,50	19,80	19,80	18,80	218,50

c)

Fig. 3. (*continued*)

3 Results of Measurement

The task of this research was to measure the electric and magnetic fields at low and high frequencies in all the locations, and to make a comparison and analysis of the obtained results, on the basis of the values prescribed by the law regulations and standards.

The objective of general implementation of this type of measurement is to detect "critical" parts of the city from the effects of EM radiation, and download specific measures to protect people in such areas.

The results of all measured electric and magnetic fields are presented graphically in Fig. 3(a–c).

The maximum value of the electric field at low frequencies is observed in primary school Jala with average value 4.5 V/m.

All other values were below 1.5 V/m.

The values of the magnetic fields at low frequencies are very low, almost negligible, as is clearly seen in the diagrams. The highest measured value was 0.820 μT, at the Faculty of Law.

The maximum values of electric field at high frequency were measured in the shopping center Robot, with average value 1.1 V/m and Chemical school, 1.2 V/m.

The maximum values of the magnetic field were measured in the central part of Tuzla city.

In addition to electric and magnetic fields, instrument for measuring EM fields at high frequencies, also has the ability to measure power density in free space in W/m^2.

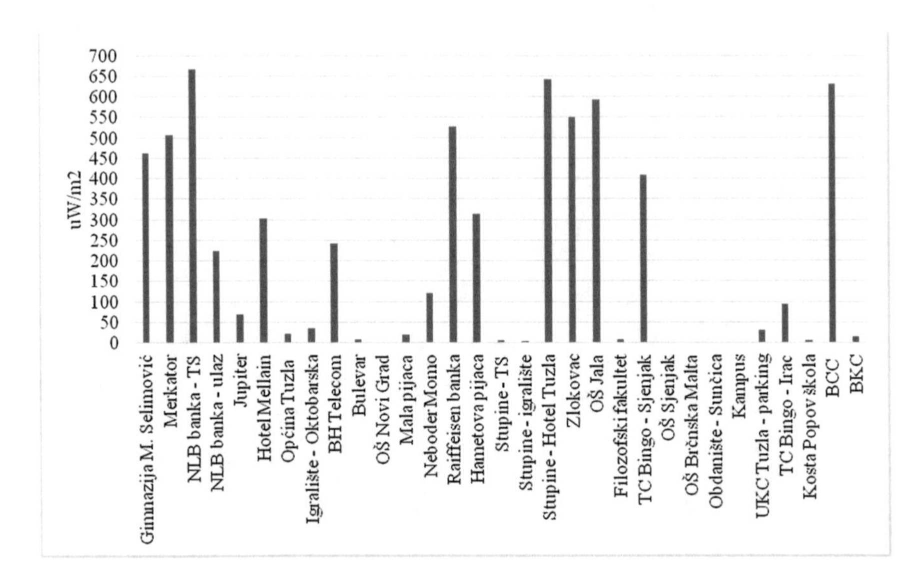

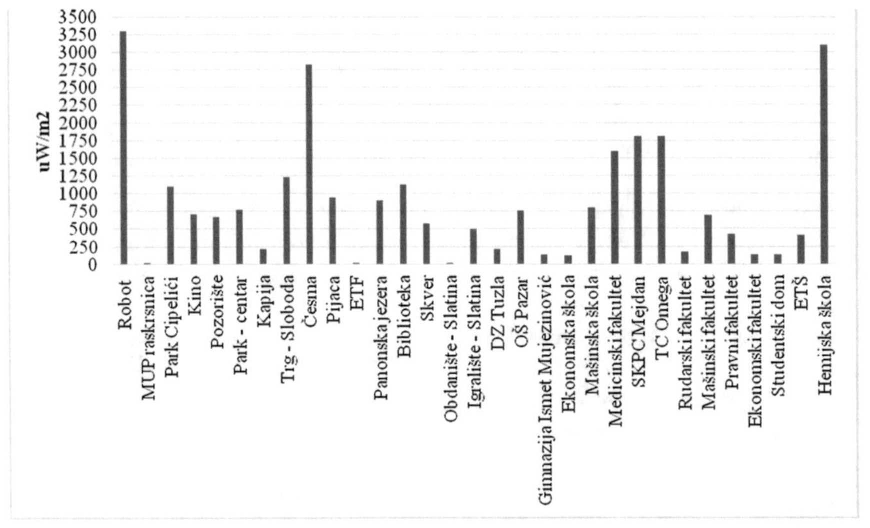

Fig. 4. Comparison of measurements results of power density at 62 locations

The power density values, measured on all 62 locations are compared and presented on diagram, Fig. 4.

All of the measured values were below those prescribed by standards and recommendations. Maximum measured value was 3250 $\mu W/m^2$.

Maximum values of electric and magnetic fields are detected near by power stations, places of crossing cable installations (power cables, telecomunication cables), switchboards. High values of RF electrical and magnetic fields 1.1–1.2 V/m and power density of 3250 μW/m2 in central part of the town are caused by rapid urbanization and a growing number of mobile telephony base stations as a result of rapid technological development.

Measurement results are in accordance with the recommendations IEC 61000-3-2, IEC 61000-3-4 i IEC 61000-3-6 recommendations on electromagnetic compatibility. They define the boundaries of the higher harmonic currents that equipment may be broadcast in the network, and that they do not cause interference with the other elements of the power system.

4 Conclusion

In this study, in order to determine the EMR pollution level in Tuzla city, EMR measurements were performed at 62 different locations at November 2017. According to over large number of measurements the maximum total LF electric field strength was 4.5 V/m and the maximum total HF electric field strength was 1.2 V/m. Although these value are below the limit suggested by ICNIRP.

Also, the maximum total LF magnetic field strength was 0.11 μT and the maximum total HF magnetic field strenght was 1.7 mA/m. Although these value are below the limit suggested by ICNIRP.

All the measurements done in the city are compared with ICNIRP safety limit values. Electric field limits (r.m.s.) defined in ICNIRP regarding the human exposure to electromagnetic fields is 30 kV/m for workers, and 10 kV/m for general public for low frequency (0 Hz to 10 kHz). Magnetic field limit (r.m.s.) defined in the standard is 1.6 mT (1280 A/m) for workers, and 0.64 mT (512 A/m) for general public.

The power density values, measured on all 62 locations are compared. All of the measured values were below those prescribed by standards and recommendations. Maximum measured value was 3250 $\mu W/m^2$.

In conclusion, careful planning in placing the electrical equipment in the buildings and parts of the city is vital. This means that the technical specification of the equipment used in the buildings, the location of electrical and high-power transmission lines, transformer stations have to be taken into account to enable safety EMF levels.

Systematic and periodic measurements are necessary to determine is the aggregate impact of all emissions in accordance with prescribed standards and norms for environmental protection. In these situations, it is very difficult to make a model of locations and realize a calculation, so that only the results obtained by measuring fully guarantee the quality assessment of the impact of electromagnetic radiation on the environment and technical equipment.

The system for monitoring of the level of electromagnetic radiation in the environment must meet three fundamental requirements: objectivity, reliability and continuity. Objectivity is achieved by the public announcement of the results of measurements whenever measurements are conducted. Reliability arises from compliance with international norms and standards relating to the measurement of electromagnetic fields, as well as exclusive use of measuring equipment calibrated (calibrated) by accredited calibration laboratories. Continuous implementation of objective and reliable measurement (24 h/365 days) allows permanent monitoring of electromagnetic radiation and maximum transparency. As the electromagnetic radiation can not be seen or feel the results of continuous measurements are of great importance for the general human population, because they represent the only objective indicator by which people can get an impression how the radiation and what is its short and long term variability, and therefore these contribute to the establishment of trust between all stakeholders.

We believe, that our study will aid the society to strengthen the protection consciousness and lead the young academicians in their future studies about the ELF measurements.

References

1. Mousa, A.: Electromagnetic radiation measurements and safety issues same cellular base stations in Nablus. J. Eng. Sci. Technol. Rev. **4**(1), 35–42 (2011)
2. Genç, O., Bayrak, M., Yaldiz, E.: Analysis of the effects of GSM bands to the electromagnetic pollution in the RF spectrum. Prog. Electromagn. Res. **101**, 17–32 (2010)
3. Habash, R.W.Y.: Bioeffects and Therapeutic Applications of Electromagnetic Energy. CRC Press, Boca Raton (2007)
4. ICNIRP Guidelines. Guidelines for limiting exposure to time - varying electric, magnetic, and electromagnetic fields (up to 300 GHz). Health Phys. **74**(4), 494–522 (1998). International Commission on Non-Ionizing Radiation Protection
5. Habash, W.T.R.: Electromagnetic Fields and Radiation. Human Bioeffects and Safety. Marcel Dekker, New York (2002)
6. Lee, J.M., Pierce, K.S., Spiering, C.A., Steams, R.D., Van Ginhoven, G.: Electrical and Biological Effects of Transmission Lines: A Review. Boneville Power Administration, Portland (1996)
7. Havas, M.: Biological effects of non-ionizing electromagnetic energy: a critical review of the reports by the us national research council and the US national institute of environmental health sciences as they relate to the broad realm of EMF Bioeffects. Annu. Rev. Environ. Resour. Rev. **8**(2000), 173–253 (2000)
8. Demirci, H.: Calculation and effect of electromagnetic fields with the human body. Marmara Master Science Thesis, University Physical Sciences Institute, Istanbul (2001)
9. Havas, M.: Radiation from wireless technology affects the blood, the hearth and the autonomic nervous system. Rev. Environ. Health **28**(2–3), 75–84 (2013)
10. Carpenter, D.O., Sage, C. (eds.).: BioInitiativereport: a rationale for biologically-based public exposure standards for electromagnetics fields (ELF and RF) (2012). www.bioinitiative.org. Accessed 20 Apr 2017
11. Levitt, B.B., Lai, H.: Biological effects from exposure to electromagnetic radiation emitted by cell tower base stations and other antenna arrays. Environ. Rev. **18**(2010), 369–395 (2010)
12. Blank, M., Havas, M., Kelley, E., Lai, H., Moskowitz, J.: International apeal: scientists call for protection from non-ionizing electromagnetic field exposure. Eur. J. Oncol. **20**(3/4) 180–182 (2015). www.emfscientist.org. Accessed 20 Apr 2017
13. Yougang, G., Lifang, Y.: Determination of dangerous region of the electromagnetic pollution caused by the electric fields around power line. In: International Conference on Communication Technology, ICCT 1998, 22–24 October Beijing, China (1998)
14. Pravilo 37/2008 o ograničavanju emisija EM zračenja. Regulatorna agencija za komunikacije BiH

Improving the Krnovo Wind Power Plant Efficiency by Means of the Lithium-Ion Battery Storage System

Filip Drinčić, Saša Mujović(✉), Martin Ćalasan, and Lazar Nikitović

Faculty of Electrical Engineering, University of Montenegro, Podgorica, Montenegro
sasam@ac.me

Abstract. Unpredictable wind nature highly affects wind power plants operation and their capability for an electrical energy production. The Krnovo wind power plant, as a new renewable energy source in Montenegrin power system, is faced with this issue. The utilization of an electrical energy storage system (EES) is proposed in this paper as a way for the problem overcoming. Through performed simulation in MATLAB Simulink software, it was concluded that connection of lithium-ion battery to the wind plant will significantly improve its efficiency. The results of simulations proves the capability of the lithium-ion battery system to maintain the value of Krnovo output power stable, making this facility a more constant power source.

1 Introduction

Due to the growing environmental pollution caused by the operation of the conventional electricity sources (mostly because of their negative influence on the atmosphere), renewable energy sources are gaining in importance. Wind power plants, as one of the fastest growing renewable energy sources, use free and inexhaustible wind energy to generate electricity. However, despite of their positive sides, as source of electricity that does not pollute the environment, when synchronizing with the electrical power system, their negative sides have to be considered: their inconsistent and unpredictable production can adversely affect the quality of the electrical energy (slower and faster voltage variations, voltage imbalance, frequency variations, occurrence of voltage flickers and harmonics, inefficiency). Variable output power of the wind power plants can also have a negative impact on the stability and reliability of the power system, and if it is connected to a weak power grid, its unsteadiness and uncertainty can cause a complete collapse of voltage.

The problems caused by connecting wind power plants to the system are not just related to the fact that the generated power changes, but for the fact that a number of variations differs from hour to hour, and that it is difficult to predict. The unpredictability of the production of a wind power plant limits their maximum participation in the production of an electricity system, and requires an increase in reserves in the system. Large variations in generated power create a problem in production planning, affecting voltage variations and stability, both by wind power plants and by the power

S. Avdaković (Ed.): IAT 2018, LNNS 59, pp. 289–302, 2019.
https://doi.org/10.1007/978-3-030-02574-8_23

system. The list of a very common problems related to wind power plants operation is given in Table 1.

Table 1. Possible negative impacts of the wind power plant on the power system

Cause	Problem
Wind impacts	Impacts of the high current
Wind speed variations	Occurrence of voltage flicker
Turning on/off the wind turbine	Occurrence of voltage flicker
Generator switching current	Voltage drop
Power converters	Occurrence of voltage harmonics

Wind speed can not be controlled, but we can control the wind power generated energy, thus its efficiency and constantinity, using electrical energy storage (EES) systems. EES transform electricity into another form of energy and use it when it is most needed for the power system.

EES play a very important role in eliminating the disadvantages of renewable energy sources, making them more efficient and more usable, through several important applications: grid stabilization, frequency regulation, transmission loss reduction, load leveling, peak shaving, capacity firming, power quality management etc. For the wind power plants, the most important type of application is capacity firming - ability of the EES to maintain the changeable output from renewable sources at a commited (firm) level for a certain period of time. Large wind power plant needs a huge EES and a good control system to firm its output and make it more consinsent, more predictable and more suitable for the power system. EES can control fluctuations of output active power and adjust reactive power.

The lithium-ion battery, as an EES with the best performance in several categories (self discharge, efficiency, charging time, capacity) is the most favourable solution for integration with wind power plants, because of its high energy density, fast voltage response, and long lifetime expectancy. This paper is based on the simulation of the control of the output power of the Krnovo wind power plant, using a lithium-ion battery, in order to increase its efficiency and role in the power system of Montenegro.

The remainder of the paper is organised as follows: In the next section (Sect. 2), the basic characteristics of the Krnovo wind power plant, its influence on the network, and the way of its connection to the power system are presented. The above data will serve as a basis for a later simulation in MATLAB Simulink software. Section 3 describes the basic principle of the operation of the lithium-ion battery, the method of connection to the system, its advantages and disadvantages. The characteristics that make this battery recommended for integration with the wind power plants are described. Section 4 offers an overview and analysis of the simulation of the connection of the lithium-ion battery to the wind power plant Krnovo, providing adequate graphical explanations. Conclusion and literature are given at the end of the paper.

2 The Krnovo Wind Power Plant

The Krnovo wind facility was put into operation in early July 2017., and it is currently the only Montenegrin wind power plant: with a total power of 72 MW, and a planned annual output of 200GWh, which is, according to estimates, sufficient to power about 50 000 households [1].

Krnovo consists of 26 asynchronous doubly-fed induction generators (DFIG), with rated power of 2.85 MW and 2.5 MW (20 × 2.85 MW + 6 × 2.5 MW). Wind generators are located at an altitude around 1500 m, with average wind speed around 5.5 to 6.5 m/s. The simplified scheme of the Krnovo wind facility is presented in Fig. 1 [1].

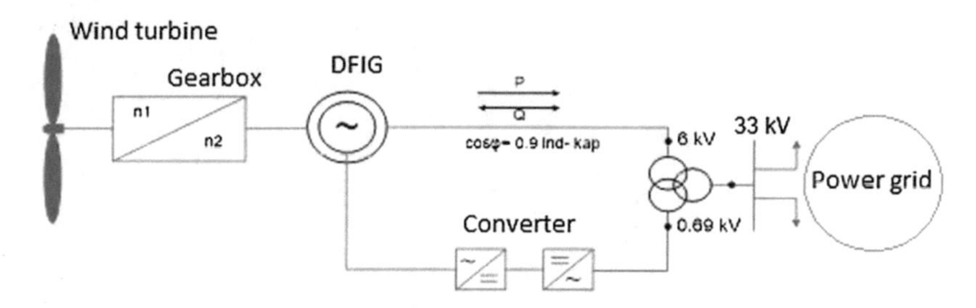

Fig. 1. Scheme of the "Krnovo" wind power plant [1]

Wind power plant can have different negative impacts on the power system, depending on the type of the wind turbine. There are two types of wind turbines: with variable and with constant rotational speed, or with the ability to adjust to wind speed changes or without that ability. In the case of wind turbines with the constant speed, the variations of produced power are directly transferred to the power system, which can cause voltage fluctuations and occurrence of voltage flickers in the power system. For the wind turbines with the variable speed, the variations are not directly transferred to the system, due to the power electronic devices which can neutralize the appearance of voltage flickers, but can cause the appearance of voltage harmonics. Using appropriate filters their impact can also be neutralized. Wind turbines on Krnovo belongs to group of turbines with ability of adjusting to the wind variations.

According to the results from the Elaborate on testing and compliance of the operation of the Krnovo, it was confirmed that all parameters concerning power quality are within the prescribed limits (voltage variations in normal operation, voltage asymmetry, power factor, voltage harmonic value), except flicker emission factors, which has occasionally appeared in values higher than allowed. However, according to the Elaborate, their impact on the network is insignificant [1]. These results are expected for this type of wind turbine. However, we can not influence on weather conditions and the forecast can not be credibly established for a longer period of time. The inability to own these informations, or the inability to plan the use of the exact amount of power from the wind power plant for a longer period, limits their efficiency and role in the power system.

By analyzing the Figs. 2, 3 and 4 obtained on the basis of the results of Krnovo wind power plant tests, it can be concluded that variations of the output power may pose a problem for its efficient use in the Montenegrin power system.

As it is mentioned above, average wind speed at Krnovo area is 5.5–6.5 m/s. Figures 2 and 3 represents typical situations on a case of one day with little wind (lower than the average), and day with much wind (higher than the average), while Fig. 4 represents the total production of the Krnovo during the trial period of work [1].

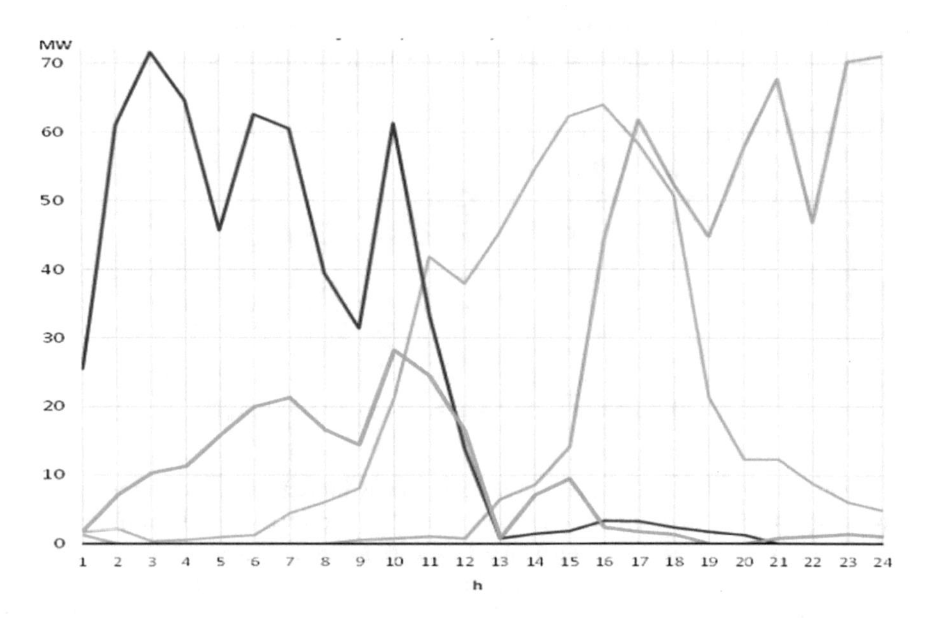

Fig. 2. Variations of "Krnovo" output power for a period of one day with little wind [1]

Forecast of the Krnovo production, based on the measurements results during trial work period, presented in Figs. 2 and 3 show significant variations in power. Different lines represents the production of Krnovo for 24 h for different days [1]. In a certain part of the day, Krnovo can achieve high production, while in the second part of the day this production is significantly lower. The total production measured at the monthly level, as shown in Fig. 4, shows large deviations between maximum and minimum production in certain months. These informations shows us that big and sudden changes in the wind speed can lead to unstable production of the Krnovo.

Taking these results into account, in order to increase the role of the Krnovo in the Montenegrin power system, this paper proposes connection of the lithium-ion battery energy storage system (Li-ion BESS), with the aim of adjusting the output power of the Krnovo to a more constant value. This type of application of the EES is called capacity firming. EES can modify variations in active power, adjust reactive power and solve problems related to the impossibility of forecasting the change of wind speed for a longer period of time.

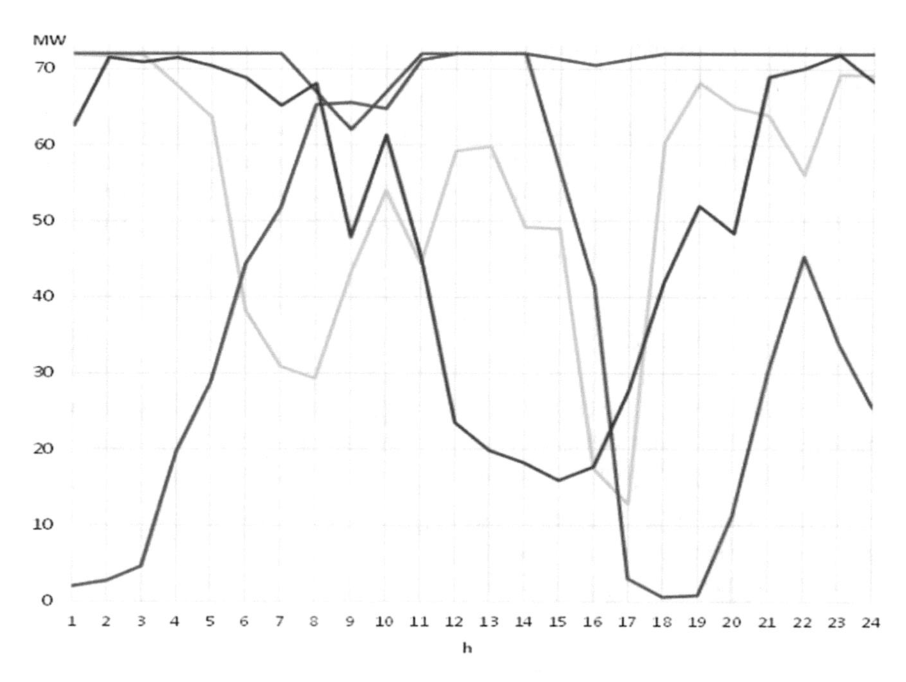

Fig. 3. Variations of Krnovo output power for a period of one day with strong wind [1]

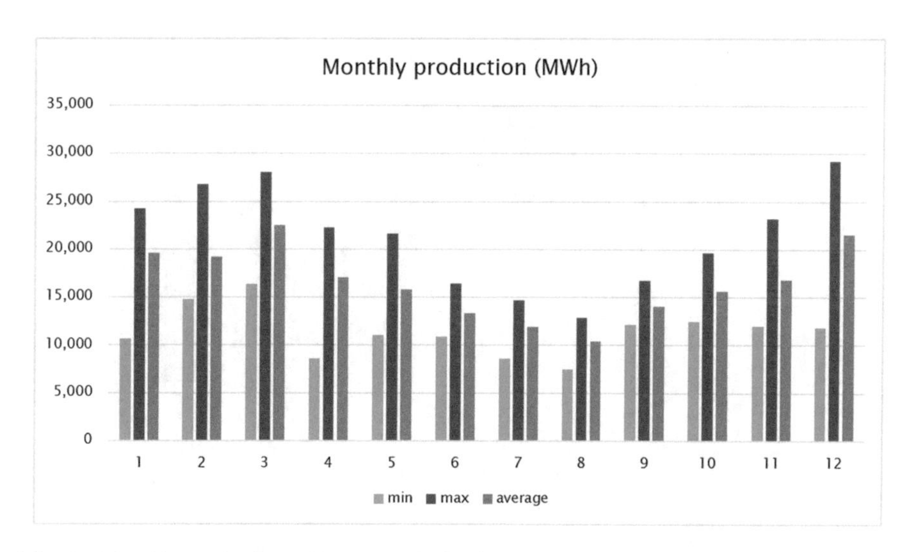

Fig. 4. Monthly production of Krnovo, obtained on the basis of the trial period of work [1]

The following technical requirements must be met for the connection of the EES at the wind power plant [5]:

(1) Long lifetime;
(2) High charge/discharge efficiency;

(3) Affordable price and payback period;
(4) Safety and reliability.

Regarding the above technical requirements, the best results have Li-ion BESS: efficiency around 90%, payback period around 10 years of use [7], lifetime around 4000 cycles [8]. Other features that distinguish the lithium battery in front of the others are power and energy density, large capacity, low self-discharge (around 0.005% per month [8]), discharge duration, etc.

3 The Li-ion BESS

BESS converts electrical into chemical energy during the process of charge, and during discharge it converts chemical to electrical energy and releases it to the system.

BESS is a complex system, which consist of 4 interconnected parts:

(1) Storage medium (li-ion battery in this case);
(2) Power conversion system (PCS);
(3) Battery management system (BMS);
(4) Balance of plant (BOP).

The scheme of the connection of BESS to the power system is presented in Fig. 5.

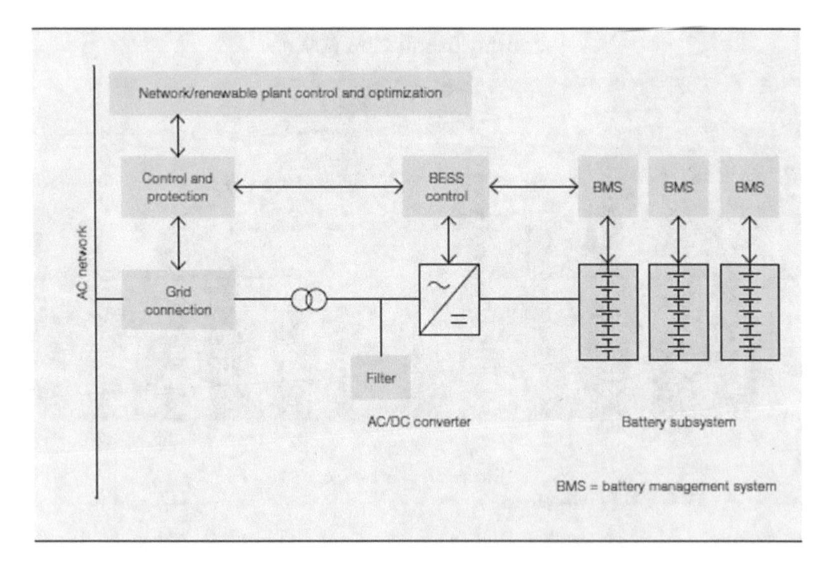

Fig. 5. Scheme of the BESS connection [3]

(1) Li-ion battery:

The battery cell consist of the following parts: electrodes (cathode$^+$ and anode$^-$), electrolyte, separator and container. Technology of the li-ion batteries is based on the lithium, which has very high reactivity and electrochemical potential, because of which the battery has higher power and energy density over other batteries. Also, lithium is the lightest metal, and because of that fact the li-ion battery has the lowest weight compared to the other types of batteries, which is important advantage. The li-ion battery scheme is shown in the Fig. 6.

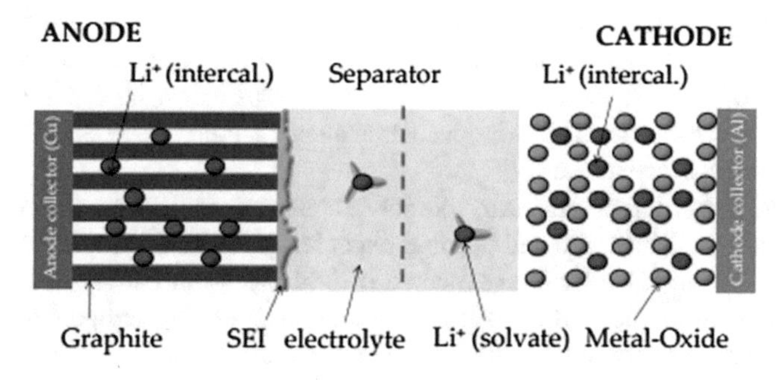

Fig. 6. Scheme of the li-ion battery [9]

During the battery process of charge, the anode receives electrons from the cathode through the external circuit (which existence is necessary in the case of li-ion battery) and the li^+ ions from the cathode are moving through the electrolyte to the anode, so that the whole system is then electrically neutral. Electrolyte serves to transfer li^+ ions during charge/discharge process. During the battery discharge process, the electrons are released into the external circuit, while the li^+ ions are moving through the electrolyte back to the cathode. The movement of the li^+ ions causes the formation of a current flow from the cathode to the anode [2].

(2) Power conversion system – PCS:

PCS controls the conversion and power flow for the battery charge/discharge process. The most common PCS solution is buck/boost converter, which adjust the power grid voltage level to the voltage level of the battery system. Another PCS solution is the bi-directional inverter, which is capable to convert DC to AC and opposite.

Another part of the PCS is the low-pass filter, which is used to mitigate the impact of the high frequency harmonics. The most commonly used is LCL filter, connected between each phase [2].

(3) Battery management system – BMS:

BMS – control unit that decides in which direction the energy flows, and unit that is responsible for battery operation. It consist of battery control and monitoring system, protection and local MicroSCADA system. Battery control unit must be at high level to avoid excessive charge or discharge of the battery cell. The most important parameters that must be controlled during battery operation are the state of charge (SOC), voltage, current, depth of discharge (DOD) and temperature.

The discharge current have to be strictly controlled at low temperatures, since the internal resistance is increasing and the avaliable capacity and voltage are decreasing drastically. While charging at low temperature, the internal resistance is increasing, which negatively affects the value of the charging current and battery voltage.

During excessive charging process there is a risk of a sudden increase in battery voltage which can lead to the battery damage. Likewise, during excessive discharging process, battery voltage can drop on too low value which can also lead to the battery damage. When the battery voltage is below 2.5 V, the discharge process have to be stopped. During the discharge process, if the SOC drops below 20% it is more difficult to operate the battery due to the high inconstancy of the voltage and the associated increase of internal resistance [5].

SOC, DOD and temperature are the parameters that can have negative impact on the battery life – calendar life and lifetime expectancy (cycle lifetime). As can be seen from the Fig. 7, the calendar life (expected life) of the li-ion battery decreases with a higher percentage of the SOC and higher optimum operating temperature. The optimum value of the battery SOC is 50%, with temperature around 20–40°C. The number of cycles decreases with a higher percentage of the DOD (Fig. 8) [2].

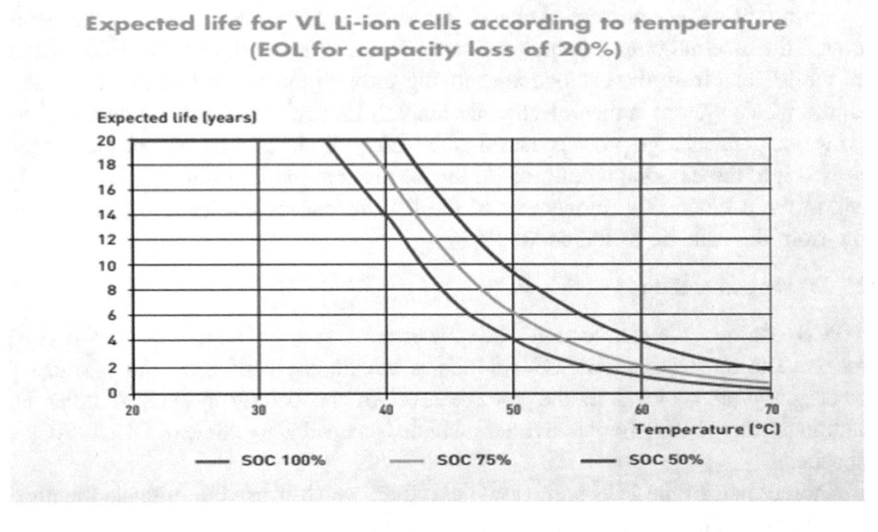

Fig. 7. Expected battery life depending on the SOC. EOL – End of life of the battery [4]

Complex control of the SOC and DOD, large safety requirements related to the control of temperature, current and voltage are the disadvantages of the li-ion over other types of batteries.

(4) Balance of plant – BOP:

The BOP system consist of transformers, switches, protective equipment, monitoring and other to ensure reliable connection of the BESS to the power plant and reliable operation.

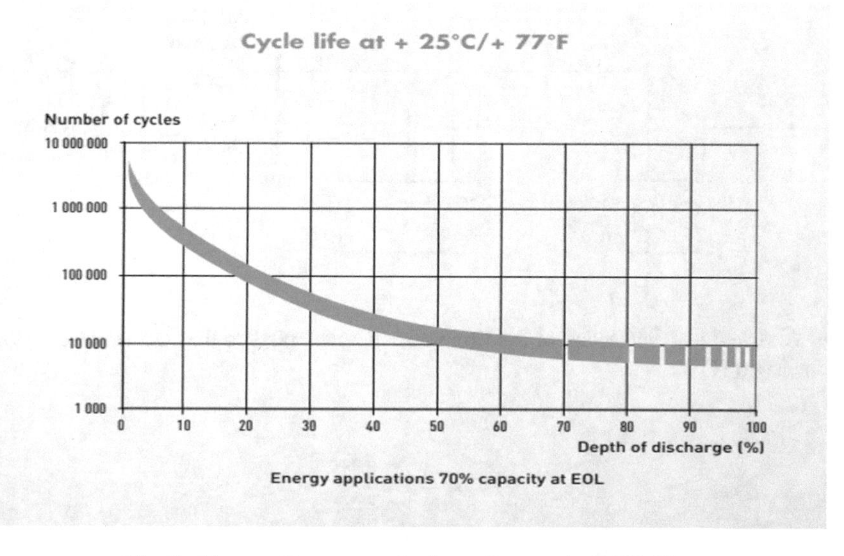

Fig. 8. Expected battery cycle depending on the DOD [4]

4 MATLAB Simulink – Simulation of the Connection of the Li-ion BESS to the Krnovo Power Plant

The effect of the Li-ion BESS connection to the Krnovo is analyzed through the simulation performed in the MATLAB Simulink software. The objective is to analyze the influence of the Li-ion BESS on the variable output power of the Krnovo and on the system's voltage stability in the case of wind variations close to the average wind speeds at the Krnovo area. In the simulation, battery system with the capacity of 20 MWh is used.

Krnovo is connected via the 33/110 kV substation to the power system of Montenegro, as it is presented in Fig. 9. Scheme from Fig. 9 is implemented in MATLAB Simulink software, which is presented in Fig. 10. Figure 11 presents scheme of the Li-ion BESS setup.

The following setup is used for Krnovo:

- Generators rated voltage: 2.85 MW/2.5 MW (20 × 2.85 MW + 6 × 2.5 MW);
- Frequency: 50 Hz;
- Power factor: 0.9 ind/cap;
- Cut-in wind speed: 3 m/s;
- Cut-out speed: 25 m/s.

The BESS parameters are:

- Total power: 20 MWh (10 × 2 MWh);
- Initial SOC: 50%;
- Rated voltage of single battery cell: 12.6 V;

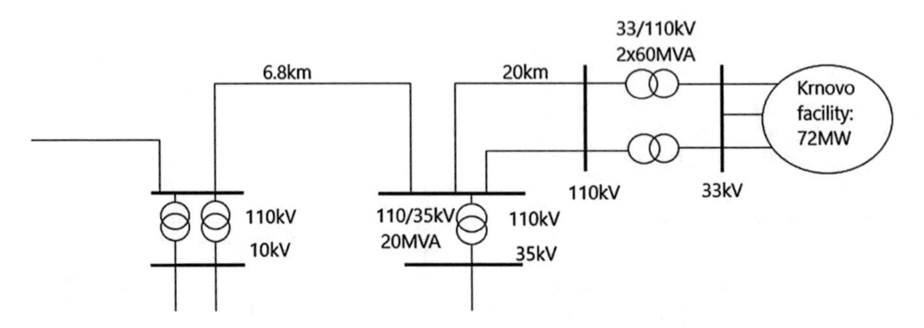

Fig. 9. Scheme of connection of the Krnovo wind power plant to the electric power system of Montenegro [1]

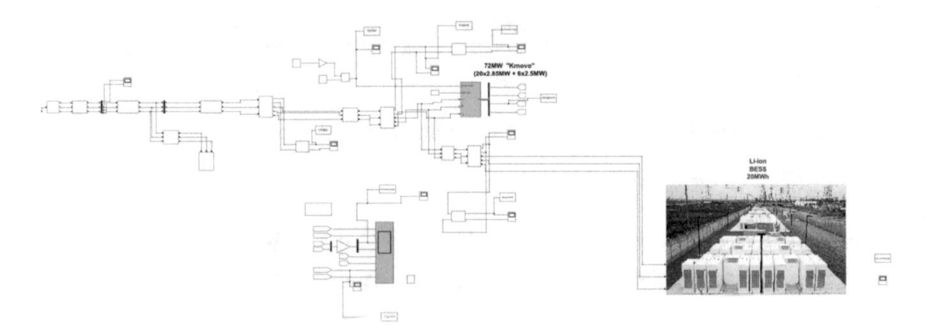

Fig. 10. Scheme of the connection of the Li-ion BESS to the Krnovo wind power plant in MATLAB Simulink software

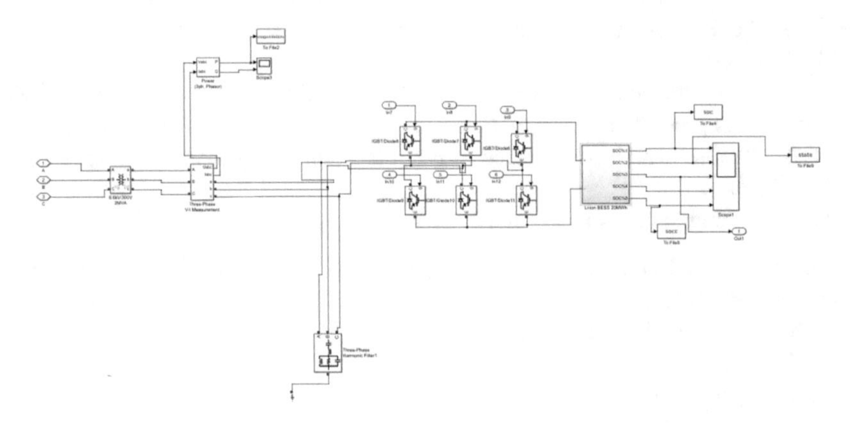

Fig. 11. Scheme of the Li-ion BESS in MATLAB Simulink software

- Battery response time: 5 s;
- PCS: bi-directional inverter + low-pass filter.

The wind speed variations are set from 4.5 m/s to 10.5 m/s, which is shown in Fig. 12. Changes in wind speeds cause variations in Krnovo output power, measured at a connection point of this facility and the power system, with a Li-ion BESS connected, as shown in Fig. 13. The output power values are changing from 40 MW to around 30 MW, at the very beginning of the simulation (Fig. 13). As the simulation progresses, the value of the output power is stabilized to a value of about 20 MW, due to the influence of the Li-ion BESS. This process corresponds to the type of the Li-ion BESS application – capacity firming. As noted earlier, the BESS maintains for a fixed period of time constant output power of the wind power plant at a specified (fixed) level. Also, at the beginning of the simulation, sudden changes in the wind speed cause voltage variations at the location of the Krnovo connection to the system, but it is also stabilized by the influence of the BESS, as shown in Fig. 14.

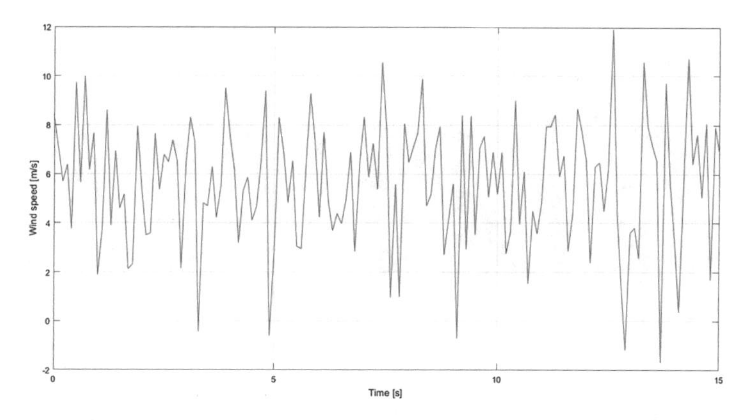

Fig. 12. Wind variations at Krnovo during the simulation

Li-ion BESS, initially set to 50% SOC, during this simulation operates in the charge mode, and it “accepts” energy during the period of voltage and power output variations, as shown in Fig. 15.

Figure 16 represents the variations of power at the point of connection of the one of the Li-ion BESS units of 2 MWh. The results from Fig. 16 shows that during the highest power variations (shown in Fig. 13), the Li-ion BESS operates in charge mode, i.e. it “takes over” the part of the Krnovo output power in order to maintain it at a more stable value.

In the case of the simulation of the operation of Krnovo facility without the Li-ion BESS, higher variations of power happens: from 40 MW to the 20–30 MW, after which the power drops rapidly to around 10–15 MW (shown in Fig. 17).

Ultimately, the conclusion is that, due to the wind power variations ranging from 4.5 m/s to 10.5 m/s, the result is destabilization of the Krnovo output power, as well as

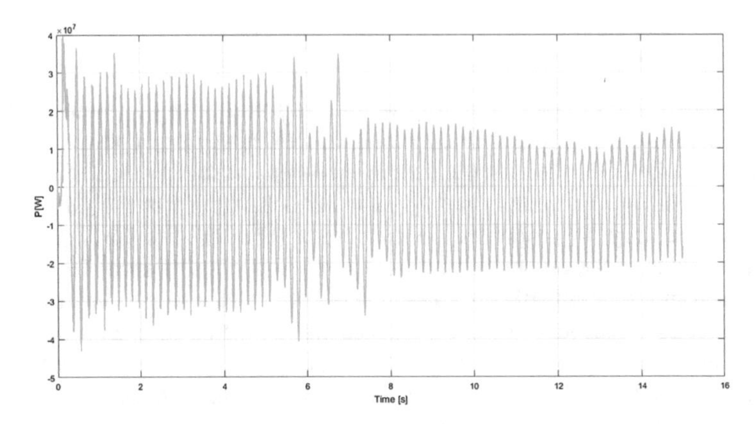

Fig. 13. Variations of the output power of the wind power plant Krnovo

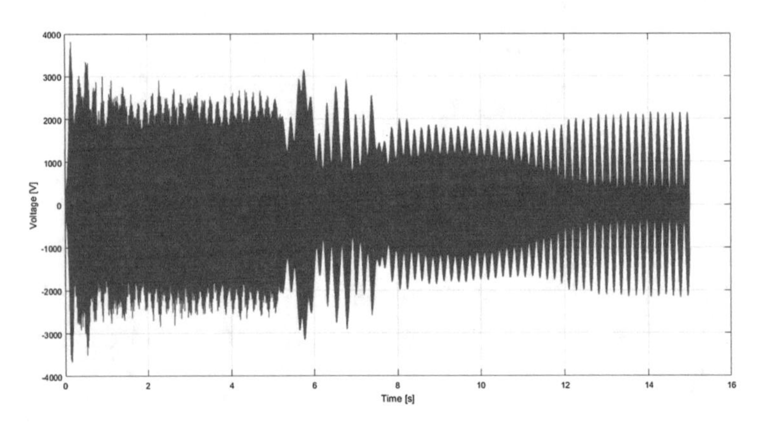

Fig. 14. Voltage variations at the point of connection of the Krnovo to the power system

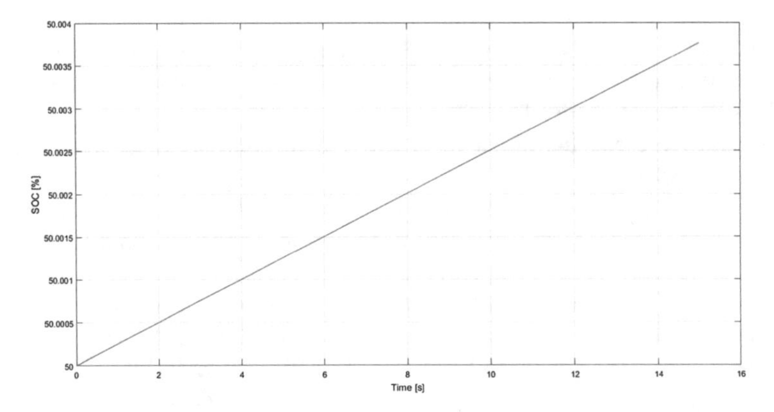

Fig. 15. SOC of the single Li-ion cell – Battery system operates in charge mode during the simulation process

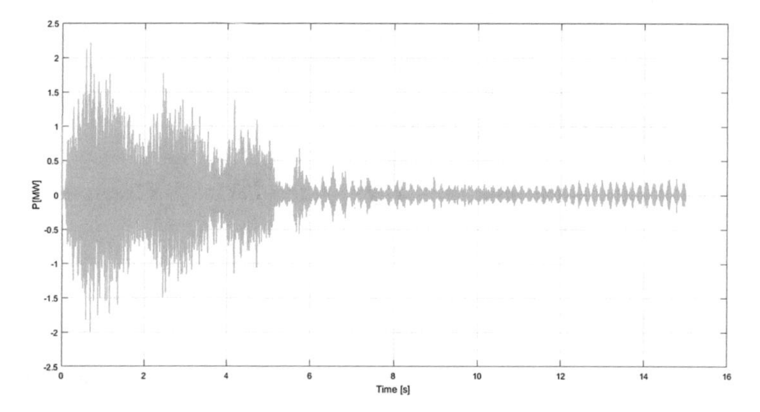

Fig. 16. Power variations at the connection point of the Krnovo and the Li-ion BESS unit

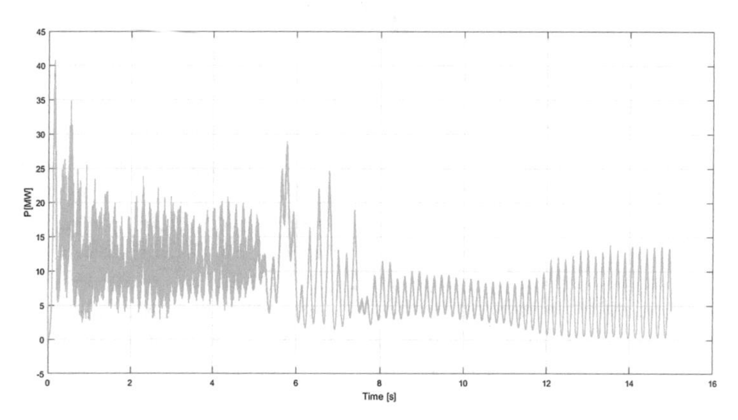

Fig. 17. Variations of the output power of the Krnovo without the connection of the Li-ion BESS

instability of the voltage at the connection point of Krnovo and Montenegrin power system. The power varies in the range of 40 MW to around 30 MW.

In order to maintain the stability of the above mentioned values, the Li-ion BESS operates in a charge mode – during huge power variations BESS receives excess of power, and thus controls the Krnovo output power. As a result of Li-ion BESS operation, the Krnovo output power is stabilized to around 20 MW, and the problem of voltage stability is also solved.

5 Conclusion

The Krnovo wind power plant can have a negative impact on the power system of Montenegro, mostly due to its inconsistent and unpredictable production, which significantly decrease its role in the power system of Montenegro.

The results presented in this paper show that the connection of the Li-ion BESS can solve the potential problems that Krnovo can make to the Montenegrin power system by controlling the variable production of the Krnovo power plant at a certain value, making it more stable source of electricity.

Acknowledgment. This work was supported by Ministry of Science of Montenegro, under the CROSSBOW research project.

References

1. Goić, R., Topčić, A.: Elaborate on testing and compliance of the operation of the Krnovo wind power plant
2. Santhanagopalan, S., Smith, K., Neubauer, J., Kim, G.-H., Keyser, M., Pasaran, A.: Design and analysis of large lithium-ion battery systems. Artech House, Norwood (2015)
3. ABB: Energy storage – the benefits beyond the integration of renewables
4. Saft: Lithium-ion battery life
5. Huang, X., Jiang, B.: Research on lithium battery energy storage system in wind power. In: 2011 International Conference on Electrical and Control Engineering. IEEE, Yichang, China (2011)
6. Diaz-Gonzalez, F., Sumper, A., Gomis-Bellmunt, O.: Energy Storage in Power Systems, pp. 129–220. Wiley, Hoboken (2016)
7. Daggett, A., Qadrdan, M., Jenkins, N.: Feasibility of a battery storage system for a renewable energy park operating with price arbitrage. In: 2017 IEEE PES Innovative Smart Grid Technologies Conference Europe (ISGT-Europe) (2017)
8. Casalini, E., Leva, S.: Feasibility analysis of storage systems in wind plants – and Italian application. In: 2017 IEEE International Conference on Environment and Electrical Engineering and 2017 IEEE Industrial and Commercial Power Systems Europe (EEEIC/I&CPS Europe) (2017)
9. Hesse, H.C., Schimpe, M., Kucevic, D., Jossen, A.: Lithium-ion battery storage for the grid —a review of stationary battery storage system design tailored for applications in modern power grids. Energies **10**(12), 2107 (2017). Department of Electrical and Computer Engineering, Technical University of Munich (TUM)
10. Saez-de-Ibarra, A., Martinez-Laserna, E., Stroe, D.I., Swierczynski, M., Rodriguez, P.: Sizing study of second life Li-ion batteries for enhancing renewable energy grid integration. In: Energy Conversion Congress and Exposition (ECCE). IEEE (2015)
11. Tomšić, Ž., Rajšl, I., Ilak, P., Filipović, M.: Optimizing integration of the new RES generation and electrical energy storage in a power system: case study of Croatia. In: 2017 52nd International Universities Power Engineering Conference (UPEC) (2017)
12. Sbordone, D.A., Di Pietra, B., Bocci, E.: Energy analysis of a real grid connected lithium battery energy storage system. Energy Procedia **75**, 1881–1887 (2015). www.sciencedirect.com
13. Bahramipanah, M., Torregrossa, D., Cherkaoui, R., Paolone, M.: Enhanced equivalent electrical circuit model of lithium-based batteries accounting for charge redistribution, state-of-health, and temperature effects. IEEE Trans. Transp. Electrif. **3**(3), 589–599 (2017)

Computer Modelling and Simulations for Engineering Applications

Modelling the Dephosphorization Process in a Swaying Oxygen Converter

Damir Kahrimanovic[1,2](✉), Erich Wimmer[3], Stefan Pirker[4], and Bernhard König[1]

[1] k1-met GmbH - Metallurgical Competence Center, Linz, Austria
damir.kahrimanovic@k1-met.com
[2] Mechanical Engineering Faculty, Process Technology Department, University of Sarajevo, Sarajevo, Bosnia and Herzegovina
[3] Primetals Technologies Austria GmbH, Linz, Austria
[4] Department of Particulate Flow Modelling, Johannes Kepler University, Linz, Austria

Abstract. Some important aspects of numerical modelling of the oxygen converter are discussed in this paper, such as multiphase flow modelling and the mass transfer of species across the metal-slag interface. Numerical models can help to interpret and understand a variety of processes occurring simultaneously in a converter. For example, converter swaying can be used to improve mixing in the liquid metal phase. This is shown for the case of dephosphorization, where phosphorus is transferred from metal to slag and vice versa.

1 Introduction

Basic oxygen furnace (BOF) is a large refractory-lined container in which liquid metal and slag are brought together in order to initialize and maintain diverse chemical reactions and the associated mass transfer across the interface, Fig. 1.

Simultaneously, the gas stirring takes place: inert gas (argon) bubbles are released from the ladle bottom in order to enhance the mixing and thus to homogenize the composition of the liquid metal. The motion and the mixing in the metal are strongly influenced by the top slag layer.

Under the most common reactions in a BOF converter are decarburization, desulphurization and dephosphorization, the latter gaining in importance in the last decades, as the remaining iron ore reserves contain more and more phosphorus. High phosphorus content in steel is responsible for brittleness, and is thus undesirable. All of these reactions are highly influenced by the oxygen activity in the ladle.

There is a vast number of analytical models for diverse processes in a ladle (see [1] and references therein). These models are based on theoretical considerations and experimental investigations. In the field of computational fluid dynamics (CFD), a huge progress has been made over the past decades as well. However, as for the steel refining in a ladle, the phenomena occurring therein have been simulated only partially, i.e. separated from each other. The reasons lie not only in the high complexity of different phenomena which occur simultaneously (compressible, multiphase flows with large variations of flow regimes and scales, chemical reactions and mass/momentum/energy

S. Avdaković (Ed.): IAT 2018, LNNS 59, pp. 305–315, 2019.
https://doi.org/10.1007/978-3-030-02574-8_24

transfer between phases, fluid-structure interaction etc.) but also in the computational costs of such simulations (calculations covering the range from few minutes to several hours of real time are needed).

In this work the dephosphorization process in a swaying oxygen converter is simulated under assumption on incompressible flow. It is to clarify whether swaying can improve the mixing in liquid metal.

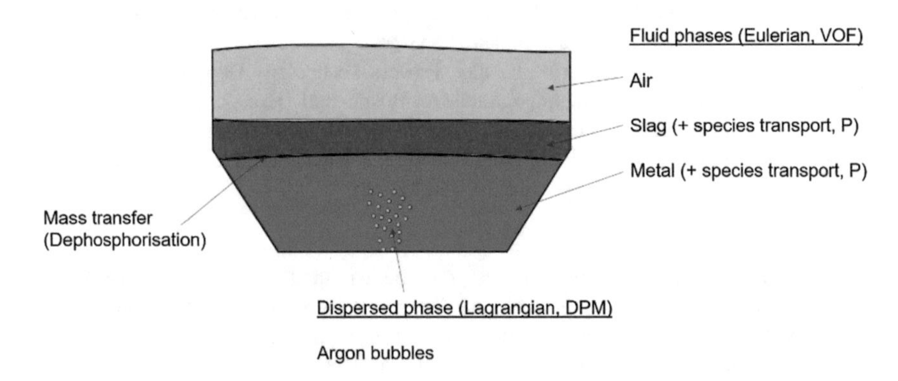

Fig. 1. Schematic illustration of a BOF converter

2 Numerical Modelling

For numerical simulations in this work the finite volume method is used. Thereby, a set of conservation equations for mass (continuity) and momentum is solved for every numerical cell in the computational domain. The continuity equation for an incompressible fluid reads as follows:

$$\nabla \cdot \mathbf{u} = 0, \tag{1}$$

Where $\mathbf{u}$ is the fluid velocity. The momentum equation can be expressed as

$$\frac{\partial(\rho \mathbf{u})}{\partial t} + \nabla(\rho \mathbf{u} \otimes \mathbf{u}) = -\nabla p + \nabla(2\mu \mathbf{D}) + \rho \boldsymbol{g} + \boldsymbol{f}_p. \tag{2}$$

$\mathbf{D}$ is the rate of deformation tensor, defined as

$$\mathbf{D} = \frac{1}{2}\left(\nabla \mathbf{u} + \nabla \mathbf{u}^T\right). \tag{3}$$

ρ and μ are the fluid density and viscosity, respectively, p is the static pressure, $\boldsymbol{g}$ is the gravitational acceleration and $\boldsymbol{f}_p$ represents the external body force from the interaction with Lagrangian bubbles, see later. A more detailed elaboration of these equations can be found in textbooks on computational fluid dynamics [2, 3].

For turbulence modelling the large eddy simulation (LES) model is used. Within this model the flow quantities are decomposed into a fully resolved part (large turbulent eddies) and an unresolved part (sub-grid scale) which needs to be modelled, [4]. The model of Smagorinsky is used to close the turbulence equations [5].

2.1 The Volume of Fluid (VOF) Model

For the simulation of Eulerian phases (liquid metal, slag, air) in an oxygen converter the VOF-model is used. Thereby, an additional equation for the volume fraction for each phase involved needs to be solved. This equation takes the following form:

$$\frac{\partial}{\partial t}\alpha_i\rho_i + \nabla\cdot(\alpha_i\rho_i\mathbf{u}) = S_{\alpha_i}, \tag{4}$$

where α_i is the volume fraction, ρ_i is the density and S_{α_i} is the mass source for the phase i.

The VOF-model belongs to the group of Eulerian models for multiphase flow simulations, but in contrast to the Eulerian Two-Fluid model it is assumed that the phases are not interpenetrating. Thus, the phases share one velocity field **u**, which means that the material properties of different phases (density and viscosity) in the momentum equation are averaged by the volume fraction α_i:

$$\phi = \sum \alpha_i\phi_i, \tag{5}$$

whereby the following must apply for every computational cell:

$$\sum \alpha_i = 1. \tag{6}$$

2.2 Species Transport

In order to include the mass transfer of phosphorus from liquid metal to slag and vice versa, additional equations for species transport in these two phases need to be solved:

$$\frac{\partial}{\partial t}\rho_i c_j + \nabla\cdot(\rho_i c_j\mathbf{u}) = -\nabla \boldsymbol{J}_j + S_{c_j}. \tag{7}$$

c_j is the mass fraction and $\boldsymbol{J}_j$ is the diffusion flux of the species j within the phase i, which arises due to the concentration gradient (the influence of temperature is omitted in this work):

$$\boldsymbol{J}_j = -\left(\rho_i D_{m_j} + \frac{\mu_t}{Sc_t}\right)\nabla c_j. \tag{8}$$

D_{m_j} is the mass diffusion coefficient of the species j, μ_t is the turbulent viscosity and Sc_t is the turbulent Schmidt number. Using the source term S_{c_j} the phosphorus mass can be transferred between metal and slag. As in the case of volume fractions for phases, the sum of mass fractions of species in a phase must equal 1.

2.3 Lagrangian Particle Tracking

Argon bubbles are simulated by calculating particle trajectories through the computational domain in a Lagrangian reference frame. For every point-particle a force balance is solved in order to get the particle position and velocity:

$$\frac{d}{dt}\mathbf{x}_p = \mathbf{u}_p, \tag{9}$$

$$\frac{d}{dt}\mathbf{u}_p = \sum \boldsymbol{f}_i. \tag{10}$$

$\mathbf{x}_p$ and $\mathbf{u}_p$ are the position and velocity of a bubble, respectively. The forces $\boldsymbol{f}_i$ acting on a single bubble include the contributions of a drag force on a spherical particle, forces due to pressure gradient and buoyancy, and the added mass force. The Basset force is neglected in these simulations. Detailed descriptions of the Lagrangian particle tracking method and the particle forces can be found in [6].

2.4 Mass Transfer Equations

Considering the mass conservation of phosphorus in the slag-metal system, the mass change of phosphorus in the metal must equal the negative mass change in the slag and vice versa. In the following expression it is assumed that phosphorus flows out of the metal and into the slag:

$$-\frac{\partial m_{PM}}{\partial t} = \frac{\partial m_{PS}}{\partial t} = \dot{m}, \tag{11}$$

where $\dot{m}$ (kg/s) is the mass which is being transferred from one phase to another, m_{PM} (kg) is the total mass of phosphorus in metal and m_{PS} (kg) is the total mass of phosphorus in slag.

The change of species (i.e., phosphorus) mass in metal (or slag, respectively) is equal to the change of metal (or slag) mass itself:

$$-\frac{\partial m_M}{\partial t} = \frac{\partial m_S}{\partial t} = \dot{m}. \tag{12}$$

m_M (kg) is the total mass of the metal phase and m_S (kg) is the total mass of the slag phase considered. The mass of phosphorus is a part of the total mass of the respective phase, which is defined by the mass concentration:

$$m_{PM} = m_M \cdot c_{PM}, \tag{13}$$

$$m_{PS} = m_S \cdot c_{PS}. \tag{14}$$

c_{PM} is the mass concentration of phosphorus in metal, and c_{PS} is the mass concentration of phosphorus in slag. Using this definition, Eq. (11) can be written as

$$-\frac{\partial(m_M \cdot c_{PM})}{\partial t} = \frac{\partial(m_S \cdot c_{PS})}{\partial t}. \tag{15}$$

The mass transfer equations for phosphorus in steel and slag are:

$$\frac{\partial m_{PM}}{\partial t} = A\rho_M k_M \left(c_{PM}^i - c_{PM}\right), \tag{16}$$

$$\frac{\partial m_{PS}}{\partial t} = A\rho_S k_S \left(c_{PS}^i - c_{PS}\right), \tag{17}$$

where A (m^2) is the area through which the flux is calculated, k_M and k_S(m/s) are the mass transfer coefficients and c_{PM}^i and c_{PS}^i are phosphorus mass concentrations in the metal and slag, respectively. By multiplying the metal and slag densities, ρ_M and ρ_S (kg/m^3), with the mass transfer coefficients k_M and k_S, one gets the density-related mass transfer coefficients k'_M and k'_S (kg/m^2s) for metal and slag, respectively, and Eqs. (16) and (17) can be written as

$$\frac{\partial m_{PM}}{\partial t} = Ak'_M \left(c_{PM}^i - c_{PM}\right), \tag{18}$$

$$\frac{\partial m_{PS}}{\partial t} = Ak'_S \left(c_{PS}^i - c_{PS}\right), \tag{19}$$

The phosphorus mass concentration in the metal and slag at the interface is defined by the phosphorus equilibrium distribution ratio:

$$L_P = \frac{c_{PS}^i}{c_{PM}^i}. \tag{20}$$

It is thereby assumed that in systems which contain any amount of phosphorus these values can never be zero. By combining Eqs. (18)–(20) and (12), it is possible to solve for the phosphorus mass concentrations at the interface:

$$c_{PM}^i = \frac{k'_S c_{PS} + k'_M c_{PM}}{k'_S L_P + k'_M}, \tag{21}$$

$$c_{PS}^i = \frac{k'_S c_{PS} + k'_M c_{PM}}{k'_S + \frac{k'_M}{L_P}}. \tag{22}$$

By putting Eqs. (21) and (22) back into Eqs. (18) and (19), respectively, the mass transfer equations for phosphorus in slag and/or metal now look more familiar:

$$\frac{\partial m_{PM}}{\partial t} = -Ak'_0 \left(c_{PM} - \frac{c_{PS}}{L_P} \right), \tag{23}$$

$$\frac{\partial m_{PS}}{\partial t} = Ak'_0 \left(c_{PM} - \frac{c_{PS}}{L_P} \right), \tag{24}$$

with the overall density-related mass transfer coefficient k'_0 (kg/m^{2}s), valid both in slag and metal:

$$k'_0 = \frac{1}{\frac{1}{k'_M} + \frac{1}{k'_S L_P}}. \tag{25}$$

2.5 Adaptive Mesh Refinement

The geometry of a BOF converter is relatively large. To make a very fine mesh everywhere would lead to large meshes and high computational times. There is a possibility to refine the mesh only in certain regions. The regions of high slag volume fractions are chosen for refinement (Fig. 2) for two reasons: the slag phase is concentrated within a small part of the computational domain, yet it is the region where the most important processes take place (i.e. mass transfer, emulsification, plume eye etc.). Thus, the mesh is refined in every cell in which the slag is present (above a certain value, here $\alpha_S \geq 0.001$). When the slag leaves that cell (which means that the slag volume concentration falls under a certain value, here $\alpha_S < 0.001$), the cell itself is coarsened again. The refinement process is done by halving the cell in all directions (one coarse cell results in eight cells for the three-dimensional case).

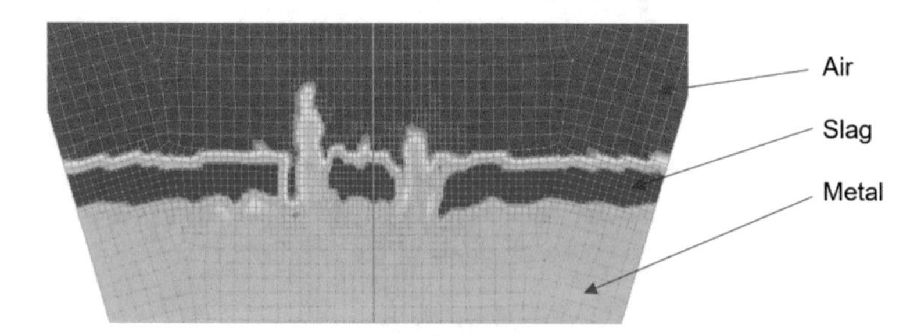

Fig. 2. Adaptive mesh refinement

2.6 Mixing Efficiency

For the monitoring of values on a sample plane which consists of unequal faces (or on a volume with unequal cells), working with the weighted values (area- or mass-weighted) is more useful than taking the ensemble average. A mass-weighted mean value of a species concentration can be expressed as

$$\bar{c} = \frac{1}{m_{total}} \sum_{j=1}^{n} \left(c_j \cdot m_j \right), \tag{26}$$

and the variance:

$$\sigma^2 = \frac{1}{m_{total}} \sum_{j=1}^{n} \left[m_j \cdot \left(c_j - \bar{c} \right)^2 \right]. \tag{27}$$

Consider the case where there are two species in a phase, and one needs to determine the mixing efficiency on a sample plane, or even for the entire domain. First, the mean value of species 1 on a sample plane at some fixed point in time is needed: $\bar{c}_1$. For example, $\bar{c}_1 = 0.3$ means that there is 30% of species 1 and 70% of species 2 at that moment. Now the actual variance for every cell in the field, σ^2, can be calculated using Eq. (27). Next, the highest (worst) variance in the field, σ_0^2, is determined by evaluating every cell and taking the maximum value. This is not the hypothetical worst possible case, which is usually used for turbulent mixing, and is not suitable for converter mixing simulations. The mixing efficiency is then

$$\eta_{mix} = 1 - \frac{\sigma^2}{\sigma_0^2}. \tag{28}$$

Thus, the mixing efficiency is a dimensionless number between 0 and 1:

- The worst mixing efficiency: $\sigma^2 = \sigma_0^2$: $\eta_{mix} = 0$.
- The best mixing efficiency: $\sigma^2 = 0$: $\eta_{mix} = 1$.

3 Numerical Set up

The simulated converter vessel has a diameter of 6.3 m and a height of 4 m. Argon is injected through eight orifices situated at a radius of 1.4 m around the midpoint at the converter bottom. For this geometry a grid consisting of 650 000 hexahedral cells is generated.

Two cases are considered for the simulations: the non-swaying and the swaying converter with the amplitude of 5° and the frequency of 1/40 Hz (which means that the converter reaches its maximum deflection of 5° in 10 s). The converter swaying is taken into account by using the moving grid.

Overall masses of metal and slag are $m_M = 250000\,\text{kg}$ and $m_S = 20000\,\text{kg}$, with the densities $\rho_M = 7900\,\text{kg/m}^3$ and $\rho_S = 2500\,\text{kg/m}^3$ and the dynamic viscosities $\mu_M = 6.5 \cdot 10^{-3}\text{Pa s}$ and $\mu_S = 0.223\text{Pa s}$, respectively.

The initial concentrations of phosphorus in metal and slag are $c_{PM0} = 0.001$ and $c_{PS0} = 0.025$, with the mass diffusivities $D_{PM} = 1.45 \cdot 10^{-8}\text{m}^2/\text{s}$ and $D_{PS} = 2 \cdot 10^{-10}\,\text{m}^2/\text{s}$, [7].

The density-based mass transfer coefficients are $k'_M = 3.95\,\text{kg/m}^2\text{s}$ and $k'_S = 1.25\,\text{kg/m}^2\text{s}$ (Eq. 25), and the phosphorus equilibrium ratio is assumed constant and set to $L_P = 288$.

The mass flow of argon bubbles is $\dot{m}_p = 0.2856\,\text{kg/s}$, with the constant argon density $\rho_p = 0.8\,\text{kg/m}^3$ and the bubble diameter $d_p = 0.005\,\text{m}$.

All calculations were performed using the commercial software package ANSYS Fluent. For simulations the VOF-LES model was used. The second-order QUICK method was chosen as an interpolation scheme for momentum, PRESTO for pressure, compressive for the VOF, and first order upwind for species transport. Further information on turbulence, multiphase and interpolation methods can be found in [8].

4 Results

In Fig. 3 the dephosphorization process in the ladle for the first 100 s is shown. The colour bar relates to the phosphorus concentration in liquid metal: as the start concentration of phosphorus is $c_{PM0} = 0.001$, and the density of metal is $\rho_M = 7900\,\text{kg/m}^3$, the red colour refers to the highest concentration of $c_{PM0} \cdot \rho_M = 7.9\,\text{kg/m}^3$.

Phosphorus concentration in metal decreases with time, as the phosphorus leaves the metal phase and enters the slag phase through the slag-metal interface. Thus, the phosphorus concentration in slag must increase, Fig. 3.

On the left side of Fig. 3 the non-swaying case is shown. The rising argon bubbles gain momentum as they move upwards. They also spread horizontally entraining the surrounding liquid and form the gas-liquid plume, which is the driving force behind the mixing behaviour in liquid metal in the case of bottom stirring. The plume is strong enough to break the slag up which strongly increases the slag-metal mixture and even brings the metal on top of the slag layer. This area is called the slag eye.

The development of slag eye is hindered in the swaying case, as is shown on the right side of Fig. 3, because gas bubbles are dispersed over the interface due to converter pivoting. For this reason the mass transfer from metal to slag is lower for the swaying case, Fig. 4.

Through the formation of slag eye, the turbulence intensity in this area is increased which results in increased slag-metal interfacial area. As the phosphorus mass transfer is directly related to the interfacial area (Eq. 23), the overall reaction rate will be higher in the non-swaying case, Fig. 5.

The rising gas-liquid plume also initializes one large toroidal flow pattern in liquid metal. It starts with the bubbles induced, upward directed flow in the middle of the converter, then the flow continues radially outwards at the slag-metal interface, downwards at the converter wall and radially inwards at the bottom. This flow is recognisable in Fig. 3 for the non-swaying case, as the mixing of phosphorus follows the flow pattern almost exactly. The concentration lowering starts at the interface, spreads radially outwards as the fluid with lower phosphorus concentration is transported in this direction, and then it gradually moves to the bottom near the converter wall.

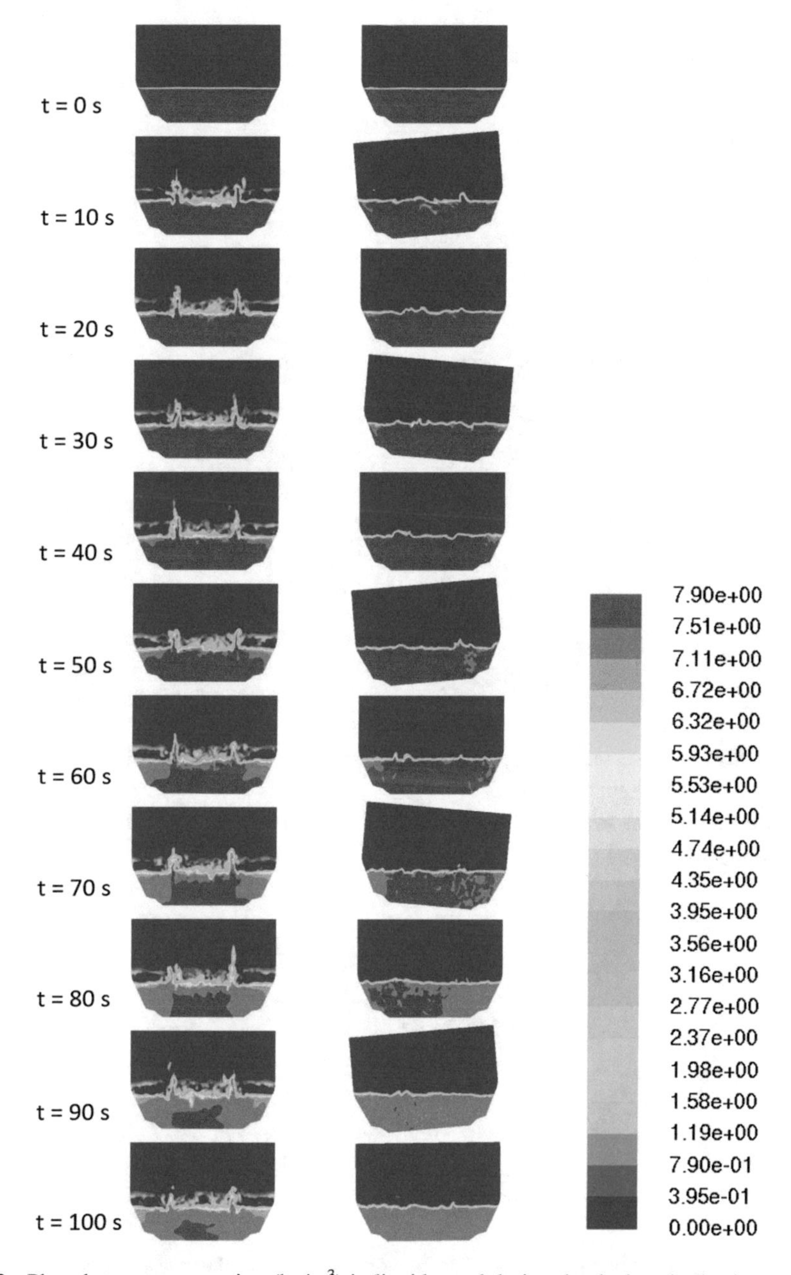

Fig. 3. Phosphorus concentration (kg/m^3) in liquid metal during the dephosphorization process. Left: non-swaying converter, Right: swaying converter

There are, however, areas with weak mixing behaviour, such as the middle of the toroidal vortex, the outer near-bottom area, and especially the area between the inert gas orifices. This is improved by the converter swaying, as is shown on the right side of Fig. 3. The toroidal flow pattern is disturbed by the alternating converter motion and

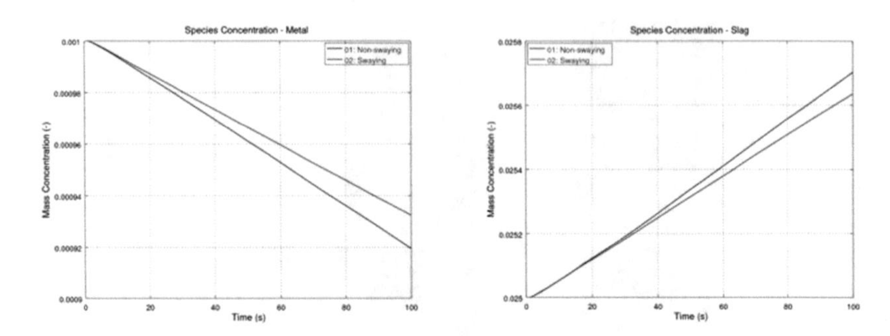

Fig. 4. Phosphorus concentration in metal and slag

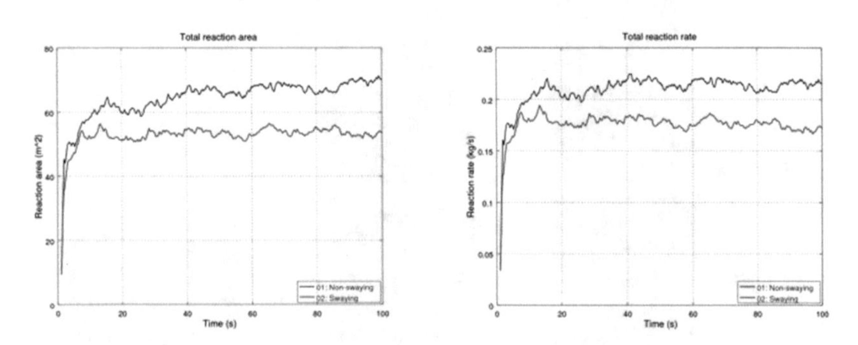

Fig. 5. Total reaction area and rate during the dephosphorization process

the liquid metal mass is pushed back and forth which results in a much faster mixing for the swaying case, Fig. 6.

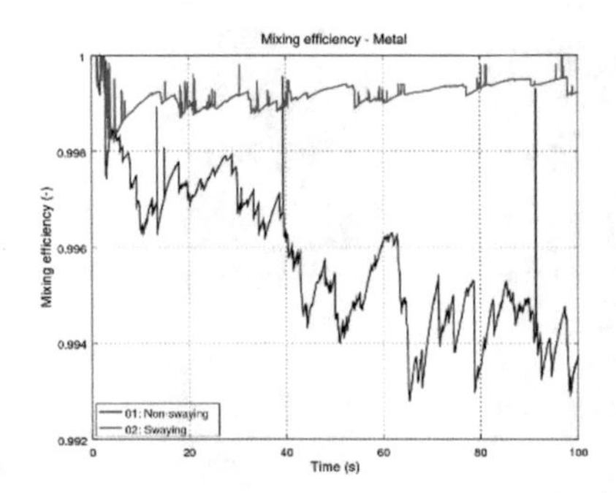

Fig. 6. Mixing efficiency for the non-swaying and the swaying case

5 Conclusion

A process of dephosphorization in a BOF converter is investigated by means of numerical simulations. Thereby, special attention was laid on the influence of converter swaying on the mixing behaviour in liquid metal. Converter swaying supress the formation of the slag eye, which in turn decreases the interfacial area between metal and slag thus decreasing the overall mass transfer. On the other hand, converter swaying strongly improves the mixing efficiency and homogenizes the phosphorus concentration much faster than in the non-swaying case.

This work represents a first step towards a complex converter model, which will incorporate many effects which are important for secondary steelmaking. Although at this state some phenomena are not yet considered (compressibility, top lance blowing, temperature influence, emulsification), many observations and comparisons can still be made, one of which is shown here.

References

1. Seetharaman, S.: Treatise on Process Metallurgy. Elsevier, Oxford (2014)
2. Ferziger, J.H., Peric, M.: Computational Methods for Fluid Dynamics. Springer, Berlin (2013)
3. Anderson, J.D.: Computational Fluid Dynamics. McGraw-Hill, New York (1995)
4. Deardorff, J.: A numerical study of three-dimensional turbulent channel flow at large Reynolds numbers. J. Fluid Mech. **41**(2), 453–480 (1970)
5. Smagorinsky, J.: General circulation experiments with the primitive equations. Mon. Weather Rev. **91**(3), 99–164 (1963)
6. Crowe, C.T., Schwarzkopf, J.D., Sommerfeld, M., Tsuji, Y.: Multiphase Flows with Droplets and Particles. CRC Press, Boca Raton (2012)
7. Poirier, D.R., Geiger, G.H.: Transport Phenomena in Materials Processing. Springer, Cham (2016)
8. ANSYS 17 Fluent Theory Guide. ANSYS Inc., Canonsburg (2016)

Bare Conductor Temperature Coefficient Identification by Means of Differential Evolution Algorithm

Mirza Sarajlić[1(✉)], Marko Pocajt[1], Peter Kitak[1], Nermin Sarajlić[2], and Jože Pihler[1]

[1] Faculty of Electrical Engineering and Computer Science, University of Maribor, Koroška cesta 46, 2000 Maribor, Slovenia
mirza.sarajlic@um.si

[2] Faculty of Electrical Engineering, University of Tuzla, Franjevačka 2, 75000 Tuzla, Bosnia and Herzegovina

Abstract. This paper describes a method for the identification of bare conductor's temperature coefficient. The authors used a numerical and real model of the conductor. The numerical model is composed of the finite elements and includes material temperature characteristics and boundary condition values. The real model represents a section of the bare conductor, on which were fixed thermosensors to measure the temperature in laboratory surroundings. Identification of the temperature coefficient was carried out by using the optimization algorithm Differential Evolution (DE), in order to reach minimum difference of the temperatures between numerical and real model.

1 Introduction

Transfer of heat energy from one area to another is known as the heat transfer [1, 2], where the heat passes from the warmer to the colder area. During the heat transfer, there is an alteration in the internal energy between two areas with atomic or molecular displacement, or by means of electromagnetic waves. Usually, the heat transfer is expressed through the following three processes [1, 2]:

- Conduction – the transfer of heat in solid materials with direct contact, without agitation of materials.
- Convection – the transfer of heat to or from a liquid or gas solid surface nearby. Convection is basically conduction with the additional possibility of heat transfer by the movements of fluids or gas.
- Radiation – the transfer of heat by means of electromagnetic waves.

This paper describes a method for the temperature coefficient identification using numerical and real model of the conductor [2]. The goal of this paper is to identify the thermal conductivity and heat transfer coefficients using optimization algorithm with an innovative and scientific approach.

The paper is organized in 7 sections. Section 2 provides a description of thermal conductivity coefficient and heat transfer coefficient. Section 3 presents the bare overhead conductor model. Section 4 describes the carried out measurements on the

S. Avdaković (Ed.): IAT 2018, LNNS 59, pp. 316–325, 2019.
https://doi.org/10.1007/978-3-030-02574-8_25

bare overhead conductor. Section 5 describes method for identification of the thermal conductivity coefficient and heat transfer coefficient using the optimization process. Section 6 present the resulting coefficients obtained by optimization process and comparison between the model calculated and measured conductor temperatures. Section 7 concludes the paper.

2 Heat Transfer Coefficients

This chapter presents thermal conductivity coefficient λ and heat transfer coefficient α.

2.1 Thermal Conductivity Coefficient λ

Figure 1 presents an example of conductivity, with two heat reservoirs, and each reservoir has a different temperature (T_1 – hot, T_2 – cold). The reservoirs are connected with a conductor of length l and cross-section S. It is necessary to consider the thermal conductivity coefficient λ, which depends on the type of material. In a steady state, the heat flow flows from a reservoir with a higher temperature to a lower temperature reservoir. Equation (1) describes the heat flow P, taking into account the thermal conductivity coefficient λ, surface S, temperature difference ΔT and conductor's length l [2].

$$P = \lambda S \frac{\Delta T}{l}. \tag{1}$$

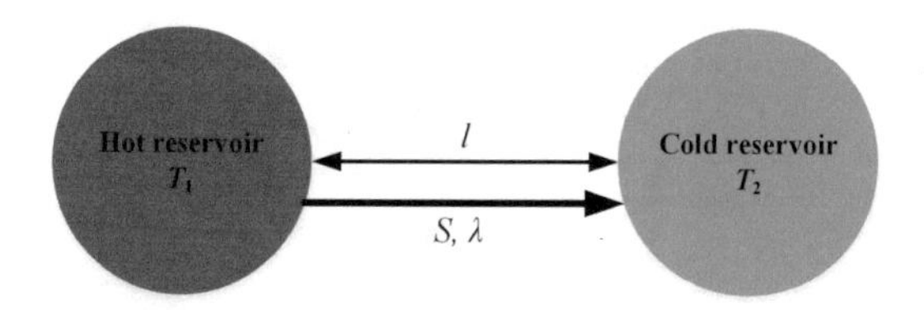

Fig. 1. Heat transfer between two reservoirs.

Thermal conductivity λ [W/m K] is a feature of material, which tells how well the material conducts heat. The thermal conductivity coefficient depends on the type of material, temperature and air pressure. Thermal conductivity is divided in three groups: solid material (metals), fluids and gas. Solid materials have the highest thermal conductivity. The conductivity decreases with additions and impurities.

2.2 Heat Transfer Coefficient α

Transfer of heat energy between a solid surface and a fluid is called convection [2]. There are two types of convection: Natural convection and forced convection. Forced convection occurs when the fluid movement is caused by a fan, pump, or some other source. The basic convection equation is:

$$j = \alpha(T_s - T_f), \tag{2}$$

where α is the heat transfer coefficient, T_s is the surface temperature and T_f is the fluid temperature.

The heat transfer coefficient α is calculated using (3):

$$\alpha = \frac{\lambda}{\delta}, \tag{3}$$

where λ is the fluid's thermal conductivity and δ is the thickness of the boundary layer between the surface and fluid.

3 Bare Overhead Conductor Model

The bare overhead conductor model is built in EleFAnT [3] using the Finite Element Method (FEM) [8], where the conductor's temperature coefficients are calculated. The conductor model is designed as a 2D problem and it is generated using a mesh generator. Figure 2 shows a 2D model of a bare overhead conductor with the finite element mesh, where the orange color shows the conductor. Table 1 indicates the specifications of the real model of a bare overhead conductor.

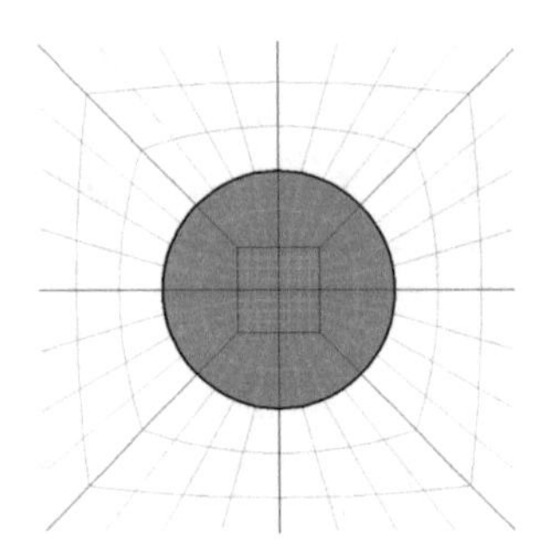

Fig. 2. 2D model of a bare overhead conductor

Table 1. Bare overhead conductor specifications [2]

Cross section [mm^2]	70
Number of wires in the conductor	7
Wire diameter [mm]	10.7
Mass [kg/km]	279
Rated voltage [kV]	24
Maximum current load at conductor's temperature 80°C, at air temperature 20 °C, IEC 1597 [A]	325
Maximum resistance of conductor at 20 °C [Ω/km]	0.434
Maximum short circuit current, 1 s [kA]	6.65
Aluminum alloy	AlMgSi

4 Measurements

Measuring elements (sensors) for measuring temperature were placed on the conductor and its surroundings. The measurements were conducted at the high voltage laboratory ICEM, Maribor. The measuring line consists of a low voltage supply, low voltage switchgear, regulation transformer, secondary disconnector, heating transformer and measuring equipment.

In the conductor, the current was 200 A, and when the temperature stabilized, the current was raised to 300 A, which created two measurement points. Figure 3 shows a bare overhead conductor with distributed thermosensors. The sensors are arranged vertically over and under the conductor on the left and right sides. The most sensors are set over the conductor, because the natural convection influence is best recorded in that way.

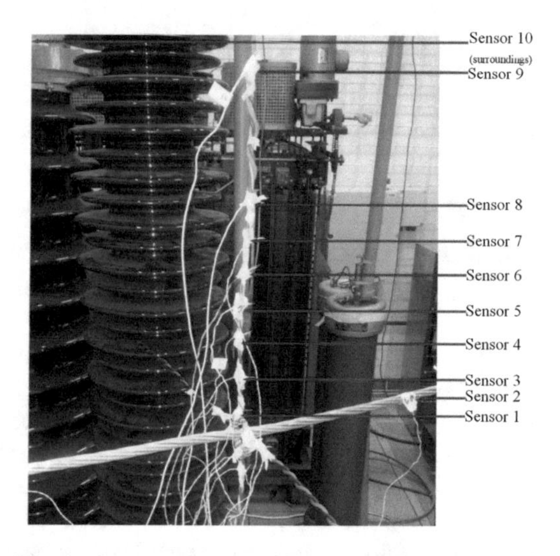

Fig. 3. Bare overhead conductor with thermosensors.

4.1 The Results of Temperature Measurements

Figure 4 shows the temperature of each sensor over time. The increase of temperature for the first half hour of measurement and the current change (dotted line in Fig. 4 shows the point of the change) can be seen. The highest temperatures have been reached on the conductor's surface. The oscillations in temperature (Fig. 4) were

Table 2. Measured temperature values at 300 A

Sensor	1	2	3	4	5	6	7	8	9	10
Temperature [°C]	117.3	57.7	46.6	34.4	32.3	28.6	27.3	26.8	26	25.5

present due to the rise of hot air layers, which are a result of the natural convection. Table 2 shows the final temperature values on the bare conductor, which are later on used in the optimization process.

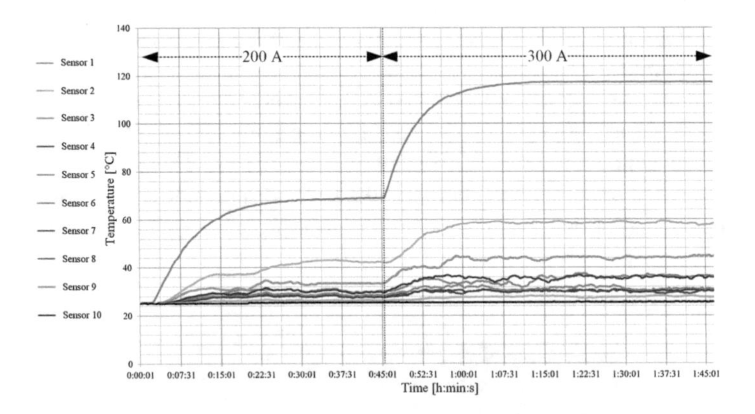

Fig. 4. Measured temperature on the bare overhead conductor.

5 Identification of the Thermal Conductivity Coefficient and Heat Transfer Coefficient Using the Optimization Process

Figure 5 shows an optimization process flow diagram for identifying the optimum parameters of the conductor's numerical model. The calculation begins with a randomly generated initial population [2, 8]. For each population member a numerical analysis is performed, which contains the calculation of the temperature field. Calculation of the magnetic field is performed only once, since the current and the power losses do not change. The current density is calculated from the magnetic field, where the skin effect current is taken into account, and power losses p, which is then used in the thermal field when calculating the temperature T. Determining the optimization parameters is performed in the optimization algorithm. A stopping condition (predefined minimal difference between measured and calculated temperature) is checked for each evaluation of the objective function. Once the stopping condition is satisfied, these optimization parameters represent the model parameters. If the stopping condition is not satisfied, the optimization algorithm makes a new set of parameters, and the whole procedure is repeated for the new iteration [2]. The process of creating potential solutions (new set of optimization parameters) is described in Sect. 5.1.

Figure 6a shows the arrangement and the value of the current density, where it can be seen that the current density is greatest on the conductor's surface ($5.097 \cdot 10^6$ A/m^2). Figure 6b shows the power losses, where it can be seen that the losses are biggest on the conductor's surface ($228 \cdot 10^3$ W/m^3).

One of the most important things about the optimization process is the objective function. In this paper, the objective function y consists of the minimum temperature

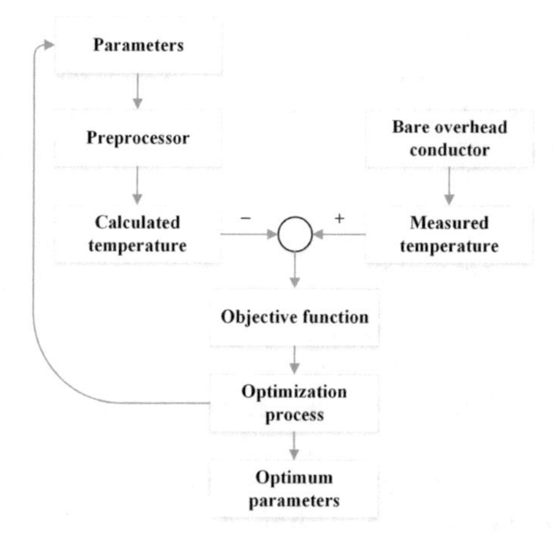

Fig. 5. Optimization process' flow diagram.

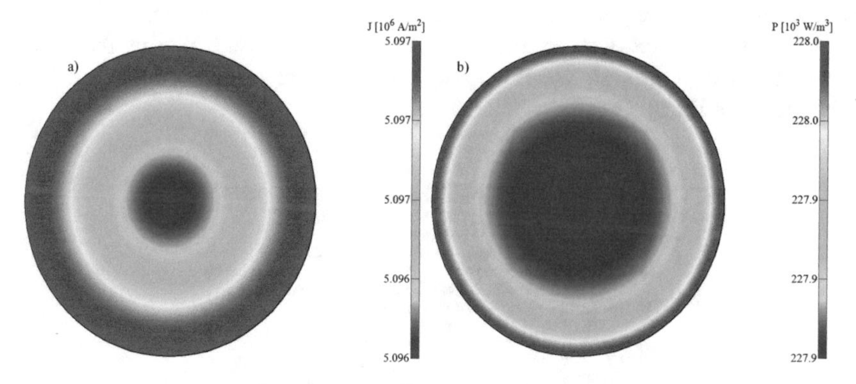

Fig. 6. The arrangement of the bare overhead conductor's: (a) current density and (b) power losses.

difference between calculated ($T_{\mathrm{i,calc}}$) and measured values ($T_{\mathrm{i,meas}}$). Index i presents the observed measurement point. The objective function y is written as follows:

$$y = 0.6 \cdot y_1 + 0.4 \cdot y_2 \tag{4}$$

where y_1 and y_2 are partial objective functions, that are written as:

$$y_1 = \frac{1}{2}\sum_{i=1}^{2} \left| T_{\mathrm{i,calc}} - T_{\mathrm{i,meas}} \right|, \tag{5}$$

$$y_2 = \frac{1}{8}\sum_{i=3}^{10} \left| T_{\mathrm{i,calc}} - T_{\mathrm{i,meas}} \right|. \tag{6}$$

The partial objective function y_1 presents two sensors, which were placed directly on the conductor. The partial objective function y_2 presents the remaining eight sensors, which were placed in the surroundings. Weights 0.6 and 0.4 in (6) were chosen based on the analysis of the influence of certain optimization parameters on the temperature calculation [2].

5.1 The Classical DE Algorithm

This subsection describes the classical DE algorithm [4, 5, 8]. The general problem with optimization algorithms is to find vector x in order to optimize $f(x)$; $x = (x_1, x_2, \ldots, x_D)$. D is a dimension of function f. The variable domains are defined with the lower and upper bounds: $x_{j,\mathrm{lower}}, x_{j,\mathrm{upper}}$; $j \in \{1, \ldots, D\}$. The initial population is chosen randomly between lower ($x_{j,\mathrm{lower}}$) and upper bound ($x_{j,\mathrm{upper}}$), which is defined for each variable x_j. The bounds are defined by the user regarding the problem.

DE is a population based algorithm and the vector $x_{i,G}$; $i = 1, 2, \ldots, NP$ is individual in that population. NP represents population size and G generation [2]. Figure 7 shows the main steps of the DE algorithm. During one generation, DE executes mutation and crossover for each vector and, thus, creates a test vector. Between the original vector and the test vector, the vector with the best value of the objective function is selected through selection. The classical DE algorithm is described in detail in the literature [5, 6].

1. Set parameters: F, CR, NP and counter of generation $G = 1$
2. Generate random initial population: $x_{i,0}$; $i = 1,2,\ldots,NP$
3. **while** stopping criterium is not satisfied
   4. **for** each vector $x_{i,G} = \{x_{1i,G}, x_{2i,G}, \ldots, x_{Di,G}\}$; $i = 1,2,\ldots NP$
      a) Choose random 3 indices r_1, r_2, r_3 within the area [1, NP]. None of these indices can not be equal to the index i.
      b) (Mutation): Generate mutation vector $v_{i,G+1}$:
      $v_{i,G+1} = x_{r1,G} + F \cdot (x_{r2,G} - x_{r3,G})$
      c) (Crossover): Generate new vector $u_{i,G+1}$

$$u_{ji,G+1} = \begin{cases} v_{ji,G+1} & \text{if } rand(j) \le CR \text{ or } j = rn(i), \\ x_{ji,G} & \text{if } rand(j) > CR \text{ and } j \ne rn(i), \end{cases}$$

      where $r(j) \in [0,1]$ j-times evaluation of randomly generated number. $rn(i) \in (1,2,\ldots,D)$ randomly chosen index.
      d) (Selection): Execute a selection scheme:

$$x_{i,G+1} = \begin{cases} u_{i,G+1} & \text{if } f(u_{i,G+1}) < f(x_{i,G}), \\ x_{i,G} & \text{otherwise.} \end{cases}$$

   End for
   $G = G + 1$
   End while
5. Display the best results

Fig. 7. The classical DE algorithm [2].

6 Optimization Results

After running the optimization algorithm, the following results are obtained: Objective function value (*bestval*), number of objective function evaluations (*nfeval*), number of iterations and CPU time of the calculation.

Table 3 shows the optimization process' control parameters and the boundaries of the searched parameters, where XV_{min} is the lower boundary and XV_{max} is the upper boundary of the searched parameter.

There are three searched parameters, that are labeled as p_1, p_2 and p_3. The parameter p_1 represents the heat transfer coefficient for air. The parameters p_2 and p_3 represent the thermal conductivity coefficient for the conductor and air, respectively.

Meanings of parameters from Table 3:

- *VTR* – Value to Reach;
- *Stopping* – number of iterations for stopping the optimization process, when the objective function does not change;
- *D* – number of parameters; *itermax* – maximum number of iterations;
- *F* – mutation factor; *CR* – crossover factor.

Table 3. Optimization process' control parameters with minimum and maximum values of the searched parameters [7]

VTR	*Stopping*	*D*	*itermax*	*F*	*CR*
1e−6	70	3	250	0.6	0.7
	p_1 [W/m² K]	p_2 [W/m K]		p_3 [W/m K]	
XV_{min}	2.5	200		0.02	
XV_{max}	20	260		0.03	

Table 4 shows the results obtained with the used optimization algorithm.

Table 4. Results after the optimization

bestval	*nfeval*	Iterations	CPU time [s]	Identified parameters		
				p_1	p_2	p_3
5.61	4865	76	8326	2.5	210	0.03

The objective function value is 5.61. The number of objective function evaluations is 4865. The number of iterations is 76. CPU time of the optimization process' duration is 8326 s. Figure 8 shows the objective function trace.

Table 5 shows a comparison between measured (T_{meas}) and calculated (T_{calc}) temperature of the bare overhead conductor model. When calculating the conductor model temperature, the values of searched parameters obtained after optimization was used. From Table 5 it can be concluded there are small deviations between calculated and measured temperatures.

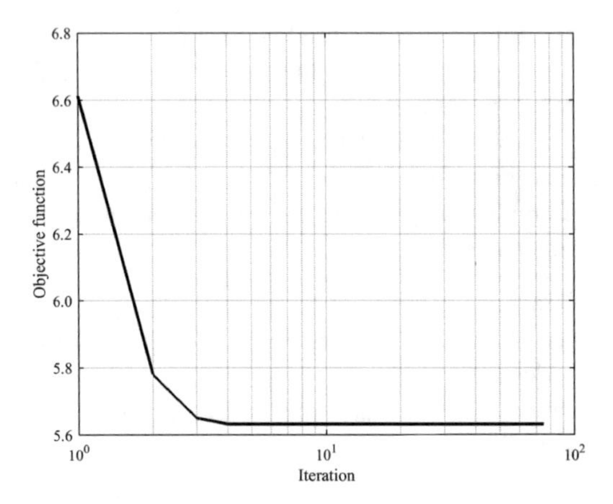

Fig. 8. The objective function trace.

Table 5. Results of the calculated temperature after the optimization process

Sensor	T_{meas} [°C]	T_{calc} [°C]
1	117.3	118.5
2	57.7	58.5
3	46.6	46.1
4	34.4	34.7
5	32.3	32.7
6	28.6	29.1
7	27.3	27.6
8	26.8	26.6
9	26	26.1
10	25.5	25

7 Conclusion

This paper describes the identification process of the bare overhead conductor's heat transfer coefficients regarding the conductor's calculated and measured temperature. The temperature measurements were carried out in the laboratory using thermosensors, which were placed on the conductor and its surroundings. The conductor model was designed as a 2D problem using the Finite Element Method.

The heat transfer coefficient for air and the thermal conductivity coefficient for the conductor and air, were determined by using the DE optimization algorithm, in order to achieve best possible agreement between measured and model-calculated values of temperature. After analyzing the obtained results, there is an exceptional matching of the calculated temperature with the measured parameters. This makes the DE algorithm suitable for identification of thermal parameters in applications as described in this article.

References

1. Bergman, T.L., Incropera, F.P.: Introduction to Heat Transfer. Wiley, Hoboken (2011)
2. Sarajlić, M., Pocajt, M., Kitak, P., Sarajlić, N., Pihler, J.: Covered overhead conductor temperature coefficient identification using a differential evolution optimization algorithm. B&H Electr. Eng. **11**, 26–35 (2017)
3. Program tools EleFAnT, Graz, Austria, Institute of Fundamentals and Theory in Electrical Engineering, University of Technology, Graz (2000)
4. Storn, R., Price, K.: Differential evolution-a simple and efficient adaptive scheme for global optimization over continuous spaces, vol. 3. ICSI, Berkeley (1995)
5. Storn, R., Price, K.: Differential evolution–a simple and efficient heuristic for global optimization over continuous spaces. J. Glob. Optim. **11**(4), 341–359 (1997)
6. Price, K.V., Storn, R.M., Lampinen, J.A.: Differential evolution: a practical approach to global optimization. Springer, Berlin, New York (2005)
7. List of thermal conductivities. https://en.wikipedia.org/wiki/List_of_thermal_conductivities
8. Sarajlic, M., Kitak, P., Pihler, J.: New design of a medium voltage indoor post insulator. IEEE Trans. Dielectr. Electr. Insul. **24**(2), 1162–1168 (2017)

Preliminary Considerations on Double Diffusion Instabilities in Two Quaternary Isothermal Systems of Biological Relevance

Berin Šeta[1](✉), Josefina Gavaldà[1], Muris Torlak[2], and Xavier Ruiz[1]

[1] Departament de Quimica Física i Inorgànica, Universitat Rovira i Virgili, URV, Tarragona, Spain
berin.seta@urv.cat

[2] Mechanical Engineering Faculty, University of Sarajevo, Sarajevo, Bosnia and Herzegovina
torlak@mef.unsa.ba

Abstract. Double diffusion phenomena plays a significant role in the transport of species when two or more driving forces are involved in the generation of buoyancy. This is the case of simultaneous gradients of temperature and concentration or concentration gradients of two or more species in multicomponent liquid mixtures. Double diffusion phenomena can be presented in different forms, which are mostly referred as fingers and overstability. The aim of this work was to establish limiting initial conditions for certain mixtures in order to control appearance of dissipative structures such as fingers and overstability in quaternary mixtures by simplifying these systems into ternary subsystem. Limits were applied for quaternary mixture of biological reference. Limits are established using analytical solution and confirmed by numerical simulations conducted in open source package OpenFOAM. Graphs of stability are obtained for certain mixtures from analytical solution along with procedure how to obtain them. quaternary mixtures region where both instabilities are presented simultaneously is found. This behaviour is not possible for ternary mixtures.

1 Introduction

Technological processes in which convection is minimized with respect to diffusion are of increasing importance in all kind of industries. This is because, compared to convection, the isothermal diffusion in liquid systems is a more controllable and reproducible process. In other words, diffusion always simplify the problems.

This work focuses on two interesting biotechnological systems related with pharmaceutical industries. Both are quaternary aqueous mixtures and the solvents are Lysozyme, Polyethylene Glycol (PEG) and Sodium Chloride in the first case and Sucrose, Sodium Chloride and Potassium Chloride in the second. Lysozyme is a model protein for protein-crystallization and antimicrobial enzymology studies, so the first quaternary mixture is important in relation to the understanding and optimizing the production of protein crystals [1]. The second mixture is also relevant because in biotechnological industries, saccharides such as sucrose are important components of formulations and salts are used to modulate the cryoprotectant action of these saccharides on biological macromolecules [2].

S. Avdaković (Ed.): IAT 2018, LNNS 59, pp. 326–335, 2019.
https://doi.org/10.1007/978-3-030-02574-8_26

But, in the Earth's gravitational field, the competition between the gradients of the three components of the mixture could unstabilize it developing convective motions [3–5]. This kind of problems have been for a long time a subject of great interest in meteorology, geology and oceanography. So, the study of the conditions under which the instability appears in complex systems and its characteristics are of great interest not only from a fundamental point of view but also from a more applied one.

The aim of the present preliminary report will be, thus, the analysis and characterization of the different kind of instabilities which could appear in laboratories on the Earth in case of the two chosen quaternary systems. In this way, with the use of these data, experimenters will have a valuable information to maintain their experiments diffusively controlled avoiding unstable conditions.

2 Problem Description

In both aqueous liquid mixtures, the non-dimensional continuity, momentum and mass transfer equations, assuming constant thermophysical properties of the mixture except for the linear variation of density with the temperature and concentration in the buoyancy terms –Boussinesq hypothesis- can be written as,

$$\frac{\partial u_i}{\partial x_i} = 0 \tag{1}$$

$$\frac{\partial u_i}{\partial t} + \frac{\partial u_j u_i}{\partial x_j} = -\frac{\partial p}{\partial x_i} + Sc\frac{\partial^2 u_i}{\partial x_j^2} - Ra_1 Sc \cdot c_1 \delta_{i2} - Ra_2 Sc \cdot c_2 \delta_{i2} - Ra_3 Sc \cdot c_3 \delta_{i3} \tag{2}$$

$$\frac{\partial c_1}{\partial t} + \frac{\partial u_j c_1}{\partial x_j} = \frac{\partial^2 c_1}{\partial x_j^2} + D_{12}^* \frac{\partial^2 c_2}{\partial x_j^2} + D_{13}^* \frac{\partial^2 c_3}{\partial x_j^2} \tag{3}$$

$$\frac{\partial c_2}{\partial t} + \frac{\partial u_j c_2}{\partial x_j} = D_{22}^* \frac{\partial^2 c_2}{\partial x_j} + D_{21}^* \frac{\partial^2 c_1}{\partial x_j^2} + D_{23}^* \frac{\partial^2 c_3}{\partial x_j^2} \tag{4}$$

$$\frac{\partial c_3}{\partial t} + \frac{\partial u_j c_3}{\partial x_j} = D_{33}^* \frac{\partial^2 c_3}{\partial x_j} + D_{31}^* \frac{\partial^2 c_1}{\partial x_j^2} + D_{32}^* \frac{\partial^2 c_2}{\partial x_j^2} \tag{5}$$

The scales of length and time used to define the dimensionless variables are, a characteristic length λ, and the relation λ^2/D_{11} respectively. In Eq. (2), $Sc = \mu/\rho_o D_{11}$ is the Schmidt number and $Ra_1 = \beta_{c1} H \lambda^3 g/\nu D_{11}$ and $Ra_2 = \beta_{c2} H \lambda^3 g/\nu D_{11}$ and $Ra_3 = \beta_{c3} \lambda^3 g/\nu D_{11}$ are the thresolutal Rayleigh numbers corresponding to the three different species considered in both mixtures. Previous model for ternary mixtures is developed [3] and extended in this work. The diffusion coefficients of Eqs. (3–5) are nondimensional, $D_{ij}^* = \frac{D_{ij}}{D_{11}}$, c_i are species concentrations, u is velocity, β_{ci} are mass expansion coefficients, g is gravitational acceleration, ρ_o is mean density of mixture, μ and ν are dynamic and kinetic viscosities respectively. The domain of integration used in the calculations has been a 5 × 5 × 10 mm parallelepipedic volume. Length normalization has been based on its height (λ = H = 10 mm) to enable, as much as possible, quick comparisons between Rayleigh and Schmidt numbers. No-slip impermeable boundary conditions for velocity and species concentration were imposed at the six walls of the domain. Initial conditions for velocity are fluid at rest and different step concentration profiles. A small flat interface has been defined in the middle plane of the domain to examine the evolution of controlled concentration gradients between top and bottom of the different species of the mixture

$$\Delta c_i = c_i^{bot} - c_i^{top}. \tag{6}$$

Being c_i^{bot} and c_i^{top} the values initially defined for the specie i in the lower and upper subdomains. Both values have been chosen here as uniformly distributed to simplify the calculations. The thermophysical properties used in all calculations are compiled in Table 1. Mention, at this respect, that we have estimated the viscosity of the mixture taking into account data of the pure species.

Table 1. Thermophysical properties used in all calculations.

	Mixture 1 [1] Lysozyme (1) + PEG2000 (2) + NaCl (3)	Mixture 2 [2] Sucrose (1) + NaCl (2) + KCl (3)
$\overline{c_1}$ (mol/l)	0.0006	0.25
$\overline{c_2}$ (mol/l)	0.05	0.5
$\overline{c_3}$ (mol/l)	0.5	0.5
ρ_0 (kg/m^3)	1035.960	1072.185
$\overline{\mu}$ (Pa·s)	3.1×10^{-3}	3.2×10^{-3}
β_{c_1} (l/mol)	3.890112	0.119998
β_{c_2} (l/mol)	0.319510	0.035861
β_{c_3} (l/mol)	0.037936	0.041234
D_{11} ($\times 10^{10}$m^2/s)	0.515	4.42
D_{12}($\times 10^{10}$ m^2/s)	0.0097	0.19
D_{13}($\times 10^{10}$ m^2/s)	0.0009	0.14
D_{21}($\times 10^{10}$ m^2/s)	2	1.0
D_{22}($\times 10^{10}$ m^2/s)	1.99	11.6
D_{23}($\times 10^{10}$ m^2/s)	0.083	0.08
D_{31}($\times 10^{10}$ m^2/s)	121	2.0
D_{32}($\times 10^{10}$ m^2/s)	21	1.40
D_{33}($\times 10^{10}$ m^2/s)	10.75	15.35

The set of governing equations, together with the corresponding initial and boundary conditions, have been solved numerically with OpenFOAM [6]. The diffusive and convective terms were discretized using second-order linear schemes while that time integration used an Euler first order implicit one. The number of mesh cells used in all calculations is 40 × 40 × 320. In the x and y directions the distribution is uniform but in the z direction the cells are slightly clustered towards the central plane to conveniently follow the instabilities which will take place in this region.

3 Results and Discussion

Three possible kind of instabilities associated with any generic quaternary mixtures, these are the so-called fingers, overstability and simultaneous fingers-and-overstability [7]. In order to detect the existence intervals of each one of these single instabilities, fingers and overstability, we use the stability diagrams of the three different ternary subsystems associated [4, 5]. In these cases and considering one of the components uniformly distributed, the corresponding ternary system can be considered as a good enough approximation of the complete mixture, at least during the initial instants of the process during which the instabilities appear and no significant gradient of the third constant component acts. Mention also that with these kind of mixtures the values of the different diffusion coefficients do not change too much if they are related with ternary or quaternary mixtures [1, 8].

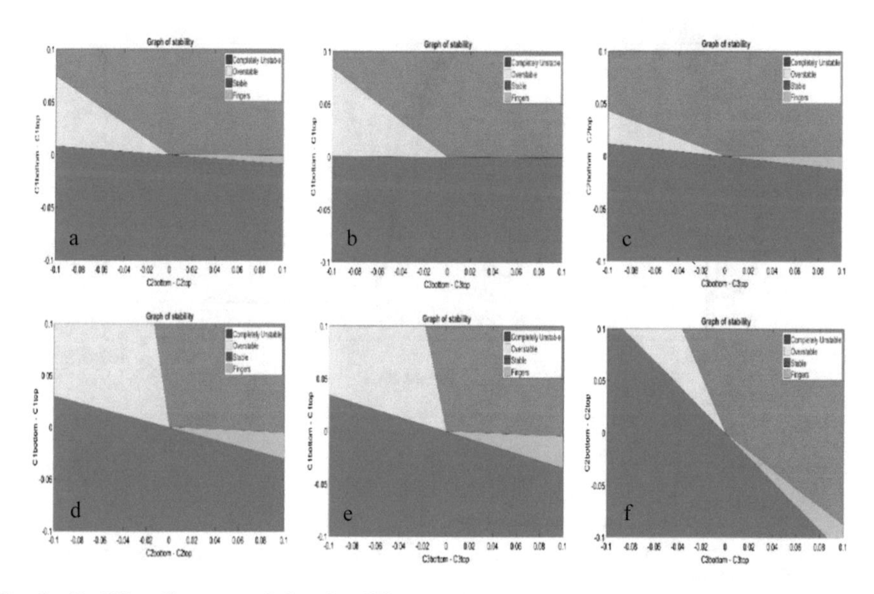

Fig. 1. Stability diagrams of the six different subsystems considered. Mixture 1 subsystems corresponds to (a), (b) and (c) cases while Mixture 2 subsystems are related with (d), (e) and (f) ones. $\Delta c_1 = 0$ in cases (a) and (d), $\Delta c_2 = 0$ for (b) and (e) and $\Delta c_3 = 0$ for (c) and (f).

In relation to the prediction of the conditions needed to the obtaining of single instabilities, Fig. 1 show the stability diagrams associated with each one of the two biological mixtures. More details about their construction can be consulted in the literature. Only say, in summary, that lines defining boundaries are a consequence of the analytical unbounded self-similar solution related with each corresponding ternary subsystem [4, 5, 9, 10]. By treating first derivates of density and seeking for its sign on the edge of boundary (overstability) or in the center of boundary (fingers) In these particular cases notice the big blue regions in which the mixture density in the top part of the parallelepipedic cell is bigger than the one in the bottom. Because this situation is gravitationally unstable these areas are labelled completely unstable. Also, big red regions corresponds to stable situations in which the mixture density in the top is lower than the one in the bottom. Experimentalists must plan their experiments in these areas to ensure pure diffusion during their experiments. Yellow regions corresponds to overstable systems and it is interesting to point out that this kind of instability is given to a greater or lesser extent in both mixtures. This is not the case of fingers in which the areas of existence are much reduced and practically vanishes in case (b), that is to say, in case of Mixture 1 and with $\Delta c_1 \approx 0, \Delta c_2 \in [0.04, 1], \Delta c_3 = 0$. It is important to mention that with the use of these simple diagrams the experimentalist could preselect a particular composition in a stable region ensuring pure diffusive conditions in their experiments.

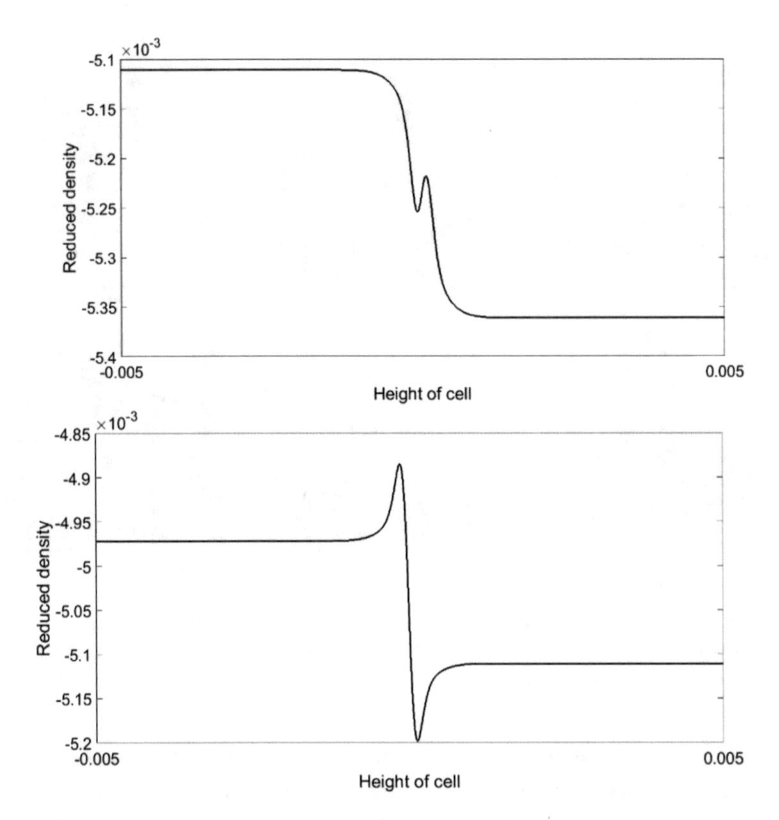

Fig. 2. Density profile along midline through height of the cell obtained using Mixture 1 with $\Delta c_1 = -0.0001$, $\Delta c_2 = 0.002$, $\Delta c_3 = 0$ on the top and $\Delta c_1 = 0.0002$, $\Delta c_2 = -0.002$, $\Delta c_3 = 0$ on the bottom

Defined the ternary diagrams and to check the correction of the previous hypothesis on the prediction of the conditions needed to obtain a determined kind of single instability, we have simulated a number of cases and always initial predictions and calculations coincides in the determination of the resulting instability. As an example, Fig. 3 shows the concentration profiles of the three different species of the corresponding quaternary Mixture 1–in particular Lysozyme (1) + PEG2000 (2) + NaCl (3) – while Fig. 2 represent density profile in the center line of the parallelepipedic domain. Density profile is obtained at the beginning of the process of diffusion while that the concentration patterns of the three different species corresponds to other instant of time in which the instability has already been developed. Concerning density profiles notice the different structure of both fingers and overstability. In the first case the density overturn is generated in the center of the domain, around the contact plane, and remains there until the mixing is complete. In the second case the instability is generated at both edges and moves to the center until the mixing is complete. Effect of overstability is visible only in the c3 profile showing a small deflection near the walls.

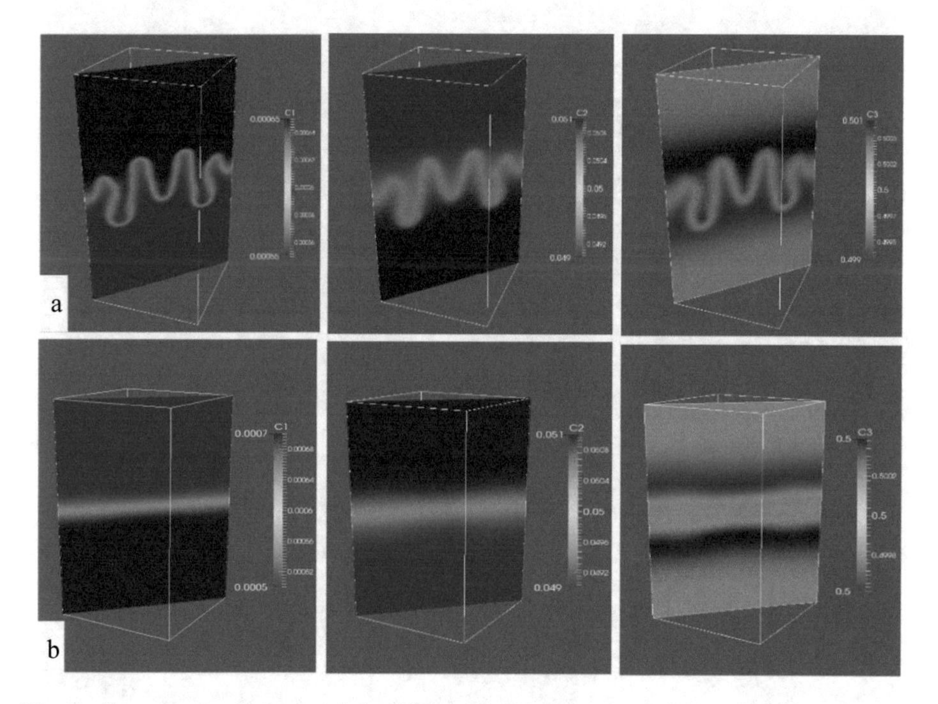

Fig. 3. Concentration patterns of the Mixture 1 for the case of fingers (a - first row) and overstability (b - second row) instability

Figure 4 first row shows two snapshots of the time dependent field associated with fingers instability. In the beginning (left side) the dynamics is constrained all along the interface until, with time, the whole domain participates in the movement (right side) by the effect of buoyancy. The same Fig. 3a shows the patterns of reduced density,

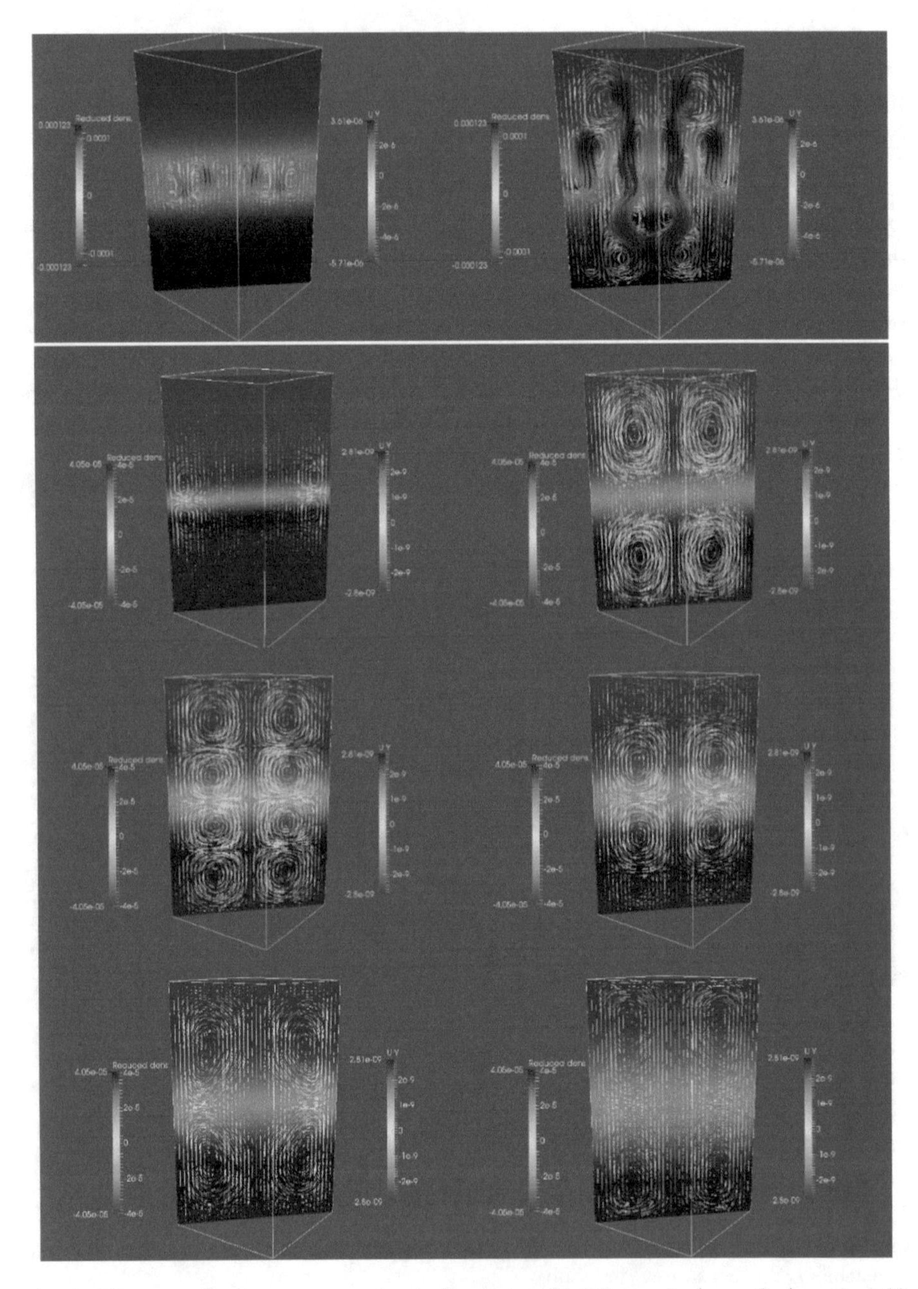

Fig. 4. First row: density patterns and velocity vectors for Mixture 1, $\Delta c_1 = 0$, $\Delta c_2 = -0.02$, $\Delta c_3 = 0.019133$, from second to fourth row: density patterns and velocity vectors for Mixture 2, $\Delta c_1 = 0$, $\Delta c_2 = -0.02$, $\Delta c_3 = 0.02278$. Time evolution is going by rows, starting from left side.

$\rho_{red} = \frac{\rho - \rho_0}{\rho_0}$. It is interesting to mention a high density in the upper region (red color) above another one of lower density (blue color) in the beginning of the process. By gravity effect this heavy layer will go down until a stable situation. The right side of the figure illustrate this point showing a density pattern in which parts of the fluid with lower density still remains in the lower part of the domain. The mixing is, thus, still not complete there. Figure 4 from second to fourth row presents a series of six sequential snapshots of the time dependent field associated with overstability. In this case, the dynamics is richer because in the beginning four small counter rotating vortexes

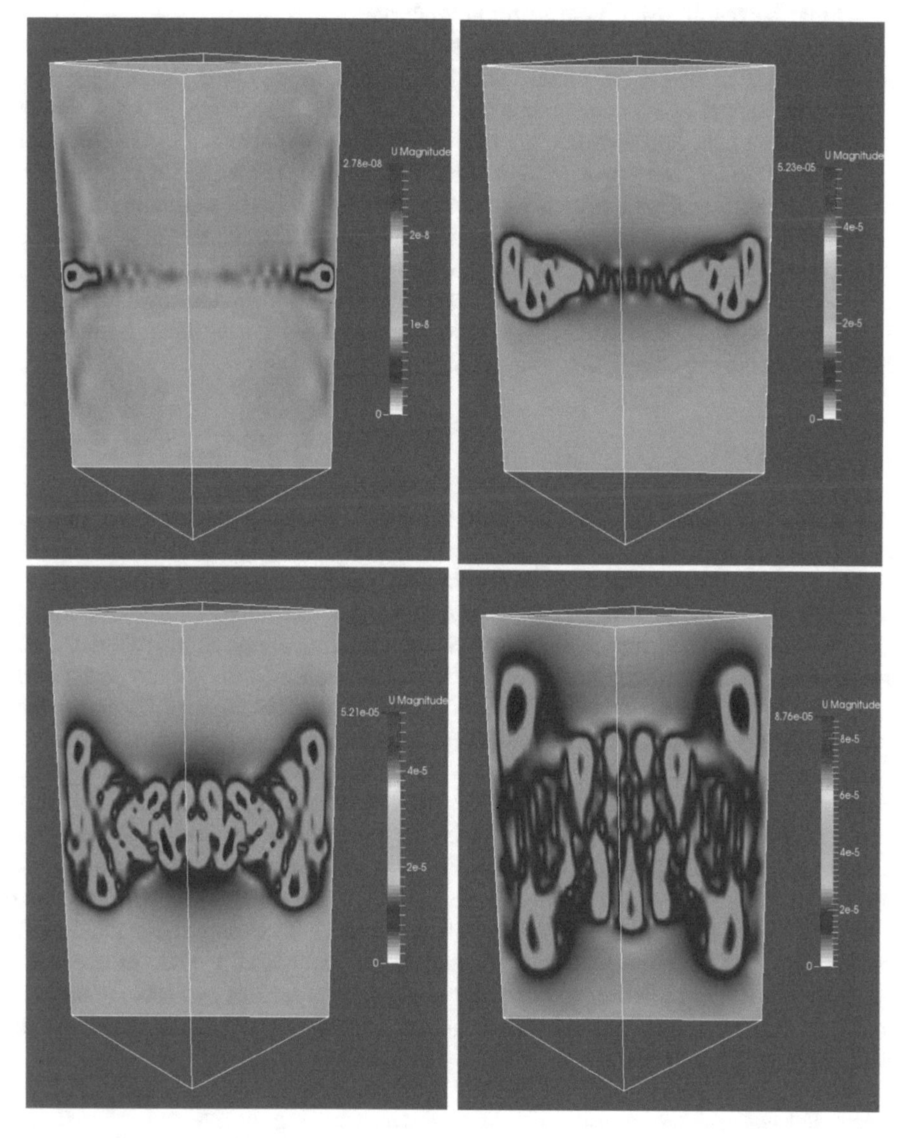

Fig. 5. Velocity patterns in case of simultaneous fingers-and-overstability.

appears near the edges. With time, these vortexes grow and moves to the top and bottom of the domain respectively generating another group of four small ones around the middle plane. These new structures continue growing at the expense of the energy of the old ones till these finally disappears. The new vortexes then remain active until the complete mixing. Mention that the values of the velocity, although small, are lower than those observed in the fingers case. Overstability dynamics is weaker than the fingers one. The same Fig. 4, second to fourth row also shows the reduced density, $\rho_{red} = \frac{\rho - \rho_0}{\rho_0}$ patterns. In this case, only a blocking region could be observed near the central plane at the beginning of the experiment. Blocking region would represent a region that blocks diffusion process for certain time.

Although the obtaining of simultaneous fingers-and-overstability is could not be obtained using in all the six ternary subsystems and corresponding stability graphs, quaternary mixtures could develop this third type of instability. In fact Fig. 5 shows the time dependent field of one particular composition which simultaneously develops both kind of instabilities. It is interesting to note that, unlike the single fingers case, the initial appearance of fingers now is on the edge, not in the center of the interface. The overstability fronts are also visible during the initial part in the upper and lower part of the domain. These suppresses the fingers in the middle part of the interface preventing their development toward the upper and lower part of the domain as in the single case. But, overstability weakens until at a certain moment it practically disappears leaving the fingers finish the complete mixing.

4 Conclusion

In this work, two different mixtures were simulated. Particular interest were put into appearance of single instabilities in quaternary systems, using double-diffusive stability graphs of ternary subsystems. It is shown that, by putting ternary subsystem into unstable range close to isodensity line and keeping concentration not considered in ternary system as constant through boundary that single instabilities (fingers and overstability) in quaternary systems appear if they appeared in ternary subsystem. However, fingers and overstability instability could not be obtained in this way, since this kind of dissipative motion is related only with quaternary systems and not with ternary ones. To obtain initial conditions that would lead to this instability, it was required to construct quaternary graph of stability. Creation of such stability graph would include analytical solution for quaternary mixtures which is complex to write. Also, graph of stability would not be plane as in case of ternary mixture but rather volume. It could be represent on 2D plots using ternary diagrams. However such representation would be limited to constrictions related with ternary diagrams. Region that represent this condition is between fingers and overstability lines intersect and isodensity line. Future work will include quaternary graphs of stability along with analytical expressions for instabilities to appear in terms of all nine diffusion coefficients of quaternary system.

Acknowledgements. The present work has been supported by grant ESP2017-83544-C3-3-P from the Spanish Ministerio de Economía y Competitividad (MINECO).

References

1. Annunciata, O., Vergara, A., Paduano, L., Sartorio, R., Miller, D.G., Albright, J.G.: Quaternary diffusion coefficients in a protein-polymer-salt-water system determined by rayleigh interferometry. J. Phys. Chem. B **113**, 13446–13453 (2009)
2. Annunciata, O., Miller, D.G., Albright, J.G.: Quaternary diffusion coefficients for the sucrose–NaCl–KCl–water system at 25 °C. J. Mol. Liq. **156**, 33–37 (2010)
3. Pallarés, J., Gavaldà, J., Ruiz, X.: Solutal natural convection flows in ternary mixtures. Int. J. Heat Mass Transf. **106**, 232–243 (2016)
4. Šeta, B., Gavaldà, J., Bou-Alí, M., Ruiz, X.: Analytical and numerical considerations about doublediffusive convection in ternary systems. Phys. Rev. E (2018, submitted)
5. Šeta, B., Dubert, D., Gavaldà, J., Bou-Ali, M., Ruiz, X.: Gravitational stability analysis on double diffusion in ternary mixtures. In: 69th International Astronautical Congress, Bremen (2018)
6. OpenFOAM User Guide Reference
7. Vitagliano, P.L., Roscigno, P., Vitagliano, V.: Diffusion and convection in a four-component liquid system. Energy **30**, 845–859 (2005)
8. Annunciata, O., Vergara, A., Paduano, L., Sartorio, R., Miller, D.G., Albright, J.G.: Precision of interferometric diffusion coefficients in a four-component system relevant to protein crystal growth: lysozime-tetra(ethylene glycol)-NaCl-H_2O. J. Phys. Chem. B **107**, 6590–6597 (2003)
9. Vitagliano, P.L., Della Volpe, C., Vitagliano, V.: Gravitational instabilities in free diffusion boundaries. J. Solut. Chem. **13**, 549–562 (1984)
10. Miller, D.G., Vitagliano, V.: Experimental test of McDougall's theory for the onset of convective instabilities in isothermal ternary systems. J. Phys. Chem. **90**, 1706–1717 (1985)

Stress Analysis of the Support for Double Motion Mechanism Inside 420 kV 63 kA SF6 Interrupter

Džanko Hajradinović[1], Mahir Muratović[2](✉), and Amer Smajkić[2]

[1] Faculty of Mechanical Engineering, University in Sarajevo, Sarajevo, Bosnia and Herzegovina

[2] EnergoBos ILJIN d.o.o. Sarajevo, Sarajevo, Bosnia and Herzegovina
mahir.muratovic@gmail.com

Abstract. Circuit breakers with voltage rating 245 kV and above are usually equipped with double motion mechanism. Double motion mechanism is needed to achieve higher opening speed which is required for withstanding higher transient recovery voltages and at the same time preserving the reliability of the driving mechanism. Double motion mechanism is required to be extremely reliable since any malfunction represents a major failure of the circuit breaker. Special attention needed to be focused on preventing the jamming of the double motion mechanism and releasing of metal particles, since both situations could lead to internal arcing. This paper presents the results of the stress analysis of the double motion mechanism supporting plates for 420 kV 63 kA SF6 circuit breaker. Additionally, the functional and reliability test results of the double motion mechanism are discussed.

Keywords: SF6 circuit breakers · Double motion mechanism
Stress analysis · Mechanical tests

1 Introduction

After initial interruption of current flow the circuit breaker contact gap must still withstand the transient recovery voltage. Dielectric performance of the high voltage SF6 circuit breaker during opening operation is mostly defined with the distance between contacts and SF6 gas density at any given time. Thus, circuit breakers with higher voltage rating (above 245 kV) must have increased contact stroke and higher opening speed. The portion of kinetic energy of the moving parts, in the total energy of the operating mechanism, is rapidly increasing since the kinetic energy increases with the product of the square of the opening speed and the mass of moving parts [1, 2]. This means a significant increase in the energy of driving mechanism which leads to higher stresses (dynamic load) and reduces overall reliability of the circuit breaker. The answer to this challenge is the double motion principle. The principle is achieved by moving two arcing contacts in opposite directions which significantly reduces the demand for the opening energy but increases the number of moving parts.

S. Avdaković (Ed.): IAT 2018, LNNS 59, pp. 336–346, 2019.
https://doi.org/10.1007/978-3-030-02574-8_27

From the point of view of the protection of the power systems from faults through the isolation of faulted parts from the rest of the electrical network, HV circuit breakers are the most important element of the system [3]. Therefore, high voltage circuit breakers are required to be extremely reliable to perform thousands of operation with no failure. The standard [4] requires 10000 operations with no failure for M2 class. The question of reliability is even more important if the double motion principle is implemented since any malfunction of that mechanism represents a major failure of the circuit breaker, possibly resulting with internal arcing. This paper presents stress analysis of the support of double motion mechanism (for 420 kV 63 kA SF6 interrupter) which, as results will show in chapter 3, has a major influence on its reliability.

2 Double Motion Mechanism for 420 kV 63 kA SF6 Interrupter

The double motion mechanism implemented on subject 420 kV 63 kA SF6 circuit breaker is shown on Fig. 1. The movable upper contact system is connected to the nozzle via a linkage system. This allows the movement of both lower and upper contact in opposite direction. In this way, the speed requirement from operating mechanism (and consequently opening energy) is significantly reduced, since the contact speed will be relative of upper contact in respect to the lower contact.

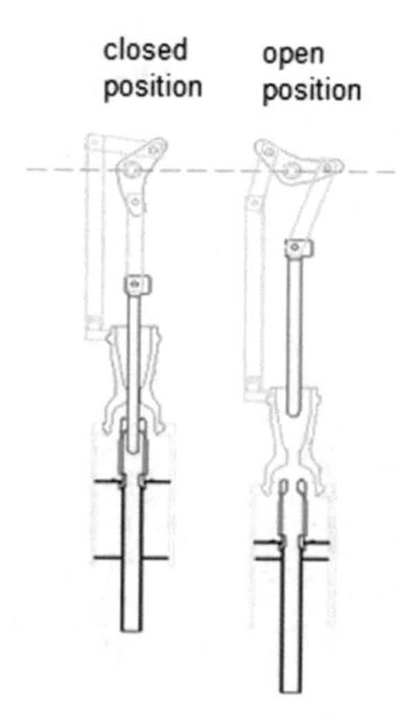

Fig. 1. Double motion mechanism for 420 kV 63 kA SF6 Interrupter [5]

The double motion mechanism shown in Fig. 1 is a non-linear double motion mechanism. The linkage system is designed in a way that it will provide a delay in the movement of upper contact at the beginning of the opening operation. Just before contact separation the upper contact will start to move (in order to increase relative opening speed) and it will continue to move until the end of operation, see Fig. 2. The delay in the movement of upper contact is required for the coordination of insulation between main and arcing contacts [6].

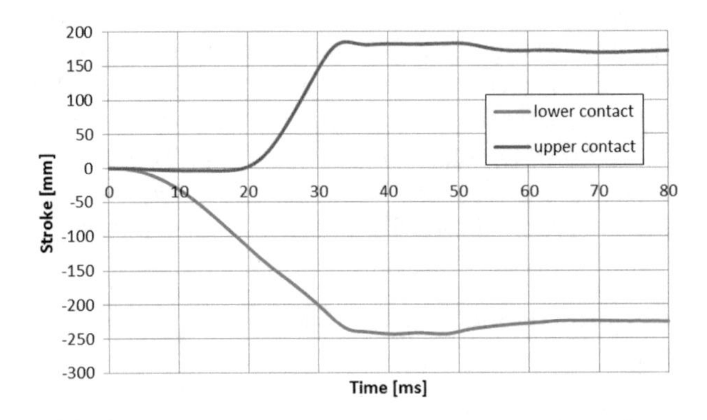

Fig. 2. Lower and upper contact stroke (measurement)

3 Functional Tests

Figure 3 shows the test model with double motion mechanism (described in chapter 2) during mechanical tests in the laboratory. The tests have been part of the overall project of development of 420 kV 63 kA GIS SF6 circuit breaker. The purpose of these tests was proving required reliability of the double motion mechanism (Fig. 4) for further tests in high power laboratory. Therefore, the test model did not include supporting insulators, compression cylinder, housing and SF6 gas. Testing was carried out in three phases with different values of the upper contact speed. The main goal of the tests was to find weak points of the design and improve it before continuing with the test until the

Fig. 3. Test model

criteria is met. Several weak points (which are corrected in the later test phases) are reported in [5]. This paper focuses on the issue with the supporting plates of the double motion mechanism. In initial design the main support of the double motion mechanism were two aluminum plates (10 mm thick) which are bolted together and on one side fixed to the insulator with four M6 bolts, see Fig. 4.

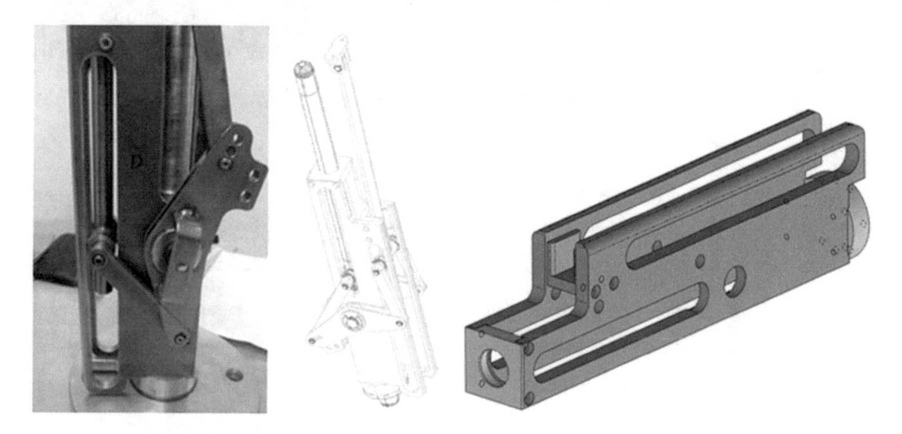

Fig. 4. Double motion mechanism: left - photo; center - 3D model; right - 3D model of supporting plates - initial design

Figure 5 (left) shows the increase in width of the upper track on both supporting plates. The increase in track width allowed hopping of the bearings and the marks of hammering of the plates have been founded also, see Fig. 5 (right). Finally, this led to the destruction of the bearings, see Fig. 6.

Another issue was high oscillations, potentially dangerous, of the supporting plates during opening operation. These oscillations were measured with a laser sensor pointed to the top of the supporting plates (opposite to the fixed side). Measured movement (oscillation) of the top of supporting plates during opening operation is given in Fig. 7.

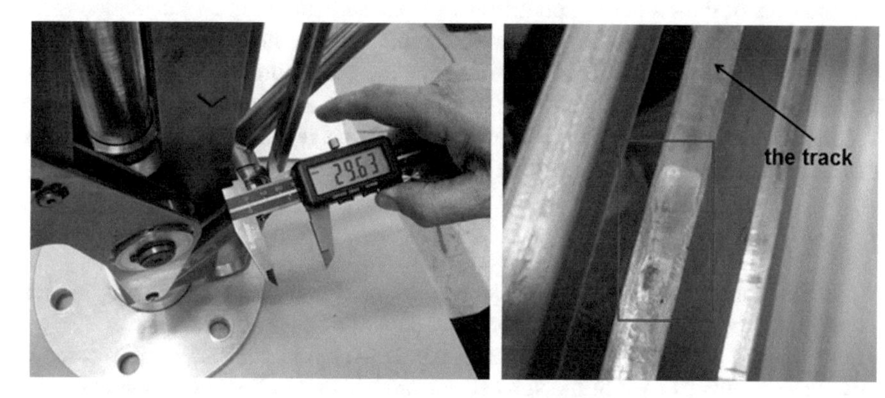

Fig. 5. Findings after mechanical tests: left - increase in the width of the track (design value is 28 mm); right - marks of hammering on the supporting plate

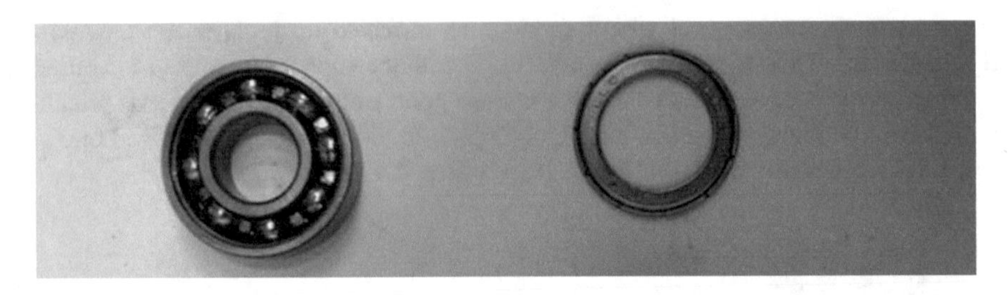

Fig. 6. Destroyed bearing

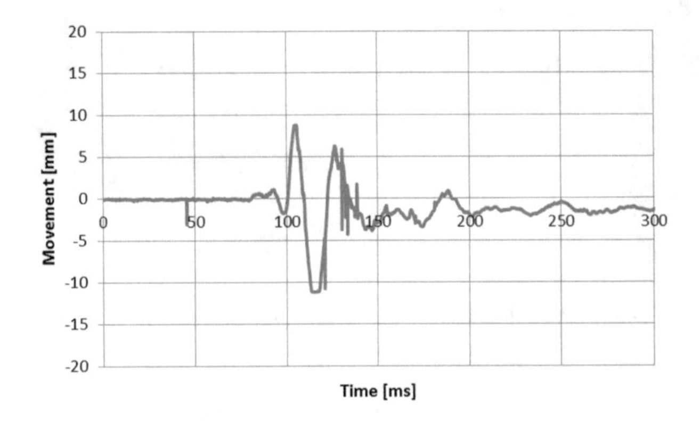

Fig. 7. Movement of the top of the supporting plates during opening operation

The cause for both issues with the supporting plates was insufficient design resistance to the dynamic load during opening operation.

4 Stress Analysis

Stress analysis of the supporting plates has been carried out based on calculation results of the deformation of supporting plates of double motion mechanism under dynamic force during opening operation. The inputs for the calculation in ANSYS have been obtained from HV CB Simulation results (the stroke and speed of upper contact). HV CB Simulation was used for the purpose of stress analysis of the supporting plates for different speeds (dynamic loads on supporting plates) of upper contact.

4.1 HV CB Simulation

HV CB Simulation provides complete simulation of high voltage SF6 circuit breakers. It is presented in [7] and more features about software application are given in [8]. In order to use HV CB Simulation results as inputs for stress analysis it is necessary to verify those results in case of the subject test model. Figure 8 shows good agreement between measured and simulated (by HV CB Simulation) values of lower and upper contact stroke.

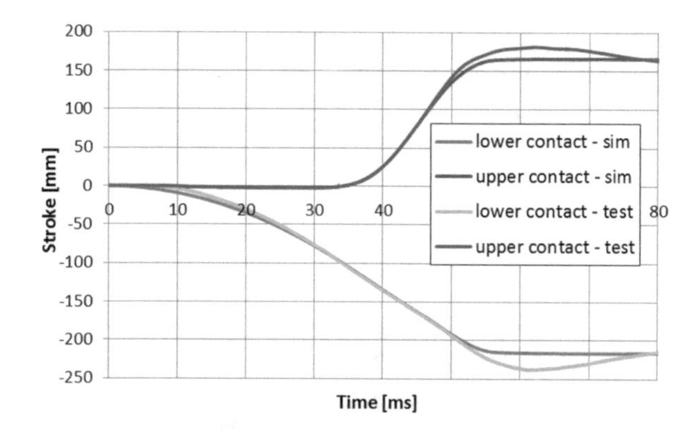

Fig. 8. Lower and upper contact stroke; comparison between measurement and HV CB Simulation results

4.2 Background for the Stress Analysis

The analysis was based on the assumption not to exceed the elastic deformation of the material. In order to take this into account, failure theory or shear-energy theory and the von Mises-Hencky theory has been used. This combination has been proved to be best used for calculation of ductile materials.

The beginning of yield of the body in a triaxial stress state is defined by the following equation:

$$2S_y^2 = (\sigma_1 - \sigma_2)^2 + (\sigma_2 - \sigma_3)^2 + (\sigma_3 - \sigma_1)^2 \tag{1}$$

Where S_y is strength and σ_1, σ_2 and σ_3 are principal stresses. If one of the main normal stresses is equal to zero than it is a biaxial or 2D stress state. The circuit breaker model can be represented with such 2D model. In order to simplify the calculation it is possible to define the larger one of the two nonzero stresses as σ_A and σ_B as the smaller one. In this way, Eq. (1) becomes:

$$S_y^2 = \sigma_A^2 - \sigma_A\sigma_B + \sigma_B^2 \tag{2}$$

For analysis and design purposes it is convenient to use von Mises stress:

$$\sigma^{vM} = \sqrt{\sigma_A^2 - \sigma_A\sigma_B + \sigma_B^2} \tag{3}$$

To provide analogy to a Cartesian coordinate system and already defined stresses σ_A and $\sigma_B$the following equations can be used:

$$\sigma_A = \frac{\sigma_x}{2} + \tau_{xy} \tag{4}$$

$$\sigma_B = \frac{\sigma_x}{2} - \tau_{xy} \tag{5}$$

Where σ_x is stress in x direction and τ_{xy} is shear stress in 2D stress state. After the substitution, the final equation can be written as:

$$\sigma^{vM} = \sqrt{\sigma_x^2 + 3\tau_{xy}^2} \tag{6}$$

4.3 Initial Design

According to the initial design of the supporting plates (Fig. 4) the material of the supporting plates was aluminum with 10 mm thickness. The plates were bolted together and on one side fixed to the insulator with four M6 bolts. In order to calculate reaction joints forces on double motion mechanism HV CB Simulation results of lower contact stroke were used (opening speed - 9.0 m/s). The worst position, in a meaning of highest forces on double motion mechanism, was identified (the maximum of lower contact acceleration) and stress analysis was performed for that position, see Fig. 9.

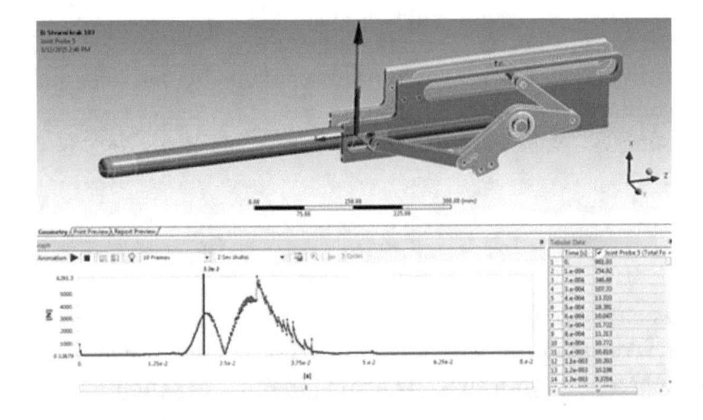

Fig. 9. Position of double motion mechanism with maximum dynamic load during opening operation of the circuit breaker

Figure 10 shows the deflection of the supporting plates and the deformation of upper track at time instant when maximum load was applied for the corresponding position of the linkage system.

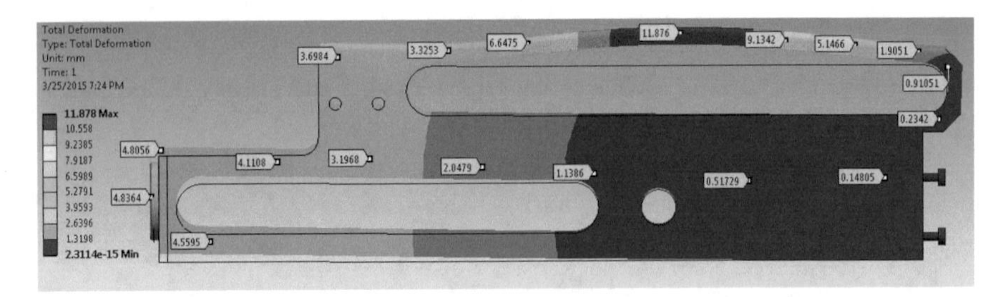

Fig. 10. Deformation of supporting plates for the time instant when maximum load was applied

Calculated deflection of the top of supporting plates (4.8 mm) was less than measured oscillation on test model (up to 12 mm) during opening operation (Fig. 7). Possible explanations can be that calculation was performed for the complete circuit breaker including supporting insulator (not for the test model). Also, it is important to have in mind that Fig. 10 show "temporary" deformation of upper track (11.8 mm maximum value) for the given time instant. Calculated plastic ("permanent") deformation of the upper track (1.8 mm maximum value) after opening is show in Fig. 11 which is comparable to the finding after mechanical test (1.6 mm, Fig. 5).

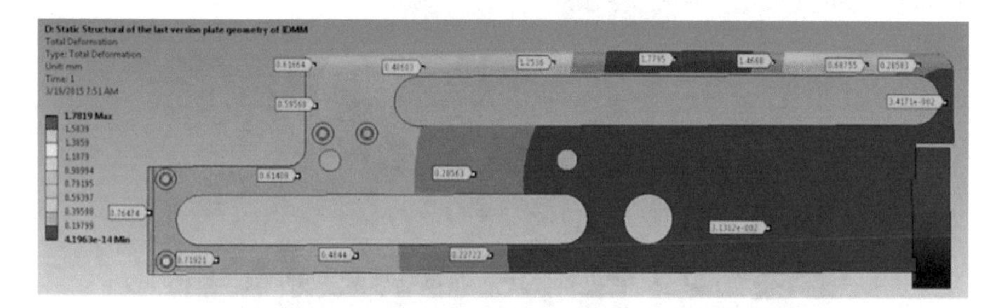

Fig. 11. Plastic deformation of supporting plates after opening operation

4.4 Improved Design

After many different variants of supporting plate design were analyzed, finally an improved design was defined.

The new, improved design included several changes:

- the material of the supporting plates was changed to steel,
- the thickness of the plates was increased to 15 mm,
- the shape of the plates was changed,
- the plates were now fixed to the insulator with six M8 bolts.

Figure 12 shows the deflection of the supporting plates and the deformation of upper track at time instant when maximum load was applied for the corresponding position of the linkage system for the improved design.

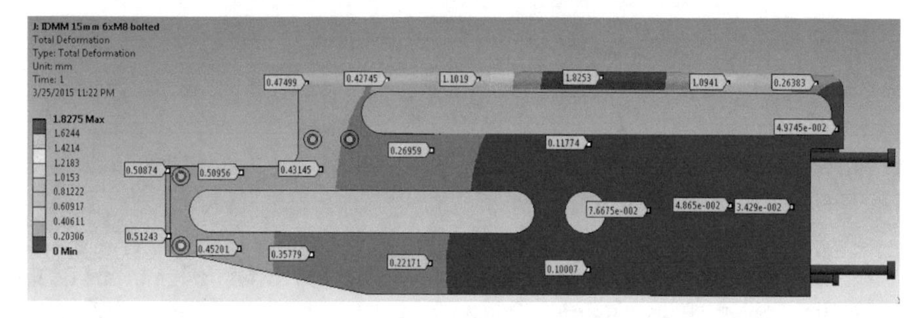

Fig. 12. Deformation of supporting plates for the improved design for the time instant when maximum load was applied

With the improved design it is clear that "temporary" deflection of top of supporting plates and deformation of upper track for the given time instant were significantly reduced (deflection of top of supporting plates - from 4.8 mm down to 0.5 mm; deformation of upper track from - 11.8 mm down to 1.8 mm).

Additional measure for the reduction of oscillation of the supporting plate was the additional support (Fig. 13) for the supporting plates on opposite side where plates are fixed to the insulator. Figure 14 gives the measured deflection of the supporting plates for the static load (applied to the top of plates in upper direction) with and without additional support in case of improved design. Practically, the additional support does not allow oscillation of the plates higher than 1 mm.

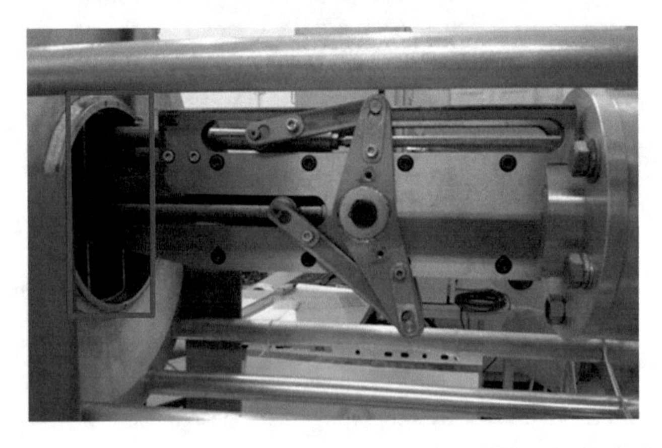

Fig. 13. Improved design of double motion mechanism with additional support (marked in red)

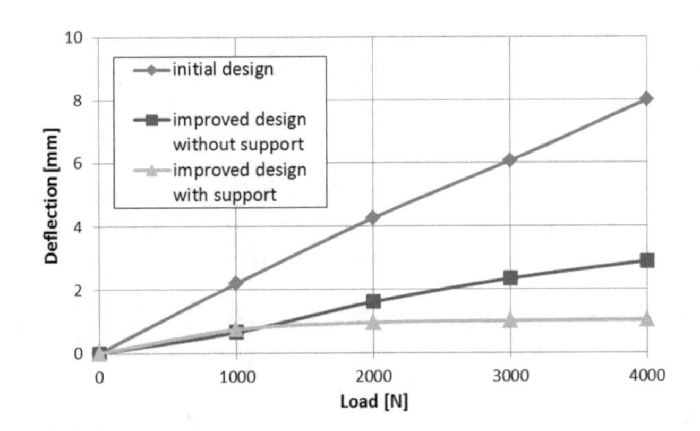

Fig. 14. Measured deflection of the supporting plates under static load applied on the top of supporting plates in upper direction

Additional mechanical tests were performed on the improved design of the double motion mechanism (with additional support) in the same conditions as before. No issues were revealed after additional testing. Figure 15 shows supporting plates (improved design) after additional mechanical tests.

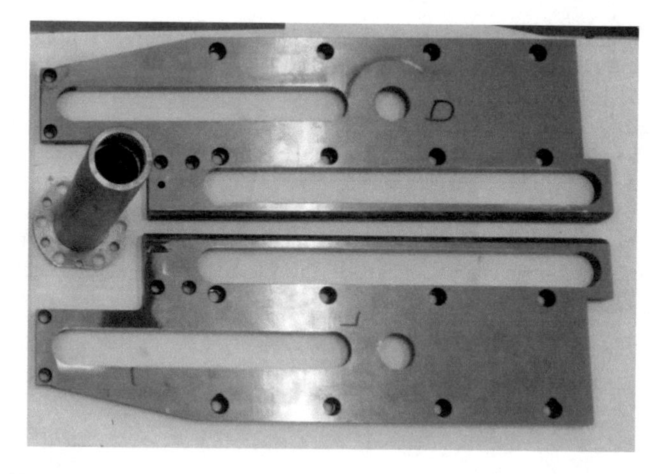

Fig. 15. Supporting plates (improved design) after additional mechanical tests

5 Conclusion

This paper presented a double motion mechanism for 420 kV 63 kA GIS SF6 circuit breaker. Double motion mechanisms are required to be extremely reliable, since any malfunction represents a major failure of the circuit breaker. Mechanical tests of the subject double mechanism revealed deformations and high oscillations of the supporting plates. Stress analysis was performed for dynamic load on the supporting plates of the double motion mechanism during operation of the circuit breaker. Stress analysis was performed in ANSYS and HV CB Simulation results were used as input. An improved design was defined based on the results of the analysis. Additional mechanical tests were performed on the improved design of the double motion mechanism in the same conditions as before. No issues were revealed after additional testing.

References

1. Kapetanovic, M.: High voltage circuit breakers, Faculty of Electrical Engineering, Sarajevo (2011)
2. Kapetanovic, M.: Visokonaponskiprekidači, Faculty of Electrical Engineering, Sarajevo (2012)
3. Muratović, M., Sokolija, K., Kapetanović, M.: Modelling of high voltage SF6 circuit breaker reliability based on Bayesian statistics. In: 7th IEEE GCC Conference and Exhibition (GCC), pp. 303–308, Doha (2013)
4. International Standard IEC 62271-100: High-voltage switchgear and controlgear – Part 100: Alternating-current circuit-breaker, Edition 2.0 (2008)
5. Smajkić, A., Kim, M.H., Muratović, M., Hajdarović, A., Gačanović, R.: Mehanička ispitivanja pouzdanosti mehanizma za dvostruko kretanje kontakata VN prekidača. B&H Electr. Eng. **9**, 20–25 (2015)

6. Bešlija, D., Muratović, M., Delić, S., Kim, K.H., Gačanović, R.: Koordinacija izolacije u VN prekidaču s aspekta sigurnosti od eksplozije, XII Session of BH K CIGRE, A3.07, Neum, Bosnia and Herzegovina (2015)
7. Ahmethodžić, A., Kapetanović, M., Gajić, Z.: Computer simulation of high-voltage SF6 circuit breakers: approach to modeling and application results. IEEE Trans. Dielectr. Electr. Insul. **18**(04), 1314–1322 (2011)
8. Muratović, M., Smajkić, A., Kim, K.H., Kim, M.H., Kapetanović, M., Ahmethodžić, A.: Criteria for successful short circuit current interruption on a real 245 kV 40/50 kA SF6 circuit breaker. In: 3rd International Conference on Electric Power Equipment – Switching Technology (ICEPE-ST), Busan, pp. 54–59 (2015)

Solving Linear Wave Equation Using a Finite-Volume Method in Time Domain on Unstructured Computational Grids

Muris Torlak(✉) and Vahidin Hadžiabdić

Mechanical Engineering Faculty, University of Sarajevo, Vilsonovo šetalište 9, 71000 Sarajevo, Bosnia-Herzegovina
{torlak,hadziabdic}@mef.unsa.ba

Abstract. In this paper, numerical solving second-order wave equation using a cell-centered finite-volume method with collocated variable arrangement on unstructured computational grids in spatial domains of arbitrary shape is discussed. A second-order accurate technique is used for discretization of spatial derivatives. Two different discretization schemes for approximation of the time derivative are employed and tested within an implicit method for time integration. Application of the numerical method is demonstrated in three simple examples.

1 Introduction

There is a variety of analytical solutions of the wave equation available in a number of standard university textbooks, for example in [1, 2]. Typically, these in most cases individual solutions are, however, tailored at specific conditions, such as those related to the number of spatial dimensions, homogeneity of the material, initial and boundary conditions etc. Many numerical methods are available in scientific literature, where the wave equation is solved in more versatile cases; for example in the works [3–5], or in the references given therein. In order to solve the wave equation for spatial distribution of a scalar quantity in time domain at various conditions, availability of one single and efficient method is desired, particularly in engineering.

In this paper, we test the finite-volume method, such as proposed by Demirdžić and Muzaferija [6], which is in most parts also described in the book of Ferziger and Perić [7]. Versatility of the method employing a unified solving approach is proved in a number of different applications [6–24]. It is used both for computation of steady-state and time-dependent problems, typically for fluid flow, heat transfer and/or solid stress. The structure of the governing equations used may yield elliptic, parabolic or (parabolic-)hyperbolic character of the problem. In a large part of time-dependent fluid flow and heat transfer problems, viscosity term and conduction term arise. Their contribution reduces the pure hyperbolic, wave-propagation character of the first-order time derivative of fluid velocity/temperature and the advection term.

On the other hand, in dynamic elastic solid-stress problems, second-order time derivative of displacements in inertial term combined with the stress tensor contribution to the surface forces delivers a second-order wave-propagation problem. However,

S. Avdaković (Ed.): IAT 2018, LNNS 59, pp. 347–356, 2019.
https://doi.org/10.1007/978-3-030-02574-8_28

most published works on application of the here applied finite-volume method in elasticity are focused on static problems, such as the works in [13, 22, 25–30], just to mention a few of them. Hence, further research on verification of the method and its performance in wave propagation analysis and solving hyperbolic problems is needed.

2 Mathematical Model

The wave propagation is described by the well-known wave equation from standard literature, either in the form of PDE or integral form, respectively:

$$\frac{\partial^2 \phi}{\partial t^2} = c^2 \left(\frac{\partial^2 \phi}{\partial x^2} + \frac{\partial^2 \phi}{\partial y^2} + \frac{\partial^2 \phi}{\partial z^2} \right), \tag{1}$$

$$\int_V \frac{\partial^2 \phi}{\partial t^2} \mathrm{d}V = \oint_S c^2 \vec{\nabla} \phi \cdot \mathrm{d}\vec{S}, \tag{2}$$

where ϕ is a generic solution variable, c is the wave propagation speed, V is the volume of the considered part of space bounded by the closed surface S.

Generic form of the wave propagation described by Eqs. (1) and (2) is applicable to a variety of physical problems, such as stretched string/membrane vibrations, pressure propagation in shallow-water, acoustic pressure waves, oscillating electric fields etc. Consequently, the variable ϕ may represent different physical properties depending on the problem under consideration, such as displacement, acoustic pressure etc.

3 Solution Methods

3.1 Analytical Solution for a 1D Problem

Equations (1)–(2) can be analytically solved for a number of simplified problems. One of these, which will be used for verification of the method in Sect. 4.1, is the problem of free, undamped vibrations of a stretched string. In that case, the solution variable ϕ represents the lateral displacement w of a string point, c is the wave propagation speed which depends on the string tension T and the specific density per unit length ρ':

$$c = \sqrt{\frac{T}{\rho'}}. \tag{3}$$

Due to string shape, the domain reduces to a 1D space, which is here adopted to be aligned with the x-axis. It is bounded by two points where the Dirichlet conditions are applied: $w_{\mathrm{b},1} = w_{\mathrm{b,n}} = 0$. Here, we consider the string having parabolic shape at the initial instant of time, which is released from the rest. Thus, the initial conditions read:

$$w_0 = ax(l - x), \quad \left.\frac{\partial w}{\partial t}\right|_0 = 0, \quad 0 \le x \le l, \tag{4}$$

where a is the auxiliary parameter determining the vertex position of the parabola, i.e. the maximum deflection: $a = 4\ w_{\max}/l^2$, and l is the span of the string. Solution of the problem can be found either by the method of separation of variables, or using Laplace transform. In doing so, one obtains [1]:

$$w(x,t) = \frac{8al^2}{\pi^3} \sum_{i=1}^{\infty} \frac{1}{(2i-1)^3} \sin\frac{(2i-1)\pi}{l}x \cos\frac{(2i-1)\pi}{l}ct. \tag{5}$$

3.2 Finite-Volume Method

The solution domain is discretized employing a number of adjoining but separate cells. The shape of cells is polyhedral. The computational points, where the solution variable ϕ is evaluated, are located at the centers of the cells. The wave equation in integral form, Eq. (2), is set up for each cell in the computational mesh, which are regarded as individual control volumes. Discretization schemes of second-order accuracy are applied. The integrals are approximated by the mid-point rule, while the gradient of ϕ is approximated at the cell-face center. Having projected it onto the cell-face normal vector, central differencing is employed using the variable values from the adjacent computational points. Non-orthogonal effect on the cell face (deviation of the face normal from the distance vector between the computational points) is taken into account using deferred-correction technique [6, 7]. The wave propagation speed c is treated as a material property. In the case of acoustic, pressure or elastic waves it is obtained from the bulk modulus and the density of the material in the control volume, analogously to Eq. (3). It is specified in the entire solution domain, assuming its piece-wise linear distribution between the computational points.

Time-dependent properties are obtained in an implicit solving procedure dividing the observed time interval into a number of, typically but not necessarily, equally sized time steps. The second derivative in time of ϕ is discretized using an appropriate differencing scheme, including the values of ϕ from the current time step m and the corresponding number of the last previous time steps, while all other values are considered in the current time step. Applying the implicit backward Euler scheme to the time derivative two times results in the following first-order accurate approximation:

$$\left.\frac{\partial^2 \phi}{\partial t^2}\right|^{(m)} \approx \frac{\phi^{(m)} - 2\phi^{(m-1)} + \phi^{(m-2)}}{\Delta t^2}, \tag{6}$$

while applying the implicit backward three-time-levels scheme [7] two times leads to the second-order accurate approximation [31]:

$$\left.\frac{\partial^2\phi}{\partial t^2}\right|^{(m)} \approx \frac{2\phi^{(m)} - 5\phi^{(m-1)} + 4\phi^{(m-2)} - \phi^{(m-3)}}{\Delta t^2}. \tag{7}$$

Implicit use of Eq. (7) in finite-volume method is similar to the Houbolt's scheme [32] which is used in numerical methods of structural dynamics such as finite element method.

Upon discretization of Eq. (2) given for each cell in the computational mesh at the considered time step, a system of linear algebraic equation is obtained. The system is solved using conjugate gradient method with incomplete Cholesky preconditioning [6], whereupon the solver advances to the next time step, and the procedure is repeated with the updated time.

4 Examples

4.1 Vibrations of a Stretched String

In this verification test, the finite-volume method results are compared with the analytical solution, Eq. (5), for stretched string vibrations.

The span of the string is l = 10 m, the wave propagation speed in the string is c = 5900 m/s, the intial conditions of the string are given by Eq. (4), and the value of the auxiliary parameter is a = 0.04 m^{-1} yielding the initial deflection of $w_{0,\max}$ = 1 m at the string midpoint. The numerical model contains 21 computational point uniformly distributed over the string span. The time step size of Δt = 10^{-5} s is chosen, with equal blending of the 1st-order and the 2nd-order time differencing, Eqs. (6)–(7). Herewith, a full oscillation period (from Mersenne's law: $T = 2\ l/c$) is reached within nearly 340 time steps. Figure 1 shows the initial shape and position with the numerical and the analytical results obtained at four evenly distributed time steps covering approximately half of the oscillation period. The agreement of the results is evident.

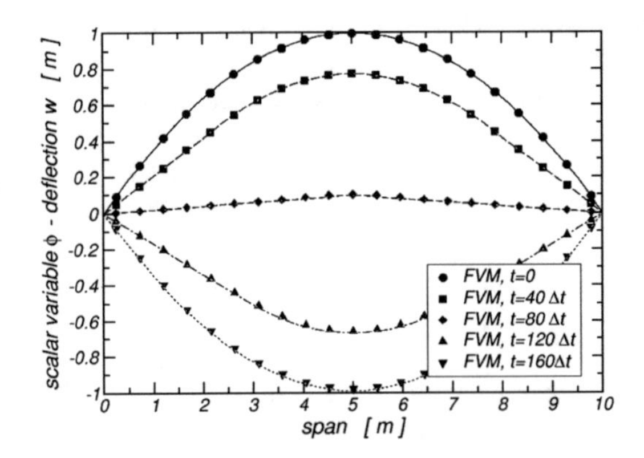

Fig. 1. Lateral deflection of the stretched string over the span length: comparison of the finite-volume-method solution (symbols) with the analytical one (lines) given by Eq. (5).

4.2 Wave Propagation in a Homogeneous Field

Here, propagation of a wave in a plane, homogeneous material is considered. The adopted solution domain is a square, see Fig. 2. At the square boundaries, the zero-gradient condition for the solution variable ϕ is applied, so they behave as symmetry lines. Thus, the calculated model can be considered as a reduced, repeating pattern of an infinite solution domain. The computational mesh is equidistant, Cartesian one,

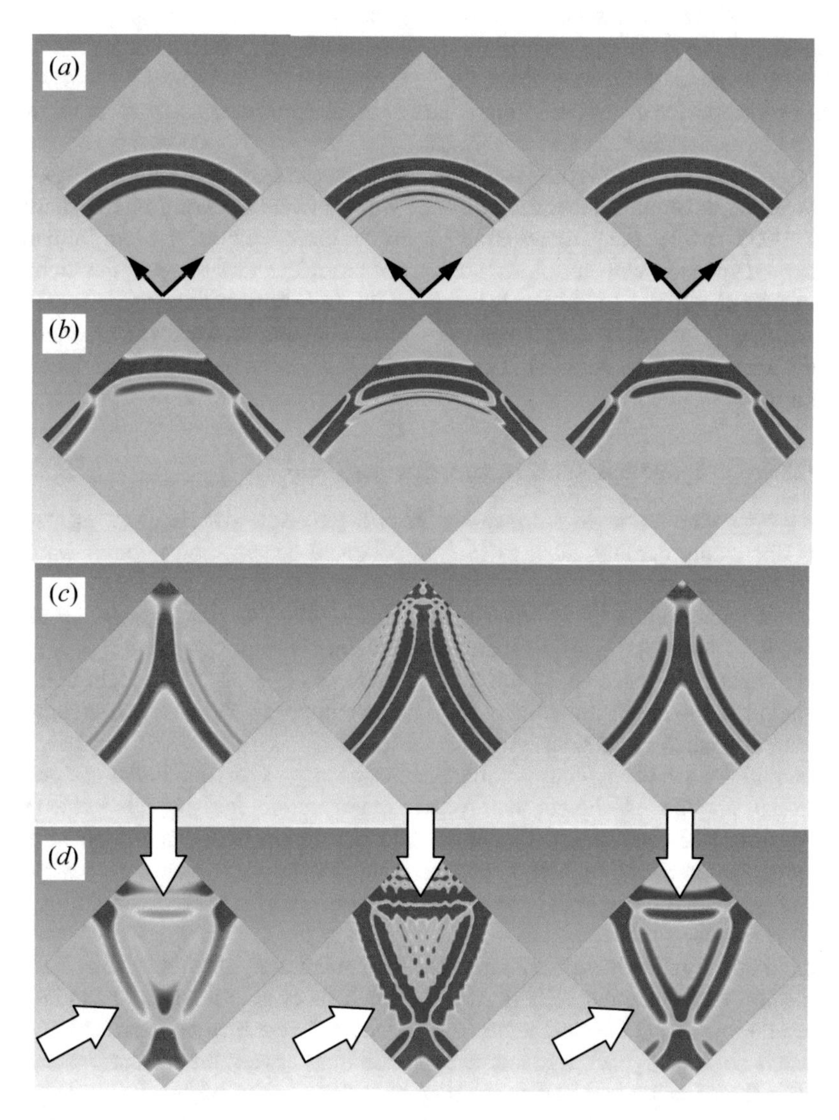

Fig. 2. Use of implicit Euler scheme (left), Houbolt scheme (middle), and a blend of the two schemes (right) at four different instants of time (from the top downwards, a–d).

where the grid lines are aligned with the square sides. It has 96 cells along each boundary.

The center of the initial, circular wave is chosen to be at the origin of the coordinate system, denoted by the black arrows in the figure. Its profile shape is described by spatial distribution of ϕ following a Gaussian function of x and y. The wave velocity at time $t = 0$ is zero.

Figure 2 shows the values of the variable ϕ at different instants of time. The results given are calculated using the first-order (left), second-order (middle) and the blended (right) time differencing scheme. At the beginning, the wave front propagates in all directions equally from the center of the wave outwards (a). After the upper boundaries are hit by the parts of the wave front, they are reflected (b), and travel back toward the origin of the coordinate system. The reflected waves collide (c) and pass through each other, creating a complex wave pattern (d).

The results obtained with the first-order accurate discretization of the time derivative show certain numerical diffusion with obvious smearing of the sharp gradients (left), while the second-order accurate discretization triggers remarkable numerical dispersion with strong oscillations of the scalar variable ϕ in the vicinity of sharp gradient regions (middle). A blend of the two schemes seems to resolve the problem (right), avoiding both smearing and oscillations to some extent, increasing thus the accuracy of the solution. This effect is clearly seen in the regions denoted by white arrows.

4.3 Wave Propagation in an Inhomogeneous Field

Wave propagation through a domain with two materials whose wave propagation speeds differ considerably, such as in case of sound propagation through water and steel, is considered.

The adopted domain is symmetric, see Fig. 3, with the size 20 m × 15 m. The dark-shaded region represents the material with larger wave-propagation speed. Here, a typical value of the wave speed for steel of 5900 m/s is used. In the light region the wave-propagation speed for water 1400 m/s is imposed. Non-reflective boundary condition is applied at all sides.

The initial wave front, whose profile is described by a Gaussian function, is parallel to the x-axis and moves with constant speed c in y-direction. Initially, it is located in the water region. The right-angled corner of the steel region is oriented opposite to the wave propagation direction. Due to considerable change of the wave speed across the interface between the materials, wave reflection with a partial transmission through the solid is expected.

Automatically generated unstructured computational grid consisting of both quadrilateral and triangular cells is used. The cell faces at both sides of the interface between the water and the steel region are matching. In the method presented here, the interface is not used to exchange boundary conditions between the regions; instead, it is treated as the other interior cell faces, they contribute to the surface integral approximation with spatially variable wave propagation speed. The wave speed at the interface is calculated by linear interpolation from the adjacent cells/regions, although some other models can be employed (e.g. geometric or harmonic averaging).

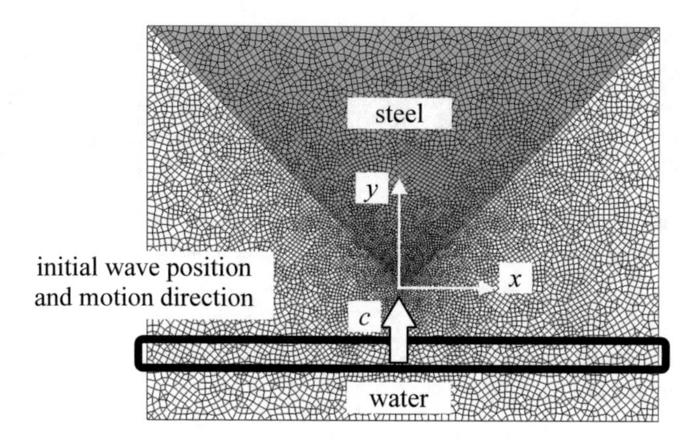

Fig. 3. Unstructured mesh in an inhomogeneous domain.

Figure 4 shows the calculated values of the solution variable ϕ in the entire domain at four different time steps, where the wave patterns are clearly recognized. The largest part of the wave is reflected by the solid interface, whereby the sign of the wave is changed (change from red- to blue-colored values). The shape of the waves and their motion direction show that the angle of incidence and the angle of reflection are the same, i.e. symmetric about the line normal to the material interface. The interface is slanted at the angle of 45° with respect to the coordinate axes. The red zone representing the wave front continues the motion in positive y-direction (the incident angle of 45°), while the blue zone representing the reflected wave continues the motion in x-direction (yielding the reflexion angle of 45°). It is also seen that the part of the wave reflected from the corner point propagates radially backwards, revealing a semi-circular shape.

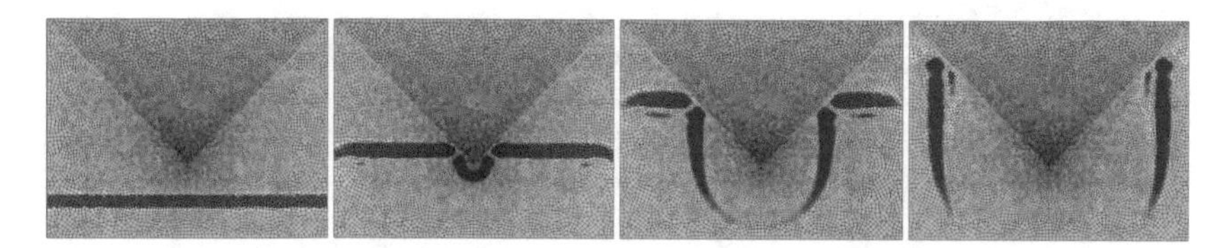

Fig. 4. Inhomogeneous media: wave propagation through the water with the scalar value ϕ from −5 (dark blue) to 5 (dark red) at the instants of time: 0 ms, 2.5 ms, 5 ms and 7.5 ms (left to right).

In Fig. 5 the calculated values of the solution variable ϕ in the solid region are presented for four different time steps. Forward radial propagation from the corner is observed. Obviously, the propagation speed found in the plot – interpreted through the distance covered by the wave front in the same period of time, and attenuation of the wave are larger in the solid domain, as expected.

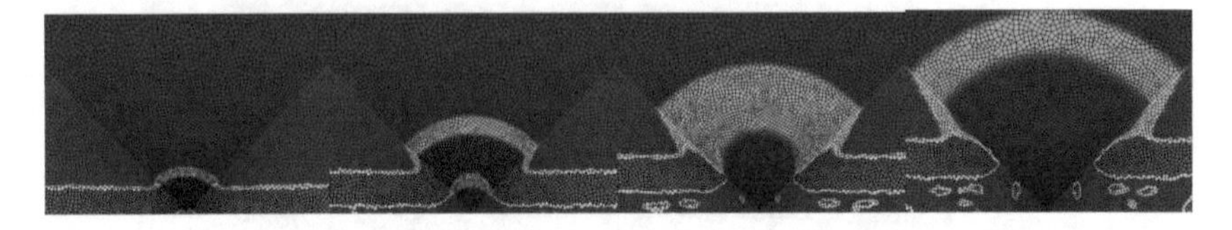

Fig. 5. Inhomogeneuous media: wave propagation through the steel with the scalar value ϕ from 0.2 (dark blue) to 0.4 (dark red) at the instants of time: 2 ms, 2.5 ms, 3 ms and 3.5 ms (left to right).

In Fig. 6 three-dimensional shape of the wave is depicted, where the previously discussed effects are also seen.

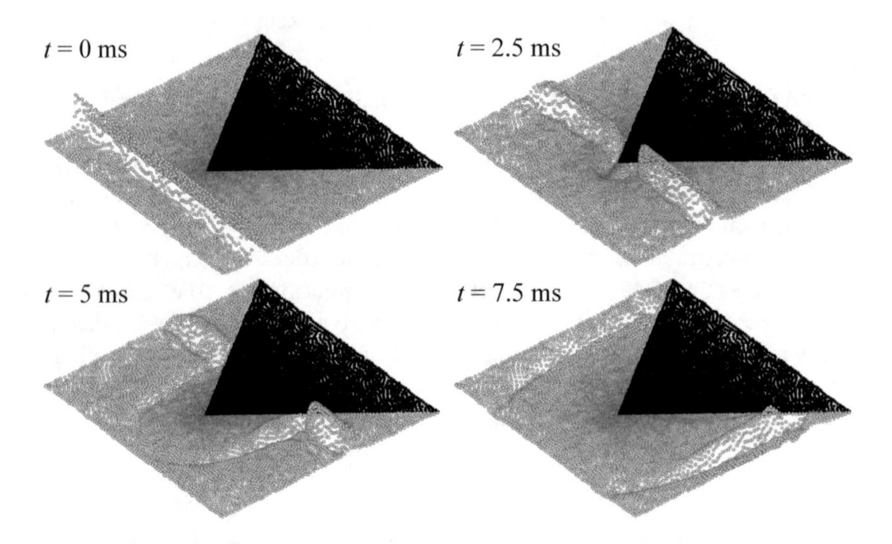

Fig. 6. Wave front through the inhomogeneuous media at different time steps.

5 Conclusions

The finite-volume method by Demirdžić and Muzaferija [1, 2] enables efficient time-dependent solution of wave-equation on unstructured grids in domains of arbitrary shapes, throughout materials with inhomogeneous properties, i.e. with spatially variable wave-propagation speed.

The 1st-order accurate time integration results in a noteable numerical diffusion, while the 2nd-order accurate scheme causes severe numerical dispersion. A remedy is found in a blend of the two time-integration schemes, keeping both stability of the low-order scheme and accuracy of the high-order one.

References

1. Kreyszig, E., et al.: Advanced Engineering Mathematics, 10th edn. Wiley, Hoboken (2011)
2. Vuković, M.: Diferencijalne jednačine 2, parcijalne diferencijalne jednačine, jedna-čine matematičke fizike. Univerzitet u Sarajevu, Sarajevo (2001)
3. Engquist, B., Majda, A.: Absorbing boundary conditions for the numerical simulation of waves. Math. Comput. **31**(139), 629–651 (1977)
4. Rowley, C.W., Colonius, T.: Discretely nonreflecting boundary conditions for linear hyperbolic systems. J. Comput. Phys. **157**, 500–538 (2000)
5. Britt, S., Tsynkov, S., Turkel, E.: Numerical solution of the wave equation with variable wave speed on nonconforming domains by high-order difference potentials. J. Comput. Phys. **354**, 26–42 (2018)
6. Demirdžić, I., Muzaferija, S.: Numerical method for coupled fluid flow, heat transfer and stress analysis using unstructured moving meshes with cells of arbitrary topology. Comput. Methods Appl. Mech. Eng. **125**(1–4), 235–255 (1995)
7. Ferziger, J.H., Perić, M.: Computational Methods for Fluid Dynamics. Springer, Heidelberg (2002)
8. Demirdžić, I., Martinović, D.: Finite-volume method for thermoclasto-plastic stress-analysis. Comput. Methods Appl. Mech. Eng. **109**, 331–349 (1993)
9. Demirdžić, I., Lilek, Ž., Perić, M.: A collocated finite-volume method for predicting flows at all speeds. Int. J. Numer. Methods Fluids **16**, 1029–1050 (1993)
10. Ivanković, A., Muzaferija, A., Demirdžić, I.: Finite volume method and multigrid acceleration in modelling of rapid crack propagation in full-scale pipe test. Comput. Mech. **20**(1–2), 46–52 (1997)
11. Muzaferija, S., Perić, M.: Computation of free-surface flows using finite volume method and moving grids. Numer. Heat Transf. – Part B **32**, 369–384 (1997)
12. Muzaferija, S., Perić, M.: Computation of free surface flows using interface-tracking and interface-capturing methods, In: Mahrenholtz, O., Markiewicz, M. (eds.) Nonlinear Water Wave Interaction, Chap. 2. Computational Mechanics Publications, Southampton (1998)
13. Wheel, M.A.: A finite volume method for analysing the bending deformation of thick and thin plates. Comput. Methods Appl. Mech. Eng. **147**(1–2), 199–208 (1997)
14. Greenshields, C.J., Venizelos, G.P., Ivanković, A.: A fluid-structure model for fast brittle fracture in plastic pipes. J. Fluids Struct. **14**(2), 221–234 (2000)
15. Demirdžić, I., Horman, I., Martinović, D.: Finite volume analysis of stress and deformation in hygro-thermo-elastic orthotropic body. Comput. Methods Appl. Mech. Eng. **190**, 1221–1232 (2000)
16. Teskerežić, A., Demirdžić, I., Muzaferija, S.: Numerical method for heat transfer, fluid flow, and stress analysis in phase-change problems. Numer. Heat Transf. B **42**, 437–459 (2002)
17. Demirdžić, I., Džafarović, E., Ivanković, A.: Finite-volume approach to thermo-viscoelasticity. Numer. Heat Transf. Part B: Fundam. **47**(3), 213–237 (2005)
18. Bašić, H., Demirdžić, I., Muzaferija, S.: Finite volume method for simulation of extrusion processes. Int. J. Numer. Methods Eng. **62**(4), 475–494 (2005)
19. Bijelonja, I., Demirdžić, I., Muzaferija, S.: A finite volume method for large strain analysis of incompressible hyperelastic materials. Int. J. Numer. Methods Eng. **64**(12), 1594–1609 (2005)
20. Bijelonja, I., Demirdžić, I., Muzaferija, S.: A finite volume method for incompressible linear elasticity. Comput. Methods Appl. Mech. Eng. **195**(44–47), 6378–6390 (2006)
21. Neimarlija, N., Demirdžić, I., Muzaferija, S.: Finite volume method for calculation of electrostatic fields in electrostatic precipitators. J. Electrostat. **67**, 37–47 (2009)

22. Cardiff, P., Karač, A., Ivanković, A.: Development of a finite volume contact solver based on the penalty method. Comput. Mater. Sci. **64**, 283–284 (2012)
23. Teskeredžić, A., Demirdžić, I., Muzaferija, S.: Numerical method for calculation of complete casting process - part i: theory. Numer. Heat Transf. B **68**, 295–316 (2015)
24. Teskeredžić, A., Demirdžić, I., Muzaferija, S.: Numerical method for calculation of complete casting process - part II: validation and application. Numer. Heat Transf. B **68**, 317–335 (2015)
25. Demirdžić, I., Muzaferija, S.: Finite volume method for stress analysis in complex domains. Int. J. Numer. Methods Eng. **37**, 3751–3766 (1994)
26. Wheel, M.A.: A geometrically versatile finite volume formulation for plane elastostatic stress analysis. J. Strain Anal. Eng. Des. **31**(2), 111–116 (1996)
27. Demirdžić, I., Muzaferija, S., Perić, M.: Benchmark solutions of some structural analysis problems using the finite-volume method and multigrid acceleration. Int. J. Numer. Methods Eng. **40**, 1893–1908 (1997)
28. Jasak, H., Weller, H.G.: Application of the finite volume method and unstructured meshes to linear elasticity. Int. J. Numer. Methods Eng. **48**, 267–287 (2000)
29. Demirdžić, I.: A fourth-order finite volume method for structural analysis. Appl. Math. Model. **40**, 3104–3114 (2016)
30. Golubović, A., Demirdžić, I., Muzaferija, S.: Finite volume analysis of laminated composite plates. Int. J. Numer. Methods Eng. **109**, 1607–1620 (2017)
31. Torlak, M.: A finite-volume method for coupled numerical analysis of incompressible fluid flow and linear deformation of elastic structures. Ph.D. thesis, Technische Universität Hamburg-Harburg, Arbeitsbereiche Schiffbau, Hamburg (2006)
32. Houbolt, J.C.: A recurrence matrix solution for the dynamic response of elastic aircraft. J. Aeronaut. Sci. **17**, 540–550 (1950)

Mechatronics, Robotics and Embedded Systems

HaBEEtat: Integrated Cloud-Based Solution for More Efficient Honey Production and Improve Well-Being of Bee's Population

Semir Šakanović(✉) and Jasmin Kevrić

International Burch University, Sarajevo, Bosnia and Herzegovina
symorgh13@gmail.com, jasmin.kevric@ibu.edu.ba

Abstract. Up to now, primary way of producing honey relied heavily on the experience and "gut feeling" of the individual beekeepers, who are utilizing traditional agricultural methods of placing the hives at locations that are predicted to be suitable for higher honey gain by visually inspecting the hives. HaBEEtat mission is to enable more efficient honey production and improve the well-being of bees by creating first integrated online platform for beekeepers, combing the hardware component enabling real time monitoring of the beehives and data analysis. The platform is composed of the hardware component containing sensors that is inserted in the beehive, accompanying web-page and smartphone application and cloud based storage for the database. Utilizing our technology, users have the opportunity to move away from the traditional model of beekeeping and transform themselves into IT enabled – data driven methods of beekeeping. This is in line with our desired outcome, which is to bring the beekeeping to the next generation (millennials) who are primarily influenced by the technological trends in the world, and make it more appealing for them thus enabling beekeeping to survive.

Keywords: HaBEEtat · Bees · Data · Analysis · Beekeeping Platform

1 Introduction

Currently there is estimated 50 million beehives [1] active in the world. The problem of traditional model is that the data collected through its utilization is usually not reliable and not recorded for future use and analysis. Integrating this model in current value chain will in future enable data driven prediction analysis on the level of the individual beekeepers, as well as big data trend analysis on the level of the entire platform. This is further enhanced by the fact that this platform aims to collect other environmental information such as temperature, humidity, geographical location and vegetation; which can be further utilized to build the dataset based on which the analysis of beekeeping [2] conditions and predictions of the seasons can be performed. Such analysis enables this platform in future to suggest the best locations for placement of beehives, by which it makes honey production more intuitive and decreases barriers of entry.

S. Avdaković (Ed.): IAT 2018, LNNS 59, pp. 359–374, 2019.
https://doi.org/10.1007/978-3-030-02574-8_29

To date, beekeepers [3] employed traditional beekeeping [4] agricultural methods such as placing hives at locations that are predicted to be suitable for higher honey gain by visually inspecting hives in order to obtain necessary information. Collected data still depends on their eye monitoring and gut predictions rather than accurate data readings that in most cases never get recorded or stored for future reference or analysis. So, our core idea is to enhance traditional honey production process and transform it to IT enabled data driven methodology and process. To make this happened our solution is designed to track beekeeping conditions based on environmental factors such as temperature, humidity, geographical location and vegetation. The concept is primarily focused on live data collection from hive sensors (integrated smart hardware component) in order to have timely and accurate information about environmental factors that are directly influencing honey production. The data collection is made available via the network of sensors that represent smart hardware component, which could be easily integrated with 10 frame Langstroth [5] hive (90% of beehives used in honey production are 10 frame Langstroth [6] hives). Data collected is further streamed to cloud based web application that further renders information through mobile and web user interface.

The primary goal is to enable any of existing beekeepers, as well as brand new beekeepers, to monitor and track beekeeping conditions measuring all available environmental data relevant to honey production. Hardware component together with online applications, both for web and mobile, are created to be extremely simple to use and user friendly, designed for offline demographics. With our approach worlds beekeepers would be able to access computed, map-based, beehive conditions based on current and historical weather and other factors such as temperature, humidity, atmospheric pressure, vegetation, honey gain, pollen, altitude, terrain, etc. By gathering and analyzing all these information, platform will be able to provide the best suitable location for hives based on historic data. In addition, it will alert beekeepers about possible production shortcomings such as irregularities within daily and weekly honey production gains as well as bee diseases and their potential spread in the area. As such, our solution will directly contribute to quality assurance of beekeeping products and the optimal use of knowledge sharing between other known solutions. The ultimate result is to enable beekeepers to produce more honey by providing an integrated support framework and path to higher outcomes replicable in the any region throughout the world.

Simply put, the main goal is to develop a cloud-based web application interoperable with hardware component (smart beehive add-on supplement), which will enable beekeepers to track current and forecasted conditions improving current process of honey production [7]. It should dramatically help beekeepers to monitor and review beekeeping conditions and to draw correlations between honey production, weather conditions and other factors. This solution will be able to provide best suitable location for hives, containment of bee diseases [8], as well as help with daily operation and maintenance. It is estimated that hives placed in optimal locations can produce 30–40% more honey. In addition, environmental data will be collected through deployed network of data sensors but also could be imported it from external sources where applicable. The web application and hardware component will be user–friendly, designed in a way so it does not require deep technical knowledge.

2 Concept and Methodology

Our solution has two components, hardware and software component. Hardware component contains the ring where the sensors and connectivity modules are housed that is placed in the existing beehive (our solution was developed for "10 frame Langstroth" hive that has 90% share of all beehives in the world [9]). This makes the ring able to gather and stream data in real time to the server. The data collected by the ring is: temperature [10], humidity [11], geolocation [12], vegetation [13], weight of honey [14] and pollen [15] in the frames (Fig. 1).

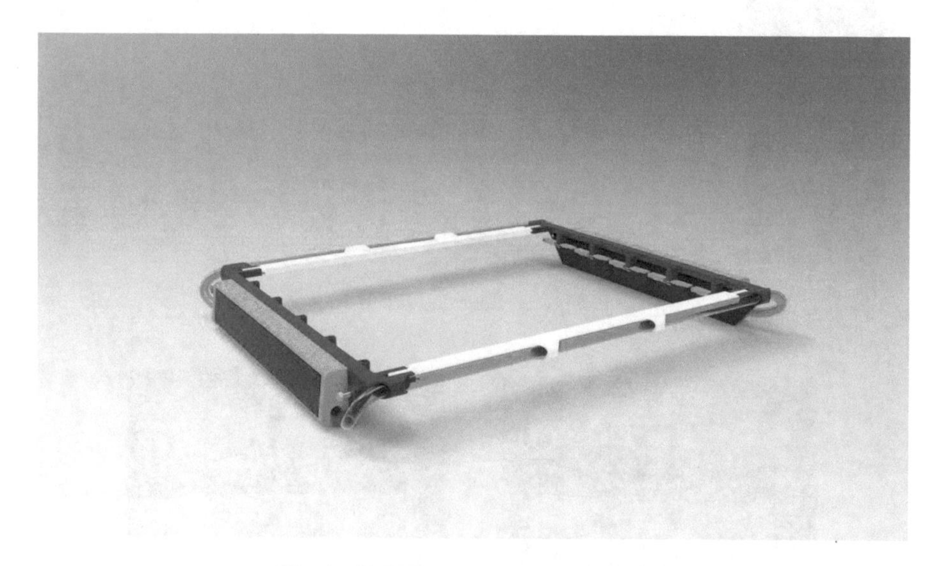

Fig. 1. HaBEEtat hardware module v3.0

On the other hand, software component revolves around the database where the data collected by the hives is stored and analysed. This is complemented with the web and mobile applications enabling to access the computed, map based, beehive conditions based on current and historical data coupled with the weather information and honey gain for their locations. In the same time, our software component includes the algorithm which analyses the data collected, and provides with the recommendations for best locations for beehives, possible irregularities and shortcomings of the location and warnings about bee diseases [16] and their potential spread in the area (Fig. 2).

Our software solution is designed in the user friendly-networking solution enabling beekeepers to create a social network where they can share their knowledge and expertise about beekeeping, bringing people together (Fig. 3).

The development of this concept started around 2 years before the date of this paper. After successful initial feasibility assessment, version 1.0 was created that was implemented in actual working environment, live bee colonies, under supervision of the professional beekeepers. The initial test ran during the entire summer season of 2016. During this time, main iterations of both hardware component and software

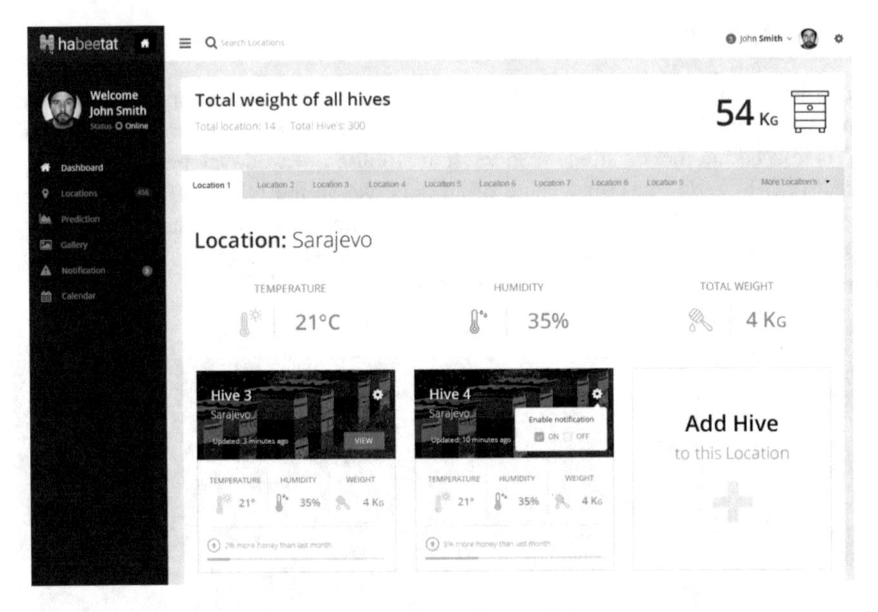

Fig. 2. User dashboard

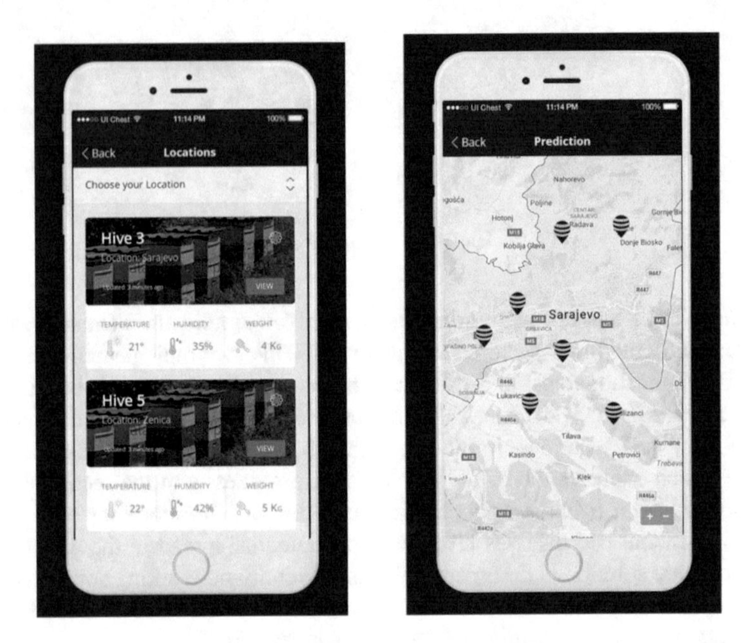

Fig. 3. Single view/map view

component were developed and tested, producing incremental improvements of our module. Summer of 2016 also generated first stream of data, every 20 min the modules would report on their current state and send data to the server. The testing was validated by the professional beekeepers, who during this time performed their regular checks of the hives according to the traditional methodology (visual inspection). The data collected by the professionals (temperature, humidity, weight of honey and pollen in the hive frames) was compared to the same data collected by the sensors, producing very reliable results that were within the margin of human error. This data was also utilized during this time for development and verification of the forecasting algorithm enabling us to further iterate on this concept.

3 Materials and Methods

Starting from initial prototype "proof of concept" until current model 3.0 we decided to partner with number of local beekeepers. Our strategy was to engage them closely and to listen to their production/maintenance pain points, which we then used as requirements to design our first prototype. Later we conducted number of use cases adopting our solution to real life conditions by setting up multiple demo hives on different location, populated with real bees colonies [17] and provided assistance and oversight from professional beekeepers.

In terms of end-users validation the real conditions testing we've completed 5 step validation process:

1. At the start our team conducted serious of interviews with local beekeepers to identify exact needs and product features in order to accurately design hardware product and software application as centralized platform for world's beekeepers;
2. Then team acquired multiple sets of live bee colonies and set up initial beehive prototypes in multiple location which were under supervision of professional bee keepers during entire summer season (2016);
3. During this time we've started testing software application and smart hardware component, which have been installed and started to generate stream of data measurements (sensors readings) from each individual hives. Data readings were provided every 20 min throughout the summer of 2016 (honey production season) when data collected were stored on our centralized server;
4. During this time visual inspection for individual hives were performed by local beekeepers and simultaneously compared to sensor data readings, such as temperature, humidity, hive frames weight measurement (honey and pollen) in reference and compared to onsite sensors data readings;
5. All data collected provided benchmark comparison between real conditions and environmental data readings. This enabled us to verify accuracy of forecasting algorithm as well as product operational flow and functional design.

4 Results

We devoted our time on testing in real life conditions. For the testing, we used two electronic boards called NodeMCU 1.0 [18] that were connected with eleven HX711 24-bit ADC, eleven Load Cells (length about 10 cm with 10 kg max weight), two

DHT-22 [19] humidity and temperature sensor, Wi-Fi router and two power banks to supply power to our boards. Eleven Load Cells are used because of design of beehive that used 11 frames (Fig. 4).

Fig. 4. NodeMCU 1.0

On each NodeMCU [20] board we had five to six HX711 24-bit ADC connected and on each HX711 24-bit ADC there was one Load Cell [21] connected that has frame attached. These two board were connected in between so that they can communicate and send data in order that one board will read the data and send, after that second board would do the same and we would see the data that was sent, on the server. We used one Wi-Fi router to provide internet connectivity, in this case we had 3G network, both NodeMCU boards had ESP8266 module that could connect on our router to get 3G network and then send all the data to the server. All this was connected on the breadboard and from the breadboard to the power banks with 10 000 mAh. Since, NodeMCU boards had microUSB connector we could easily connect Power Banks to the boards and with that had enough power to supply boards and everything that was connected. Power Banks were also connected to the home electricity so that we could charge them whole time since we made our boards to send data every 10 min, and with that the power consumption was so huge that we needed to have power banks recharging all the time for our boards to be able to have enough power (Figs. 5, 6 and 7).

Our testing period showed us that our electronic does not disturb bees nor harm them. We didn't see any bad behavior during our testing with electronic. What we saw is that we needed something that consume less power and something that is less in size (smaller electronic and less electronic components).

After summer time was spent in testing electronic components such as sensors, load cells etc. We decided that we should use this period to make our own hardware board which would have all the necessary inputs and outputs that we needed for our HaBEEtat module. We called our first board "HaBEEtat board 1.0". Board had ATmega328P microprocessor that was the brain of the board and all the inputs and

Fig. 5. DHT-22 (temperature and humidity sensor)

Fig. 6. HX711 24-bit ADC

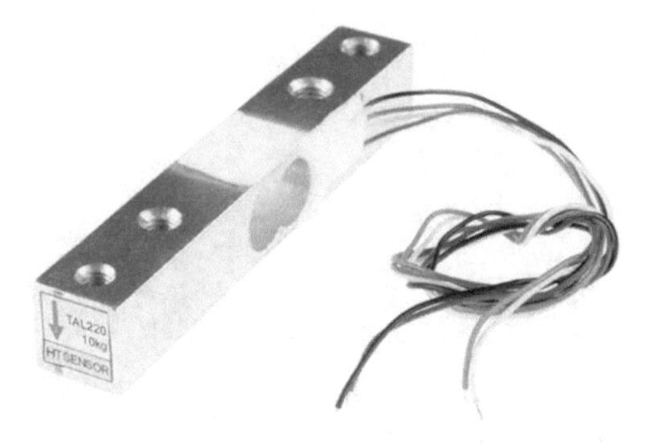

Fig. 7. Load Cell 10 kg

outputs were connected and controlled by this microprocessor. Ten HX711 24-bit ADC for 10 frame Langstroth hive were integrated on the board which then were connected with load cells. This time we used small load cells for our HaBEEtat module. Our board could be easily programed using Arduino software since we used the microprocessor that Arduino board was using. This time we had Lithium Ion 18650 batteries [22] as our power supply, also we tested with Energizer Ultimate Lithium AA batteries which were the strongest AA batteries. We conducted test with eight Ultimate Lithium AA batteries, four batteries for the board and other four batteries for the 3G modem. From the test, we could see and analyze how long these batteries will last and we got really good results and we were impressed by them. We also made our software that runs on ATmega328P microprocessor less power consumption, we made the code that would run on low power when the board is not reading the sensors. So, what we did is that board would turn on, 3G modem would turn on, board would connect on our 3G modem, after that board would read the data and would send to the server and then board would enter into deep sleep mode which reduced our power consumption to very minimum, since everything else was shut down, and just small amount of power was used to keep microprocessor alive. With this, we also reduced our electronic effect on bees. Another test was done with Lithium Ion 18650 batteries [23]. We tried with four batteries connected and with two batteries connected. From this test, we realized that using these batteries we can have almost unlimited power source if we use solar panels to charge these batteries, this was also one of the benefits over Energizer Ultimate Lithium AA batteries [24], since you can't charge Energizer Ultimate Lithium AA batteries. So, we made a set-up that consist of two Lithium Ion 18650 batteries, two battery chargers, two solar panels, 3G modem and HaBEEtat Board 1.0. Two solar panels were connected to the solar chargers for charging batteries with nice and steady

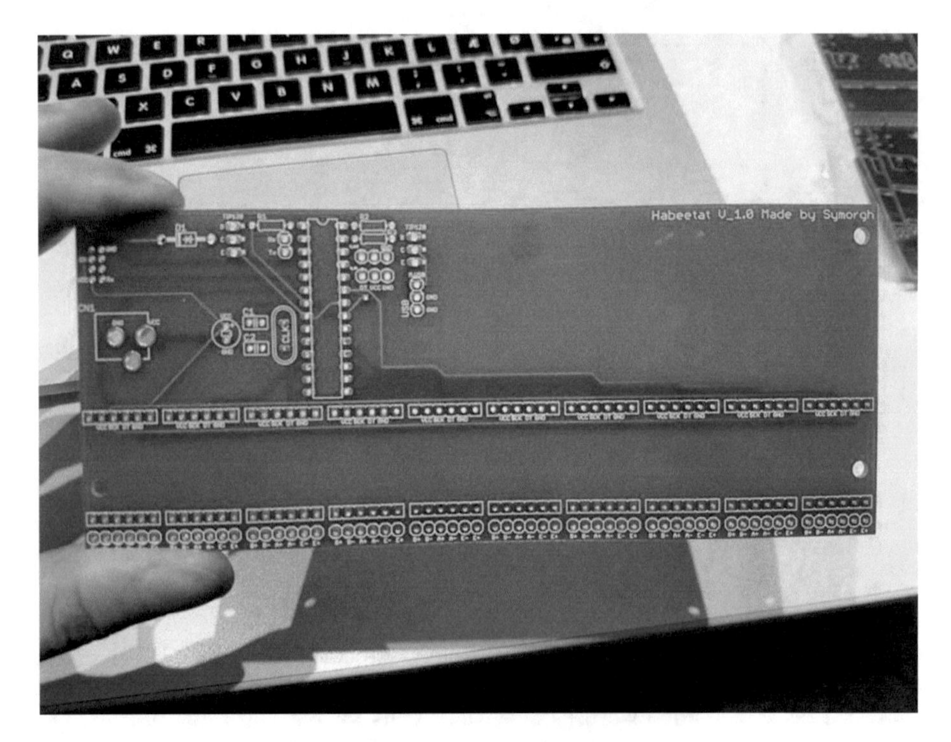

Fig. 8. Board v1.0 (without components)

voltage and then batteries connected to the board. We conducted more tests and made small design changes on board, so now we had version 1.1 of our HaBEEtat board. We made tests with version 1.1 with both kind of batteries and the board was running without any problems. From the tests, we learned a lot and we used that knowledge to make software running on the board even better, we reduced more power consumption and made better readings (Fig. 8).

After getting more and more knowledge from the testing, we came up with solution to reduce size of the board by half, with this design it will reduce the size of the box

Fig. 9. Board v1.0 (assembled)

that electronic would go, also the cost of making the board. We ordered five of these boards to make tests. These boards were version 2.0. This time we could also make tests with real time environment because the bee season started. Again, from testing we improved our code that was running on our microprocessor. This board was half the size from the first board but it contained same parts as version 1.0 and version 1.1. We also tested different type of modems, since we had ability to test in real environment with real conditions, we decided that ZTE MF70 modem was the best from the modems that we used and tested. This modem had 3G network but it supported 2G and frequencies below 2G (Figs. 9, 10, 11 and 12).

After testing with board version 2.0 we made hardware design of our new board version 3.0. This board was half the size of board version 2.0 but it used two multiplexers and only one HX711 24-bit ADC (Figs. 13, 14, 15 and 16).

So, we reduced the size of the board by half and also reduced HX711 24-bit ADC [25] from ten to only one. With this we cut the cost of the board again, the cost of electric components and we cut power consumption since we had less electronic components on this board version 3.0. We tested version 3.0 and it was working really good. Since we had multiplexers we needed to change the code. We made it even better and even more precise readings then before. After conducting tests, we made small changes on version 3.0 for better performances. Currently we are testing our HabBEEtat modules on the field and tracking all data and readings.

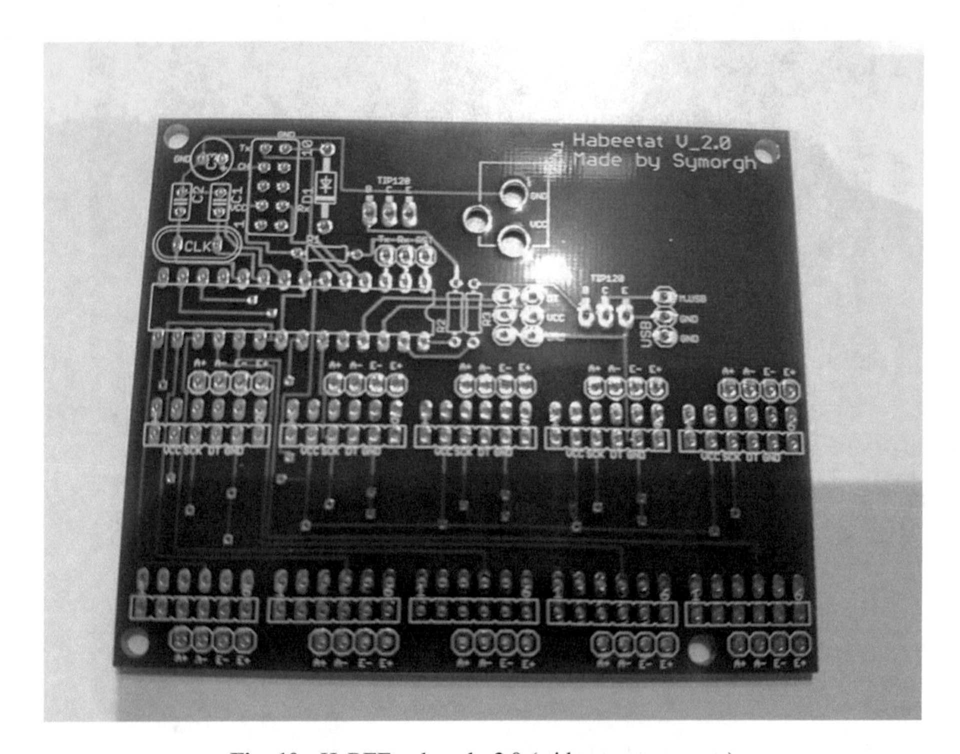

Fig. 10. HaBEEtat board v2.0 (without components)

Fig. 11. HaBEEtat board v2.0 (assembled)

Fig. 12. Difference between v2.0 and v1.0

5 Discussion

The challenge of tackling the bee depopulation is not only present in the EU but it is a truly global issue, especially considering the role bees play in the food chain. Bees are natural pollinators of the plants that produce most of our food, without them that job has to be performed by hand, which reduces the biodiversity and increases the costs. Greenpeace report on bee depopulation from 2014 has produced an infographic (shown in Fig. 17) breaking down the differences between the biodiversity of our food chain with bees involved and without bees. As we can see the difference bees play is huge and thus the impact of their extinction is catastrophic.

Thus, the problem of bee depopulation has been researched in the past in order to uncover specific solutions to it. EU wide research performed for EU reference laboratory for bee health which had the task of uncovering the state of play of honey-bee colony losses in each of the participating member states, named "EPILOBEE". The results showed that during the winter seasons of 2014 the losses of honey-bee colonies in the participating member states ranged from 2.4% to 15% with an average of 7.9% loss in one single season.

As such the situation portrayed on the infographic above seems more and more of a reality if no action is to be taken. However, EU has already started tackling the issue by

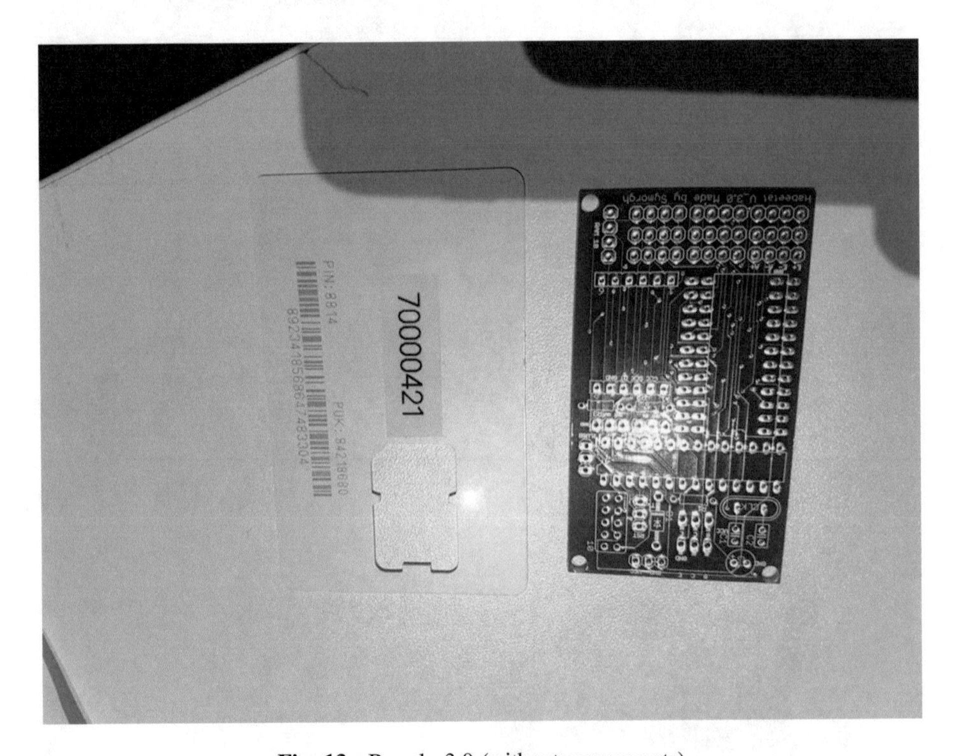

Fig. 13. Board v3.0 (without components)

Fig. 14. HaBEEtat board 3.0 (assembled)

Fig. 15. Difference between v3.0 and v2.0

Fig. 16. Difference between v3.0 and v1.0

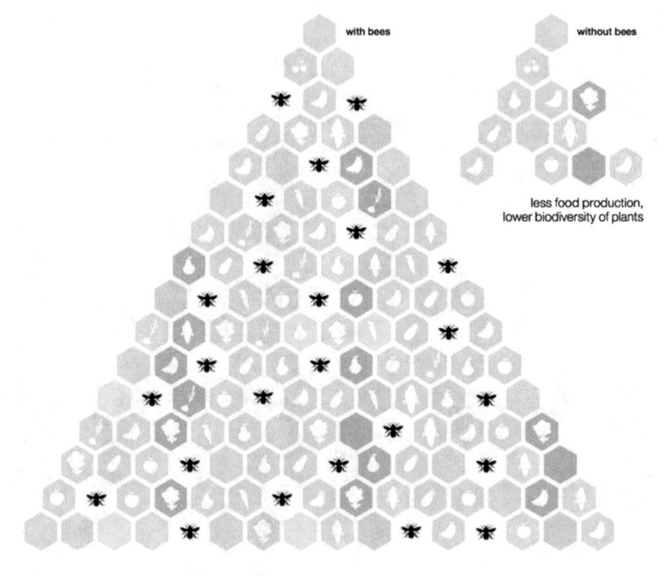

Fig. 17. Greenpeace infographic

introducing temporary bans and controls on pesticides that are harmful for bees. This, coupled with the Greenpeace research proving that reduction of industrial farming practices helps reduce mortality of bees, are the first steps in uncovering the solution to this issue. HaBEEtat project has been envisaged as the possible tool for helping the cause of saving the bees. Our solution provides direct answer to the above described through multiple ways.

6 Conclusion

Firstly HaBEEtat, if adopted in sufficient quantity, can serve as the data gathering tool for future research in the topic of bee preservation. Our wide data sets are prime for big data analysis into the topic, evaluating different hypothesis and testing different solutions to the challenge.

Secondly, our solution provides an early warning system able to detect the harmful substances utilized near the spots that beekeepers can consider placing their hives to. This enables both the beekeepers to keep their losses low, and for the EU officials to be notified of the breach of the bans on using the bee harmful substances. This, will surely reduce the bee depopulation in the short run as the beekeepers will naturally avoid the harmful sites since they do not want to lose their bee colonies.

Lastly, our solution is directly in line with the practices of reducing the industrial farming [26]. Our solution makes it easier for individual users to set up their honey productions, coupled with reducing the food losses and waste product generating it enables more and more producers to break even on their production. Considering that the beekeepers can sell their honey to the producers who bulk up the honey they collect form the individual beekeepers in the finished products that are distributed to the market. Considering that our solution brings the beekeeping to the general population and makes it easier for them to start and maintain local productions, it is essentially moving away from large scale industrial farming, and moving towards the many small-scale productions that are performed in more sustainable way.

References

1. Hasluck, P.N.: Beehives and bee keepers' appliances (1911)
2. Conte, Y.L.: Beekeeping and science. In: Beekeeping: From Science to Practice (2017)
3. Vanengelsdorp, D.: Next steps: outreach and problem solving with beekeepers. In: 2016 International Congress of Entomology (2016)
4. Crane, E.: Beekeeping (1999)
5. Langstroth, L.L.: Langstroth on the hive and the honey-bee, a bee-keeper's manual (1853)
6. Smith, O.D.: Langstroth, the "Bee Man" of Oxford. Ohio State Archaeological and Historical Society Columbus (1948)
7. Baglio, E.: Honey: processing techniques and treatments. In: Chemistry and Technology of Honey Production. SpringerBriefs in Molecular Science (2017)
8. Sturges, A.M.: Bee diseases. In: Bee World (1928)
9. Kasangaki, P., Chemurot, M., Sharma, D., Gupta, R.K.: Beehives in the world. In: Beekeeping for Poverty Alleviation and Livelihood Security (2014)

10. Jankowsky, H.-D.: Body temperature and external temperature. In: Temperature and Life (1973)
11. Moisture and humidity measurement in industrial plants. In: Industrial Moisture and Humidity Measurement (2013)
12. Progri, I.: Introduction to geolocation of RF signals. In: Geolocation of RF Signals (2011)
13. Smith, A.D.: A study of the reliability of range vegatation estimates. In: Ecology (1944)
14. Bailey, L., Ball, B.V.: The honey bee. In: Honey Bee Pathology (1991)
15. Shivanna, K.R., Rangaswamy, N.S.: Pollen collection. In: Pollen Biology (1992)
16. Archer, G.J.S.: Immunity and bee disease. In: Bee World (1928)
17. Crane, E.: Some effects of latitude on honey bee colonies. In: Bee World (2005)
18. Kühnel, C.: Building an IoT Node for Less Than 15 Dollars NodeMCU and ESP8266. Skript Verlag Kühnel, Altendorf (2015)
19. Bogdan, M.: How to use the DHT22 sensor for measuring temperature and humidity with the arduino board. ACTA Universitatis Cibiniensis **68**(1), 22–25 (2016)
20. Škraba, A., Koložvari, A., Kofjač, D., Stojanović, R., Stanovov, V., Semenkin, E.: Streaming pulse data to the cloud with bluetooth LE or NODEMCU ESP8266. In: 2016 5th Mediterranean Conference on Embedded Computing (MECO) (2016)
21. Olmi, G.: Load cell training for the students of experimental stress analysis. In: Experimental Techniques (2015)
22. Zhang, S.S., Xu, K., Jow, T.R.: Charge and discharge characteristics of a commercial LiCoO2-based 18650 Li-ion battery. J. Power Sources **160**, 1403–1409 (2006)
23. Wu, H., Yuan, S., Jiang, L., Yin, C., Miao, W.: Electro-thermal modeling and cooling optimization for 18650 battery pack. In: 2016 IEEE Vehicle Power and Propulsion Conference (VPPC) (2016)
24. Blomgren, G.: Lithium batteries at energizer. In: Proceedings of the Conference on Thirteenth Annual Battery Conference on Applications and Advances (1998)
25. Redmayne, D.: 24-bit ADC measures from DC to daylight. In: Analog Circuit Design (2015)
26. Hood, E.: Industrial farming: implications for human health, with peter thorne. In: Environmental Health Perspectives (2009)

PID-Controlled Laparoscopic Appendectomy Device

Abdul Rahman Dabbour(✉), Asif Sabanovic, and Meltem Elitaş

Faculty of Engineering and Natural Sciences, Sabancı University, Istanbul, Turkey
{dabbour,asif,melitas}@sabanci.edu

Abstract. Minimally invasive surgery is a surgical method, which boasts many advantages over regular surgeries, such as decreasing the risks involved by minimizing the incision area, thus reducing the risk of infection compared to invasive surgeries. Laparoscopic surgery tools built for this purpose are mostly singular in function, which means that it requires multiple incisions for multiple tools or changing tools using the same incision during the operation. This project attempts to motorize an affordable multifunctional mechanical surgical tool prototype. The tool is designed using SolidWorks and controlled using MATLAB/Simulink. Three motors are used to motorize the multifunctional laparoscopic tool and their control architectures made it more precise and more accurate for noninvasive operations. It is shown that with some physical modifications and simple PID control, the multifunctional laparoscopy tool can be controlled and modified for the robotic-assisted surgery. Possible future improvements include attachment of the cameras and wireless control for the tele-operational applications.

1 Introduction

Laparoscopic surgery, widely known as minimally invasive surgery, is accomplished using special tools to aid the surgeon in accomplishing relatively complex tasks in very small incisions. Laparoscopic operation is low-risk, minimally invasive procedure that facilitates recovery periods of operations thanks to its small incisions (0.5–1.5 cm). It also provides less pain, small scars, quick recovery, and short hospital stays upon the surgery, hence, it has low-level risk of hospital-acquired infections, especially the surgical site infections [1–3]. Laparoscopic operations use relatively long surgical tools (40–45 cm), which are inserted through trocars into the abdomen. One of incisions is used for the light and the camera insertion. Surgeon operates using the visual feedback obtained from the camera. Forceps, hooks, scissors, dissectors and probes are among the most commonly used surgical instruments in the laparoscopic operations. Removal of the appendix, gallbladder, and parts of the intestines are among the various procedures that laparoscopic surgery is often performed [4–6].

While laparoscopic surgery presents less risk of complications for the patients, the operation is more complicated compared to open surgeries due to its limited working area. Surgeons have limited motion capability, reduced tactile sensation and depth perception; therefore their dexterity is decreased [7, 8].

S. Avdaković (Ed.): IAT 2018, LNNS 59, pp. 375–382, 2019.
https://doi.org/10.1007/978-3-030-02574-8_30

In order to overcome these limitations and provide more ergonomic operation conditions to the surgeons, some improvements have been recently performed [9]. Multi-functional laparoscopic tools have been developed to avoid complications arise due to changing tools such as gas leakage from the abdomen, focusing the same tool-tip during the operation and readjusting camera properties [10, 11]. Moreover, these tools decreased the operation time. Their designs made these instruments superior to be used in the robotic-assisted surgeries where only one robotic arm might control several tools. Hence, the necessity of using multiple robotic arms to mount the surgical instruments will disappear and more economic and surgeon-friendly robots might be developed thanks to modernization of laparoscopic instruments [12].

In this study, we present a motorized-laparoscopic-surgery tool that has been designed as a multi-functional laparoscopic instrument for appendectomy operations [10]. As mentioned above, appendectomy is an operation involving the removal of the appendix, usually after a patient is diagnosed with appendicitis, the infection of the appendix [13]. Scissor, endo-loop, endobag are three main laparoscopic instruments, which are commonly used in appendectomy. The procedure starts with inserting a 10–12-mm diameter trocars (port, tubing) in to the incisions. One of them is allocated for the laparoscopic camera to transfer the desired part of the operation area on the screen. Using the other trocars, laparoscopic instruments are placed through the abdominal cavity to isolate appendix from the vessels and fat tissue. Next, endoloop squeezes the appendix, and then endobag covers the appendix. Scissor cuts the appendix through the endloop-knot, and the appendix is taken out in the endobag through the trocar. This process takes 30–40 min. The multi-functional appendectomy device developed by Elif et al. [10] combined the scissor, endoloop and endobag in one laparoscopic instrument that is dedicated for the appendectomy. As a result, appendectomy becomes fast, easy, safe, stable and physically less tiring for the surgeon.

Here, we motorized and controlled the multi-functional laparoscopic appendectomy device. We performed MATLAB simulations for position control of the scissor, endoloop and endobag. Our results presents PID-based position control for the motorized-multi-functional laparoscopic instrument.

2 Methods

Design of the multi-functional laparoscopic appendectomy device was performed using Solidworks (2015) as previously reported in [10]. Although our results will present the MATLAB/Simulink results of the scissor (main body), endoloop, and endobag, Fig. 1 shows the schematic of the multi-functional laparoscopic appendectomy device based on Solidwork designs.

The SolidWorks files were transferred into MATLAB/Simulink (2017b) files to be able to control virtual motors, gears and the tips of the laparoscopic tools. Figure 2 shows the position control of the multi-functional laparoscopic appendectomy device in the MATLAB/Simulink.

To define the relation of a revolutionary gear with a prismatic link the rack and pinion constraint was defined, Fig. 3 illustrates the control block diagram of the endobag in Simulink.

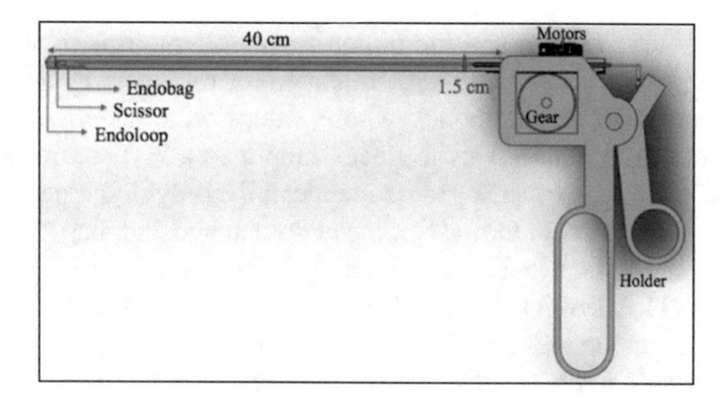

Fig. 1. Design of the motorized, multi-functional laparoscopic appendectomy device.

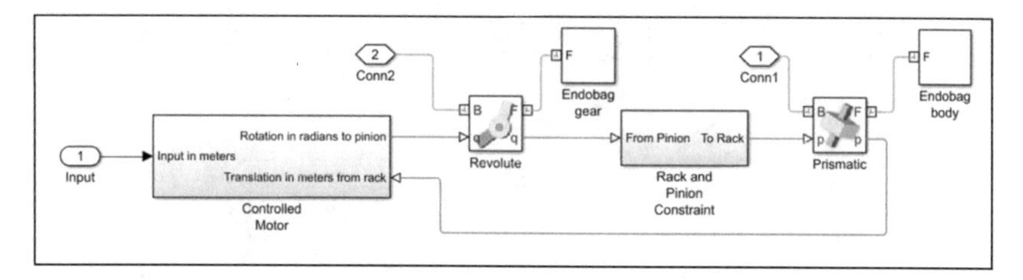

Fig. 2. Representation and control of endobag in MATLAB/Simulink (2017b).

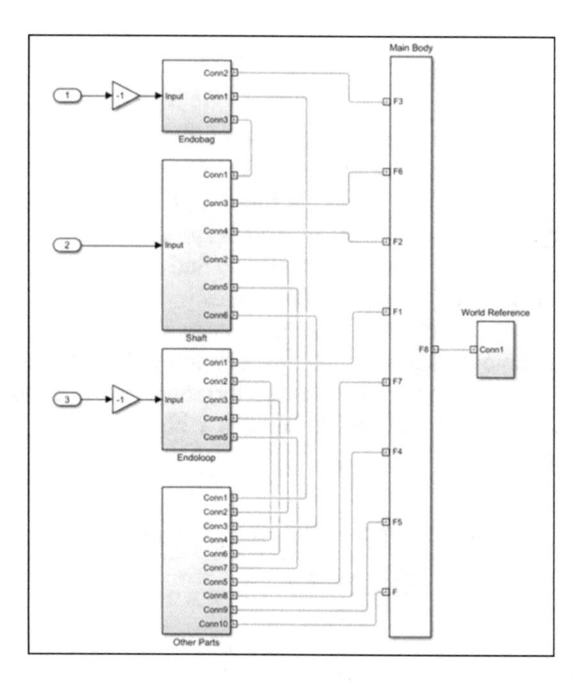

Fig. 3. Position control block diagram of the multi-functional laparoscopic appendectomy device in the MATLAB/Simulink.

The rack prismatic motion and the pinion revolute motion is tightly linked in the Simulink model [14]. A servomotor was simulated for each gear by adding a feedback loop to a simulated DC motor model, whose values were taken from the Herkulex DRS-0201 Smart Servo motor. This feedback loop uses a PID controller as illustrated in Fig. 4. The control loop of each gear is identical; the only difference is the gear ratio used for the gear-link pairs. In the case of the endoloop and endobag, they are identical. The gear ratio of the scissor (main invasive body) is different in size, and its cylindrical constraints in MATLAB/Simulink (2017b) need to be changed to have no target states for the kinematics to be calculated without an error. To tune the PID controllers, MATLAB/Simulink's internal auto-tune function was used.

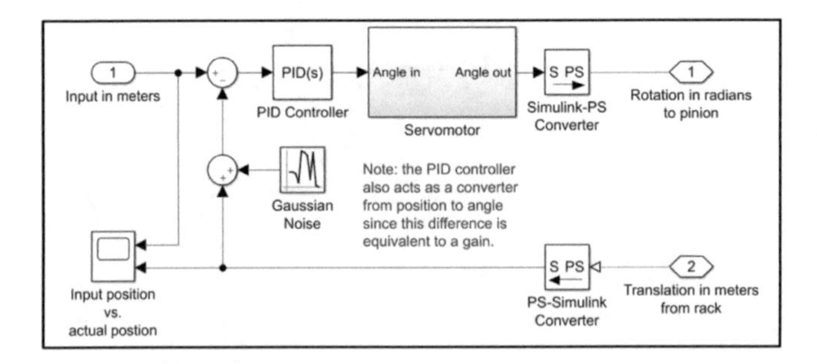

Fig. 4. Representation of the motor control architecture in Simulink.

3 Results

Position control of the endoloop, endobag and scissor were tested applying a 1-cm step response both in the absence and presence of Gaussian noise (Mean: 0, variance: 10^{-8} m). Using MATLAB's PID auto-tune function, two sets of values were found.

In set A, we attempt to minimize the overshoot and the settling and rise times, to the extent that the simulation allows as to. In set B, we find sub-optimal PID values that are uniform for all parts and are within feasible ranges.

Table 1 shows set A PID controller parameters for each part of the simulated motorized, multi-functional laparoscopic appendectomy device. Table 2 and Figs. 5 and 6 present the step response characteristics of the endoloop, endobag and scissor. Table 3 shows the parameters of set B. Table 4, with Figs. 7 and 8, show the step response characteristics of set B.

Table 1. Set A of PID values for part motors

Part	Proportional (P)	Integral (I)	Derivative (D)	Filter coefficient (N)
Endobag	2315	7625	173.8	8780
Endoloop	2315	7625	173.8	8780
Main body	663.2	2255	48.71	8232

Table 2. Step response characteristics of set A

Part	Overshoot (%)	Settling time (ms)	Steady state error (mm)	Rise time (ms)
Endobag	2.68	40.8	0.00	14.9
Endoloop	2.68	40.8	0.00	14.9
Main body	2.03	317	0.00	16.3

Table 3. Set B of PID values for part motors

Part	Proportional (P)	Integral (I)	Derivative (D)	Filter coefficient (N)
Endobag	319.8	1857	13.64	1305
Endoloop	319.8	1857	13.64	1305
Main body	95.93	557.1	4.092	1305

Table 4. Step response characteristics of set B

Part	Overshoot (%)	Settling time (ms)	Steady state error (mm)	Rise time (ms)
Endobag	3.59	612	0.00	142
Endoloop	3.59	612	0.00	142
Main body	3.59	612	0.00	142

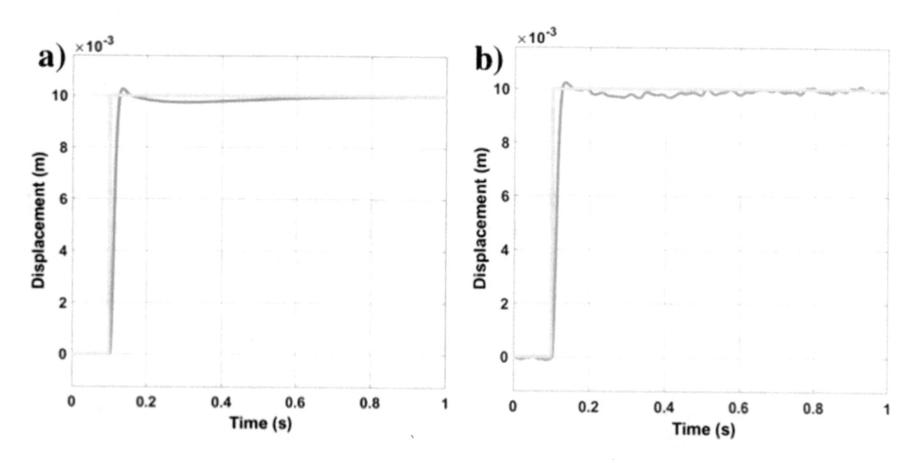

Fig. 5. For set A, position control for the endoloop and endobag. Step input is applied as reference (yellow), and positions of the tips are presented (blue). (a) Without noise, (b) with noise.

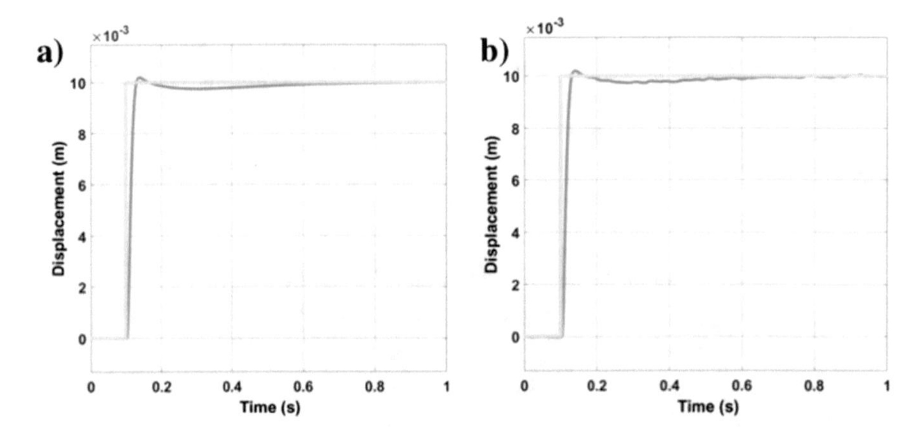

Fig. 6. For set A, position control for the scissor (main body). Step input is applied as reference (yellow), and positions of the tip is presented (blue). (a) Without noise, (b) with noise.

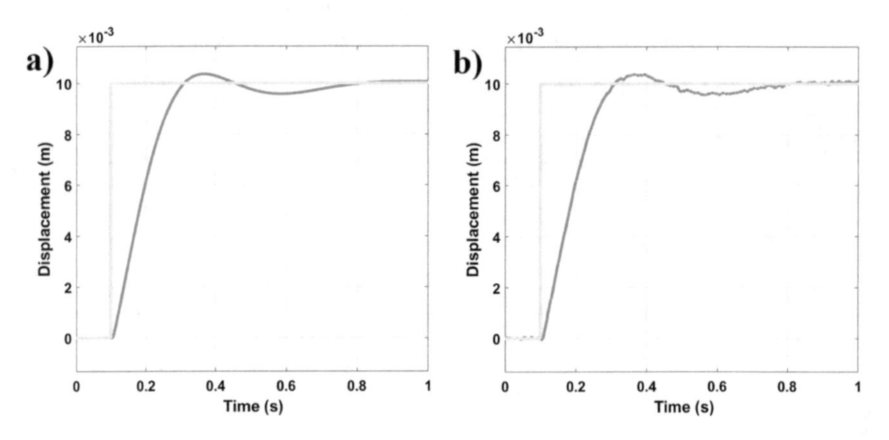

Fig. 7. For set B, position control for the endoloop and endobag. Step input is applied as reference (yellow), and positions of the tips are presented (blue). (a) Without noise, (b) with noise.

4 Discussion

Here we present the simulation results for the position control of the motorized laparoscopic surgery tool. Still most of the laparoscopic surgery tools are manual. Our preliminary results show that a robust, fast and accurate position control of the laparoscopic surgery tool. Previously, moving the gear manually controlled the position of the tool-tips [10]. The PID control is one of the simplest and most widely used control architectures for industrial tools in automation. It is also convenient for the control of surgical tools in medicine. Furthermore, it is economic to implement. Stability, accuracy, repeatability and reliability of the tools can be achieved applying various control architectures according to microenvironment of the operating area and

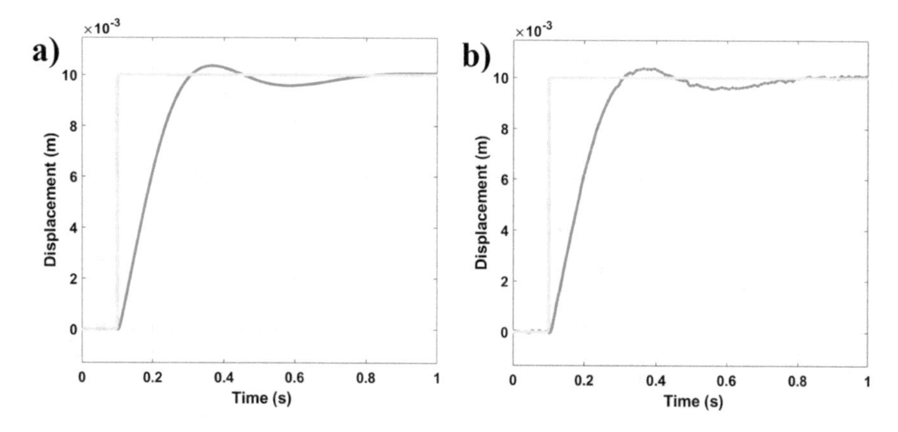

Fig. 8. For set B, position control for the scissor (main body). Step input is applied as reference (yellow), and positions of the tip is presented (blue). (a) Without noise, (b) with noise.

the task of the operation. The results also show response performance comparable to that of surgeons, with surgeons usually having an accuracy in the μm to mm range [15, 16]. Today robotic systems are still very expensive to be able to part of every operating room in hospitals [12]. Therefore, development of cheaper, portable and user-friendly robotic systems or tools will be very valuable. Automated robotic-assisted surgery in conjunction with artificial intelligence will significantly contribute to the surgery and medicine in near future.

References

1. Clarke, H.C.: Laparoscopy—new instruments for suturing and ligation. Fertil. Steril. **23**(4), 274–277 (1972)
2. Hunter, P.: The cutting edge, a synergy of modern surgical techniques and science improves patient survival and recovery. EMBO Rep. **8**(11), 999–1002 (2007)
3. Alexander, J.I., Hull, M.G.: Abdominal pain after laparoscopy: the value of a gas drain. Br. J. Obstet. Gynaecol. **94**(3), 267–269 (1987)
4. Mirhashemi, R., Harlow, B.L., Ginsburg, E.S., Signorello, L.B., Berkowitz, R., Feldman, S.: Predicting risk of complications with gynecologic laparoscopic surgery. Obstet. Gynecol. **92** (3), 327–331 (1998)
5. Jimenez-Rodríguez, R.M., Segura-Sampedro, J.J.: Laparoscopic approach in gastrointestinal emergencies. World J. Gastroenterol. **22**(9), 2701 (2016)
6. Bhandarkar, D., Mittal, G., Shah, R., Katara, A., Udwadia, T.E.: Single-incision laparoscopic cholecystectomy: how I do it? J. Minimal Access Surg. **7**(1), 17–23 (2011)
7. van der Westebring Putten, E.P., Goossens, R.H., Jakimowicz, J.J., Dankelman, J.: Haptics in minimally invasive surgery – a review. Minim. Invasive Ther. **17**(1), 3–16 (2008)
8. Gallagher, A.G., McClure, N., McGuigan, J., Ritchie, K., Sheehy, N.P.: An ergonomic analysis of the fulcrum effect in the acquisition of endoscopic skills. Endoscopy **1**(7), 617–620 (2007)
9. Rosser, J.C., Rosser, L.E., Savalgi, R.S.: Skill acquisition and assessment for laparoscopic surgery. Arch. Surg. **132**, 200–204 (1997)

10. Taşkın, E., Kurt, E., Elitaş, M., Tansuğ, T.: Çok fonksiyonlu apendektomi cihazı, Otomatik Kontrol Ulusal Toplantısı (2017)
11. Frecker, M.I., Schadler, J., Haluck, R.S., Culkar, K., Dziedzic, R.: Laparoscopic multifunctional instruments: design and testing of initial prototypes. JSLS **9**, 105–112 (2005)
12. da Vinci Surgery. http://www.davincisurgery.com/. Accessed 10 July 2017
13. Sallinen, V., Mentula, P.: Laparoscopic appendectomy. Duodecim **133**(7), 660–666 (2017)
14. Budyans, R.G., Nisbett, J.K., et al.: Shigley's Mechanical Engineering Design, vol. 9. McGraw-Hill, New York (2008)
15. Podsędkowski, L.R., Moll, J., Moll, M., Frącczak, L.: Are the surgeon's movements repeatable? An analysis of the feasibility and expediency of implementing support procedures guiding the surgical tools and increasing motion accuracy during the performance of stereotypical movements by the surgeon. Kardiochirurgia i torakochirurgia polska=Pol. J. Cardio-Thorac. surg. **11**(1), 90 (2014)
16. Riviere, C.N., Rader, R.S., Khosla, P.K.: Characteristics of hand motion of eye surgeons. In: Proceedings of the 19th Annual International Conference of the IEEE Engineering in Medicine and Biology society, vol. 4, pp. 1690–1693. IEEE (1997)

Radial Basis Gaussian Functions for Modelling Motor Learning Process of Human Arm Movement in the Ballistic Task – Hit a Target

Slobodan Lubura[1(✉)], Dejan Ž. Jokić[2], and Goran S. Đorđević[3]

[1] Faculty of Electrical Engineering, University of East Sarajevo, East Sarajevo, Bosnia and Herzegovina
slubura@etf.unssa.rs.ba

[2] International Burch University, Sarajevo, Bosnia and Herzegovina
dejan.jokic@ibu.edu.ba

[3] Faculty of Electronic Engineering, University of Niš, Niš, Serbia
goran.s.djordjevic@elfak.ni.ac.rs

Abstract. Mathematical tool for modelling motor learning of human arm movement in the ballistic task – hit a target is described in this paper. Proposed tool is used for quantification of the subject's ability to learn motor control of their arm movements in the ballistic task after training. Conducted research showed that the key role in the ballistic task – hit the target had velocity profiles of the arm/joystick movement. Therefore, the velocity profiles have been an object of refined analysis and modelling performed for the purpose of determining whether motor learning of humans arm movement is possible or not. Radial Basis Gaussian Functions (RBGF) are used as a tool for analysis and modelling, because they can reveal behaviour of human's arm movement around more local points or in the more stages of movement. The proposed tool is verified by conducted experimental analysis. The experimental analysis was performed as practice of 50 subjects in the ballistic task – hit a target.

Keywords: Radial Basis Gaussian Function · Human arm · Motor learning Ballistic task

1 Introduction

The study of motor control is essentially the study of sensor-motor transformations. For example, during arm movement or during interaction between arm and an object that changes the dynamics of the movement, the human motor control system must coordinate a variety of forms of sensory and motor data. Such data is generally provided in different formats and may refer to the same entities but in different coordinate systems. This means that motor system should be a bridge between different coordinate systems or correlation between different data types in order to close sensor-motor control loop [1–3]. Moreover, it is clear that human motor system simultaneously controls dynamic object (arm) and the interaction of this object with the environment. The interaction between human arm and joystick in the ballistic task – hit a target, or more precisely, the tool for modelling motor learning process in ballistic task is described in this paper.

S. Avdaković (Ed.): IAT 2018, LNNS 59, pp. 383–393, 2019.
https://doi.org/10.1007/978-3-030-02574-8_31

Test equipment for evaluation of motor learning process in the ballistic task – hit a target is shown in Fig. 1.

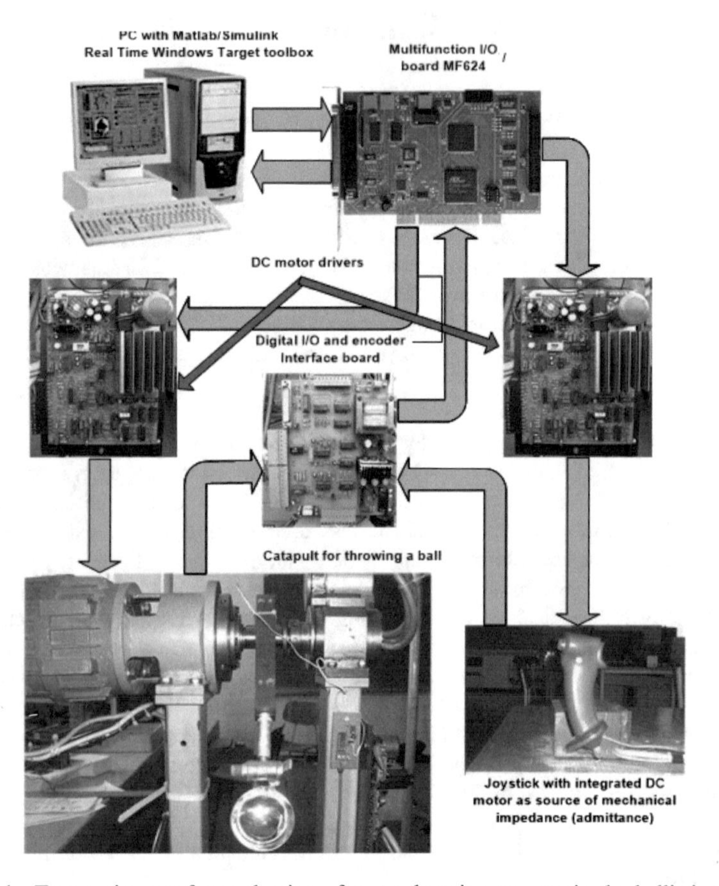

Fig. 1. Test equipment for evaluation of motor learning process in the ballistic task

At beginning, in early experimental phase, fifty subjects participated in the training. Their task was to throw a number of balls into a target (box) over an obstacle. By moving a joystick, their arm movements were transferred further to the catapult for throwing balls. The success of the throws was quantified by binary logic and the number of successful hits during the entire training of twenty series in the early experimental phase was recorded. Simultaneously, position, velocity and acceleration of arm and catapult movements were recorded for further analysis. But, the question which arose was which tool to use for analysis of arm movement in order to reveal whether motor learning occurred and at which stage of arm movement motor learning process is dominant. If the model of motor learning process of human arm movement is known, it can be integrated in external control structure (joystick with integrated DC motor as source of variable mechanical impedance) in order to boost success of the task – hit a target. This hypothesis was proved in [9]. Conducted analyses confirmed that velocity profiles of human arm movement were more variables (carried out more information of movement) during training phase and therefore these profiles were used for quantification of motor learning process of human arm movement in the ballistic task – hit a target. Refined analysis of human arm movement aimed at finding the model of motor learning of human arm movement was carried out by using a Radial Basis Gaussian Functions (RBGF).

2 Modelling Arm Movement with RBGF

The main feature of radial basis functions is symmetrical monotonous rise or fall in regard to its central point. One type of radial basis functions are RBGF which can be expressed as follows:

$$h(x) = e^{-\frac{(x-x_c)}{2\sigma^2}} \tag{1.1}$$

where x_c is centred point of RBGF and σ is width of RBGF [4, 5]. Single isolated RBGF with parameters $x_c = 0$ and $\sigma = 1$ is shown on Fig. 2.

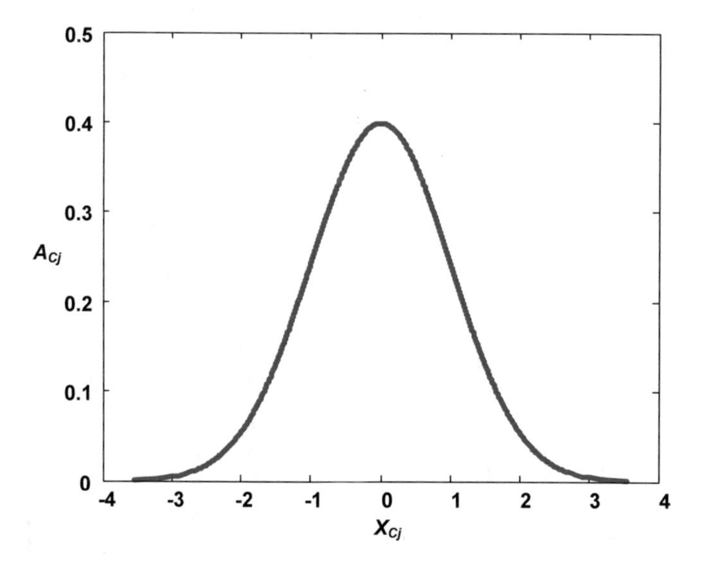

Fig. 2. Isolated RBGF with parameters $x_c = 0$ and $\sigma = 1$

By virtue of its locality (narrowness), the maximum value is obtained around its central point, as shown in Fig. 2. Feature of RBGF locality was used for parsing arm movement in several phases in order to check whether motor learning process is existent in each phase of arm movement, and discover in which phase of movement it is dominant. The first question which arose was what the number of RBGFs i = 1, 2…n is for correct parsing and modelling the arm movement in the ballistic task with a predefined error. Also, finding correlation between individual RBGF was being particular research task. Motor learning process of arm movement was proposed as linear regression of more individual RBGFs $h_j(x)$ = 1, 2… n, as:

$$f(x) = \sum_{j=1}^{n} A_j h_j(x) \tag{1.2}$$

Where A_j are weighted coefficients of regression. Given functional dependency allows approximation (fitting) of wide class non-parametric functions such as position, velocity and acceleration profiles of humans arm movement by simply choosing a weighting coefficients A_j and rarely one of the parameters of individual RBGF x_{cj} and σ_j.

Let's suppose that $h_j = e^{-\frac{(x - x_{c_j})}{2\sigma_j^2}}$ represents more RBGFs on segment [0,1], as it is shown in Fig. 3. Each RGBF h_j locally approximates non-parametric functional dependency around centred point x_{cj}.

Consequently, the linear regression (1.2) on the entire segment [0, 1] can be written in the form of the sum of the weight coefficients A_j and individual lGRBF h_j, as follows:

$$f(x) = \sum_{j=1}^{n} A_j h_j = \sum_{j=1}^{n} A_j e^{-\frac{(x - x_{c_j})}{2\sigma_j^2}} \tag{1.3}$$

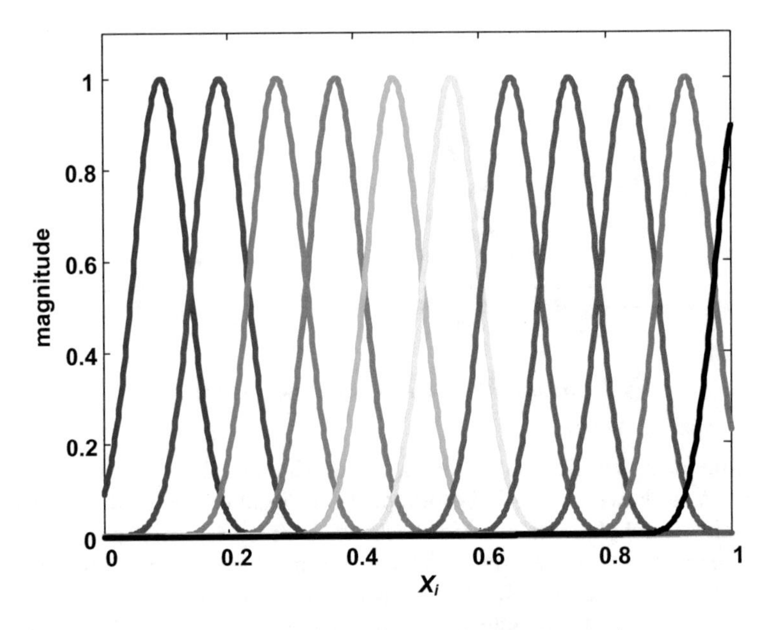

Fig. 3. Class of shifted GRBF h_j for linear regression

It is obvious that in the given Eq. (1.3) the four parameters n, A_j, x_{cj}, and σ_j are unknown and should be determined in order to be able to approximate non-parametric functional dependency with predefined uncertainty. In other words, proposed linear regression (1.3) is actually tool for modelling motor learning process. Further in the paper will be shown that only two (A_j, x_{cj}) of four (n, A_j, x_{cj}, σ_j) parameters are sufficient for quantification of existence of motor learning process of human arm movement in the ballistic task – hit a target. In order to solve the mentioned task and determine four unknown parameters in Eq. (1.3), the MATLAB user-defined function ***lsqcurvefit*** has been used [6–8]. The key part in the script is the following function:

[f_gauss_coeff(i_dir,:),resnorm(i_dir,:),data_fit_errors]= lsqcurvefit(@fa_gauss_new,init_condtns,time,fit_data),

After execution of the function ***lsqcurvefit,*** following output parameters were obtained:

- Estimated value of:
  - weighting coefficients A_j (*f_gauss_coeff*),
  - central points of individual RBGF x_{cj};
- Quadratic residue norm, (*resnorm*),
  $resnorm = \sum f(coeff, xdata - ydata)^2$
- Fitting error of approximation in each point x_i, y_i for estimated values of RBGF x_{cj} and A_j, and predefined value of n and σ_j (*f_gauss_parameters*).

The *quadratic residue norm* and *fitting error* are primarily used to estimate the required numbers ***n*** of RBGF in the Eq. (1.3) that are used to fit non-parametric functions with an allowable fitting error. Verification procedure of proposed mathematical tool is described in next section of this paper.

3 Parameter Selection for RGBF

As it is mentioned earlier, a key role in the ballistic task – hit a target have velocity profiles of human's arm movement and these velocity profiles have been the subject of modeling in order to quantify the motor learning process of the human's arm movement by using the statistical parameters [10–16].

On the Fig. 4(a) and (b) are shown the velocity profiles of the human arm/joystick movement during one series of the training, for the early (Fig. 4a) and late (Fig. 4b) stage of the training. The subject who performed the movement and a series of movements were chosen arbitrarily.

In the given linear regression (1.3) two of four free parameters n, A_j, x_{cj} and σ_j are fixed, n is required number of RBGF and σ_j is RBGF width.

Reason for fixing the first parameter is due to the uniformity of the model and the second parameter σ_j – RBGF width is fixed in order to maintain "locality". This means that each individual RBGF is the most "responsible" in the vicinity of his center point x_{cj}. Also, it should be noted that there is a reciprocal relationship between the number of RBGF and their width σ_j. If a smaller number of RBGFs is chosen, in order to keep out approximation error within the permissible limits, the width of the RBGF must be

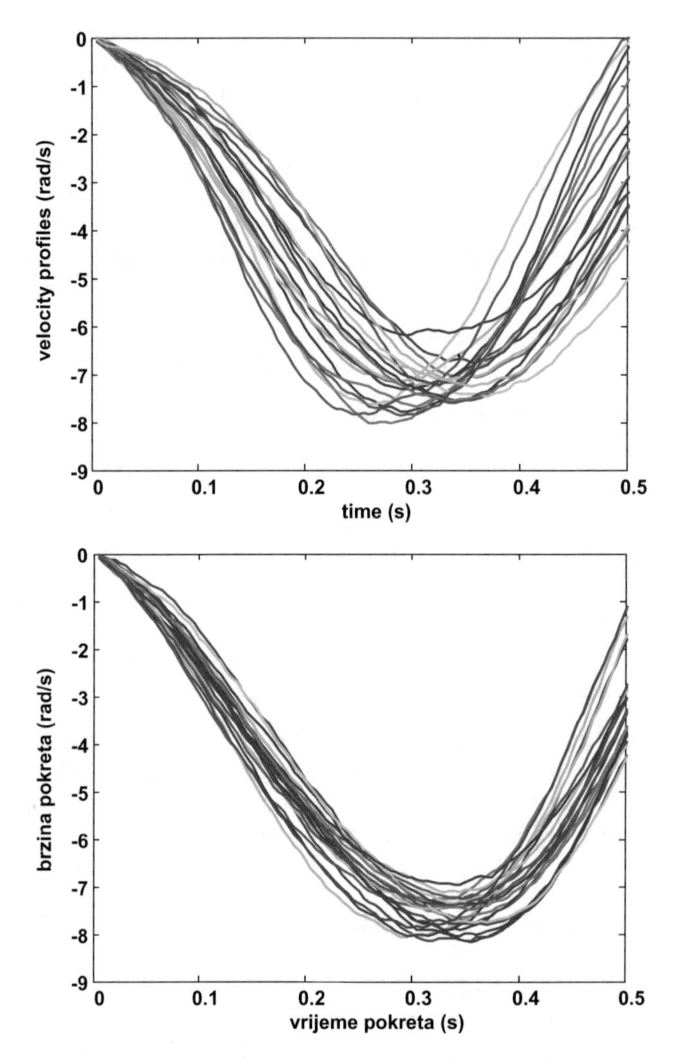

Fig. 4. (a) Velocity profiles of the human arm movement in the early stage of training (b) Velocity profiles of the human arm movement in the late stage of training

greater and vice versa. The conducted analysis gave the answer to the question for choosing the number of RBGF and their width σ_j.

To evaluate the number of RBGF needed for parametric model of human arm movement in the ballistic task and simultaneously deploy motor learning model, one training series of randomly selected subject who participated in the experimental phase of the research is observed.

The procedure of determining the number of RBGF and their width was carried out through following iterative steps:

- Set the required number of RBGF: n,
- Set RBGF width: σ_j,

- Analyze the output parameters of the function ***lsqcurvefit.m*** (resnorm- fitting error),
- If the fitting error is greater of the predefined value the number of RBGF must be increased.

This proposed procedure can be modified by fixing the number of RBGF (n) in one iteration and changing their width σ_j. The described procedure for determination number of RBGF and their width has been fully performed in MATLAB.

At beginning of analysis, the required number of RBGFs was $n = 5$ and RBGFs with two widths $\sigma_j = 0.15$ and $\sigma_j = 0.1$ were initially chosen.

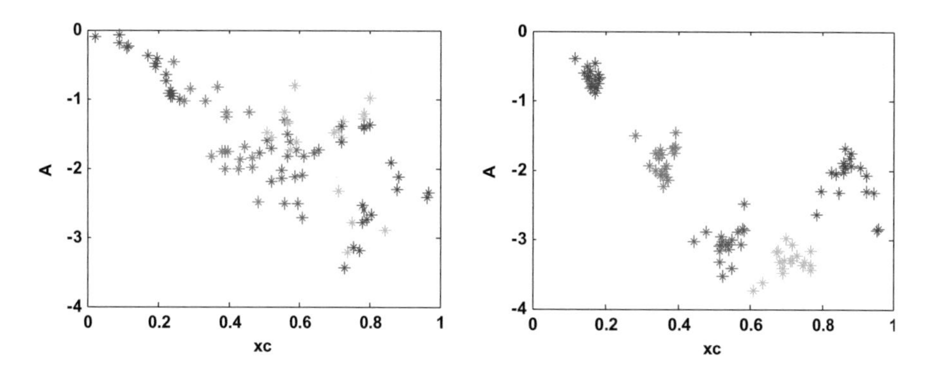

Fig. 5. (a) Positions x_{cj} and A_j, for $n = 5$ and $\sigma = 0.15$ (b) Positions x_{cj} and A_j, for $n = 5$ and $\sigma = 0.1$

4 Extraction of Features

On the Fig. 5(a) and (b) position of centers x_{cj} amplitude A_j for a given number of RBGFs n and two different widths σ_j are shown. It is obvious, that if RBGF (Fig. 5a) has too wide width σ_j, then locality of RBGF will be loosed, hence making it difficult to quantify motor learning process. This means that it will be difficult to separate individual phases of the arm movement and will be difficult to determine at which stage of the arm movement the motor learning process is dominant.

In the next stage of the analysis, Fig. 6(a) and (b), the required number of n RBGFs is increased to $n = 6$, with two same widths $\sigma_j = 0.15$ and $\sigma_j = 0.1$. The increased number of RBGFs compared to the previous case in the linear regrresion (1.3), together with the reduction RBGF width σ_j, caused a better grouping of individual RBGF parameters A_j and $x_{cj,}$ in other words, the better localization of individual phases of the arm movement with the same or lower approximation error.

A similar procedure for the analysis of the required number n of RBGFs for the linear regrresion (1.3) was continued with $n = 7$, with two same widths $\sigma_j = 0.15$ and $\sigma_j = 0.1$.

In order to make a general conclusion about choice of the required number $\boldsymbol{n}$ of RBGFs and their width σ_j, besides the conducted analyses of the grouping of RBGF parameters, i.e. the localization properties of individual RBGF, additional analyses of

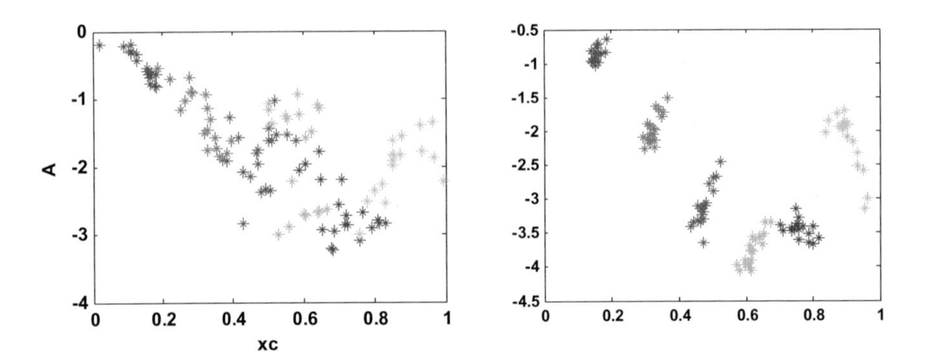

Fig. 6. (a) Position x_{cj} and A_j, for $n = 6$ and $\sigma = 0.15$ (b) Position x_{cj} and A_j, for $n = 6$ and $\sigma = 0.1$

the fitting error of the linear regression (1.3) in the form of the sum of squared errors (resnorm) was carried out.

In Fig. 7 are shown fitting errors for $n = 5$, $n = 6$, $n = 7$ and for four RBGF widths of interest: $\sigma_{j1} = 0.1190$, $\sigma_{j2} = 0.1121$, $\sigma_{j3} = 0.106$ and $\sigma_{j4} = 0.1$.

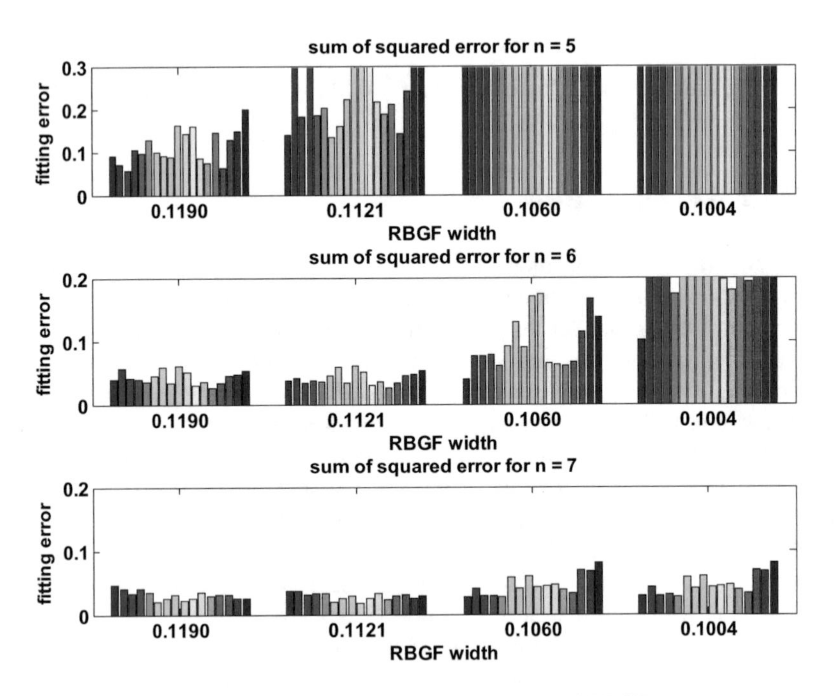

Fig. 7. Squared fitting error vs. number of RBGFs

Each vertical strip in the Fig. 6 represents a sum of squared fitting error of individual human arm movement velocity profile during the observed throw series. At the beginning of analysis, for required number of RBGFs n and their width σ_j, a reasonable assumption that fitting error should be less than of 0.1, i.e. *resnorm* = *0.1* was made. The results of the conducted analyses are shown in Fig. 7. Which show an interval in which the RBGF widths σ_j should be changed in order to assure RBGF locality around the center x_{cj}, which is necessary for a refined analysis of the motor learning process of the arm movement in the ballistic task.

As a result of the analyses, following RBGF width intervals of interest were obtained: $\sigma_{j1} = 0.1190$, $\sigma_{j2} = 0.1121$, $\sigma_{j3} = 0.106$ and $\sigma_{j4} = 0.1$. In the beginning of the analysis, the minimum required number of RBGFs $n = 5$ and maximum width $\sigma_{j1} = 0.1190$ were chosen in order to group amplitude of each individual RBGF near their center x_{cj} and retain RBGF locality.

For some velocity profiles, a sum of squared error exceeded the set *resnorm* = 0.1. As the RBGF widths σ_j were decreasing, the fitting errors exceeded over set value *resnorm*. In the conducted iterations, RBGFs with lower widths $\sigma_{j2} = 0.1121$, $\sigma_{j3} = 0.106$, $\sigma_{j4} = 0.1$ and $n = 5$ were selected. As a result, fitting error increased over set value *resnorm* = 0.1 which was expected. Further analysis was continued with a larger number of RBGFs $n = 6$ with maximum width $\sigma_{j1} = 0.1190$ and fitting errors for all velocity profiles were reduced below set value *resnorm*. The analysis was continued for the number of RBGFs $n = 7$ and maximum width $\sigma_{j1} = 0.1190$. In this case, the fitting

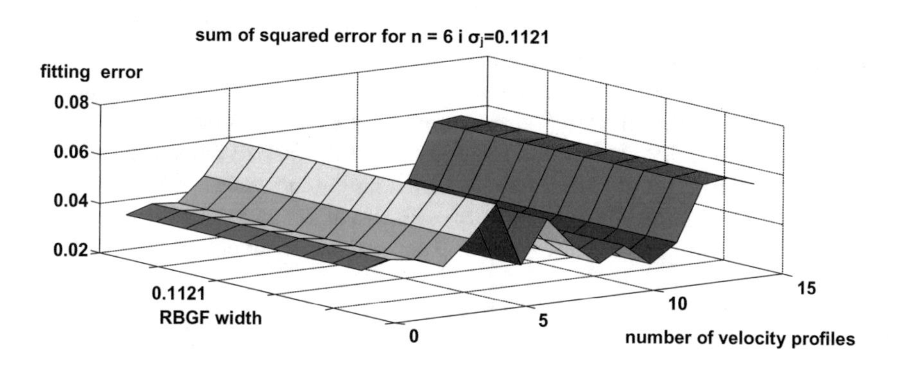

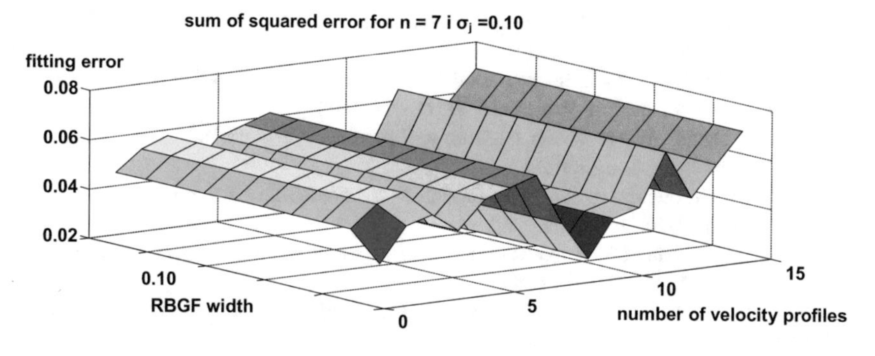

Fig. 8. Comparative characteristics of the fitting error *resnorm* for two combinations of interest: the number of RBGFs n vs. RBGF width σ_j

error decreased even more, but the feature of the RBGF locality was lost and this width σ_{j1} was rejected as unacceptable. The feature of RBGF locality has also been adverse for RBGF widths $\sigma_{j2} = 0.1121$ and $\sigma_{j3} = 0.106$, so only for width $\sigma_{j4} = 0.1$ acceptable locality was obtained. On the other hand, however, by reducing the width of the RBGFσ_j and increasing their locality the sum of squared fitting error is increased. The result of the conducted analysis is reducing the choice of possible combinations between the number of RBGF and their width into two combinations: $n = 6$, $\sigma_{j2} = 0.1121$ and $n = 7$, $\sigma_{j4} = 0.1$, as it is shown in Fig. 8.

From Fig. 8, it is evident that both combinations of the number of RBGFs n and width σ_j yield similar results. The combination $n = 6$, $\sigma_{j2} = 0.1121$ gives slightly better results, and this combination was chosen for parametric modeling of motor learning process of human arm movement (motor learning model) in the ballistic task.

5 Conclusion

Therefore, as a conclusion of the performed analysis, we can point out the following: from the four free parameters n, A_j, x_{cj} and σ_j in the functional approximation (1.3), two are fixed $n = 6$ and $\sigma_{j2} = 0.1121$. The other two parameters A_j and x_{cj} are variables for each fitted velocity profile of arm movement during a single throw series, and represent a sufficient set for modeling the arm/joystick movement (derive the model of motor learning) in the ballistic task – hit a target. This conclusion is drawn from the sum of squared error (resnorm) in the described procedure of fitting the velocity profiles. Therefore, based on the above given parameters, it was possible to obtain a parametric motor learning model of the arm/joystick movement in the ballistic task – hit a target. In order to additionally confirm the hypothesis that the RBGF parameters A_j and x_{cj} represent a sufficient set of parameters for modeling the movement of the arm/joystick in the individual training phase, a statistical estimate of the parameters A_j and x_{cj} of the proposed motor learning model was made. This analysis is not described in this paper.

References

1. Jordan, I.M., Wolpert, M.D.: Computational motor control. In: The Cognitive Neurosciences. MIT Press, Cambridge (1998)
2. Kawato, M., Wolpert, D.M.: Internal models for motor control. In: Glickstein, M., Bock, R. (eds.) Sensory Guidance of Movement, Novartis (1998)
3. Paul, D.R., Daniel, W.M.: Motor learning and prediction in a variable environment. Curr. Opin. Neurobiol. **13**, 1–6 (2003)
4. Orr, L.J.M.: Introduction to Radial Basis Function Networks. Centre for Cognitive Science, University of Edinburgh (2003)
5. Bishop, C.: Improving the generalisation properties of radial basis function neural networks. Neural Comput. **3**(4), 579–588 (1991)
6. Martinez, L.W., Martinez, R.A.: Exploratory Data Analysis with MATLAB. Chapman & Hall/CRC Press, Boca Raton (2005)
7. Martinez, L.W., Martinez, R.A.: Computational Statistics Handbook with MATLAB. Chapman & Hall/CRC Press, Boca Raton (2002)

8. More, J.J.: The Levenberg-Marquardt algorithm: implementation and theory. In: Watson, G. A. (ed.) Numerical Analysis. Lecture Notes in Mathematics, vol. 630, pp. 105–116. Springer, Heidelberg (1977)
9. Slobodan, L.: Solving dynamic tasks in complex systems with the human-machine interface by using the motor learning model. PhD thesis, Faculty of Electrical Engineering, University of East Sarajevo, East Saraejvo (2009)
10. Mussa-Ivaldi, F.A., Bizzi, E.: Motor learning through the combination of primitives. Phil. 7ims. R. SIX. Lond. R (2000)
11. Motulsky, H., Christopoulos, A.: Fitting models to biological data using linear and nonlinear regression. GraphPad Software (2003)
12. Schaal, S.: Dynamic movement primitives – a framework for motor control in humans and humanoid robotics. In: AMAM (2003)
13. Schaal, S., Peters, J., Nakanishi, J., Ljspeert, A.: Learning movement primitives. In: International Symposium on Robotics Research (ISRR 2003), Springer Tracts in Advanced Robotics. Springer, Ciena (2004)
14. Shadmehr, R., Mussa-lvaldi, F.A.: Adaptive representation of dynamics during learning of a motor task. J. Neurosci. **74**(5), 32083224 (1994)
15. Thoroughman, A.K., Shadmehr, R.: Learning of action through adaptive combination of motor primitives. Nature **407**, 742 (2000)
16. Wang, T., Đorđević, S.G., Shadmehr, R.: Learning the dynamics of reaching movements results in the modification of arm impedance and long-latency perturbation responses. Biol. Cybern. **85**, 437–448 (2001)

An Open and Extensible Data Acquisition and Processing Platform for Rehabilitation Applications

Sehrizada Sahinovic[1(✉)], Amina Dzebo[1], Baris Can Ustundag[2], Edin Golubovic[3], and Tarik Uzunovic[1]

[1] Faculty of Electrical Engineering, University of Sarajevo, Sarajevo, Bosnia and Herzegovina
{ssahinovic1,adzebo2,tuzunovic}@etf.unsa.ba
[2] Department of Computer Engineering, Istanbul Technical University, Istanbul, Turkey
ustundag16@itu.edu.tr
[3] Inovatink, Istanbul, Turkey
edin@inovatink.com
https://www.inovatink.com

Abstract. Recently we witnessed a great deal of progress in the field of medicine, as well as treatments that improve the patient therapy and care. However, physiotherapy and rehabilitation fields still face the challenges of treating patients in remote regions. Considering that, developing a data acquisition and processing platform that collects data of rehabilitation movements at home can play a key role in the success of a patient's recovery process. The designed system is composed of three main parts: wearable sensor capable of collecting movement data with 3 axial accelerometer, gyroscope and magnetometer sensors, central hub for processing and a cloud system which is used as a link between the therapist and patient. The system was tested for purpose of monitoring rehabilitation exercises usually done during recovery from an elbow fracture. Experimental results have shown that the system presented in this paper gives successful results for rehabilitation applications.

1 Introduction

Technological advancements in the last decade have resulted in remarkable progress of patient treatment in physiotherapy and rehabilitation. Nevertheless, physiotherapy and rehabilitation fields still face challenges associated with the healthcare delivery. Currently, healthcare delivery is centralized to hospitals and treatment centers which poses challenge of access for patients in remote regions and produces the economic challenges associated with hospitalization. Healthcare procedures such as diagnosis and determination of progress are currently based on the infrequently sampled tests which poses challenge with incomplete assessment and late discovery of illness. Finally, present physiotherapy procedures are reactive. Nevertheless, latest advancements in technology regarding the miniature wearable sensors, advanced computation devices, signal processing techniques and the Internet of Things (IoT) are promising to shift the

S. Avdaković (Ed.): IAT 2018, LNNS 59, pp. 394–406, 2019.
https://doi.org/10.1007/978-3-030-02574-8_32

physiotherapy delivery model to decentralized, patient-centric, continuously monitored and, most importantly, preventive.

Patients undergoing physiotherapy typically perform corrective movements under the supervision of therapist. Therapist tracks the patient's progress and devises the new set of exercises according to the progress. The exercises are done on the daily basis for several weeks or even months depending on the severity of injury and patient's physical condition [1]. With the help of advanced sensing and communication and information technologies, the physical therapy can be performed outside of the hospitals and rehabilitation centers, inside patient's home, also referred to as telerehabilitation [2–4]. The quality of telerehabilitation can be increased using wearable sensors that precisely quantify the body movement [4–6]. The use of advanced signal processing techniques allows for automatic movement recognition and provision of timely feedback in an autonomous fashion [7, 8]. Using IoT technology therapists can have full access to processed motion data and automated reports regarding the patient's exercises [9–12] that allows them to make timely and precise decisions regarding the patient's health.

Telerehabilitation research deals with the studies of topic in three directions [13], patient satisfaction with developed systems [14], therapist acceptance of developed systems [15] and technical issues of the telerehabilitation systems [16]. Literature review of telerehabilitation shows that the most popular telerehabilitation applications are upper and lower limb rehabilitation [17], knee surgery rehabilitation [18], rehabilitation for patients with motor impairment [19] and rehabilitation for patients with chronic pain [20], spinal cord injuries [19] and neurological injuries [21, 22].

This paper introduces a system for data acquisition and processing of motion data during physical therapy. The system consists of a wearable sensor that records specific movements during the rehabilitation process, central processing hub that receives the data from wearable, performs local processing on the data and communicates the movement data to the cloud system with storage capabilities and portal where therapists and patients can track the progress of rehabilitation. The most prominent characteristics of developed system are its openness and extensibility. The system is based on the readily available, low cost and open source technology. All the results of this study, including technical details and production ready hardware and software materials can be found on project online repository [23].

The rest of this paper is organized as follows: Sect. 2 compares several data acquisition systems for rehabilitation. Section 3 describes design and implementation of the system. The experimental results are presented in Sect. 4. Section 5 provides final conclusions and directions for future work.

2 Analysis and Comparison of Data Acquisition and Processing Systems for Telerehabilitation

This section analyzes four telerehabilitation solutions previously published in the literature. Main reasoning behind such analysis is to understand the requirements for building of a telerehabilitation system and to gain insight into deficiencies of the current systems. Four telerehabilitation systems used in this analysis [24–27] are intended for monitoring of patient movement during the rehabilitation exercises.

Literature survey has shown that telerehabilitation systems are commonly composed out of three main building blocks: sensors that collect movement data during rehabilitation, a processing hub that processes the collected data and arranges it into proper format for therapist and patient to track the progress and optionally, in the case of Internet connected systems, a cloud system to be used as a link between the therapist and patient and for storage and further processing of patient's data.

Authors in [24] developed a wearable system for upper arm rehabilitation of stroke patients. The wearable consists of optical linear encoder as sensing element, accelerometer, digital signal controller for signal processing and wireless radio transceiver operating on 2.4 GHz frequency. The wearable is used to measure the orientation of the arm using accelerometer signals. Authors designed base station that has radio frequency circuitry capable of receiving data from wearable and an USB adaptor for data transfer to PC. The data collection, organization, processing and visualization is done on a PC. The cloud system and mobile data processing was mentioned as future directions in this work. Authors share fair amount of details regarding the biomechanical model of an arm, however, besides the architectural explanation, details about hardware and software are not detailed in this work.

An extensible platform for wearable motion sensing, sensor research and IoT applications called BlueSense is represented by [25]. This platform is not primarily made for any specific telerehabilitation application but due to its design can be adopted for such purpose. The wearable is designed based on the microcontroller unit (MCU), motion sensor and Classic Bluetooth communication. Wearable runs on battery and has USB interface for programming and configuration. Additionally, wearable device features SD memory card for long term monitoring. The interface of this wearable is done using a Bluetooth dongle and a PC. The wearable extensibility is provided through customized set of connectors which allow for attachment of additional sensors and other circuitry. The cloud system is not discussed in the context of this work, however, the authors intend this system to work with either smart phone or PC, which can then be extended for cloud operation.

In [26] authors developed wearable sensing system for human motion monitoring in physical rehabilitation. The system is designed such that wearable sensors act as slaves in master-slave wireless network. The master module is interfaced to PC and intuitive graphical user interface was designed to assist therapists and researchers to easily track the progress of patients. Wearable device consists of tri-axial accelerometer, microcontroller and an RF module for wireless data transmission. Wearable device also hosts connectors for system extensibility. This system does not have dedicated cloud system, as it is case with previous systems, there is possibility to extend PC software with addition of cloud client for telerehabilitation system. The system details are not open, however authors give fair amount of details about the communication protocol in master-slave wireless communication system.

Authors in [27] created telerehabilitation system based on wireless motion capture sensors to be used as a personalized home-based rehabilitation system for elderly people. Developed system consists of wearable devices and a base station that is connected to PC. Wearable devices are attached to the part of the body that needs to be monitored. Wearable devices include connectors for system extensibility. The readings from inertial and magnetic sensors are then continuously collected and sent to the base station via wireless communication. Base station forwards the information to the PC where real time calculation of 3D orientation of the module is done and results are delivered in the form of pitch, roll and yaw angles. The system is not connected to cloud.

Table 1 summarizes the considered systems [24–27] with regards to common design criteria for telerehabilitation systems based on the wearable devices.

Table 1. Design criteria for telerehabilitation system

Criteria	[24]	[25]	[26]	[27]
Wireless connection	2.4 GHz proprietary	Classic bluetooth	2.4 GHz proprietary	2.4 GHz proprietary
Integrated sensors	Optical linear encoder, accelerometer	Accelerometer, gyroscope, magnetometer	Accelerometer	Accelerometer, gyroscope, magnetometer
Wearable dim. (mm)	–	30 × 30	84 × 52	46 × 46
On-board processing	Sensor measurement	Orientation calculation	Sensor measurement	Sensor measurement
Openness	✗	✗	✗	✗
Extensibility	✗	✓	✓	✓

Current systems found in literature are designed as wireless wearable devices that communicate with central hub. Analyzed systems are connected to PC which limits the mobility of the entire system and imposes extra cost to the system. Most common deficiency of the current systems is the lack of cloud communication and integration of IoT technology. Another deficiency of the current systems is that they are not open to engineers and researchers. In the light of the analysis, we present our solution in the next section. The developed solution follows main design criteria of most of the systems found in literature while mitigating common deficiencies of the present systems. Designed system is cloud connected, open, extensible and low cost.

3 Design and Implementation of an Open and Extensible Data Acquisition Platform for Rehabilitation Applications

Architecturally the designed system, shown in Fig. 1, is composed of three main parts: wearable device, central hub for processing and doctor/patient portal. Wearable collects movement data and sends it via Bluetooth Low Energy (BLE) protocol to central hub. Central hub is implemented using Raspberry Pi 3 (RPi3). RPi3 has an on-board BLE module, able to receive data sent out from wearable device. Doctor/patient portal is cloud-based software where achieved results can be visualized and rehabilitation progress can be tracked.

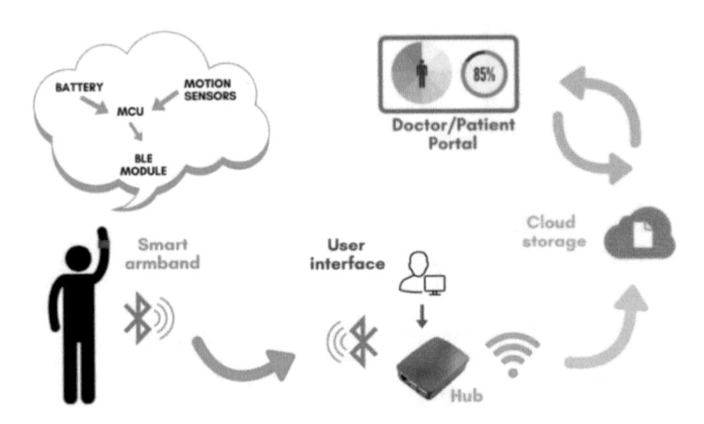

Fig. 1. Designed telerehabilitation system

3.1 Wearable Sensor

The success and applicability of the system presented in this paper depends on the reliable and accurate measurement of patient's motion during exercise. The motion data is measured with the help of custom built wearable device. The architectural block diagram of the designed wearable is shown in Fig. 2. The main building blocks of wearable device are motion sensor, BLE module, microcontroller, battery, battery charger, voltage converters, extensibility interface and USB interface for programming and charging.

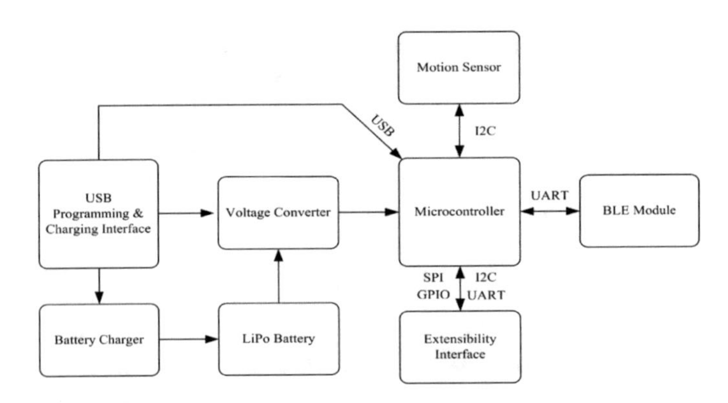

Fig. 2. Wearable device architectural block diagram

Motion sensor is Invensens's MPU9250 [28], a multi-chip MEMS module that consists of 3-axis accelerometer, gyroscope, compass and a dedicated processing unit. The chip provides raw and processed sensor values through dedicated inter-integrated circuit (I2C) bus interface.

Data acquisition and on-board processing is done by low power ARM Cortex M0 microcontroller unit (Atmel's SAMD21G) [29]. Microcontroller reads sensor data through I2C interface, additionally processes signals to obtain Euler angles from quaternions received from digital motion processor on MPU9250. Next to data acquisition and processing, microcontroller handles the communication between wearable and hub through interfacing of BLE Module. Used BLE Module is Microchip's RN4871, a module that integrates Bluetooth 4.2 baseband controller, on-board Bluetooth stack, digital and analog inputs/outputs (I/O), and RF power amplifier into compact form factor. Module and microcontroller exchange data through Universal Asynchronous Receiver Transmitter (UART) interface. RN4871 has a private generic attribute profile named Transparent UART that simplified the data transfer between the microcontroller and processing hub [30]. Due to the simplified implementation of Transparent UART, any serial data sent into BLE module's UART is transferred to the connected peer device via Transparent UART Bluetooth service. When data is received from the peer device over the air via Transparent UART connection, this data outputs directly to UART. Wearable device is powered by LiPo battery through system voltage converter. A single cell 220 mAh LiPo is used and system voltage is designed as 3.3 V. Battery is charged through USB friendly integrated LiPo Charger [31]. The extensibility interface is designed to accommodate additional sensors in easy fashion. Extensibility allows the use of wearable in several different application by accommodating additional sensor. The extensibility interface was designed in Mikroelektronika's Click format [32] to make use of easily accessible over 130 different sensors [33]. The microcontroller can be programmed via both JTAG and USB interface. Wearable has visual indication via two LED's, blinking green once the firmware is in regular execution and steady blue once the connection to peer is established. Wearable enclosure is made out of 3D printed ABS plastic (see manufacturing files for details [34]). By the means of elastic straps the wearable can be attached to any part of the body such as upper and lower extremities, torso, neck and head. Implemented wearable device is shown in Fig. 3.

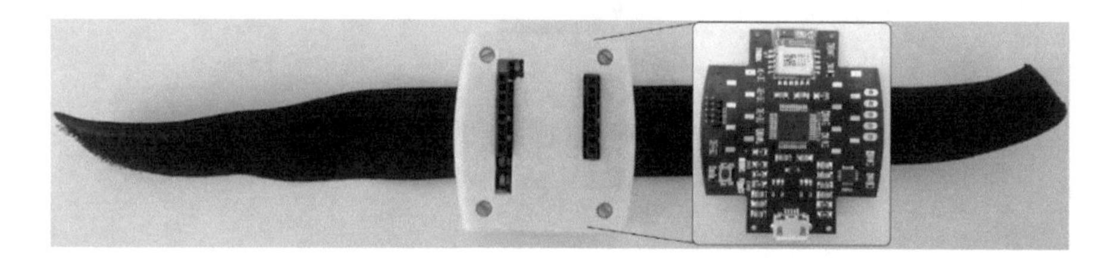

Fig. 3. Wearable device implementation

Wearable sensor firmware is written using Arduino software framework [35]. Firmware is designed such that sensor values are read periodically and sent to connected peer via BLE module. RN4871 Bluetooth module is configured in transparent UART mode. The sample period and other configurations are done in firmware. Firmware starts with MCU initialization and application configuration. The firmware is written in the form of finite state machine. Execution starts in Idle state. Once the periodic data is available from sensor, Sample state is entered and sensors values are obtained via I2C interface. If sampling fails, the program returns to Idle state and waits for new sample (this rarely happens due to temporary sensor error). After data is read from sensors firmware enters Serialize state where the data from individual sensors is serialized into a message string to be sent via Bluetooth. After serialization program enters Send state where data is sent to BLE module. After sending of data is done the firmware enters Receive state where configuration data received through BLE is read and wearable configuration is updated. Designed firmware is fully open and modifiable [36].

3.2 Acquisition and Processing Hub

Prior to the exercise the wearable device is turned on and sending of data is started. Patient places the wearable on the part of the body that is being rehabilitated. On the other hand, central hub, scans the nearby devices and connects to the desired wearable. This is done in the initial setup of the system. For the purpose of BLE communication Bluetooth Generic Attribute Profile software development kit (GATT SDK) for Python is used [37]. The hub software is implemented using Python 3 language and codes are fully available online [38]. Pseudocode below shows the procedure for initialization:

```
install gatt
install python3-dbus
gattctl –discover
gattctl --connect [MAC address]
```

Once the wearable is connected, the device gets paired for easier and automatic reconnection. After initialization procedure RPi3 can start to receive movement data sent by wearable. RPi3 discovers previously discussed Transparent UART service on the wearable and subscribes to the updates through notification mechanism native to BLE protocol. In this setting RPi3 gets notified every time new data is available from wearable. This process is further clarified through pseudocode given below.

```
import gatt
access armband
access related service (see Github for details)
access related characteristic (see Github for details)
turn notify on
receive data and save it to file
```

After the exercise is finished the received and stored data is sent to cloud for storage and possibly further analysis. It’s assumed that RPi3 is connected to Internet via WiFi or Ethernet. Once useful data about patient’s rehabilitation workout is stored on cloud it can be processed for further usage on Doctor/Patient portal. Doctor can use this portal as admin with the right username and password, i.e. doctor can assign special exercises for patient, manage the data about the patient and generally has more authority than a patient who with their credentials only see personal progress and prescribed exercise plan.

Every wearable can be personalized, i.e. every patient gets a different wearable with its unique Bluetooth address which can be related to the patient’s personal information like name, gender, date of birth, type of injury etc. Also, different movements during the exercise have their own characteristics, for example how fast the movement was, how correct, when it was made and similar. All these metadata about patient and exercise can be used for data processing update and easy retrieval. For example, doctor can do a search on the portal for all female patients under the age of 25 with wrist injury by 2017 and then see the most successful rehabilitation, all this with the help of metadata.

At the current state of development, the doctor/patient is not fully implemented so it is not possible to share further details. As soon as the development finished the implementation details will be available online [23].

4 Experimental Results

The system was tested as a system for monitoring rehabilitation exercises with an elbow fracture injury. Rehabilitation exercises in elbow injuries are divided into two groups: stretching exercises and strengthening exercises. Stretching exercises can be done at the beginning of physical therapy, while strengthening exercises can only be done when stretching is nearly painless. The goal of these exercises is to accelerate healing by expanding a range of motion.

Figure 4 [39] summarizes the exercises performed in experiment including wrist flexion and extension, forearm pronation and supination and active elbow flexion and extension. Stretching exercises can improve flexibility and a range of motion, as well as increase circulation. However, strengthening exercises strengthen muscles and help to increase the speed, as well as load that can be put on the elbow.

Strengthening exercises that were performed are [39]:

- wrist flexion, extension and radial deviation strengthening,
- forearm pronation and supination strengthening,
- wrist extension with broom handle.

Wearable device is recording acceleration and angular velocity data with the sampling frequency 50 Hz. Depending on the time required to perform the exercise, all acceleration and angular velocity data recorded at that time interval will be sent over Bluetooth. Thereafter, these data are filtered using a low pass filter to remove noise from the signals. Figure 5 shows filtered accelerometer and gyroscope data while recording rehabilitation exercises for an elbow fracture injury. Data was recorded on a healthy, young female patient without an elbow injury. As we can see in Fig. 5, from the behavior of signals, it is possible to conclude type of performed exercises and from the parameters of signals, it is possible to know whether the exercise was performed in the correct way.

When signals in Fig. 5(b) and (g) are compared, it is seen by the signal form that the movement is forearm pronation and supination. The same conclusion can be made if the signals of flexion and extension rehabilitation exercises, shown in Fig. 5(c) and (h), are compared. By comparing exercises accelerations, it is apparent that the exercise shown in Fig. 5(h) did not reach its maximum.

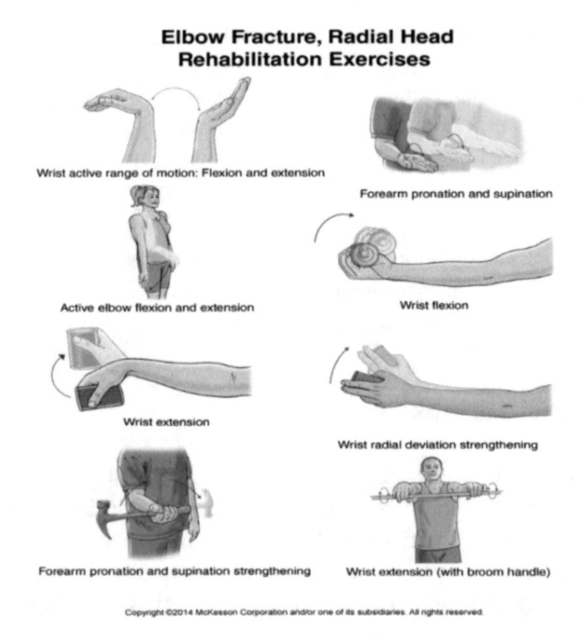

Fig. 4. Elbow fracture, radial head rehabilitation exercises [39]

In addition, from the displayed signals, the doctor can know exactly how long a patient has been doing a certain exercise and how long he has held his arm in a certain position, and whether there are visible improvements in exercise speed and range of motion. The experimental results have shown that designed shows successful results for characterization of movement in rehabilitation applications.

5 Discussion and Conclusion

In this paper the data acquisition and processing system for telerehabilitation is presented. Presented system architecturally consists of three elements: a wearable sensor that records specific movements during the rehabilitation process, a central processing hub that on, one hand receives the data from wearable and performs local processing on the data, and, on the other hand communicates the movement data to the cloud system with storage capabilities and professional portal where therapists and patients can track the progress of rehabilitation. The most prominent characteristics of developed system are its openness, extensibility and low cost. The system is based on the readily available and open source technology.

Wearable sensor hardware is made open source [23]. Main MCU used in the implementation of wearable supports the Arduino Software Framework which gives an opportunity for implementation of custom algorithms through available open source code. The wearable device is easily extended to accommodate over 130 readily available sensors through extensibility interface in Mikroelektronika's Click format [32]. Central hub is implemented using Raspberry Pi 3, the most popular computation

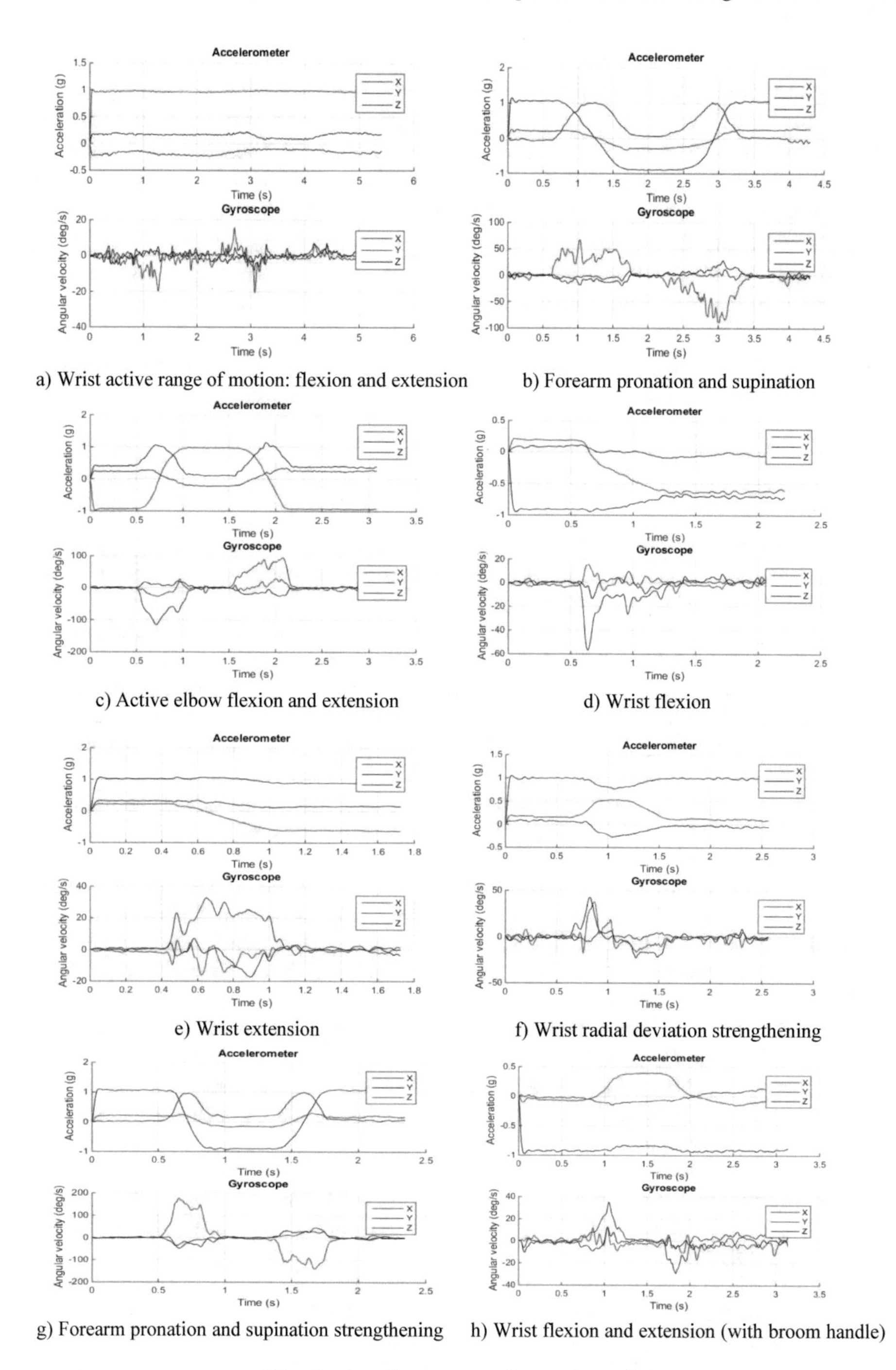

a) Wrist active range of motion: flexion and extension

b) Forearm pronation and supination

c) Active elbow flexion and extension

d) Wrist flexion

e) Wrist extension

f) Wrist radial deviation strengthening

g) Forearm pronation and supination strengthening

h) Wrist flexion and extension (with broom handle)

Fig. 5. Accelerometer and gyroscope data

platform in maker and prototyping community. Hub software is open source [23]. The upper limit of total cost of hardware for developed system is 99$.

The functionality of the system has been demonstrated by performing some of the common exercises used in rehabilitation of elbow fracture. The sensor data is collected at 50 Hz and transmitted to RPi3 in real time. The future work will consist of full implementation of doctor/patient portal, development of graphical user interface for the system that can be displayed on handheld display that will be attached to RPi3 and development of machine learning and advanced signal processing algorithms for automatic exercise recognition and determination of patient's progress.

Acknowledgments. Authors would like to acknowledge Inovatink (www.inovatink.com) for providing material and operational support in realization of this project.

References

1. O'Sullivan, S.B., Schmitz, T.J., Fulk, G.: Physical rehabilitation. FA Davis (2013)
2. Theodoros, D., Russell, T., Latifi, R.: Telerehabilitation: current perspectives. Stud. Health Technol. Inform. **131**, 191–210 (2008)
3. Castro, H., Cha, E., Provance, P.G.: Home-based physical telerehabilitation in patients with multiple sclerosis: A pilot study. J. Rehabil. Res. Dev. **45**(9), 1361 (2008)
4. Thiers, A., Orteye, A., Orlowski, K., Schrader, T.: Technology in physical therapy. In Proceedings of the International Joint Conference on Biomedical Engineering Systems and Technologies, vol. 5, pp. 500–505. SCITEPRESS-Science and Technology Publications, Lda, March 2014
5. Levene, T., Steele, R.: The Quantified self and physical therapy: the application of motion sensing technologies. In: Proceedings of the International Conference on Compute and Data Analysis, pp. 263–267. ACM, May 2017
6. Patel, S., Park, H., Bonato, P., Chan, L., Rodgers, M.: A review of wearable sensors and systems with application in rehabilitation. J. Neuroeng. Rehabil. **9**(1), 21 (2012)
7. Hadjidj, A., Souil, M., Bouabdallah, A., Challal, Y., Owen, H.: Wireless sensor networks for rehabilitation applications: challenges and opportunities. J. Netw. Comput. Appl. **36**(1), 1–15 (2013)
8. Zhou, H., Hu, H.: Human motion tracking for rehabilitation—a survey. Biomed. Sig. Process. Control **3**(1), 1–18 (2008)
9. Caporuscio, M., Weyns, D., Andersson, J., Axelsson, C., Petersson, G.: IoT-enabled physical telerehabilitation platform. In: 2017 IEEE International Conference on Software Architecture Workshops (ICSAW), pp. 112–119. IEEE, April 2017
10. Bilic, D., Uzunovic, T., Golubovic, E., Ustundag, B.C.: Internet of things-based system for physical rehabilitation monitoring. In: 2017 XXVI International Conference on Information, Communication and Automation Technologies (ICAT), Sarajevo, Bosnia and Herzegovina, pp. 1–6 (2017)
11. Dobkin, B.H.: A rehabilitation-internet-of-things in the home to augment motor skills and exercise training. Neurorehabil. Neural Repair **31**(3), 217–227 (2017)
12. Maksimović, M., Vujović, V.: Internet of things based e-health systems: ideas, expectations and concerns. In: Handbook of Large-Scale Distributed Computing in Smart Healthcare, pp. 241–280. Springer, Cham (2017)

13. Sevcenco, A.M., Li, K.F.: Motion tracking and learning in telerehabilitation applications. In: 2012 Seventh International Conference on Broadband, Wireless Computing, Communication and Applications (BWCCA), pp. 420–427. IEEE, November 2012
14. Moffet, H., Tousignant, M., Nadeau, S., Mérette, C., Boissy, P., Corriveau, H., Marquis, F., Cabana, F., Belzile, É.L., Ranger, P., Dimentberg, R.: Patient satisfaction with in-home telerehabilitation after total knee arthroplasty: results from a randomized controlled trial. Telemed. e-Health **23**(2), 80–87 (2017)
15. Liu, L., Miguel Cruz, A., Rios Rincon, A., Buttar, V., Ranson, Q., Goertzen, D.: What factors determine therapists' acceptance of new technologies for rehabilitation–a study using the Unified Theory of Acceptance and Use of Technology (UTAUT). Disabil. Rehabil. **37**(5), 447–455 (2015)
16. Agostini, M., Moja, L., Banzi, R., Pistotti, V., Tonin, P., Venneri, A., Turolla, A.: Telerehabilitation and recovery of motor function: a systematic review and meta-analysis. J. Telemed. Telecare **21**(4), 202–213 (2015)
17. Wang, Q., Markopoulos, P., Yu, B., Chen, W., Timmermans, A.: Interactive wearable systems for upper body rehabilitation: a systematic review. J. Neuroeng. Rehabil. **14**(1), 20 (2017)
18. Han, S.L., Xie, M.J., Chien, C.C., Cheng, Y.C., Tsao, C.W.: Using MEMS-based inertial sensor with ankle foot orthosis for telerehabilitation and its clinical evaluation in brain injuries and total knee replacement patients. Microsyst. Technol. **22**(3), 625–634 (2016)
19. Meijer, H.A., Graafland, M., Goslings, J.C., Schijven, M.P.: A systematic review on the effect of serious games and wearable technology used in rehabilitation of patients with traumatic bone and soft tissue injuries. Archives of physical medicine and rehabilitation (2017)
20. Goncu-Berk, G., Topcuoglu, N.: A healthcare wearable for chronic pain management. Design of a smart glove for rheumatoid arthritis. Des. J. **20**(Suppl. 1), S1978–S1988 (2017)
21. Tran, V., Lam, M.K., Amon, K.L., Brunner, M., Hines, M., Penman, M., Lowe, R., Togher, L.: Interdisciplinary eHealth for the care of people living with traumatic brain injury: a systematic review. Brain Injury **31**(13–14), 1701–1710 (2017)
22. Dobkin, B.H.: Rehabilitation strategies for restorative approaches after stroke and neurotrauma. In: Translational Neuroscience, pp. 539–553. Springer, Boston (2016)
23. https://github.com/inovatink/ws-hardware
24. Lim, C.K., Chen, I.M., Luo, Z., Yeo, S.H.: A low cost wearable wireless sensing system for upper limb home rehabilitation. In: 2010 IEEE Conference on Robotics Automation and Mechatronics (RAM), pp. 1–8. IEEE, June 2010
25. Roggen, D., Pouryazdan, A., Ciliberto, M., BlueSense: designing an extensible platform for wearable motion sensing, sensor research and IoT applications. In: International Conference on Embedded Wireless Systems and Networks (2017)
26. González-Villanueva, L., Cagnoni, S., Ascari, L.: Design of a wearable sensing system for human motion monitoring in physical rehabilitation. Sensors **13**(6), 7735–7755 (2013)
27. Macedo, P., Afonso, J.A., Alexandre Rocha, L., Simões, R.: A telerehabilitation system based on wireless motion capture sensors. In: PhyCS (2014)
28. https://www.invensense.com/wp-content/uploads/2015/02/PS-MPU-9250A-01-v1.1.pdf
29. https://cdn-shop.adafruit.com/product-files/2772/atmel-42181-sam-d21_datasheet.pdf
30. http://ww1.microchip.com/downloads/en/DeviceDoc/50002466B.pdf
31. https://github.com/inovatink/ws-hardware/blob/master/ws_v2_schematic.pdf
32. https://www.mikroe.com/click
33. https://www.github.com/inovatink/ws-hardware/blob/master/ws_v2_external_pinout.png
34. https://github.com/inovatink/ws-hardware/tree/master/3D%20Drawings
35. https://www.arduino.cc/en/main/software

36. https://github.com/inovatink/ws-firmware-v1
37. https://github.com/getsenic/gatt-python
38. https://github.com/inovatink/ws-rpi3-hub
39. https://www.summitmedicalgroup.com/library/adult_health/sma_radial_head_fracture_exercises/. Accessed 17 Apr 2018

Information and Communication Technologies

Smart Home System - Remote Monitoring and Control Using Mobile Phone

Merisa Škrgić[1], Una Drakulić[2], and Edin Mujčić[1(✉)]

[1] Faculty of Technical Engineering Bihac, University of Bihac, Bihać, Bosnia and Herzegovina
merisa.skrgic@gmail.com, edin.mujcic@unbi.ba
[2] Faculty of Electrical Engineering Tuzla, University of Tuzla, Tuzla, Bosnia and Herzegovina
una.drakulic@fet.ba

Abstract. Today, when technology is getting more and more progressive, there is growing demand for automation. In automation systems we can include home automation systems. Within home automation systems the most commonly embedded systems are: security system, temperature control system, lighting control system and etc. Control of these systems are most commonly realized using wireless communication. In this paper is described smart home system controlled by the ATmega2560 microcontroller. Smart home system beside microcontroller, also has a SIM900 GSM module that enables remote control of a smart home using SMS messages. The smart home security system consists of: security system based on IR sensors located on windows and security system based on password protection located at the front door. In the event of a security breach, an alarm with visual and audible warning signals is activated and an SMS message about the security breach is sent to the user. The lighting control system is controlled by SMS message. The user sends SMS messages that turn on/off lights in individual rooms of the smart home. The user also has the ability to turn on/off individual lighting groups in the smart home. The motion sensor is used to detect movement outside the smart home. When the motion sensor is activated, outdoor lighting turn on. The use of systems based on microcontrollers and GSM, according to research, represents a serious competition to earlier systems in terms of price, reliability and ease of use.

Keywords: Smart home · ATmega2560 · GSM module · Security system
Lighting control system

1 Introduction

Today, systems of smart or intelligent home are increasingly used, so more research on this topic has emerged. A smart home has a built-in main control system. This main system is able to integrate several different subsystems: heating system, climate system, security system, lighting system, electronic devices control system, alarm system, etc. Term 'smart home' means something interesting and it is wise decision to integrate a smart home system. One of the huge advantages of these types of systems is that they significantly reduce energy consumption.

S. Avdaković (Ed.): IAT 2018, LNNS 59, pp. 409–419, 2019.
https://doi.org/10.1007/978-3-030-02574-8_33

Smart home systems are most usually realized using microprocessors and microcontrollers [1–3].

The occurence of microcontrollers and microprocessors are regarded as one of the greatest technical achievements that characterized twentieth century. Main difference between microcontrollers and microprocessors is that they are the first optimized for speed and performance with computer programs, while microcontrollers are optimized towards integration of a large number of circuits real-time control, mass production, low cost and low power consumption. Microcontrollers are also more resistant on variation of voltage, temperature, humidity, vibration, etc. Huge advantage is reflected in the fact that can be programmed, beside Assembler, and in high-level programming languages: C, Pascal, Basic, etc. This increases number of users who can write programs and thus also apply. They are used in a wide variety of modern devices such as: robots, telecommunication devices, satellites, cars, measuring instruments, mobile phones, cameras, etc. Also they are widely used in many home devices such as washing machines, microwave ovens, breadmakers, etc. Today on market there are few major manufacturers microcontrollers which in its production program have different microcontroller familys. The most popular of them are Intel, Motorola, Atmel and Microchip [4–6].In this paper is used microcontroller ATMega2560 which has very good characteristics.

Mobile phone is often used for remote monitoring and control of smart home. Most of mobile phones are using Wi-Fi, Bluetooth or SMS message. Wi-Fi requires smart phone and internet connection, while Bluetooth has limit distance [7]. Because of these requirements, we have used SMS message for remote monitoring and control of smart home system. In this case is only required GSM network [8, 9].

In this paper is described smart system for monitoring and control of smart home using mobile phone. We have used SIM900A GSM/GPRS module that has possibility to send/receive SMS messages and calls [10]. The same way it can be used in GPRS mode for connecting on internet and to run different applications for collecting and control of data.

Smart home system consist of four smart subsystems. The first smart system is motion detection system used to detect motion when approaching the house. The second smart system is remote control system that is used for indoor and outdoor lighting control. The third smart system is system based on password protection located at the front door. After house owner enter the correct password, front door automatically are opening. The fourth smart system is security system based on IR sensors located on windows. In the event of a house security breach, an alarm with visual and audible warning signals is activated and SMS message about the security breach is sent to the house owner.

All these smart systems have the option of upgrading and the main system has ability to include new smart subsystems.

2 Design and Implementation of Smart Home System for Remote Monitoring and Control Using Mobile Phone

In this part of paper is described design and implementation of smart home system for remote monitoring and control using mobile phone.

For control smart home system is used microcontroller ATMega2560. Because easier implementation and connecting with other components and modules we used Arduino board with microcontroller ATMega2560. For programming microcontroller ATMega2560Arduino IDE free software is used.

For easier implementation of the projected system, the model of the house (see Fig. 1) is made. House model is made from wood except windows and front door which are made from glass. For outdoor lighting is used LED stripes and for indoor lighting is used LED diodes.

Fig. 1. Model of the house

The main part of this system is the Arduino board with ATMega2560 microcontroller on which are connected other components and modules: GSM/GPRS module, LEDs, LED strips, IR transmitters, IR receivers, LCD display, buzzer, keyboard and other components necessary for the correct work of the system.

On Fig. 2 is shown functional block diagram of smart home system. On this diagram power system is not included.

This project is consist of several different systems which are:

1. Motion detection system
2. Remote control system
3. Security system based on password protection
4. Security system based on IR sensors.

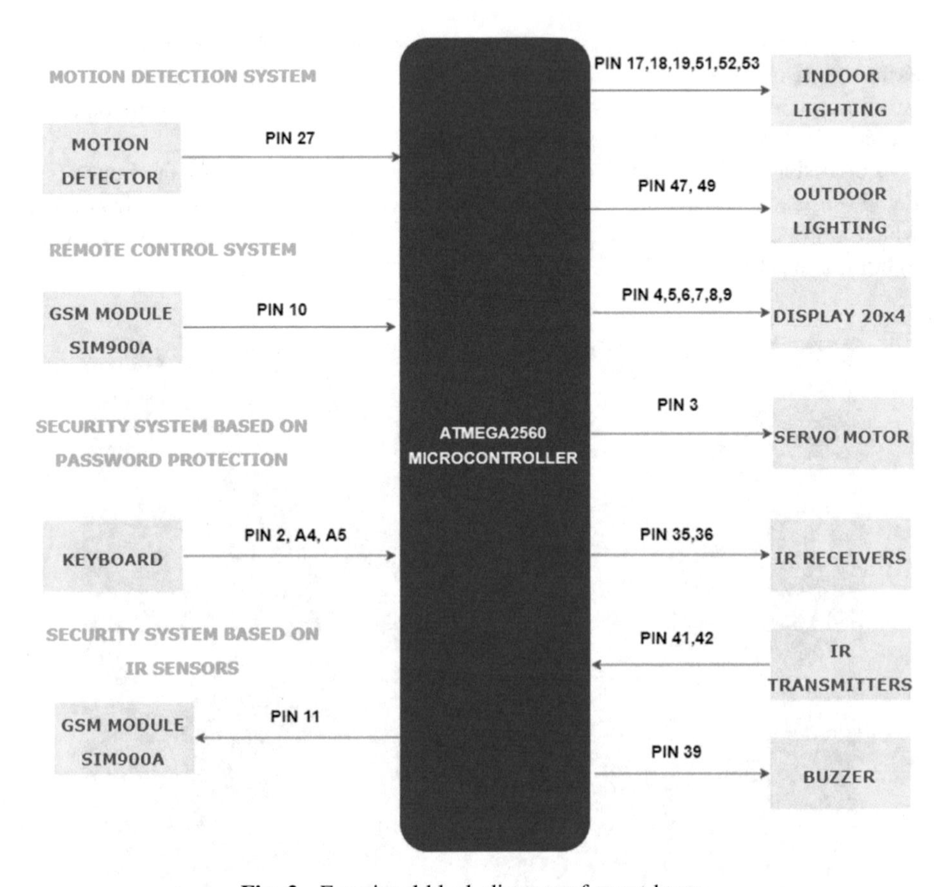

Fig. 2. Functional block diagram of smart home

2.1 Motion Detection System

The motion detection system is consists of motion sensor, ATMega2560 microcontroller and outdoor lighting. That system is shown in Fig. 3.

Fig. 3. Functional block diagram of motion detection system

The working principle of motion detection system is: HC-SR501 PIR (Passive Infrared) sensor detects movement in immediate area [11]. Output signal from PIR sensor is forwarded to input pin 27 of Arduino board. ATMega2560 microcontroller is processing received information and acts on output pins 47 and 49. On these output pins are connected LED strips for outdoor lighting over power amplifier. The outdoor lighting can be controlled using and SMS message. The problem of dual control of outdoor lighting is solved in program.

2.2 Remote Control System for Lighting

Remote control system for lighting turn on and turn off indoor and outdoor lights using SMS messages. That system is shown in Fig. 4.

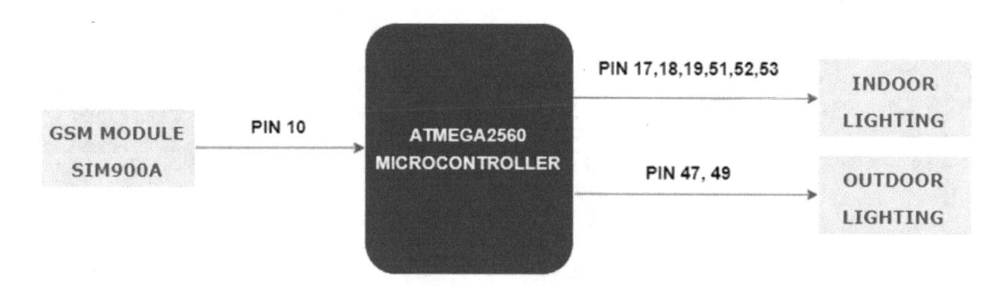

Fig. 4. Functional block diagram of remote control system for lighting

Using SIM900A GSM module enables remote communication with mobile phone. House owner sends SMS message with specified instruction. SIM900A GSM module receives SMS message and forward it to microcontroller. The ATMega2560 microcontroller is checking received information, and if it is correct, makes the decision to turn on/off the desired light.

2.3 Security System Based on Password Protection

Security system based on password protection is located at the front door and consist of keyboard, LCD display 20 $\times$ 4, ATMega2560 microcontroller and servo motor. That system is shown in Fig. 5.

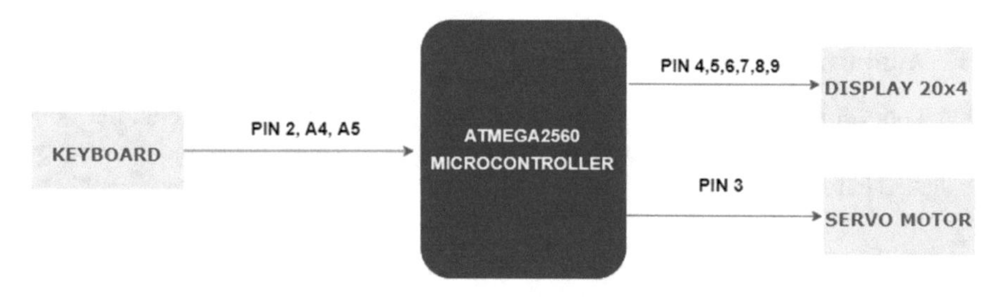

Fig. 5. Functional block diagram of security system based on password protection

The password consists of a four-digit number entered using the keyboard located near the front door. The user enters the numbers which are displayed at the same time on the LCD display. If the password is correct on the LCD display appears message '*Dobrodosli !*' and servo motor is opening the door. Otherwise, if password is incorrect on LCD display appears message '*Lozinka netacna !*'. After 10 s the ATMega2560 microcontroller is reset and on LCD display appears message '*Unesite lozinku* !'.

2.4 Security System Based on IR Sensors

Security system based on IR sensors consists of SIM99A GSM/GPRS module, ATMega2560 microcontroller, two IR transmitters and two IR receivers which are located on the sides of the windows.

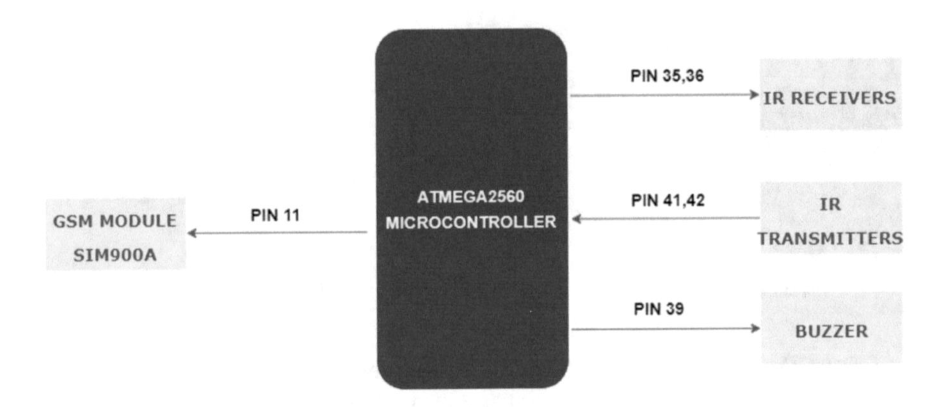

Fig. 6. Functional block diagram of security system based on IR sensors

On Fig. 6 is shown functional block diagram of security system based on IR sensors. In the case of house security breach the signal from IR transmitter to the IR receiver is interrupted. Based on that information the ATMega2560 microcontroller process and forwards that information to SIM900 GSM/GPRS module. This module sends SMS message '*Neko je u vasoj kuci* !' to house owner, i.e. information about house security breach. Also in that case, the ATMega2560 microcontroller activate buzzer sound.

2.5 Appearance of the Final System

After connecting all the components, which are located in basement of the house model, smart home system for remote monitoring and control using mobile phone is finished. That is shown in Fig. 7.

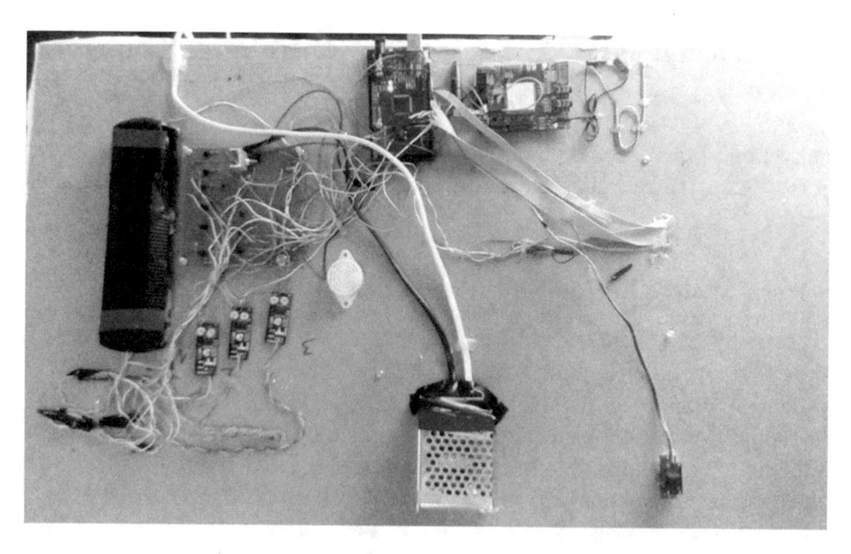

Fig. 7. Appearance of finished home control system

On Fig. 8 is shown model of the house for experimental analysis of projected control system.

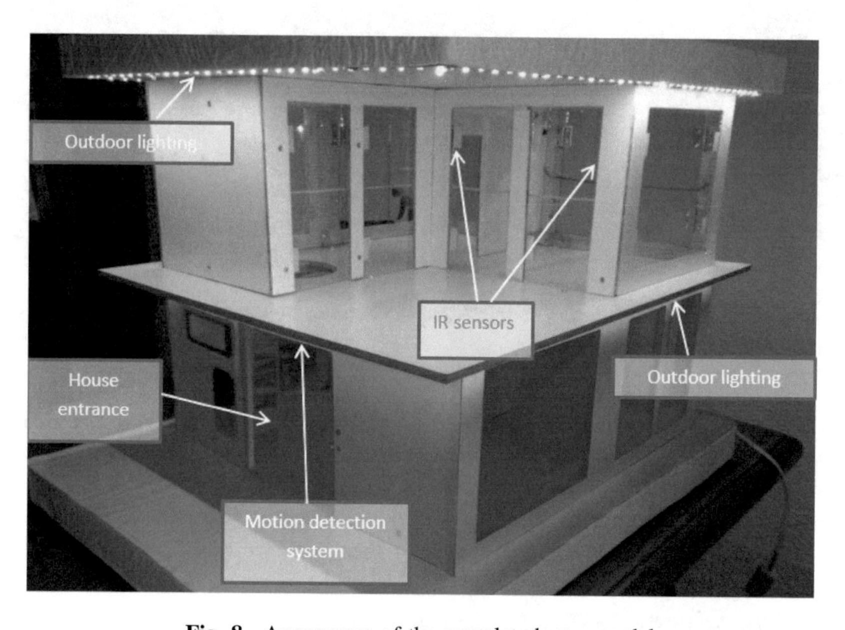

Fig. 8. Appearance of the complete house model

In the continuation of this paper, an experimental work analysis of the projected system is perform.

3 Experimental Work Analysis of Projected Home Control System

In this section of the paper is described experimental work analysis of projected home control system. The experimental work analysis is performed for each system individually.

3.1 Work Analysis for Motion Detection System

This system detects the movement from the front of the house. When motion detection sensor activate, outdoor lighting will turn on, i.e. white and blue LED strips will turn on. Outdoor lighting can also be turn on/off using SMS messages. After the experimental analyses of the projected system, it can be concluded that:

If the lighting turn on using the detection motion system, lighting will stay turn on 5 s (this time is set in the code), and then it will turn off. In the other case, if lighting turn on using SMS message, then lighting will stay turn on until it turn off using SMS message. In this case the motion detection system is not active.

Fig. 9. (a) The motion sensor is not active and the outdoor lighting is off (b) Motion sensor is active and outdoor lighting is on

On Fig. 9a is shown case when the lighting is off, i.e. when is not detected movement nearby. On Fig. 9b shows case when lighting is on, i.e. when is detected movement in front of the house.

Based on the experimental analysis for motion detection system we can conclude that projected system is working properly.

3.2 Work Analysis for Remote Control System

The remote control system is used to turn on and off indoor and outdoor lighting using SMS messages. This way of lighting control is quite interesting, practical and very useful. For the experimental analysis of this system we will use standard commands programmatically defined to turn on/off individual lighting in the house. In our case, with the home owner phone we will send an SMS message *'#SobaxP'* that is used to turn on lighting on the ground floor. After sending an SMS message, lighting on the ground floor turned on. Turning on ground floor lighting using the SMS message is shown in Fig. 10.

Fig. 10. Turning on ground floor lighting using the SMS message

Based on the experimental analysis for remote control system we can conclude that projected system is working properly.

3.3 Experimental Analysis for Security System Based on Password Protection

In this subsection is performed experimental analysis for security system based on password protection. We will begin experiment analysis of this system by entering an incorrect password.

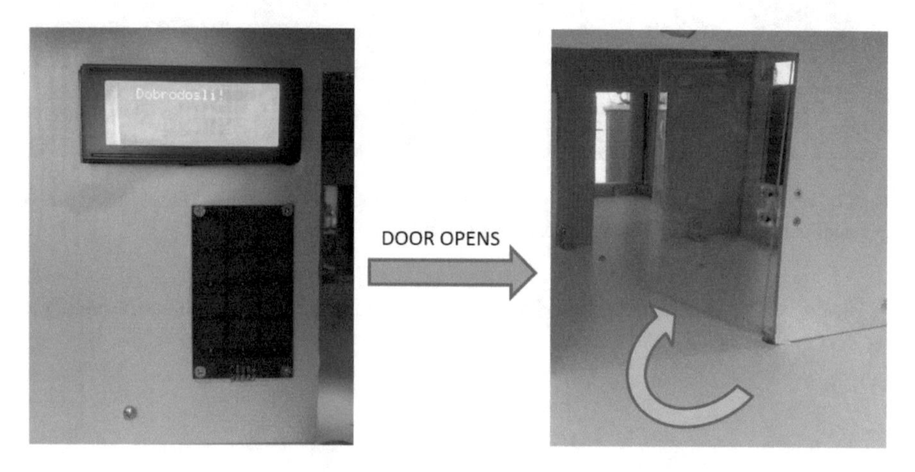

Fig. 11. Entering the correct password and opening the front door

If an incorrect password is entered on the LCD screen, the message '*Lozinka je netacna!*' appears. After 10 s the ATMega2560 microcontroller resets and on LCD display appears message '*Unesite lozinku !*'. If the correct password is entered, the message '*Dobrodosli!*' appears on the LCD screen. After that, the servo motor is opening the front door. This case is shown in Fig. 11.

Based on the experimental analysis, it can be concluded that the security system based on password protection works properly.

3.4 Experimental Analysis for Security System Based on IR Sensors

Security system based on IR sensors consists of two IR transmitters and two IR receivers located from the side of the window. The IR transmitter, i.e. the IR diode, continuously sends the signal to IR transmitter. In case of interruption of this signal (unauthorized input through the window) an alarm is triggered.

For an experimental analysis of this system, we will simulate an unauthorized entry into the house. That is shown in Fig. 12(a).

In our case, obstacle is the hand that interrupts the transmission of the IR signal. A few seconds after an obstacle occurs, a sound alarm is activated. Shortly after, the home owner receives SMS message. The content of the SMS message is shown in Fig. 12(b).

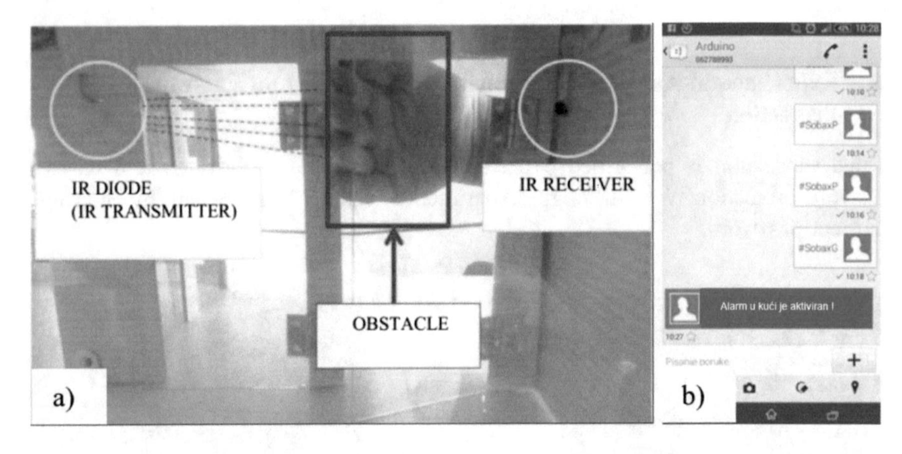

Fig. 12. (a) Simulation of the obstacle between the IR transmitter and IR receiver (b) Received SMS message about house security breach

Based on the experimental analysis, it can be determined that the functionality of this system is correct.

4 Conclusion

In this paper is described system for remote control and monitoring of the smart home using mobile phones. For realization, the Arduino board with industrial ATMega2560 microcontroller, that has very good characteristics (big speed processing, safety, large number of analog and digital pins etc.), has been used. Using this microcontroller, it is possible to achieve good system control and connect a big number of external components. For communication with user the SIM900 GSM GPRS module is used. The implemented systems consist of: motion detection system, remote control system, security system based on password protection and security system based on IR sensors. For each of these systems an individual experimental analysis has been performed, on which can be concluded that the projected system is working properly. We can also conclude that the projected system can, with minimal changes, be applied to the real system. Possible future upgrades of the projected system are: remote control of heating and cooling, remote control of roller shutter, automatic ventilation, remote turn on/off of electronic devices etc.

References

1. Nadira, Đ., Mujčić, E.: Primjena ZigBee tehnologija u razvoju pametnih kuća. In: 9th International Scientific Conference on Production Engineering, RIM (2013)
2. Clinckx, N., Baffalio, Y., Duplan, A., Ferrand, B.: Smart Home: Hope or Hype. Greenwich Consulting, International Management Consulting for Telecommunications, Media, Utilities and Insurance (2013)
3. Huang, Y., Huang, Ch., Huang, K., Liu, K.: Energy optimization approaches for smart home applications, energy optimization approaches for smart home applications, Zakopane, Poljska, str. pp. 344–349 (2012)
4. Mujčić, E., Drakulić, U., Škrgić, M.: Advertising LED system using PIC18F4550 microcontroller and LED lighting. In: Internacionalni simpozij o inovativnim i interdisciplinarnim aplikacijama savremenih tehnologija (IAT)(BHAAAS), Neum (2016)
5. Mujčić, E., Drakulić, U.: Upotreba android operativnog sistema i mikokontrolera ATMega2560 za upravljanje 3D-LED matričnim displejem. In: 11th International Scientific Conference on Production Engineering, RIM (2017)
6. Technical Specification Microcontroller ATMega 2560. http://www.mantech.co.za/datasheets/products/A000047.pdf. Accessed 21 Apr 2018
7. Lodhi, D.K., Vats, P., Varun, A., Solanki, P., Gupta, R., Padey, M.K., Butola, R.: Smart electronic wheelchail using arduino and bluetooth module. IJCSMC **5**, 433–438 (2016). Monthly Journal of Computer Science and Information Technology, India
8. Miguelez, C.G.: GSM operation onboard aircraft. European Telecommunications Standards Institute, ETSI White Paper No. 4, France (2007)
9. Zamzami, A.A., Pramana, J., Putra, E., Zainudin, A.: Reliability analysis of GSM network using software defined radio-based system. In: International Electronic Symposium IES, At Surabaya (2015)
10. Li, X., Yuan, Q., Wu, W.: Implementation of GSM SMS remote control system based on FPGA. In: Information Science and Engineering ICISE 2010, China. IEEE Xplore (2011)
11. Urfaligu, O., Soyer, E.B., Toreyin, B.U.: PIR-sensor based human motion event classification. In: Signal Processing, Communication and Applications Conference, Turkey, 2008. IEEE Xplore (2008)

Development of Educational Karate Games with the Help of Scenes and Characters from the Popular Cartoon Series

Jasna Hamzabegović[1(✉)] and Mirza Koljić[2]

[1] Faculty of Technical Engineering, University of Bihac, Bihac, Bosnia and Herzegovina
jasna.hamzabegovic@unbi.ba

[2] Pedagogical Faculty, University of Bihac, Bihac, Bosnia and Herzegovina
mirzak@live.com

Abstract. For today's younger populations, computer games are an integral part of life, and the virtual world of their heroes is a natural environment. The use of a video game for educational purposes provides interesting ways of presenting facts, acquiring knowledge and the ability to apply knowledge in the virtual world, improving the process of learning itself and bringing students closer to learning goals. In addition to computer games, popular cartoons such as: Sponge Bob, Pokemon, Dragon Ball and others, have a significant impact in the process of growing up of young people.

Karate as a sport is increasingly popular in the world and the interest in this martial art is very high. The popular animated series that children and teenagers are watching are certainly contributing to this. One of them is Dragon Ball Z, which some people judge, but others point to its positive effects.

One such positive effect is shown in this paper. It is shown how characters and scenes from the animated series can be used in the development of educational computer games to learn very complex names of body parts, poses and actions of karate. The prototype of this educational computer game was developed in the popular programming technology Windows Presentation Foundation. This paper presents how the characters and scenes from the animated series are used in the WPF animation system.

Keywords: Video games · Animated series · Dragon ball Z · Karate
Programming technology windows presentation foundation
Prototype of educational computer game for karate

1 Introduction

The form of teaching and learning through a game goes back to the distant past. Games are fun, motivational, and creative. Regardless of whether it's a game outside, a social game or a game on a computer, the game teaches us new facts, develops skills, improves the sociological aspect and forms a way of thinking. The game inspires imagination and creativity. Today's younger population plays computer games and experiences the virtual life and the world of its heroes on a daily basis.

S. Avdaković (Ed.): IAT 2018, LNNS 59, pp. 420–429, 2019.
https://doi.org/10.1007/978-3-030-02574-8_34

The use of video games for educational purposes offers interesting ways to present knowledge and the ability to apply knowledge in the virtual world by facilitating the process of learning and bringing students closer to the goals of the learning itself.

Games have a high presence in informal and formal learning segments. Unfortunately, in regular education, games are still often seen as a frivolous activity and the potential of learning through games remains undiscovered [1].

The use of educational applications is very convenient in teaching the rules and techniques of various sports disciplines. One such discipline, which is increasingly popular with children and teenagers, is the martial art of karate. This paper presents the prototype of an educational application for teaching basic karate stances. This application was created by combining WPF programming technology and scenes from the popular cartoon Dragon Ball.

2 Computer Games and Animated Series

The game represents the basic activity of a child through which it is being tought about the outside world. Throughout the game, it learns about social relations as well as rules of behavior. The global advancement of technology has led to a flood of computer games. The primary purpose of computer games is entertainment for players, but also earnings for game creators. Adult, computer or video games are generally called games. This stems from the fact that they are perceived exclusively as children's things, something that adults are not interested in, something that is simple and childish [2].

But is that really true?

2.1 The Role of Computer Games in Contemporary Societygames

Modern games are neither simple nor childish either by their technological capabilities or by the audience that prefers them. We are witnesses to the fact that they are becoming increasingly complex and require more and better players' skills and knowledge. It is therefore wrong to call them "games". Playing on a computer was in its beginning a game reserved for adolescents. Today, the target audience of game makers is not just children and teenagers, but adults, who according to various research spend even more time playing computer games than children.

Nowadays, the industry of computer/video game development is very popular in the world. It is bigger than the film industry. It is estimated that Star Wars: The Old Republic, an online game published in 2011, cost between $ 150 and $ 200 million. The Grand Theft Auto V, which came out two years later, allegedly cost $ 265 million. The most expensive video game ever made is Destiny (Destiny) whose action is set to 700 years in the future and which deals with a war between humans and aliens. The development of this game cost $ 500 million. This game is also more expensive than the Most Expensive Hollywood Blockbuster Avatar, which cost $ 75 million [3].

In addition to games intended for entertainment, games are also used for training and education, but also for various competitions and measurements of forces, both physical, mental, and creative, for the purpose of organizing numerous world competitions.

2.2 Animated Film

Animated films create an illusion of motion by a series of thumbnails, each one slightly different. The production of short animated films became a real industry at the beginning of the last century, with them being produced to be featured in cinemas.

This form of entertainment was present in the process of growing up of almost every person. Animated films are there to fill our free time, to entertain us, and to teach us something new. Nowadays, the number of animated films is increasing with every day. New characters are born, and children choose which characters they want to watch. It is not unusual for children to watch current animated films, not because they are attracted to them, but to fit in. Because the vast majority of their peers are watching them, they can then participate in conversations and discussions during a break between classes or during extracurricular gatherings. There are, however, those kids who identify themselves with a character from a favorite animated series and thus become more confident, brave, wiser.

Many children often, if not everyday, place themselves in animated films and become heroes, princesses, warriors or singers for a moment [4].

Today's most popular animated films are: Ben 10, Winx, Bratz, Pokemon, Yu-Gi-Oh, Dragon Ball Z and others. Most pedagogues condemn these films as they strive for battles and destruction. Of course, in these animated films, there may be some positive messages like helping others and the importance of friendship [5].

2.3 Dragon Ball Z

Definitely one of the most popular, but also one of the most condemned animated series among children is surely the Dragon Ball Z. The boys are fascinated by the characters, their strength and invicibility. In their fantasies they experience individual scenes from the series that make them feel safer in themselves, learn how to establish control, but also how to defend themselves. Although many pedagogues challenge and attack it, this series (if dosed to a population of a certain age) can contribute to greater self-confidence, better control, and self-defense. This animated series, which has over 600 episodes, has prompted many boys and girls to enroll in karate courses.

3 Martial Arts

Fighting skill is a system of traditions and exercises whose original purpose was the preparation of the body and mind for the battle. In its modern form it is used to maintain physical readiness, meditation, and build a person. The martial art literally means the art of war, but is usually referred to as the "art of fighting" and thus represents a system of codified practice and the tradition of fighting. All martial arts have similar goals: physically defeating another person or defending themselves or others against physical threats. Martial arts are both art and science. Many world cultures have historically created their systems for training and transferring martial arts, but currently the most popular and widespread martial arts systems are those originating from Asian countries, primarily Korea, Japan, China, Thailand and the Philippines [6].

Although martial arts are usually considered as a system of physical exercises whose martial movements resemble shadow combat, they are of particular importance in the mystical or spiritual dimension. Some examples of Asian martial arts are: aikido, jiujitsu, karate, ninjutsu, chuan fa, kung fu, Tai Chi Chuan, kendo, taekwondo, judo.

3.1 Karate

Karate is a martial art of the 16th century that is being practiced today as a sport. In the literal translation from the Japanese language means KARA - empty (naked), TE - hand (hand), DO - way, that is, how to defend from the attacker with bare hands (空手道: からてどう - "the way of empty hands").

Karate is a martial art that uses all parts of the body in combat. In the basic technique, individual strokes, blocks, stances, etc. are learned. Kate is a special set of accurately determined movements (strokes, stances and blocks). Fighting (Kumite) can be free or arranged by agreeing to knock out the blow, and by blocking that blow.

Techniques that are processed in karate school are the basis on which more complex elements are built in other stages of training. There are more than 150 different strokes, blocks and stances in the karate sport, each of which has a unique name, and is performed in its own way. In addition, there are various techniques for sweeping, throwing opponents, clutches and jumps, or techniques that manifests themselves from the air. All these names would be easily and rapidly mastered through a fun educational application whose prototype was developed for the needs of this paper.

4 Windows Presentation Foundation

Windows Presentation Foundation (WPF) is a graphical subsystem for Windows. WPF is designed for .NET Framework. When it comes to the .NET framework, it refers to a system within the operating system, developed by Microsoft developers so that independent developers can easily develop their programs. By doing so, the .NET Framework is a package of the final program code required by application programmers for common programming issues, including a virtual machine that runs executing programs written for the .NET Framework.

The .NET Framework software framework supports several programming languages that provide language interoperability, so each program language can use a code written in another programming language. For the needs of an educational interactive application for karate, whose prototype we developed, the C # programming language was used. The .NET Framework class library provides user interface, data access, cryptography, web application development, numerical algorithms, and network communications. When creating our application, we combined the .NET Framework classes with our own code.

4.1 WPF Animation System

Animations are an essential part of the WPF model. Timers aren't needed to make them active. Instead, they are created declaratively, configured using one of several classes and by mouse movements they are placed in a drive that is simulated or downloaded without writing a program code. Animations are easy to integrate into ordinary WPF windows and pages. An animated button or figure that is set or simulated to move through the active window can be stylized, can get a focus, or can be clicked on it, which triggers a typical event processing code.

Often, animation is considered a series of frames. To perform an animation of this type, the frames are displayed one by one. Basically, WPF animation is a simple way of modifying values at a certain time interval. It can't add or remove an existing element in the created animation, but it can manipulated the time interval in which certain elements appear, and therefore this "shortcoming" can be easily circumvented. A wide range of animated effects can be created using the common features that each element supports.

In general, there are two types of animations - those that gradually change the value of an attribute from the initial to the last value, and those whose property value changes sharply from one value to another. Examples of the first category which uses linear interpolation to smoothly change the values are DoubleAnimation and ColorAnimation, and the animations of this category are also used in the prototype of an educational application for karate [7].

5 Prototype of an Educational Computer Game for Learning Basic Movements in Karate (Karateka)

Karateka is designed as an educational computer game with a variety of animations that simulate the performance of techniques in karate. Each animation is accompanied by an explanation and a name, and as such visualizes and facilitates the learning of karate techniques. The advantage of Karateka is that it puts the user in a position to visualize techniques, and thereby adopt and implement them.

There are a large number of video tutorials for learning karate on the Internet. Karateka distinguishes itself from them by being created as an interactive educational game in which a user can actively take on the role of a fighter, and with the help of a mouse click, he can activate the execution of a particular technique and manage the created character.

5.1 Karateka GUI

The game itself has a rich graphical user interface. It was created in the form of a karate school where, in the temple of martial arts, you can choose several different options in which the trainers are different characters. One character teaches users the names and basics of karate, the other explains the execution of certain techniques, simulates the fight and urges the user to interactively participate, while the third one checks the acquired knowledge through various questions in order to determine to what extent the user has overcome the presented content. Figure 1 shows the initial window of this educational game.

Fig. 1. The initial window of the educational computer game KARATEKA

After entering the temple of martial arts, the user is given the opportunity to select the characters who talk to him and clarify their role in teaching karate (Fig. 2).

Fig. 2. A window in which the user is presented with the choice of a guide

By selecting the character that guides the user through the simulation of the fight and performance of basic movements and techniques, a window in Fig. 3 is obtained.

The active window at this time offers the possibility of simulating hand and foot blows, which are activated by the push of the button, and the selected character

Fig. 3. The appearance of the window in which the fight is simulated

performs the technique on the other fighter, with the user having the ability to play and manage the fighter.

By selecting a central character at the entrance to the martial arts temple, the user gets the opportunity to learn the name of body parts in Japanese, and through animation he visualizes the techniques of blows and blocks and adopts their names. (Fig. 4).

Fig. 4. Learning body parts

By clicking on any part of the body, animations appear and explain the names of the corresponding body parts in Japanese.

As Karateka is an educational game aimed at teaching and mastering the skills of karate, the essential part of the game is the ability to check knowledge through questions in the form of an interactive quiz (Fig. 5).

Fig. 5. Testing knowledge through the game

Questions are asked in the form of pictures. The user had previously adopted them through the animation by simply playing the game. Now in the quiz, in the given field the user should enter the names of the blows or body parts he learned. In addition to the above, there is also the ability to "drag and drop" and join certain offered names and randomly generated images in order to get the correct answer. The overall knowledge quiz process is carried out by the figure shown in Fig. 5 which gives a comment on the given answer, and it is possible to seek help in case of not knowing the correct answer.

5.2 Animation System in the Game Karateka

Karateka is mostly made by animations. The reason for this is the nature of the computer game itself, which aims to teach the user about the basics of karate, but also to give him the opportunity to learn through fun. Through the game karateka, a large number of different types of animations are presented, starting with constant communication with a character that plays the role of the guide to other characters who perform martial arts. The communication with the characters is achieved by alternating text messages that appear above the marked character, and give the impression that the character speaks, i.e. is talking with the user (Fig. 6).

Figure 7 shows an animation that simulates the fight by performing a particular technique. It is achieved through the alternating display of a frame in which each person has distinct characters.

Fig. 6. Guide talks through text messages

Fig. 7. Alternate activation of certain frames with different characters

6 Conclusion

Although we live in a time of high technology, the game is still the best instrument for the development of mental and psychomotor abilities because it guarantees fun. It develops creativity, ability to think and reason, logical conclusion, and solve obstacles in order to achieve a certain goal.

Computers, tablets and smartphones represent the natural environment of today's young population. Accordingly, the game moved to these devices. Today we witness the great popularity of the video games. In the process of growing up, besides computer games, animated series and films also occupy a significant place. Karate as a sport is very popular in the world. Certainly, popular animated series that children look forward to are also contributed to this. One of them is the Dragon Ball which some people judge, while others point to its positive effects.

One positive effect is shown in this paper. It is shown how characters and scenes from the animated series can be used in the development of an educational computer game to learn the very demanding names of parts of bodies, stances and actions of the martial art of karate.

Through this work we dealt with the use of modern programming technology in important segments of life and development of a young person such as: entertainment, sports and education. The result is Karateka, a prototype of an educational computer game that can certainly help the traditional learning and training of karate.

References

1. Pivec, M.: Igra i učenje: Potencijali učenja kroz igru. Edupoint časopis (2006). http://edupoint.carnet.hr/casopis/49/clanci/1.html. Accessed 10 Jan 2018
2. Klopke elektronske zabave. Pogled (2006). http://pregled.com/print_preview.php?id_nastavak=796&grupa=zdravlje. Accessed 18 Jan 2018
3. Destiny – a non-gamer's guide to 2014's biggest game. The Guardian (2014). https://www.theguardian.com/technology/2014/sep/08/destiny-non-gamers-guide-bungie-rpg. Accessed 18 Jan 2018
4. Jovanovac, I.: Utjecaj crtanih filmova na djecu predškolske dobi (2017). https://zir.nsk.hr/islandora/object/ufzg%3A329/datastream/PDF/view. Accessed 20 Jan 2018
5. Fegeš, K.: Manipulativni sadržaji i poruke u animiranim filmovima za djecu (2016). http://cultstud.ffri.hr/kultura-u-akciji/osvrti/558-manipulativni-sadrzaji-i-poruke-u-animiranim-filmovima-za-djecu. Accessed 18 Jan 2018
6. Horton, N.: Japanese Martial Arts. Summersdale Publishers Ltd. UK (2005). https://www.e-reading.club/bookreader.php/134614/Horton_-_Japanese_Martial_Arts.pdf. Accessed 18 Jan 2018
7. MacDonald, M.: Pro WPF in C# 2008: Windows Presentation Foundation with .NET 3.5, pp. 729–732. Apress. New York (2008). Accessed 20 Jan 2018

A Platform for Human-Machine Information Data Fusion

Migdat Hodžić(✉)

International University of Sarajevo, Sarajevo, Bosnia and Herzegovina
mhodzic@ius.edu.ba

Abstract. The overall goal in this concept paper is to present an innovative and rigorous mathematical methodology and an expert self learning data fusion and decision platform, which is scalable and effective for a variety of applications. This includes (i) Interface design that incorporates the understanding of how both machines and humans fuse soft and hard data and information, and (ii) Forming a shared perception and understanding of the environment between the human and the machine, which supports human decisions and reduces human soft and decision making errors. With this paper we continue our research on Uncertainty Balance Principle (UBP) which is at the core of our soft hard data fusion and decision making strategy. The proposed methodology can be employed in the context of Artificial Intelligence (AI) and Machine Learning (ML) applications, such as banking risk assessment, Block Chain peer to peer systems, ecological and climate modeling, social sciences, econometrics, as well as defense applications such as battle management.

Keywords: Human machine data fusion · Probabilistic data · Possibilistic data

1 Introduction

At the heart of any data fusion is a notion of uncertain data. According to IBM 2012 report [18]. "Today, the world's data contains increasing uncertainties that arise from such sources as ambiguities associated with social media, imprecise data from sensors, imperfect object recognition in video streams, model approximations, and others. Industry analysts believe that by 2015, 80% of the world's data will be uncertain". By 2018 these estimates have grown beyond 80%. In the modern world humans and machines both produce huge amounts of hard and soft and uncertain data, and they require common understanding and shared perception to make decisions based on these data, and in the process maximize benefits of the human-machine team. It is often common that machines integrate hard data from multiple sensors (sources, or hard statistics), at different time periods using various probabilistic methodologies to obtain meaningful information. This is considered hard data fusion. The main idea is to associate different pieces of data based on the likelihood that they represent the same or related events, people, and objects [1]. On the other hand, humans produce soft subjective data most of the time, which can not be modeled by probabilistic means. Also, humans make decisions based on all available data, both hard and soft. Hence in general data fusion process, one of the human's key roles is to act as possible soft data

S. Avdaković (Ed.): IAT 2018, LNNS 59, pp. 430–456, 2019.
https://doi.org/10.1007/978-3-030-02574-8_35

source as well as a decision maker, supporting manual, semi automated or automated reasoning techniques by using visual, audio, graphics, numerical and other pattern recognition and semantic reasoning [2]. In a view of this, a complete data fusion system is composed of computers (machines) which combine hard and soft data from multiple sources, plus an interface to present the combined and any additional information to a human operator who views, encodes, and interprets the data to make a decision. The machine fuses and represents the data in some optimal way enabling the human to make effective decisions. This process optimizes human-machine team and it should produce significant benefits for the human-machine team. Our methodology implements the following objectives:

(O1) Concept of an effective Human Operator - MachinE Team Graphical User Interface, HOME Team GUI, supporting a range of hard and soft data types stored in corresponding data bases, as well as a smart soft data learning feature to reduce Human Operator (HO) soft errors.
(O2) Hybrid data fusion based on a combined hard data and soft data (hence hybrid), with two filters, Hard Fusion Filter (HFF) and Soft Fusion Filter (SFF), combined for improved confidence levels for both data types. The HFF uses two kinds of hard data, the original sensor fused, plus human generated soft data which is Uncertainty Aligned, UA for short, to hard data. Similarly SFF uses two kinds of soft data, the original human produced soft data plus sensor data which is Uncertainty Aligned (UA) to soft data. The two data filters and UA rely on our concept of Uncertainty Balance Principle (UBP) and its inverse (UBP^{-1}) [13–15].
(O3) Decision process based on an idea of optimizing some Utility Function (UF) using several hybrid data distance measures, following HO soft input and fusion with the available hard data. The UF could be Local (for specific sensor data) or Global HOME Team decision, LD or GD respectively. Human soft input(s) in effect optimize the UF presented back to the human operator. After several iterations the HO makes some decision and the overall result is a joint Human-Machine decision process and improved HOME Team benefit. HOME Team GUI supports both (O2) and (O3) objectives. We assume that the hard data is obtained by some probabilistic (statistical) method, whereas soft data, uncertain and subjective, can be modeled by fuzzy (possibilistic) distributions. This is often done in literature [3–6, 12]. What is new in our approach is as follows:

(i) HOME Team GUI soft learning ability, via HO repeated soft inputs (over certain period of time, minutes, hours, days, weeks or longer), hence reducing HO subjective errors in the process.
(ii) Soft and hard data uncertainty alignment via UBP and UBP^{-1} without loss of data information, plus soft and hard data filters for hybrid data fusion.
(iii) Improved human decision process by HO soft input and viewing hard data and UF value, assisted by the machine which supplies soft templates, fuses the data, and calculates UF value for the HO ahead of soft inputs or some decisions.
(iv) Ever increasing smart soft and hard data bases suitable for additional mining and discovering new data correlations for HOME Team benefits continuous

improvement. We believe our approach offers significant flexibility as well as mathematical rigor and learning ability for a variety of real life applications.

2 Home Team GUI Concept

Figure 1 shows HOME Team GUI action flow chart, while Fig. 2 has a conceptual view of the GUI. Figure 3 shows structure of various data bases behind the GUI. In our proposed solution, HOME Team GUI can be implemented on a large or smaller personal screen with touch ability, with all functionality distributed in a user intuitive manner [7, 12]. The human can view the data, produce soft inputs, move icons around the screen, reconfigure, invoke data menus, align the data, update data bases, and make decisions. Our HOME Team GUI:

(i) Supports a variety of different classes and sources of hard data (Visual, Picture, Audio, Semantic, Geo, Graphical, Intelligence, Tracking). Various attributes of these data could be analyzed and incorporated into the GUI data bases.

(ii) Presents hard sensor data to HO practically obtained or simulated in a desired form. This data resides in Hard Data Base (HDB), and it will be simulated or updated with real life sensor, machine or statistical hard data. The HDB supports data types from (i) or even more types. The HO can review hard data ahead of any HO soft input and any HO decisions.

(iii) Presents pre-stored soft data templates to the HO. This data resides in Soft Data Base (SDBase), with an option for the HO to fine tune the templates (making a soft input) after interpreting the data. Over time this produces a learning effect for soft data, and reduces HO soft errors (as in training). The templates are based on preliminary knowledge about various data types in (i) as well as how HO handles and potentially fine tunes each type of soft template data. They are in the form of various fuzzy data distributions, such as triangular (TFNs), trapezoidal or some general convex functions.

(iv) Accepts fine tuned soft input from the HO. The original templates will be held unchanged (or they could be updated at a future time) and the fine tuned HO soft inputs are also stored for next fusion cycle and further fine tuning. Various menus will be available for this.

(v) Supplies the original data as well as UA data from both data bases to HFF and to SFF which perform hybrid data fusion, resulting in reduced data errors. UA data act as inputs to appropriate filters together with the original data. The recalculated (re-fused) data is presented back to the HO together with UF value, and HO can choose to make another soft input or make any decisions, and so on, resulting in an iterative learning process over time.

(vi) Allows HO to make a timely decision if ready and fused data supported. The decision is based on optimizing average UF value. The decision may be a partial one, for a particular data type in (i), for example related to exceptionally good or bad data, i.e. LD. Or it could be the final GD, if HO believes the overall data supports this action. This way HO acts as a hybrid computer and the decision made is based on a shared vision and perception between the machine and the human, hence maximizing HOME Team Benefits.

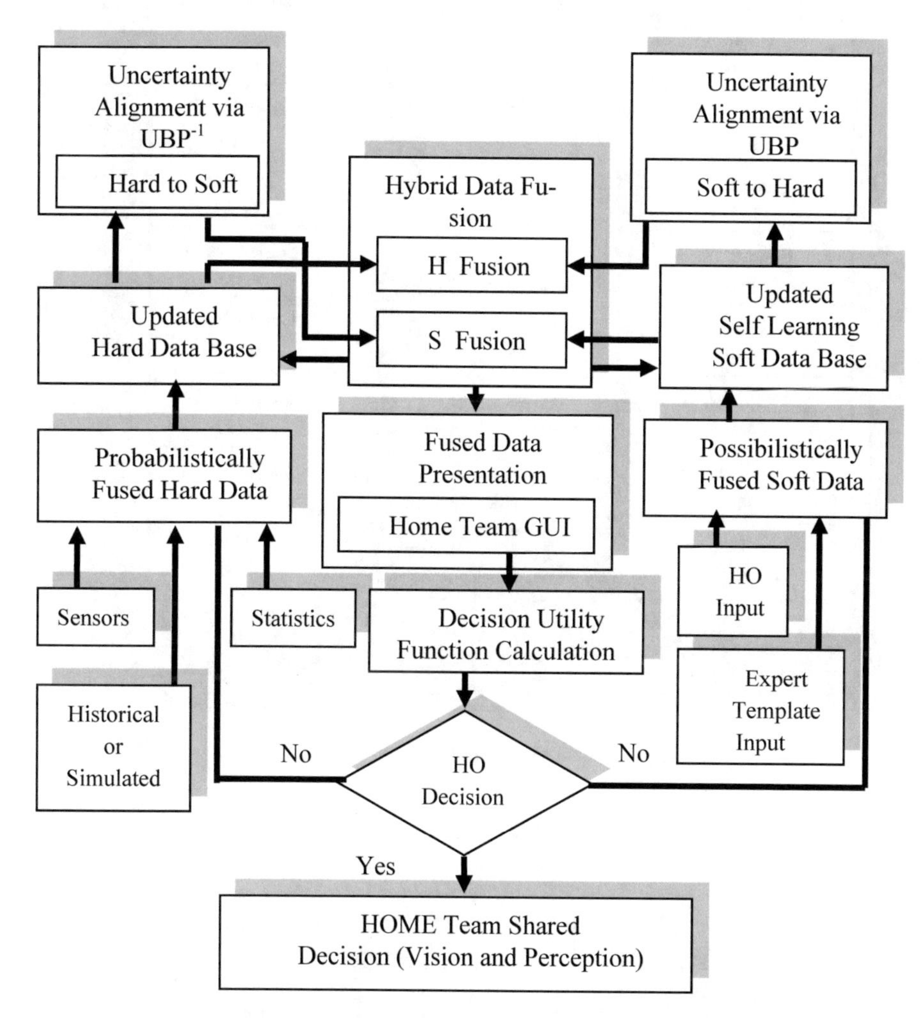

Fig. 1. HOME team GUI time and action flow chart

3 Uncertain Data

Source of data uncertainty maybe in the data origin, data precision, objectivity of subjectivity of them, timing, sensor or machine type producing the data, and many other reasons. In this paper we deal with two types of uncertain data, i.e. hard (machine, objective) and soft (human, subjective) which may come in various shapes and forms. Other definitions of what is meant by hard and soft data exist as well [19, 20].

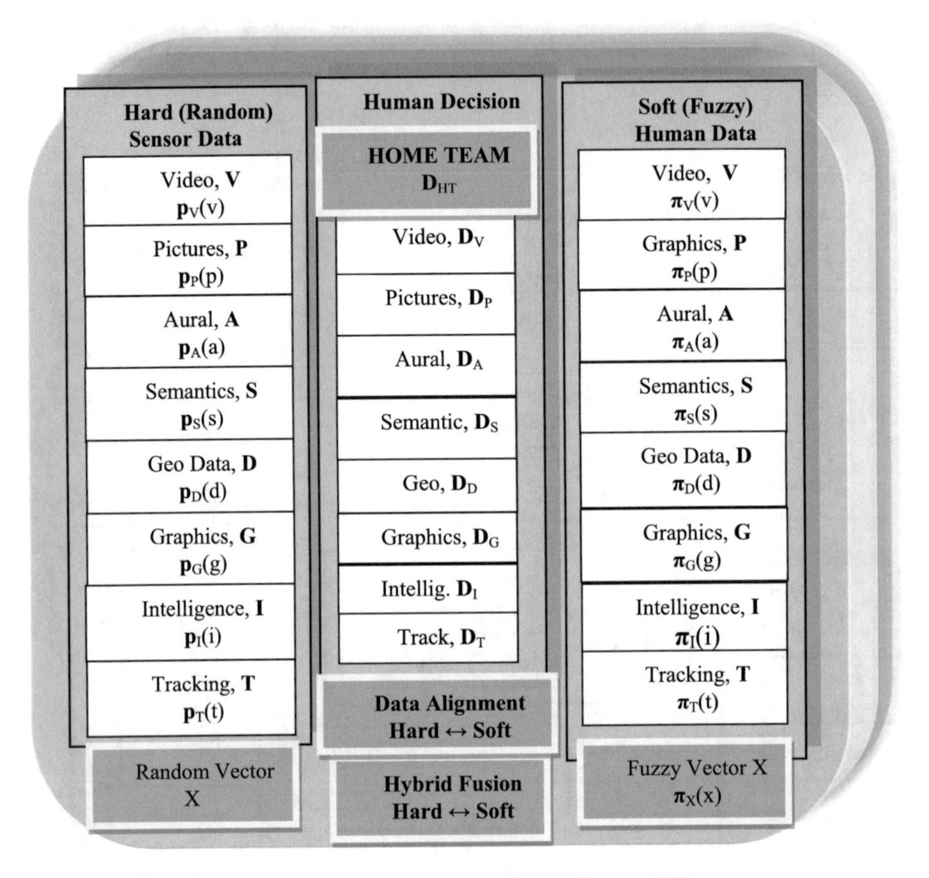

Fig. 2. Conceptual view of HOME team GUI

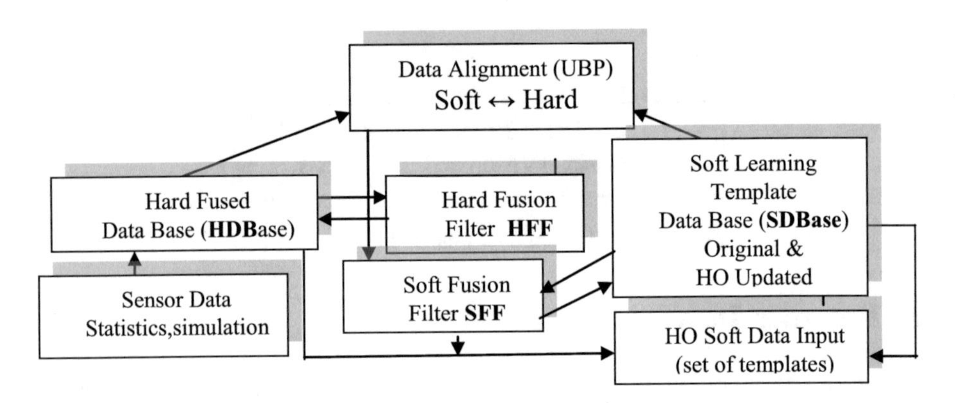

Fig. 3. Hard and soft data base structure

3.1 Data Types

Per Fig. 2, HOME Team GUI may contain a variety of data types such as:

$$\begin{aligned} X &= \{\mathbf{V}\text{isual},\ \mathbf{P}\text{icture},\ \mathbf{A}\text{udio},\ \mathbf{S}\text{emantic},\text{Geo}\ \mathbf{D}\text{ata},\ \mathbf{G}\text{raphics},\ \mathbf{I}\text{ntelligence},\mathbf{T}\text{racking}\} \\ &= \{\mathbf{V},\mathbf{P},\mathbf{A},\mathbf{S},\mathbf{D},\mathbf{G},\mathbf{I},\mathbf{T}\} \end{aligned} \tag{1}$$

These data can be soft (subjective) or hard (objective); see Table 1. The two are intuitively related and associated. We make the association notion very precise mathematically via our UBP and useful practically. The data can be one dimensional or multivariate. Hard and soft data are **p** (probabilistic) and $\boldsymbol{\pi}$ (possibilistic) distributed, and $\boldsymbol{\pi}$ is equivalent to fuzzy membership $\boldsymbol{\mu}$ [11]. Data in (1) span wide range of situations and point to feature range, which is a base for soft templates. Some of the data may not be hard relevant, some may not be soft relevant, some are both. The vector X can split into two vectors, one fully hard, one fully soft, they could overlap or be of the same size, some of the distributions in Fig. 2 may not be of interest. For example we can have two vectors with **G** data common to both parts:

$$X = X_{\text{soft}} + X_{\text{hard}} = \{\mathbf{V},\mathbf{P},\mathbf{S},\mathbf{G},\mathbf{I}\} + \{\mathbf{A},\mathbf{D},\mathbf{G},\mathbf{T}\} \tag{2}$$

hence X_{soft} and X_{hard} are "related". In the future we will address the issue of **p** and $\boldsymbol{\pi}$ consisting of a set of related distributions. In this paper we assume no relation and the distributions split into product of individual distributions (see also comment at the end of this document).

Table 1. Soft and hard data designations

Soft data ↔ human operator	Hard data ↔ sensor (machine, statistics)
Subjective	Objective
Express valuation	Express measure
Possibilistic methodology	Probabilistic methodology
Fuzzy models	Random models
Distribution $\boldsymbol{\pi}_X(x) = \boldsymbol{\mu}_X(x)$	Distribution $\mathbf{p}_X(x)$

Figure 2 shows random data vector X with the overall probability distribution $\mathbf{p}_X(x)$, and the individual distributions, whichever one is of interest:

$$\begin{aligned} \{\mathbf{X}\} &= \{\mathbf{V},\mathbf{P},\mathbf{A},\mathbf{S},\mathbf{D},\mathbf{G},\mathbf{I},\mathbf{T}\} \sim \{\mathbf{p}_X(x)\} \\ &= \{\mathbf{p}_V(v),\mathbf{p}_P(p),\mathbf{p}_A(a),\mathbf{p}_S(s),\mathbf{p}_D(d),\mathbf{p}_G(g),\mathbf{p}_I(i),\mathbf{p}_T(t)\} \end{aligned} \tag{3}$$

The small letters {v, p, a, s, d, g, i, t} are specific values for {V, P, A, S, D, G, I, T}. This data is stored in HDBase, generated by sensor data or hard statistics obtained from a specific source (banking for example), or by Monte Carlo simulation (Fig. 3). The examples of probabilistic distributions with means (m_1, m_2, m_3) and maximums (a_1,

a_2, a_3) are in Fig. 4. The 1st and 2nd moments are included, as well as other relevant statistics. Most often they are Gaussian. Other distributions can be considered, including multivariate ones. The distributions can be of data itself, or some attribute of it [8, 12]. For example, tracking data **T** can have direct distributions, whereas intelligence **I** data distributions may be of some function of (weighted) data features. Each data type may result in different distribution. Modeling these distributions for a class of HOME Team autonomous systems is one of the key items in this methodology. These systems can come from defense, banking, financial, ecological, climate, or social, medical and other models. Note that the area under each distribution is 1, and that the probability of an event that **G** data randomly distributed per $\mathbf{p}_G(g)$, is between a and b, is:

$$\mathbf{P}(\mathbf{G}) = \int \mathbf{p}_G(g)dg, \text{ integrated over a to b} \tag{4}$$

3.2 Hard (Random) Data

There may be more than one source of hard data. Our aim is to have a single hard data distribution before Uncertainty Alignment with the soft data. Hence there may be a need for an interim hard data fusion, as in Fig. 1, which uses a probabilistic method. There are many methods as reported in [3, 4, 6, 10]. Well known result for the sum of two independent random variables $z = x + y$, with the corresponding distributions $\mathbf{p}_X(x)$ and $\mathbf{p}_Y(y)$ is a convolution:

$$\mathbf{p}_Z(z) = \int \mathbf{p}_X(z - y)\mathbf{p}_Y(y)dy = \int \mathbf{p}_X(x)\mathbf{p}_Y(z - x)dx \tag{5}$$

3.3 Soft (Fuzzy) Data

Figure 5 also shows fuzzy data vector X for the same type of data in (1), with the distribution function π_X and individual ones, if all of interest:

$$\begin{aligned}\{\mathbf{X}\} &= \{\mathbf{V}, \mathbf{P}, \mathbf{A}, \mathbf{S}, \mathbf{D}, \mathbf{G}, \mathbf{I}, \mathbf{T}\} \sim \{\pi_X(x)\} \\ &= \{\pi_V(v), \pi_P(p), \pi_A(a), \pi_S(s), \pi_D(d), \pi_G(g), \pi_I(i), \pi_T(t)\}\end{aligned} \tag{6}$$

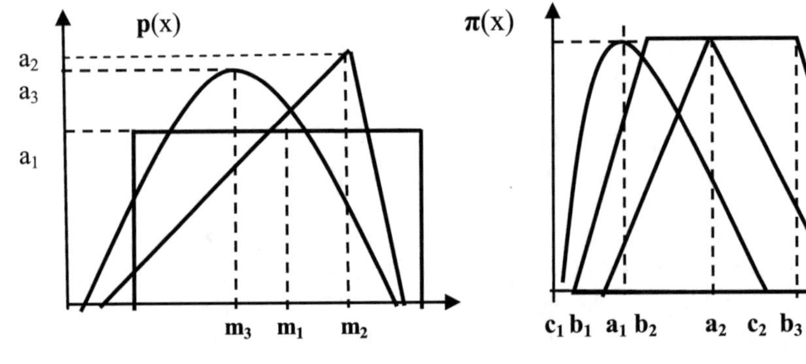

Fig. 4. Hard (random) data

Fig. 5. Soft (fuzzy) data

Figure 5 shows three types of fuzzy numbers (triangular, trapezoidal, other). Their critical values are ($a_1a_2a_3$), ($b_1b_2b_3b_4$) and (c_1c_2) and are numeric or semantic, depending on data type. The π maximum value is 1, and the areas under the curves are not 1, in general. Initial fuzzy templates are stored in SDBase. For example we can have soft data with the following critical values:

$$
\begin{aligned}
&\text{Video} = \mathbf{V}(\text{reject, consider, accept, consider, reject})\\
&\text{Picture} = \mathbf{P} \text{ and Graphics} = \mathbf{G}(\text{not present,unrecognizable, recognizable})\\
&\text{Audio} = \mathbf{A}(\text{not present, reject, not clear, consider, accept}),\ \text{ or Audio} = \mathbf{A}(5., 9., 1.)\\
&\text{Semantic} = \mathbf{S}(\text{not present, garbled, unclear, acceptable, very clear})\\
&\text{Intelligence} = \mathbf{I}(\text{nonexistent, not clear, acceptable, very clear})\\
&\text{Geo Data} = \mathbf{D}(0, 15, 25), \text{Tracking Data} = \mathbf{T}(0, 35, 50, 75, 89)
\end{aligned}
\tag{7}
$$

Each is with its own π. As an example, data type **A** can be described by both semantic and numeric values. These distributions could be one dimensional or multivariate [8]. HO can fine tune initial soft templates (updating π), choosing semantics or numerics out of a list of choices. HO can also use other expert's soft inputs. This could be implemented as a drop down menu for each data type. Figure 6 shows semantic choices. Time for this HO action can be minutes (if time critical) or weeks (banking credit decisions). We believe this new approach reduces HO errors, and increases effectiveness of HOME Team. The problem which human errors pose for soft data and hard/soft fusion process is well described in [1, 2, 16]. We believe that our approach and SDBase built-in- learning feature diminishes this issue, and produces better hybrid fusion process. The SDBase can be an exceptionally powerful tool for any autonomous system for which specific data tuning is performed. There may be more than one source of soft data, to be fused to produce single possibilistic distribution for hybrid soft hard fusion [8]. SDBase fuzzy numbers can be added, subtracted, multiplied, divided. One can find their max/min values and various max/min convolutions [8]. Figures 7, 8 and 9 show an example of two TFN's min ($\wedge$) and max ($\vee$) operations.

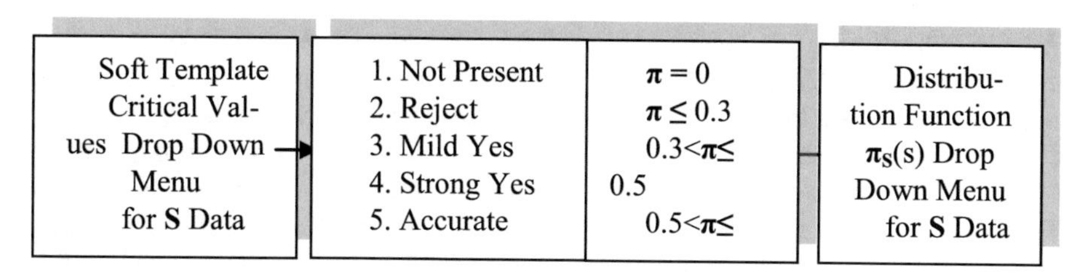

Fig. 6. Drop down menus concept for semantic data

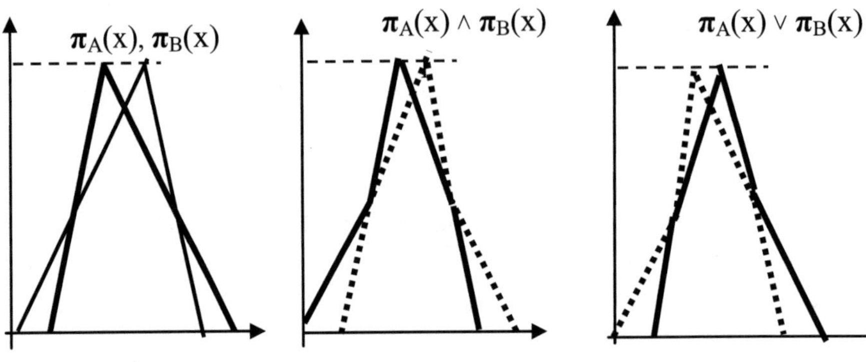

Fig. 7. Fuzzy numbers **Fig. 8.** Fuzzy min **Fig. 9.** Fuzzy max

4 Uncertainty Aligned Data

Here we consider aligning (transforming from one to another) soft and hard data as describing the same or related phenomenon, or even completely unrelated. Both types could have been obtained from single sources each or a result of corresponding data fusion. In either case we have single hard and single soft data distributions available. Then we proceed to Uncertainty Align (UA) two distributions. There may be three different situations related to these distributions, namely related to their uncertain variable x, let us call it ProPos variable x, i.e. Probabilistic and Possibilistic uncertain variable. These three situations are: (i) two distributions can have equivalent variable x, (ii) two distributions can have related variable x and (iii) two distributions can have completely unrelated or independent variable x. The methods to UA these two distributions are different. Which method is employed will depend on the task at hand, in particular what is, if any, uncertainty relationship between the data. Independent data will be handled differently than the dependent ones. All of the methods are various Data Fusion methods.

4.1 Hybrid Data Method

In order to bridge the gap between the notions of random (hard) and fuzzy (soft) data, and pursue proposed hybrid fusion, one should combine them in some useful way. One method is to define a hybrid number H [8]. We consider a fuzzy number F, with $\pi_F(x)$. It can be shifted to the left/right by some random value r, distributed with $\mathbf{p}_R(r)$, where R is a random variable. Then F moves randomly according to the law $\mathbf{p}_R(r)$. The pair (F, R) = H is called hybrid number. Note that both F and R preserve their original properties and no individual information is lost in forming H. The effect of adding R to F is that F has distribution $\mathbf{p}_R(r)$. We can write H in two forms:

$$\mathrm{H} = \mathrm{F}[\pi_\mathrm{F}(\mathrm{x}), \mathbf{p}_\mathrm{R}(\mathrm{r})] \quad \text{or} \quad \mathrm{H} = \mathrm{F} + \mathrm{R} \tag{8}$$

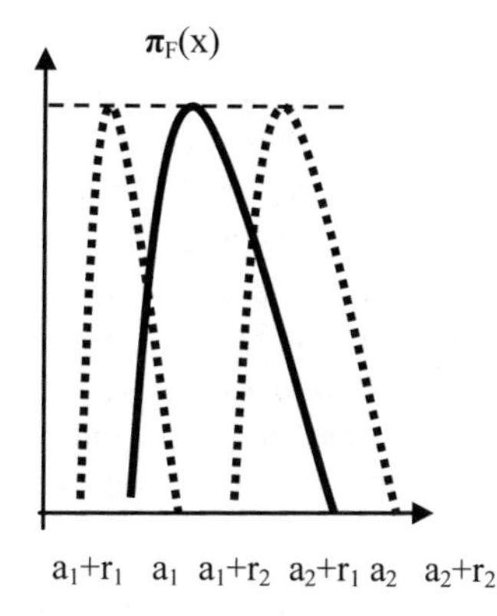

Fig. 10. Random fuzzy

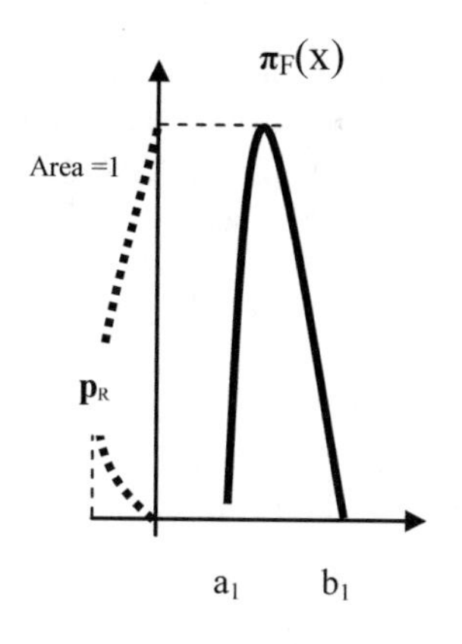

Fig. 11. Fuzzy random

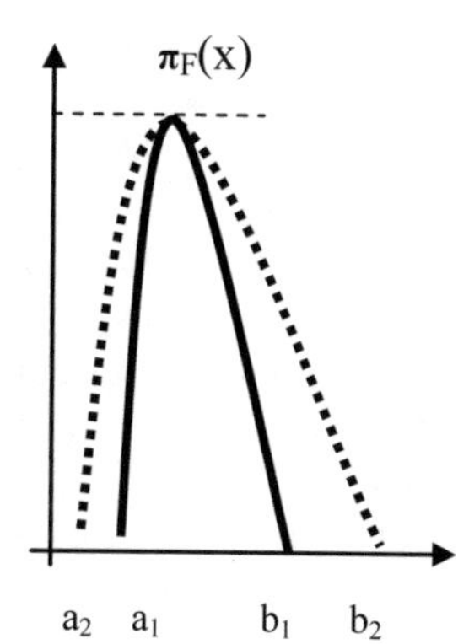

Fig. 12. Type 2 fuzzy

where "+" indicates "adding" F and R. If R has a range (r_1, r_2), $r_1 < 0$ and $r_2 > 0$, Fig. 10 shows H extreme positions. We call H a random fuzzy number. Note that $\mathbf{p}_R(r)$ may or may not be convex. Figure 11 shows a fuzzy random number where π is randomly distributed itself between 0 and 1, according to $\mathbf{p}_R[\pi_F(x)]$. Here $\mathbf{p}$ indicates how probable any interval of π is. Finally, Fig. 12 shows a fuzzy number of Type 2 where the critical intervals (a_2, a_1) and (b_1, b_2) are themselves fuzzy numbers [8]. Let us define two hybrid numbers H_1 and H_2:

$$H_1 = F_1 + R_1,\ H_2 = F_2 + R_2 \tag{9}$$

with the distributions π and $\mathbf{p}$. $H_1 + H_2$ can be treated as a hybrid convolution:

$$H = H_1 + H_2 = (F_1 + R_1) + (F_2 + R_2) = \{F_1(+)F_2, R_1[+]'R_2\} = (F,\ R) \tag{10}$$

where (+) corresponds to fuzzy max/min convolution for addition, and [+] to random sum/product convolution for addition [8]. The interpretation for the above addition is given by the distributions $\pi_{H1}(x)$, $\pi_{H2}(y)$, $\mathbf{p}_{R1}(x)$, $\mathbf{p}_{R2}(y)$ relations:

$$\begin{aligned} \pi_F(z) = \ \pi_{H1(+)H2}(z) &= \vee[\pi_{H1}(x) \wedge \pi_{H2}(y)] \\ z &= x + y \end{aligned} \tag{11}$$

$$\mathbf{p}_R(z) = \int \mathbf{p}_{R1}(z - y)\mathbf{p}_{R2}(y)dy = \int \mathbf{p}_{R1}(x)\mathbf{p}_{R2}(z - x)dx \tag{12}$$

where $\vee$ and $\wedge$ are max-min operations. The integral is replaced by a sum for discrete data. One can also use expectation E_π of a hybrid number H, i.e. $E_\pi(H)$. The distribution function π_E of the expectation of H = (F, R) is π_F shifted by the expectations of R:

$$\pi_E(x) = \pi_F + E_\pi(R) \tag{13}$$

This way we average random influence on a fuzzy number by mean value shift. Other operations over hybrid numbers can be also defined [8]. Another fuzzy concept is Sheaf Of Fuzzy Numbers (SOFN). If we have N observations of the same phenomenon, each resulting in a fuzzy number F_i, i = 1, 2, …, N, then the set of F_i forms a SOFN whose confidence interval expectation results in an averaged interval of confidence limits. Similarly, if each confidence interval is not equally weighed, but with some probability, than an appropriate mean value of the SOFN can be defined on a set of hybrid numbers as well. This can also be employed in hybrid fusion. Let us assume (3, 5, 8) TFN, plus Gaussian random number (3, 4) = (mean, variance), plus a hybrid number (1, 6, 8) + (1, 5) with (1, 6, 8) TFN and Gaussian part (1, 3). The sum of all the parts is:

$$\{(3,5,8)+(3,4)+[(1,6,8)+(1,5)]\} = [(4,11,16)+(4,9)] = [(8,15,20)+(0,9)] \tag{14}$$

where we used addition properties of fuzzy and independent Gaussian numbers, using Eqs. (8) to (12). The result is a TFN (8, 15, 20) with the $\sqrt{9} = 3$ spread. Hence we have:

$$\begin{aligned}(8-3,\ 15-3,\ 20-3) &= (5,12,17) \leq (\mathrm{a,b,c}) \leq (8+3,\ 15+3,\ 20+3) \\ &= (11,18,23)\end{aligned} \tag{15}$$

This method can be employed for manipulating various hybrid numbers. In the case of our data filters, hard HDF and soft SDF, the above described methods could be employed as one option.

4.2 Simple (Ad Hoc) Uncertainty Alignment

Each data in (1) may have both random **p** and fuzzy π distributions. See (2) as well. The UA is performed on both types of data and the idea is not to loose any information in the process, or loose a minimum. Following the UA, each data type is stored in a proper updated data base (Fig. 6) together with the original data. Besides Hybrid method described above, one can also use an ad hoc UA method, which in some cases may cause loss of useful information.

(i) Soft to Hard UA_{SH}. An ad hoc (and somewhat a superficial) idea for fuzzy to random UA_{SH} is to normalize fuzzy to unit area under $\pi(x)$. If doable, this would be mathematically correct but it may not make sense probabilistically. Figures 13 and 14 indicate general idea. Each fuzzy distribution $\pi(x)$ would need to be normalized to obtain unit area to represent it as a probabilistic distribution $\mathbf{p}_X(x)$. The problem with this approach, loosely speaking, is in a view of a fact that any possibility $\mathbf{\Pi}_X(x)$ is always greater than or equal to a corresponding cumulative probability and it is not clear if such a normalization will uphold that principle [19, 20]. We call this Complementarity Principle (CP) as described first in general terms by Zadeh in [11] and given as:

$$\prod\nolimits_{\mathbf{X}}(\mathrm{x}) \geq \mathbf{P}_{\mathrm{X}}(\mathrm{x}) \tag{16}$$

This Principle also intuitively describes level of agreement between possibilistic and probabilistic distributions, i.e. $\pi(x)$ and $\mathbf{p}(x)$ [11]. Here $\mathbf{\Pi_X}(x)$ is not equivalent to $\pi(x)$ in general, it is understood as an upper limit to the corresponding $\mathbf{P}_X(x)$, per Eq. (16) [19, 20].

(ii) Hard to Soft UA_{HS}. Random data probability density function $\mathbf{p}_X(x)$ can be normalized to 1, and then in turn it can be related to fuzzy data. This is mathematically correct but it may cause some loss of information [19, 20]. Figures 15 and 16 show this procedure for three UA_{HS}. Other approaches have been described as summarized in [19, 20] which also uphold (16) above, which should work in both directions, i.e. UA_{SH} and UA_{HS}. In various literature our Uncertainty Alignment is generally known as Uncertainty Transformation. It seems to us that the term Alignment may be a better choice than Transformation due to different nature of uncertainty description by fuzzy and by random methodologies. In any case that is a minor point and a matter of personal preference. The key is in the way that each type of data is made into another type with a mathematical rigor as well as reasonable level of intuitiveness and usability for practical applications.

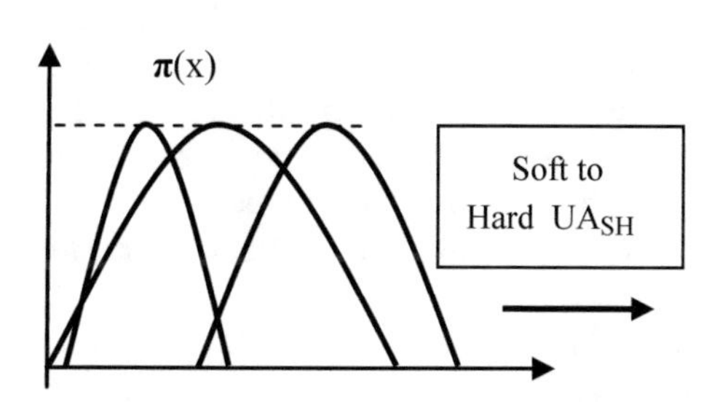

Fig. 13. Original fuzzy data

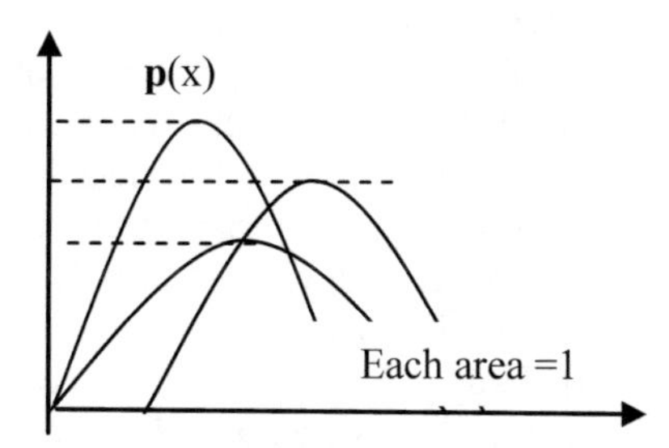

Fig. 14. Fuzzy to random UA

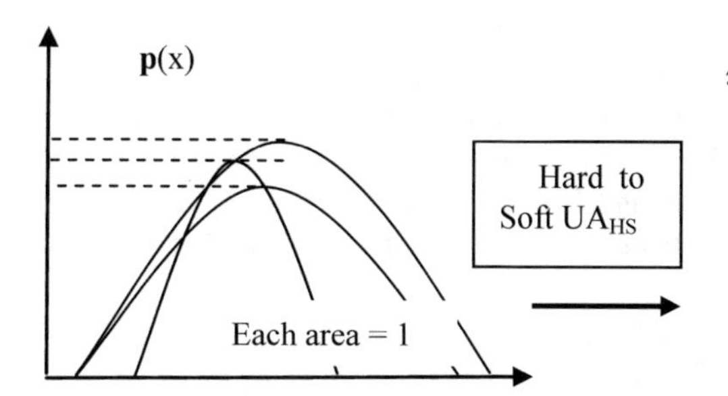

Fig. 15. Original random data

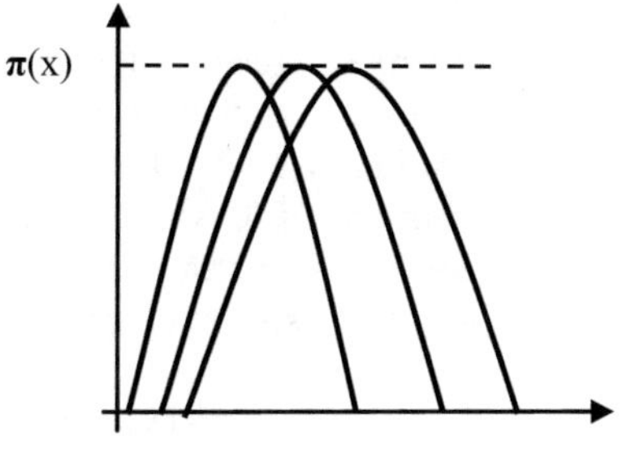

Fig. 16. Random to fuzzy UA

4.3 Uncertainty Balance Principle

(i) Soft to Hard UA_{SH}. Another methodology which is much more intuitive and rigorous mathematically is based on our Uncertainty Balance Principle (UBP), which satisfies CP idea [13–15]. Our approach to implement UA_{SH} is to use cumulative probability $\mathbf{F}_X(x) = \mathbf{P}[X \leq x] = \int \mathbf{p}_X(u)du$, instead of probability density function. This methodology produces probabilistic distribution from a fuzzy one and it satisfies CP in the process [13–15]. This is applicable to any convex $\pi(x)$. As stated in [11] for semantic data but applicable elsewhere, "A fuzzy proposition in a natural language may be translated into a procedure to compute probability distribution of a set of its attributes". This is what UBP does, and in this case it produces a random (probabilistic) distribution which corresponds to a given fuzzy distribution. The random distribution is not unique, and our method produces a class of such distributions. Which one will be chosen depends on an expert opinion which produced the original fuzzy (possibilistic) distribution. See also [19, 20]. The essence of our UBP is in the following simple relation [14–15]:

$$\pi_{\mathbf{X}}(x) \geq \Delta\mathbf{P}_X(x) \tag{17}$$

from which a family of cumulative, as well as density function (as derivatives, if they exist), can be obtained, which also satisfy Complementarity Principle (16) and performs Soft to Hard UA_{SH}. Once the alignment is done one can proceed and perform hard data fusion which will depend on the nature of corresponding ProPos variable x. Note that in some cases original and then aligned soft variable x may nor be the same as the corresponding hard variable x. They may be related or independent. This will determine what kind of hard data fusion is employed. This is what our Hard Fusion Filter (HFF) does (see below).

(ii) Hard to Soft UA_{HS}. Our Uncertainty Balance Principle can be applied in the reverse direction as well, i.e. UA from the original hard (random, probabilistic) data to a soft (fuzzy, possibilistic). Hence the "inverse" UBP i.e. UBP^{-1}. This is a bit less clear as far as the method's applicability compared to UBP itself for soft to hard UA. See also [19, 20]. We believe UBP^{-1} can be instrumental in soft data base learning and fine tuning feature, whereas the data base is populated initially with the best fuzzy distributions for various type of soft data and their features (soft templates), and as new hard data comes along, or new expert soft opinion is solicited over time, these inputs to SDBase also improve via UBP and UBP^{-1}. This is also in accordance with our idea of HOME Team approach whereas machine and Human Operator improve each other's performance and in particular reduce Human Operator soft errors over time. The essence of UBP^{-1} is summarized as:

$$\begin{aligned} \Delta\pi_{\mathbf{X}}(x) + \mathbf{P}_X(x) &\leq 1 \text{ for } \mathbf{P}_X(x) \leq 0.5 \\ \Delta\pi_{\mathbf{X}}(x) + \mathbf{P}_X(x) &\geq 1 \text{ for } \mathbf{P}_X(x) \geq 0.5 \end{aligned} \tag{18}$$

from which a family of $\pi_{\mathbf{X}}(x)$ can be obtained which satisfy Complementarity Principle (16) and performs Hard to Soft UA_{HS}. Similar comment as far as ProPos variable in hard data case applies here as well. The soft data fusion is performed by our Soft Fusion Filter (SFF) (see below).

5 Hybrid Data Fusion

The random and fuzzy data, UA procedures, and their hybrids produce HDBase and SDBase, Fig. 6. Each consists of two parts, one with the original data, one with UA (via UBP or UBP^{-1}) data of the other type. Data bases feed two fusion filters, HFF and SFF. We can treat soft data or hard data as new measurements for both HFF and SFF (after HO soft input or Expert input and UA_{SH}). New hard sensor data come continuously (from sensors, statistics or via simulation, Fig. 4) at some rate higher than HO irregular or regular soft input rate. Similarly, SFF takes UA_{HS} hard data and fuse it with the original HO soft data. The result is two hybrid filters with hard or soft outputs or individual pure hard and pure soft filters running only on the native data. All are compared in HO decision process via appropriate UFs. One way to fine tune these procedures is to use known historical data with the known results. These data can be used to test effectiveness of a particular UA and fusion method, especially in a view of decision making process.

5.1 Hard Fusion Filter (HFF)

Basic HFF structure is shown in Fig. 17, and its time operation in Fig. 18, with one UA soft data per three hard data. This is arbitrary and other data timing can be assumed as well. That will depend on the practical frequency with which the data is obtained, and it could be also asynchronous with the original hard data. The upper part is hard data, D_i, and lower UA_{SH} HO (or expert) data, D_j^H. With no soft input, HFF operates on hard data only. At the HO soft input time, new UA soft data is used to produce fused estimate. In both cases filtering method is of Kalman type predicting between the measurements, and correcting at the measurement times, producing some sort of (optimal or suboptimal) conditional distributions $\mathbf{p}(x/Z_H)$ [17]. Native sensor hard only output (or corresponding statistics) will be record as well in HDBase. Monte Carlo simulation can be used to access the quality of filtering methodology, as well as use of historical data. For the purposes of calculating various probabilities (such as for defense, banking or other applications), our interest is more on a posteriory distributions $\mathbf{p}(x/Z_H)$ rather than uncertain variable estimates $x_H^* = E_p(x/Z_H)$. This is an interesting point which makes the methodology different than the standard Kalman Filtering method. We are more interested in the resulting probability distribution rather than some specific independent variable value estimate. For soft hard data fusion and corresponding risk assessment the key is to produce a measure of change in random (probabilistic) distribution with UA added soft (fuzzy distribution) data which also satisfies CP in the UA process. Note that UA be performed using one of the methods described above.

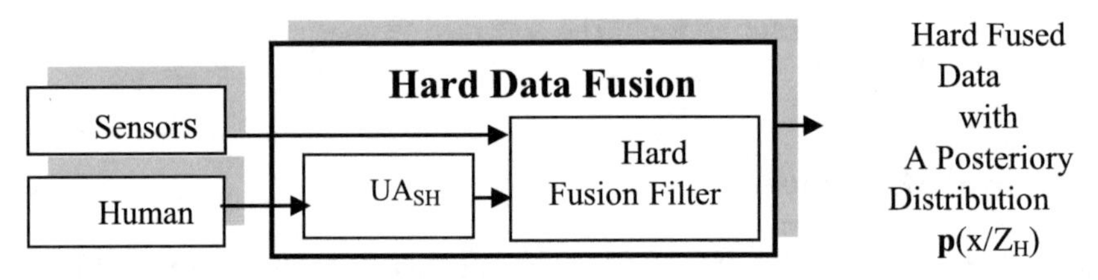

Fig. 17. Hard fusion filter structure

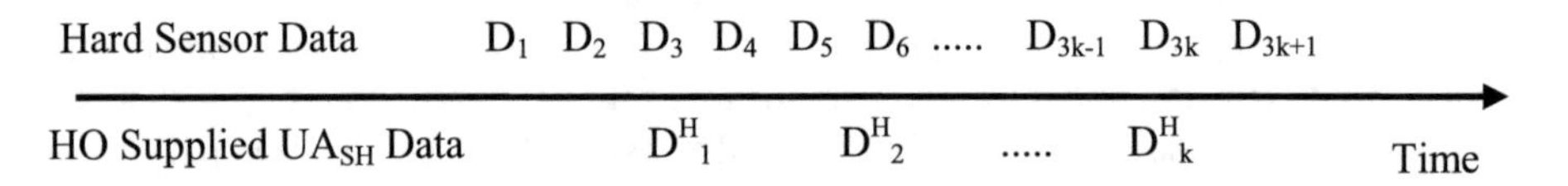

Fig. 18. Hard fusion filter time operation

5.2 Soft Fusion Filter (SFF)

SFF structure and time operation are shown in Figs. 19 and 20, with one soft per three UA data. The upper part corresponds to HO soft data, D_i, lower to sensor UA_{HS} data, D^S_j. As in HFF, we can have any regular or irregular timing between two data types. At the update times, the HO soft data is used to produce new optimal fused estimate, based on some conditional like distribution $\pi(x/Z_S)$ [8]. Other filtering methods based on fuzzy numbers convolution can be also used. Fuzzy distance measures will be used to access the effectiveness of the filtering method [8]. Native HO soft only output will be recored as a reference in SDBase. Different methods may be better suited for different data types. This will also be compared by a proper Monte Carlo simulation. Specific soft filtering techniques will be discussed in a future paper. The SFF and HFF outputs are to be used by HO Decision Making process (Fig. 2) as described below.

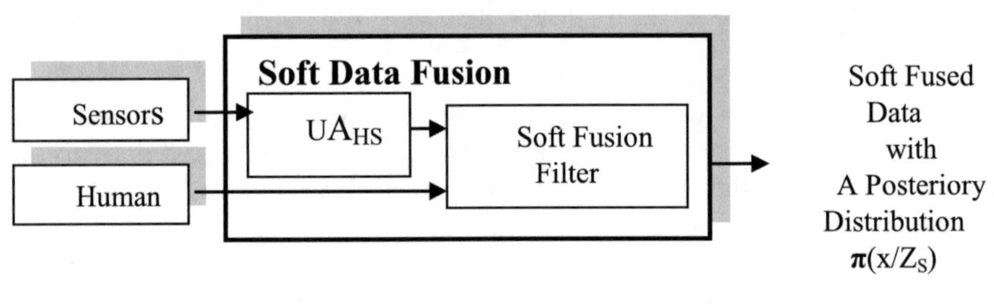

Fig. 19. Soft fusion filter structure

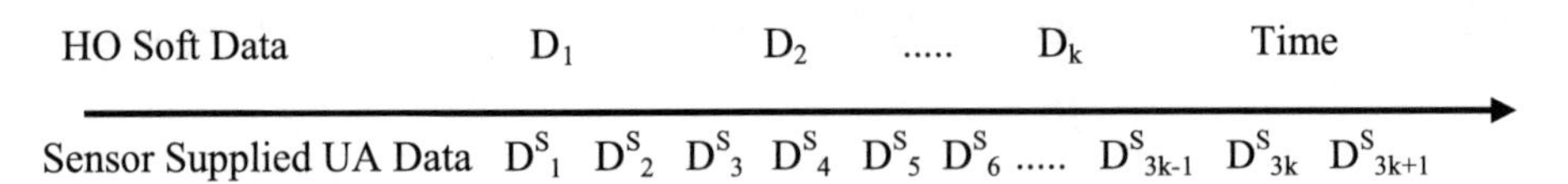

Fig. 20. Soft fusion filter time operation

6 Integrated Decision Process

Much of the information for HO decisions are based on possibilistic (fuzzy) rather than probabilistic (random) data. Figures 1, 2, and 3 indicate HO decision flow and related hybrid data fusion, which is in turn based on fuzzy and random data and their UA. The HO decisions are based on (i) viewing the hybrid data, (ii) HO fine tuned soft input, (iii) observing the value of utility function (**UF**), and when optimal (HO comfortable, following HO soft input) (iv) making a decision. This process can be simulated. Optimal will mean minimal averaged UFs which we base on a certain data distance measures. Figure 21 indicates general problem if no data fusion is done, and where the decisions are made solely based on either hard or soft data alone. Figure 22 indicates decision making when data fusion and data alignment are employed, which produces fused hard and soft data. In general case we should expect better decision made with a properly done soft to hard (hybrid) data fusion. Figure 23 indicates learning feature in our approach when hard to soft data fusion is performed.

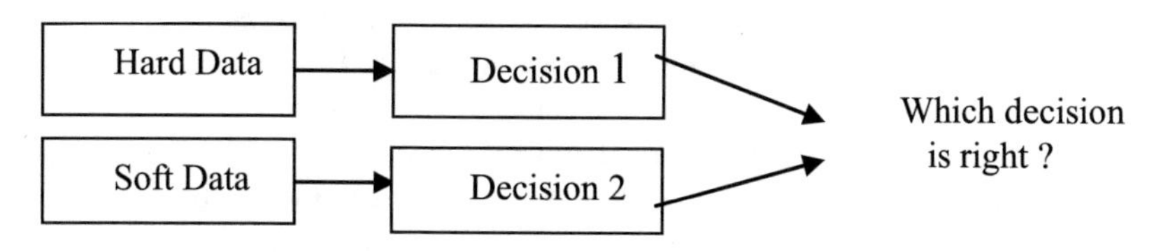

Fig. 21. Decision making without data fusion

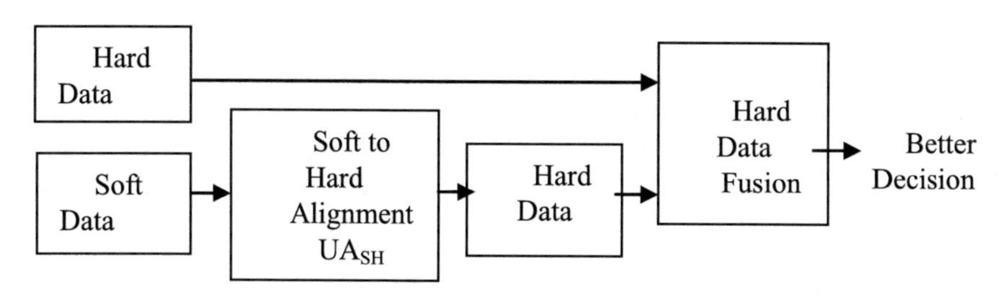

Fig. 22. Decision making with soft to hard data fusion and UA_{SH}

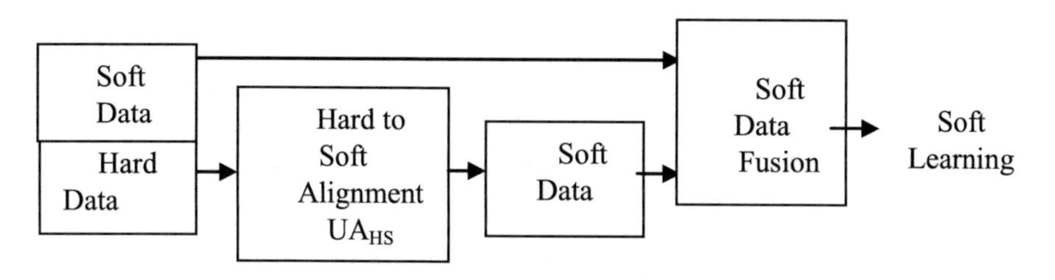

Fig. 23. Decision making with soft to hard data fusion and $UA_{HS6.}$

We define (i) **Group Utility** function and a number of (ii) **Local Utilities**, per data in (1). Each UF can be used individually or combined into a weighted sum. The extreme poor data situations may result in big biases and unacceptable fuzzy and random data variances, making it difficult for HO to make a decision (less certainty means harder decision making, and vice versa). Similarly one can consider a situation for very reliable data, or when the time is of the essence and a decision can or must be made. HO can eliminate some data for a while until quality improves, or give some data an utmost credence. The HO fine tunes soft data templates in the process, which improves decision making, and in turn HOME Team benefits. The HO would use rich and intuitive HOME Team GUI for data entry and decision making.

6.1 Utility Function

The ultimate goal is to improve some form of HOME Team benefit UF. The UF could be a group UF (final HO decision across all data) or local UF (related to utility of individual data). In Fig. 2 group decision is noted as $\mathbf{D}_{HT}$, and individual decisions by $\mathbf{D}_V$, $\mathbf{D}_P$, $\mathbf{D}_A$, etc. In general case we can denote group utility by U(x, **D**) and local by $U_i(x, \mathbf{D}_i)$, where $x \epsilon X$ is an uncertain state vector of interest, in our case vector of data in (1), or it could be just x_i (Video, Aural, Tracking, etc.). The $\mathbf{D}_i \epsilon \mathbf{D}$ (**D** = allowable decisions), is the decision related to the i-th data. As the process evolves, measurements $z \epsilon Z_H$ (subscript H for Hard) are obtained about the vector x. Based on the $z \epsilon Z_H$, HO inputs soft data "z" and eventually makes a decision **D** = **D**(z) which boils down to HO accepting minimized expected UF value, and pressing GUI "decision icon", which represents HOME Team agreement (benefit). One can consider both random and fuzzy utilities, according to choice of UBP or I_UBP, correspondingly.

6.1.1 Random Data Group Utility

We assume one HO and N data types collected in a vector x as in (1). Since x is a random variable with the distribution **p**(x), we seek to determine random expected utility. Once the measurements $z \epsilon Z_H$ are received (Z_H is allowable random space), we can calculate posterior distribution $\mathbf{p}(x/Z_H)$ (super Bayesian) or some reasonable approximation of it and use it in expected utility calculation:

$$E_H[U(x, \mathbf{R_G})/Z_H] = \int U(x, \mathbf{R_G})\mathbf{p}(x/Z_H)dx, \text{ or } E_H[U(x, \mathbf{R_G})/Z_H] = \sum U(x, \mathbf{R_G})\mathbf{p}(x/Z_H) \tag{19}$$

for continuous or discrete time case, respectively.. The expected group utility can be minimized by a proper choice of group (i.e. HO) decision $\mathbf{R_G}$ [9, 10], which is practically equivalent (directly or indirectly) to HO soft input. The Machine simply calculates the averaged UF and presents it back to the HO who compares averaged minimized values and chooses when to make a decision. With a priory and a posteriory distributions we can solve the minimization problem:

$$\mathbf{R}_\mathbf{G}^* = \arg\ \min_\mathbf{R} \int U(x, \mathbf{R_G})\mathbf{p}(Z_H/x)\mathbf{p}(x)dx \tag{20}$$

Figure 2 shows $\mathbf{R}_{\mathbf{G}}^{*} = \mathbf{D}_{\mathrm{HT}}$. A priory random distributions $\mathbf{p}(x)$ are shown in Fig. 2, and a posteriory are simulated in HF Data Base, Fig. 3. In real applications this is supplied by an external hard sensor data filter which uses $\mathbf{p}(x)$, and it generates $\mathbf{p}(x/Z_H)$ and $\mathbf{p}(Z_H/x)$. In Gaussian case, Kalman Filter generates optimal 1st and 2nd conditional moments. One of the key items is to determine $\mathbf{p}(x)$ for a set of data (or their features) in (1). Monte Carlo simulation can generate sample distributions and compare with the corresponding filtered data. One can also use historical data for training purposes and fine tuning.

6.1.2 Random Data Individual Utility

We can also consider individual i-th Bayesians for which $\mathbf{p}(x/Z_H)$ is maximized by local decision choice $\mathbf{R}_I$, i = 1, 2, …, N:

$$\mathbf{R}_{\mathbf{I}}^{*} = \arg\min_{R} \sum w_i E_H[U_i(x, \mathbf{R}_{\mathbf{I}})/Z_H] = \arg\min_{R} \sum w_i \int U_i(x, \mathbf{R}_{\mathbf{I}})\mathbf{p}(Z_H/x)\mathbf{p}(x)dx \tag{21}$$

where $0 < w_i < 1$, $\sum w_i = 1$. The other option is to center the utility around some security level $C_H(i)$ which "safeguards" i-th data interests:

$$\mathbf{R}_{\mathbf{I}}^{*} = \arg\min_{R}\left\{\sum w_i[|E_H[U_i(x, \mathbf{R}_{\mathbf{I}})/Z_H] - C_H(i)|]^{\gamma}\right\}^{1/\gamma} \tag{22}$$

for γ between $-\infty$ (max-min solution) and $+\infty$ (min-max solution) and also for $\gamma = 1$ as a special case. Referring to Fig. 2, we can have $\mathbf{R}_{\mathbf{I}}^{*} = \mathbf{D}_{\mathbf{V}}$, or any other type. One problem with local utilities U_j is their different scale. For a similar scale, the minimization problem is solvable [10]. Note also that we used full data vector x above, but if all data types in (1) are independent (decoupled sensors), we could use some combination of vector components x_i. If this is so, we can save some computation time by using a subset instead of full vector x. As in group utility case, one can resort to Monte Carlo simulation to determine sample statistics. One can also use historical data for training purposes and fine tuning.

6.1.3 Nash Solution for Random Data

When U_i's are not of the same scale and can not be compared, one can resort to Nash product [10] of individual utilities,

$$\mathbf{R}_{\mathbf{N}}^{*} = \arg\min_{\mathbf{R}} \prod_i |E_H[U_i(x, \mathbf{R}_{\mathbf{N}})/Z_H] - C_H(i) \tag{23}$$

where $C_H(i)$ has no effect on the solution, it just relates to a local maximum. The same Monte Carlo comment and historical data apply as well. This would correspond to another group utility. Note that all of the above UF's would use UA via our UBP or other methodology which produces fused hard data, via HDF.

6.1.4 Fuzzy Data Group Utility

Similar to random data, we assume one HO and N data types collected in a vector x as in (1). Since x is also a fuzzy variable with the distribution $\pi(x)$, we seek to determine fuzzy expected utility. Once the (HO soft) "measurements" $z \in Z_S$ are received, we can calculate fuzzy posterior distribution $\pi(x/Z_S)$ (similar to super Bayesian, [11]) and use it in expected utility calculation via appropriate fuzzy convolution "*", see (11) as well:

$$E_S[U(x, \mathbf{F})/Z_S] = U(x, \mathbf{F}) * \pi(x/Z_S) \tag{24}$$

The expected fuzzy group utility can be minimized by a proper choice of fuzzy decision $\mathbf{F_G}$,

$$\mathbf{F_G^*} = \arg\min_{\mathbf{F}} U(x, \mathbf{F_G}) * [\pi(Z_S/x)\pi(x)] \tag{25}$$

which can be solved once a priory and a posteriory fuzzy distributions are given. A priory fuzzy distributions $\pi(x)$ are shown in Fig. 2. Unlike random distributions $\mathbf{p}(x)$ which are from an outside source or from our simulation, determining $\pi(x)$ is a key task in this Proposal. For fuzzy $\pi(x/Z_S)$ and $\pi(Z_S/x)$ we use SLT Data Base, and assume the distributions as a posteriory after HO acts upon received data and fine tunes the soft data accordingly. The tuned data will be in general different than the original soft data, and hence a posteriory distributions can be evaluated. Following UA, fuzzy data filtering is done in SFF and optimal fuzzy estimates are obtained. Monte Carlo simulations are run to obtain sample statistics. In a classic paper [9] fuzzy decision D is considered to be an intersection of fuzzy Goals and Constraints, $D = G \cap C$, which corresponds to minimum between π_G and π_C. We will also address this angle in our simulations.

6.1.5 Fuzzy Data Individual Utility

For N fuzzy observations of the same phenomenon, we form a sheaf of fuzzy numbers (SOFN) [8]. Various operations over a SOFN can be performed, including fuzzy expectation E_S (S for soft). We now consider i-th individual data for which $\pi(x/Z_F)$ is minimized by individual decision choice $\mathbf{F}_I$, i = 1, 2, …, N:

$$\mathbf{F_I^*} = \arg\min_F \sum w_i E_S[U_i(x, \mathbf{F_I})/Z_S] = \arg\min_F \sum w_i U_i(x, \mathbf{F_I}) * [\pi(Z_S/x)\pi(x)] \tag{26}$$

where $0 < w_i < 1$, $\sum w_i = 1$. The other option is to center the utility around some security level $C_S(i)$ which "safeguards" i-th data interests:

$$\mathbf{F_I^*} = \arg\min_F \left\{ \sum w_i [|E_S[U_i(x, \mathbf{F_I})/Z_S] - C_S(i)|]^\gamma \right\}^{1/\gamma} \tag{27}$$

for γ between $-\infty$ (min/max solution) and $+\infty$ (max/min solution) and also for $\gamma = 1$ for the absolute value of the expected utility. Other comments from random section apply as well.

6.1.6 Nash Solution for Fuzzy Data

As for the random data, U_i's are not of the same scale and can not be compared, we can resort to Nash product of individual utilities,

$$\mathbf{F}_{\mathbf{N}}^{*} = \text{arg min}_{\mathbf{F}} \prod\nolimits_{i} |E_F[U_i(x, \mathbf{F}_{\mathbf{N}})/Z_S] - C_S(i) \quad (28)$$

where $C_S(i)$ has the same role as in random case. Similar to comment for hard data UF's, we note that I_UBP or other described methods can be used to produce SDF soft data output. As stated earlier Monte Carlo simulation as well as historical hard and soft data can be used to fine tune and improve the quality of both hard and soft data.

6.1.7 Hybrid Data Group Utility

We described a variety of hybrid data earlier. They can be random fuzzy or fuzzy random type or fuzzy Type 2, Figs. 10, 11 and 12. All the comments made for random and fuzzy utilities apply here as well. The form of a specific utility function $\mathbf{H}_G$ may vary with the data type.

Hybrid Data Individual Utility. $\mathbf{H}_I$, similar comments apply.
Nash Solution for Hybrid Data. $\mathbf{H}_N$, similar comments apply.

The above equations use general UFs. Next we describe specific UFs based on various distance measures for both fuzzy and random data.

Table 2. General utility function choices

Type	Random data	Fuzzy data	Hybrid data
Group	$\mathbf{R}_G$	$\mathbf{F}_G$	$\mathbf{H}_G$
Individual	$\mathbf{R}_I$	$\mathbf{F}_I$	$\mathbf{H}_I$
Nash	$\mathbf{R}_N$	$\mathbf{F}_N$	$\mathbf{H}_N$

6.2 Utility Function Choices

With the above general utility presentation, we now look at several specific options for the UFs, both random and fuzzy. Note that weighted sums of the above utilities (Table 2) can be defined in the HO decision process. First we present basic hybrid numbers formation. Let us assume we have a hard random number R_H and UA number R_{SH} originated from SLT Data Base or from HO fine tuning input. Both correspond to the same event such as target tracking data. We can have two situations. First, we assume reasonably that the hard data is Gaussian and that UA data R_{SH} after several fine tunings and averaging will also exhibit Gaussian like properties per Central Limit Theorem. Hence both are described by their means and variances. The other possibility is that both are of some random distribution and we would convolute them to obtain resulting distribution. On the fuzzy side we have the original soft data F_S and UA associated F_{HS}. We can now define a hybrid number H in couple of ways:

$$H = (R_H + R_{SH}) + (F_S + F_{HS}) = R + F, \text{or } H = (R_H + F_S) + (R_{SH} + F_{HS}) = H_1 + H_2 \quad (29)$$

As an example, we can form a difference between measured and UA data:

$$\Delta H = H_1 - H_2 = R_H - R_{SH} + F_S - F_{HS} = \Delta R + \Delta F \quad (30)$$

The utility functions can be some suitable functions of ΔH, ΔR, or ΔF. Few general options for the group, individual and Nash utilities for both random and fuzzy data are described next.

6.2.1 Dissemblance Index (DI)

Two hybrid numbers H_1 and H_2, are separated by:

$$\delta(H_1, H_2) = [(1/2)/(\beta_2 - \beta_1)]\, d(H_1, H_2) \quad (31)$$

which is a normalized distance, and the interval $(\beta_2 - \beta_1)$ surrounds non zero values of H_1 and H_2, with $d(H_1, H_2)$ some distance measure such as [8]:

$$d(F_1, F_2) = d_L(F_1, F_2) + d_R(F_1, F_2), \text{and } d_L(F_1, F_2) = |(a_1, b_1)|, d_R(F_1, F_2) = |(a_2, b_2)| \quad (32)$$

with intervals of confidence (a1, a2) and (b1, b2). Integrating from 0 to 1 for π we obtain:

$$\delta(F_1, F_2) = [(1/2)/(\beta_2 - \beta_1)]\int d(F_1, F_2)da = \delta(F_1, F_2) = [(1/2)/(\beta_2 - \beta_1)]\int(|a1 - b1| + |a2 - b2|)da \quad (33)$$

Utility function can be defined as $\delta(F_1, F_2)$ and would need to be minimized by HO action.

6.2.2 Fuzzy Optimum (FO)

Fuzzy part of a hybrid number gets shifted by random mean value. This way we can calculate their fuzzy optimums. For two hybrid numbers $H_1 = F_1 + R_1$ and $H_2 = F_2 + R_2$, their random expectations are two fuzzy numbers:

$$E(H_1) = H_1^* = F_1 + E(R_1),\ E(H_2) = H_2^* = F_2 + E(R_2) \quad (34)$$

and we can calculate their sum min and max, as with any other fuzzy number. These values can be compared for $E(H_{FS})$ and $E(H_{UA})$ and their difference (utility) minimized.

6.2.3 Degree of Consistency (DC)

DC between densities **p** and π [11] can be defined as:

$$DC = \pi_1 \mathbf{p}_1 + \pi_2 \mathbf{p}_2 + \ldots + \pi_N \mathbf{p}_N = \sum \pi_i \mathbf{p}_i \quad (35)$$

DC can be viewed as a measure of closeness of fuzziness and randomness. We can compare DC_{FS} against DC_{UA} and use the difference as the utility function.

6.2.4 Error Norm (EN)

EN for ΔH, ΔR and ΔF can be used as the utility function:

$$U^*(x, \mathbf{R_G}) = E_R[U(x, \mathbf{R_G})/Z] = E_R(\|\Delta H\|) = E_R\left(\left[\Delta R^2 + \Delta F^2\right]^{1/2}\right) \quad (36)$$

6.2.5 Least Square (LS)

For H_1 and H_2 we can define utility to be minimized as:

$$U^*(x, \mathbf{R_G}) = (H_1 - H_2)^T Q(H_1 - H_2) \quad (37)$$

6.2.6 Mean Deviation (MD) and Variance Deviation (VD)

1st and 2nd order deviations for fuzzy numbers are notions similar to mean and variance (1st and 2nd moments) for random data, and they carry similar advantages and disadvantages as far as information loss [8]. In the case of Gaussian random data the two moments suffice for the full description of $\mathbf{p}_{Gauss}(x)$. One can use difference between 1st order deviations for two hybrid (fuzzy) numbers as a measure of their closeness. Similarly for 2nd order deviation.

6.2.7 Information Content (IC)

IC of a fuzzy number is another useful concept. The fuzzy IC is equivalent to $\pi(x)$, $I(x) = \pi(x)$ [11]. The narrower (more constrained) $\pi(x)$ is, more information useful for decision making it conveys. We can use size of the 1st order deviations as the measure of fuzzy narrowness [8].

6.2.8 Other Utilities

We can consider quadratic fuzzy agreement index, min-max, and other operations over hybrid numbers [8]. For the random numbers, we consider their means and variances. See Fig. 1 for a general concept of hybrid decision process for any Utility. Some optimally weighed utility combination will be analyzed.

6.3 Additional Comment

Besides one dimensional distributions $\mathbf{p}$ and π, one can consider multivariate distributions as well, with multiple attributes (visual, aural, geo, semantic, or numeric). In general for N related attributes A_i we can write the total distribution:

$$\prod\nolimits_{A1(x)A2(x)\ldots AN(x)} (u_1, u_2, \ldots, u_n) = \pi_F(u_1, u_2, \ldots, u_N), u_1, u_2, \ldots, u_N \in U \quad (38)$$

where $A_1(x)$, $A_2(x)$, …, $A_N(x)$ is a set of attributes, or it could be understood as data types in (1). If the attributes are unrelated as we assume here (Eq. 3) then the problem decomposes into N individual distribution $\pi_{Fi}(u_i)$ products:

$$\prod\nolimits_{A1(x)A2(x)\ldots AN(x)}(u_1,\ u_2,\ \ldots,\ u_N) = \prod\nolimits_i \pi_{Fi}(u_i) \tag{39}$$

Similar comment applies to distribution **p**. This is very important area which affects the fusion and decision results.

7 Implementation and Future Work

The general Platform goals are stated in Introduction Section. The implementation consists of various tasks for achieving stated objectives. These are

(i) Machine Human Interface Analysis
(ii) Modeling Data Distributions
(iii) Soft and Hard Data Bases Design
(iv) Uncertainty Alignment Algorithms
(v) Hybrid Fusion Methodology
(vi) Decision Making Procedure Design
(vii) HOME Team GUI Concept Simulation

There may be few design iterations until an optimal concept of HOME Team GUI is obtained. These iterations serve dual purpose to (1) optimize our concept in general, and to (2) optimize running of a HOME Team system prototype in real life and real time for a specific application. The real life system operation would use similar iterations to fine tune HOME Team system performance continuously, with the soft-hard fusion, smart learning data bases, hard-soft data alignment, and Human Operator decision making process. Performance improvement of HOME Team design, as well as an increasing data base for added data mining and further improvements are important keys to our approach. Next we briefly describe all listed tasks above.

7.1 Machine Human Interface Analysis

In this task we will analyze several representative Machine-Human interfaces. We will also look into how the HO makes the decisions and produces any required soft input. In each of the examples, HO acts in many ways but there are some common and core action features (soft data). From the set of these core features we will extract the essential ones for soft data template modeling. Task 1 is split into several subtasks, depending on the application at hand. Some examples of the applications are:

1. UAV (Quadcopter) plus Human Operator, mission decision
2. Banking credit expert vs hard statistical data, credit decision
3. Battle commander and battlefield data, battle related decision
4. Government official and ecological data, policy decision
5. Political party and elections, strategy decision
6. Corporate War Room, corporate decision
7. Block Chain and crypto coin mathematics and prediction
8. Social sciences analysis and predictions, decisions
9. Ecology and climate modeling and predictions, decisions

Table 3. Data Type and interface example matrix

	Visual	Picture	Audio	Semantic	Geo	Graphs	Intel	Track
UAV	X	X			X	X	X	X
Bank	X			X	X	X	X	
Battle	X	X	X	X	X	X	X	X
Ecology	X	X		X	X	X	X	X
Party	X		X	X	X	X	X	X
Corp.	X			X	X	X	X	X
BChain				X		X		
Soc. Sc.	X	X		X		X		
Ecology		X		X	X	X		X

Table 3 shows an example with data in (1) produced by specific HOME interfaces above. Two outcomes from this Task come to mind, i.e.

(i) Description of chosen Machine-Human systems, with the emphasis on the interface requirements, unique and common features, any implicit data alignment, HO decision making process, as well as soft-hard data aspect of each example. These items will be listed for each case and for data types in (1), if applicable.
(ii) Soft and hard feature descriptions (semantic and numeric) of the data types in (1) and the examples above. This will serve as a starting point for Task 2, where we will determine corresponding random **p**(x) and fuzzy π(x) distributions for a range of data in (1) and given interface examples above in Table 3.

7.2 Modeling Soft and Hard Data Distributions

In this Task we model a priory random distributions **p**(x) and fuzzy distributions π(x) shown in Fig. 2. For **p**(x) we assume certain sensor errors (or statistics obtained) and treat them as Gaussian distributions. The HF Data Base, contains hard data and update algorithm (of Kalman Filter type) to simulate new sensor data and calculating optimal hard data estimates with a posteriory distributions **p**(x/Z_H) which the algorithm produces. These are used in Tasks 4, 5 and 6. The a priory fuzzy distributions π(x) (and the corresponding soft templates) are deduced from Task 1. The corresponding a posteriory fuzzy distributions π(x/Z_S) are determined as the overall process is fine tuned after HO makes soft data inputs generally different than a priory fuzzy templates. Both of these will be stored in SLT Data Base. Task 2 is split into four subtasks:

(i) Determining a priory hard data distributions p(x) based on Monte Carlo simulated sensor data or real hard statistics and assumed sensor error statistics.
(ii) Determining a posteriory hard data distributions **p**(x/Z_H) with Monte Carlo sensor data Z simulation or real statistical data and the corresponding Kalman Filter like estimation algorithm
(iii) Determining a priory soft data distributions π(x) for a specific application
(iv) Determining a posteriory soft distributions π(x/Z_S) based on HO soft input Z where HO acts as a “soft sensor” after viewing simulated hard data.

7.3 Soft and Hard Data Bases Design

Figure 3 shows various data bases and their relations with other key system parts. There are two data bases, hard fused (HF) and soft learning template (SLT) data base. This Task is about designing these data bases, establishing data formats, their relations, built in hard data update algorithm, soft templates as well as fine tuned HO soft data. Task 3 is split into three subtasks:

(i) Design of HF Data Base. It consists of two parts, (A) the original simulated hard data as well as (B) UA_{SH} HO soft data (Figs. 3 and 22)
(ii) Design of Soft Learning Template (SLT) Data Base. It consists of three parts, (A) initial soft templates, (B) HO soft updates, and (C) UA_{HS} hard data (Figs. 3 and 23)
(iii) Data Base Updates (sensor data simulation in HF Data Base and soft updates based on Human soft input).

7.4 Uncertainty Alignment Algorithms

Our data fusion is based on availability of both hard and soft data in the original and UA forms. This Task defines and implements UA methods, which take soft/hard data inputs and produce the opposite type, i.e. hard/soft, making sure all their properties are preserved. We will implement two UA_{HS} and two UA_{SH} algorithms and compare their effectiveness using Monte Carlo simulations. Task 4 consists of two subtasks:

(i) Design of Hard-to-Soft uncertainty alignment (UA_{HS}) algorithms
(ii) Design of Soft-to-Hard uncertainty alignment (UA_{SH}) algorithms.

7.5 Hybrid Fusion Methodology

Once new hard data and HO soft input are received, they undergo UA process and are stored in proper data bases. Next they are combined with previous cycle data as inputs to HFF and SFF which produce updated optimal hard and soft estimates, combined into optimal hybrid number. We use conditional distributions $\mathbf{p}(x/Z_H)$ and $\pi(x/Z_S)$ as optimality methods. The estimate errors are reduced and as such are presented back to the HO who decides to make another soft input or make final or local data driven decision. We have:

(i) Design of Hard Fusion Filter (HFF)
(ii) Design of Soft Fusion Filter (SFF).

7.6 Decision Making Procedure Design

This Task is about designing specific Group, Individual and Nash strategies for both fuzzy and random numbers. We use various fuzzy measures such as (i) Dissemblance Index (DI), (ii) Fuzzy Optimum (FO), (iii) Information Content (IC), (iv) Quadratic Fuzzy Agreement Index (QI), (v) Mean Deviation (MD) and Variance Deviation (VD), (vi) Degree of Consistency (DC), (vii) Error Norm (EN), and (viii) Least Square (LS).

Items (vi), (vii) and (viii) also apply to random data, in addition to mean and variance measures. The following are specific tasks:

(i) Design of random data Utility Functions
(ii) Design of fuzzy data Utility Functions
(iii) Design of hybrid data Utility Functions.

7.7 HOME Team GUI Concept Simulation

This Task is a final one in Phase 1 and it is aimed at integrating all of the above items, simulating and testing our concept on a limited number of data types and HOME Team interfaces in Table 3. We will follow conceptual GUI design (Figs. 2 and 3). The design will be user friendly and built on a general purpose computer to test our approach, including HO soft inputs with an appropriate graphical drop down menus. The subtasks and deliverables are:

(i) Conceptual Design of HOME Team GUI
(ii) GUI Integration (Data Bases, Data Alignment, Fusion Filters, Decision Algorithms).

7.8 Current Research and Development

At this point in our research and development we are well advanced in the second application mentioned in Sect. 7.1 above, Banking Credit Risk assessment, whereas both soft and hard data base are formed from local Bosnia and Herzegovina banking market, plus a much large data base from Malaysia which is in the design. The ultimate goal is to produce an Expert System to improve HO (Banking credit expert) decision making ability, and also reduce his/her soft data errors and subjective credit related judgments. All the theoretical work related to data fusion, hard and soft data filters, including uncertainty alignment are completed and partially reported in [13–16].

8 Conclusion

This conceptual paper presents new methodology for soft hard data modeling, fusion and filtering with the aim to support human decision making process, as well as a generic soft hard data fusion for other applications besides decision making. The pair Human Operator and Machine are treated as a joint HOME Team which cooperates to assist in data fusion and decision making, aided by a user friendly GUI for entering, presenting, fusing and data learning. Various applications can be handled by this approach which improves human effectiveness and also reduces human soft errors by way of data learning feature. Currently methodology is being used to handle bank risk assessment improvement as well mathematical modeling in the area of Block Chain and crypto coin mathematical infrastructure foundation.

References

1. Jenkins, M.P., Gross, G.A., Bisantz, A.M., Nagi, R.: Towards context aware data fusion: Modeling and integration of situationally qualified human observations to manage uncertainty in a hard [plus] soft fusion process. Inf. Fusion **21**, 130–144 (2015)
2. Hall, D.L., McNeese, M.D., Hellar, D.B., Panulla, B.J., Shumaker, W.: A cyber infrastructure for evaluating the performance of human centered fusion. In: Proceedings of the 12th International Conference on Information Fusion, pp. 1257–1264. IEEE (2009)
3. Nachouki, G., Quafafou, M.: Multi-data source fusion. Inf. Fusion **9**, 523–537 (2008)
4. Khaleghi, B., Khamis, A., Karray, F.O., Razavi, S.N.: Multisensor data fusion: a review of the state-of-the-art. Inf. Fusion **14**, 28–44 (2013)
5. Smirnov, A., Levashova, T., Shilov, N.: Patterns for context-based knowledge fusion in decision support Systems. Inf. Fusion **21**, 114–129 (2015)
6. Wozniak, M., Grana, M., Corchado, E.: A survey of multiple classifier systems as hybrid systems. Inf. Fusion **16**, 3–17 (2014)
7. Kurc, T., Janies, D.A., Johnson, A.D., Langella, S., Oster, S., Hastings, S., Habib, F., Camerlengo, T., Ervin, D., Catalyurek, U.V., Saltz, J.H.: An XML-based system for synthesis of data from disparate databases. J. Am. Med. Inform. Assoc. **13**(3), 289–301 (2006)
8. Kaufmann, A., Gupta, M.M.: Introduction to Fuzzy Arithmetic, Theory and Applications. Van Nostrand Reinhold (1985)
9. Belman, R.E., Zadeh, L.A.: Decision making in a fuzzy environment. Manag. Sci. **17**(4), B141–B164 (1970)
10. Durrant-Whyte, H.: Multi Sensor Data Fusion. University of Sydney (2001)
11. Zadeh, L.A.: Fuzzy sets as basis for a theory of possibility. Fuzzy Sets Syst. **1**, 3–28 (1978)
12. Sumari, A.D.W., Ahmad, A.S.: Design and implementation of multi agent-based information fusion system for decision making support. ITB J. ICT **2**(1), 42–63 (2008)
13. Hodzic, M.: Fuzzy to random uncertainty alignment. SEJSC **5**(1), 58–67 (2016)
14. Hodzic, M.: Uncertainty balance principle. PEN **4**(2), 17–32 (2016)
15. Hodzic, M.: Soft to hard data transformation using uncertainty balance principle. In: International Symposium on Innovative and Interdisciplinary Applications of Advanced Technologies, IAT 2017. Advanced Technologies, Systems, and Applications II, pp. 785–809. Springer, January 2018
16. Brkić, S., Hodžić, M., Djanić, E.: Fuzzy logic model of soft data analysis for corporate client credit risk assessment in commercial banking. ICEI, Tuzla, Bosnia and Herzegovina, December 2017
17. Leon-Garcia, A.: Probability, Statistics, and Random Processes for Electrical Engineering. Pearson, London (2008)
18. IBM: Global Technology Outlook (2012). www.IBM.com
19. Dubois, D., Prade, H.: Possibility Theory. Plenum, New York (1988)
20. Dubois, D., Prade, H.: Possibility theory and its applications: where do we stand? IRIT-CNRS, Universit´e Paul Sabatier, 31062 Toulouse Cedex 09, France (2011)

Soft Data Modeling via Type 2 Fuzzy Distributions for Corporate Credit Risk Assessment in Commercial Banking

Sabina Brkić[1(✉)], Migdat Hodžić[2], and Enis Džanić[3]

[1] American University in Bosnia and Herzegovina, Tuzla, Bosnia and Herzegovina
sabinabrkic@outlook.com
[2] International University Sarajevo, Sarajevo, Bosnia and Herzegovina
mhodzic@ius.edu.ba
[3] Univerzitet u Bihaću, Bihać, Bosnia and Herzegovina
enis.dzanic@bih.net.ba

Abstract. The work reported in this paper aims to present possibility distribution model of soft data used for corporate client credit risk assessment in commercial banking by applying Type 2 fuzzy membership functions (distributions) for the purpose of developing a new expert decision-making fuzzy model for evaluating credit risk of corporate clients in a bank. The paper is an extension of previous research conducted on the same subject which was based on Type 1 fuzzy distributions. Our aim in this paper is to address inherent limitations of Type 1 fuzzy distributions so that broader range of banking data uncertainties can be handled and combined with the corresponding hard data, which all affect banking credit decision making process. Banking experts were interviewed about the types of soft variables used for credit risk assessment of corporate clients, as well as for providing the inputs for generating Type 2 fuzzy logic membership functions of these soft variables. Similar to our analysis with Type 1 fuzzy distributions, all identified soft variables can be grouped into a number of segments, which may depend on the specific bank case. In this paper we looked into the following segments: (i) stability, (ii) capability and (iii) readiness/willingness of the bank client to repay a loan. The results of this work represent a new approach for soft data modeling and usage with an aim of being incorporated into a new and superior soft-hard data fusion model for client credit risk assessment.

1 Introduction

Fuzzy set theory and fuzzy logic were introduced by Zadeh in 1965 who was almost single-handedly responsible for the early development in this field. Fuzzy Set Theory is a mathematical theory for describing impreciseness, vagueness and uncertainty. The first fuzzy logic framework is referred to as Type 1 Fuzzy Logic Sets and Systems. As an extension of his theory of fuzzy sets and fuzzy logic [17], the "Theory of Possibility" was developed by Zadeh in 1978 [19] in which he explained that possibility distributions were meant to provide a graded semantics to natural language statements by

S. Avdaković (Ed.): IAT 2018, LNNS 59, pp. 457–469, 2019.
https://doi.org/10.1007/978-3-030-02574-8_36

interpretation of membership functions of fuzzy sets as possibility distributions. He introduced the concept of possibility fuzzy distributions, contrary to random and probabilistic distributions, and noticed that what is probable must initially be possible, but not vice versa. In [19] Zadeh wrote that in dealing with soft data, encountered in various fields, the standard practice was to rely almost completely on probability theory and statistics and he stressed out that those techniques could not cope effectively with those problems in which the softness of data is non-statistical in nature. Soft data encounter predominance of fuzziness. Author's rationale for using fuzzy logic for soft data analysis "rests on the premise that the denotations of imprecise terms which occur in soft database are for the most part fuzzy sets rather than probability distributions" [20]. The difference between probability and possibility is that the concept of possibility is an abstraction of our intuitive perception while concept of probability depends on likelihood, frequency, proportion or strength of belief.

Fuzzy logic has been utilized in various industry areas such as, in artificial intelligence, computer science, control engineering, decision theory, expert systems, logic, management science, operations research, pattern recognition and robotics [23]. Considering risk assessment, many studies of fuzzy logic have appeared in different business areas such as information security, software development, ground water nitrate risk management, system failure, civil hazardous materials, natural hazards, bank, etc. [21].

The main criticism of Type 1 Fuzzy Logic Systems is in its limited capability to directly handle data uncertainties [14], considering that the membership grade in the fuzzy set is expressed exactly i.e. making it a crisp value. Therefore Type-2 fuzzy sets and systems are introduced by the inventor of fuzzy sets [18] which generalize Type-1 fuzzy sets and systems so that more uncertainty can be tackled. The main strength of type-2 fuzzy logic is its ability to deal with the second-order uncertainties that arise from several sources [9] and are preferred over Type 1 fuzzy systems in highly uncertain environment to better handle uncertainty [22]. Considering Type 2 Fuzzy Systems advantages references [22, 13] generated lot of interest in the research community. Type 2 models offer more modeling flexibility in various practical applications.

The purpose of this study is to design and develop possibility distribution modeling of soft data used for corporate client credit risk assessment in commercial banking by applying Type 2 fuzzy membership functions (distributions) for the purpose of developing a new expert decision-making fuzzy model for evaluating credit risk of corporate clients in a bank. Terms fuzzy and possibility distributions are used interchangeably. The paper is an extension of previous research conducted on the same subject which was based on Type 1 fuzzy distributions. Our aim in this paper is to address inherent limitations of Type 1 fuzzy distributions so that broader range of banking data uncertainties can be handled and combined with the corresponding hard data, which all affect banking loan decision making process.

Expert sample is created ad hoc with a commercial bank in Bosnia and Herzegovina that was willing to take part in this project at this initial phase. We are now in a process of adding data from other local banks for the purpose of expanding the relevant soft database. Top senior credit risk assessment experts from this bank were interviewed and they have provided all information about the process, data processing and inputs used for credit risk assessment. Experts have provided inputs for generating universe of discourse, as well as the number and description of membership functions related to each soft variable. Data processing is done by listing all identified soft variables and by mapping their membership values into membership functions based on inputs from interviewed experts. Results of Type 1 fuzzy distributions from previous research [3] are incorporated in this study.

The results of this work represent a new approach for soft data usage/assessment with an aim of being incorporated into a new and superior soft-hard data fusion model by applying the method of Uncertainty Balance Principle [6, 7] for the purpose of creating a new decision-making fuzzy model of credit risk assessment that will assist bank managers in identifying credit risk factors and improve evaluation of the corresponding default risks of their loan applicants. Design and development of Type 1 and Type 2 possibility distributions of soft data/variables used for corporate client credit risk assessment serve as critical steps in this process.

In this paper, we first present a general overview of credit risk assessment in commercial banking. Following section provides an overview of the results of this study based on which Type 2 fuzzy distribution model of identified soft variables is developed, used by the bank for assessing the credit risk of a corporate loan applicant. Finally, we make conclusions and give directions for future research.

2 Credit Risk Assessment in Commercial Banking

As it was described in our previous research, credit risk is one of the largest risks faced by commercial banks and it is assuming increased importance in a changing regulatory regime and quite volatile market conditions. Risk analysis techniques are powerful tools that help professionals manage uncertainty and can provide valuable support for decision making. These techniques can be either qualitative or quantitative depending on the information available and the level of detail that is required [2]. Quantitative techniques rely heavily on statistical approaches while qualitative techniques rely more on judgment than on statistical calculations.

The complex and uncertain nature of loan processing has enforced banks to make loan decisions by utilizing experienced lending officers to perform the essential tasks and evaluations. A loan officer has to fully understand the level of risk a loan would entail and thus has to understand and assess the following: the financial position, repayment ability and strength of the company, whether the company has a sound record of credit worthiness, work history, what is applicants experience and management skills, does the company have a sound business plan which demonstrates his/her understanding of the business and his/her commitment to the success of the business, is company's cash-flow solid and stable, willingness to repay debt and many more. Such analysis incorporates not only the economic data but also the qualitative information concerning the borrower. Data which is subject to this analysis can be classified as hard and soft data. Hard data is usually objective, they express a measure and thus are measurable, quantitative and crisp, while soft data is linguistic, qualitative, subjective and non-measurable.

Besides loan officers, banks usually use various types of scoring models to assess credit risk of a borrower before disbursing a loan. Scoring models were initially introduced to standardize the decision making process and to increase the transparency of a bank's business. They are usually estimated with historical data and statistical methods. Scoring models generally do not follow the Basel II regulatory capital framework definitions since their primary aim is not to fulfill the supervisory requirements but to provide internal decision support. Credit scoring could also be considered as a data mining technique, introduced in 1950s, and since then many methods for applying this technique for credit scoring have been proposed. They can, in general be classified as hard and soft models of data mining [12, 16]. Soft techniques use fuzzy logic compared to crisp in case of hard techniques. In order to overcome shortcomings of credit scoring models many researches have suggested the use of hybrid methods, which use the advantages of various models. Thus, a single model may not be sufficient in order to identify all the characteristics of the data [10]. An example of such model is a soft version of traditional multi-layer perceptrons which is proposed as an alternative classification model, using the unique soft computing advantages of fuzzy logic in which instead of crisp weights and biases, fuzzy numbers are used in multi-layer perceptrons for better modeling of the uncertainties in financial markets [11].

Statistical theory offers a variety of methods for statistical risk assessment. In general, such statistical models use the borrower's characteristic indicators (usually data from financial statements) and (if possible) macroeconomic variables which were collected historically and are available for defaulting (or troubled) and non-defaulting borrowers. Depending on the statistical application of this data, various methods can be used to predict the performance of a borrower. These methods have a common feature in that they estimate the correlation between the borrowers' characteristics and the state of default in the past and use this information to build a forecasting model. The Internal Rating Based Approach (IRBA) of the New Basel Capital Accord allows banks to use their own rating models for the estimation of probabilities of default (PD) as long as the systems meet specified minimum requirements [1]. Most common statistical methods for building and estimation of such models are Regression Analysis, Discriminant Analysis, Logit and Probit Models, Panel Models, Hazard Models, Neural Networks, Decision Trees [5].

Traditional risk models are based on probability and classical set theory which are widely used for assessing market, credit, insurance and trading risk. However, many risks still cannot be analyzed sufficiently by applying classical probability models because of lack of sufficient experience data, lack of knowledge and vagueness, as well as complex cause-and-effect relationships that are inherent in certain risk types. Many authors believe that the best way to solve obstacles in facing with any type of uncertainties is by utilizing fuzzy logic and theory of possibility. It provides a mathematical advantage to capture the uncertainties associated with human cognitive processes, such as thinking and reasoning. "Fuzzy logic is a superset of conventional (Boolean) logic that has been extended to handle the concept of partial truth, truth values between completely true and completely false" [4]. By applying fuzzy logic most variables of a model are described in linguistic terms which makes fuzzy logic models more intuitively similar to the human reasoning. For risks that do not have a proper quantitative probability model, a fuzzy logic system can help model the cause-and-effect relationships, assess the degree of risk exposure and rank the key risks in a consistent way, considering both the available data and experts opinions [15].

3 Soft Data Modeling via Type 2 Fuzzy Distributions

With an aim to eliminate the uncertainty of Type 1 fuzzy distribution results we have extended our previous research [3] to Type 2 fuzzy distributions and have created a comprehensive database of soft data fuzzy Type 1 and 2 based on inputs provided by interviewed experts. In this section we show the results of Type 2 fuzzy distributions which are used by the targeted bank for the purpose of credit risk assessment of corporate clients.

We conducted a series of interviews with credit risk specialists from a local bank which resulted in a database of soft data used for credit risk assessment of corporate clients. It contains 12 main soft variables that have been analyzed from the perspective of five possible outcome/states per variable, generating a total of 60 fuzzy distributions per expert, as well as per type of fuzzy logic sets and systems (i.e. 60 possibility distributions from the perspective of Type 1 and 60 per Type 2 fuzzy distributions). All identified soft variables can be grouped in following segments: stability, capability and readiness/willingness of the client to repay a loan. Each of these segments have a variety of impact on the assessments going from low impact to medium and high. Considering the resulted number of possibility distributions we are not able to show all results, thus here we consider only some examples of the results with a focus on comparison of different perception of the same variable by different expert, provided in Figs. 1, 2, 3, 4, 5, 6, 7, 8, 9, 10, 11 and 12. Moreover, due to confidentiality we do not disclose estimation results that have been given by the bank experts but we are instead showing graphical illustration of the possibility distribution results. The examples of results shown here are chosen so that they can illustrate various types of distributions e.g., triangles, trapezoids, S-membership function, Z-membership function etc.

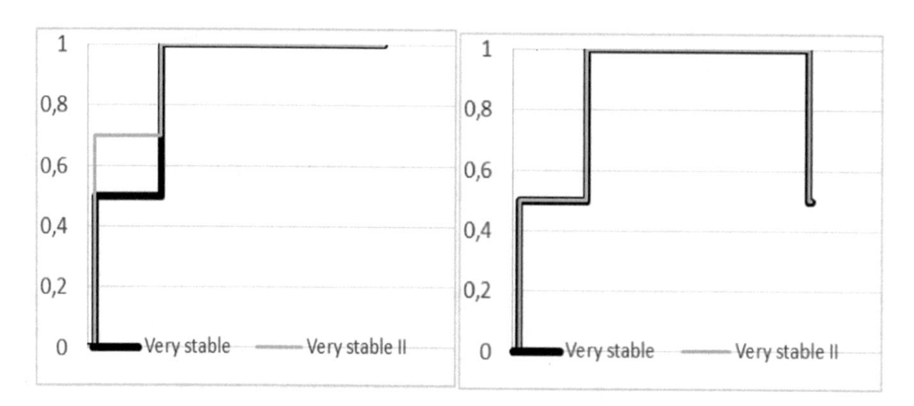

Fig. 1. Very stable Type 1 and Type 2 fuzzy distributions with comparison of perception between two experts of Stability/Capability of the loan applicant based on company size considering its total assets

Fuzzy membership functions examples of Type 2 Stability and capability of a company assessed considering company's size based on its total assets, total income, as

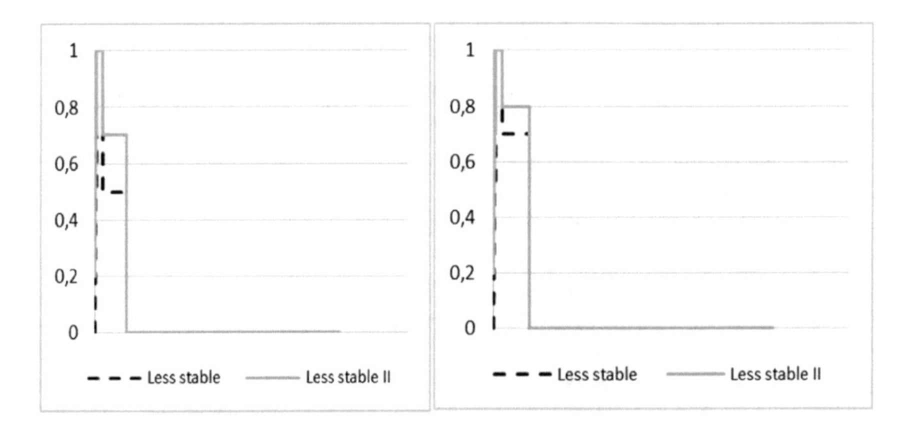

Fig. 2. Less stable Type 1 and Type 2 fuzzy distributions with comparison of perception between two experts of Stability/Capability of the loan applicant based on company size considering its total income

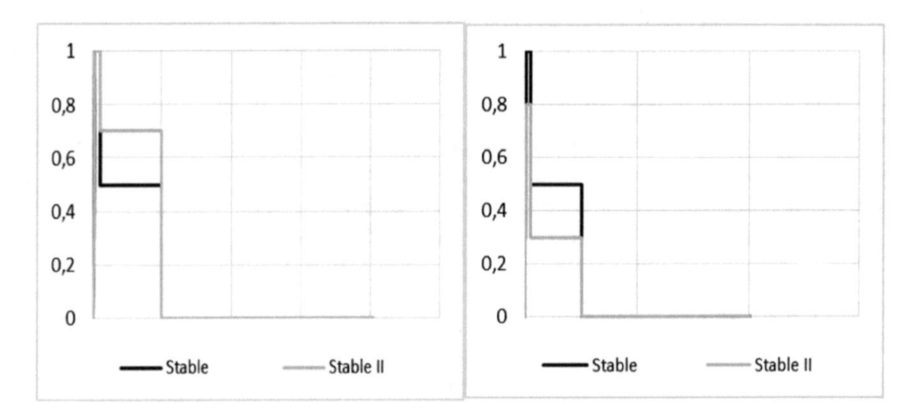

Fig. 3. Stable Type 1 and Type 2 fuzzy distributions with comparison of perception between two experts of Stability/Capability of the loan applicant based on company size considering its total number of employees

well as total number of employees, are shown in Figs. 1, 2 and 3. The results for example in Fig. 1 show a shift in Type 2 membership function in case of one expert, while no shift in case of the perception of another expert for the same variable. In examples shown in Figs. 2 and 3 there is a clear shift of membership function assessed by both experts in case of Type 2 evaluation.

Figure 4 demonstrates comparison between less stable Type 1 and Type 2 fuzzy distribution of Stability of the loan applicant considering number of years the company is doing business, as well as comparison in the perception of two experts from the same bank. In case of the first expert Type 2 membership function is shifted to the right while there is an increase in the membership grade in the case of the second expert.

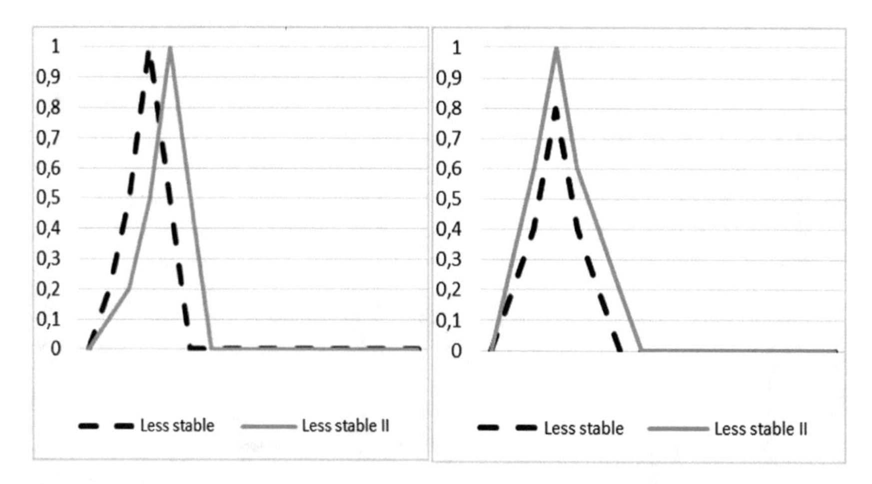

Fig. 4. Less stable Type 1 and Type 2 fuzzy distributions with comparison of perception between two experts of Stability of the loan applicant considering number of years the company is doing business

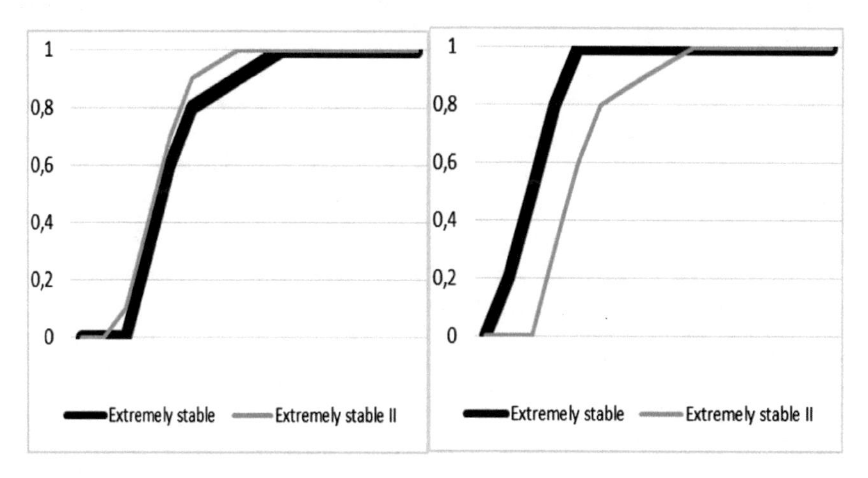

Fig. 5. Extremely stable Type 1 and Type 2 fuzzy distributions with comparison of perception between two experts of Stability of the loan applicant considering number of years it operates profitably (considering operating income)

Extremely stable Type 2 membership function, shown in Fig. 5, is shifted to the left in case of one expert's perception, while in the case of another expert the membership function is shifted to the right.

Figures 6, 7 and 8 demonstrate similar shapes of trapezoid possibility distributions in case of different expert evaluation but are shifted in case of Type 2 evaluation in all shown examples.

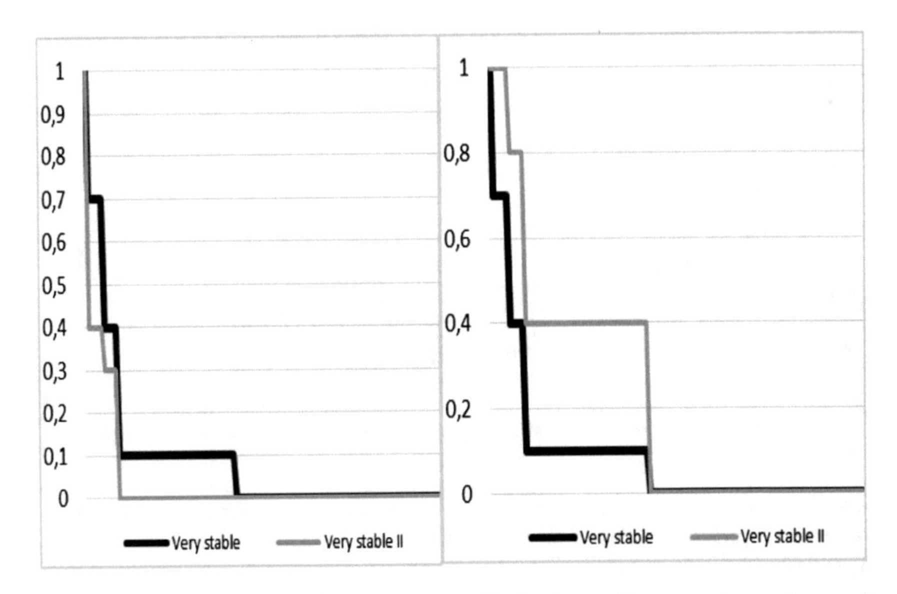

Fig. 6. Very stable Type 1 and Type 2 fuzzy distributions with comparison of perception between two experts of Stability of the loan applicant considering number of days of blocked bank accounts in the last year

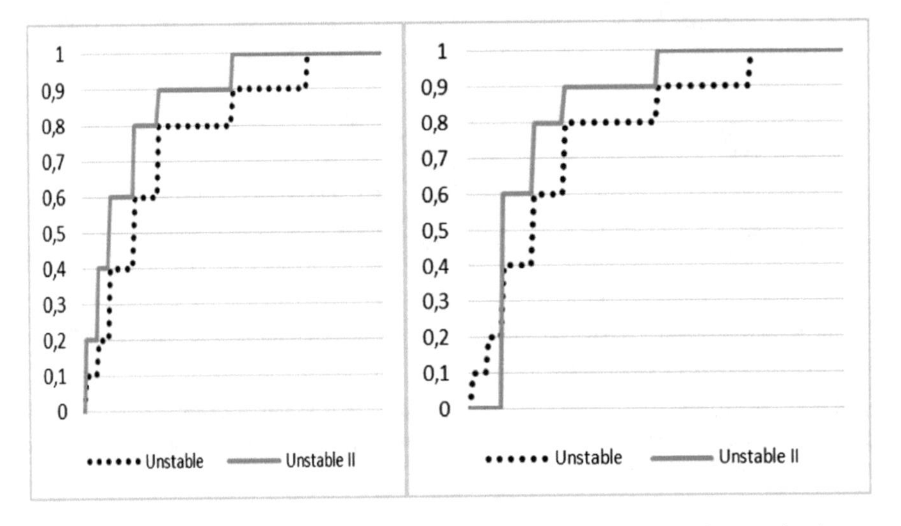

Fig. 7. Unstable Type 1 and Type 2 fuzzy distributions with comparison of perception between two experts of Stability of the loan applicant considering company's repayment history in the bank (if already a client)

Example of less stable Type 1 and Type 2 fuzzy distributions of Stability of the loan applicant considering future development of the company, shown in Fig. 9, express different shapes of membership function between two experts from the same

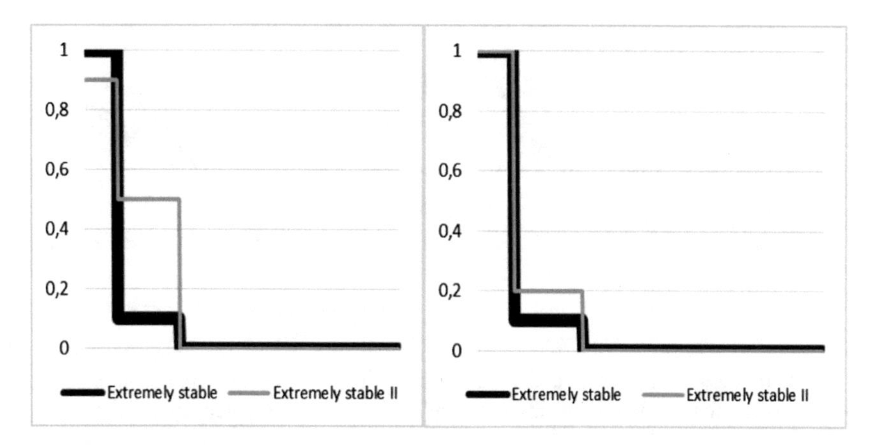

Fig. 8. Extremely stable Type 1 and Type 2 fuzzy distributions with comparison of perception between two experts of Stability of the loan applicant considering company's worst regulatory classification found in the Central Credit Registry

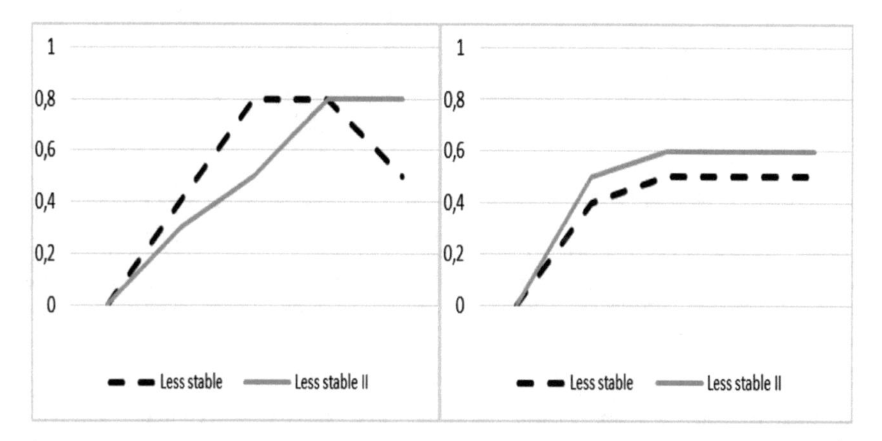

Fig. 9. Less stable Type 1 and Type 2 fuzzy distributions with comparison of perception between two experts of Stability of the loan applicant considering future development of the company

bank. Results of expert evaluation shown in left graph of Fig. 9 indicate shift to the right along the universe of discourse, while results from another expert shown in the right part of the Fig. 9 indicate a shift upwards along the membership grade.

In case of the Stability of the loan applicant considering company's competition extremely stable Type 1 and Type 2 fuzzy distributions demonstrate in both cases a Z-membership function. However, in the perception of one expert Type 2 membership function is expressed mostly through a slight shift to the right, while an increase in the membership grade is expressed by another expert for the same variable in case of Type 2 estimation.

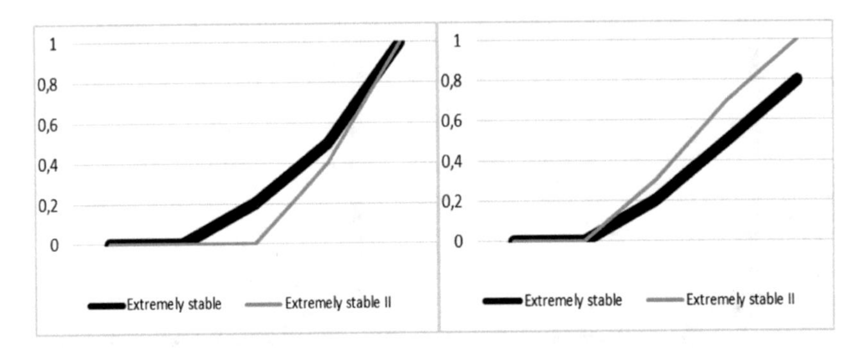

Fig. 10. Extremely stable Type 1 and Type 2 fuzzy distributions with comparison of perception between two experts of Stability of the loan applicant considering company's competition

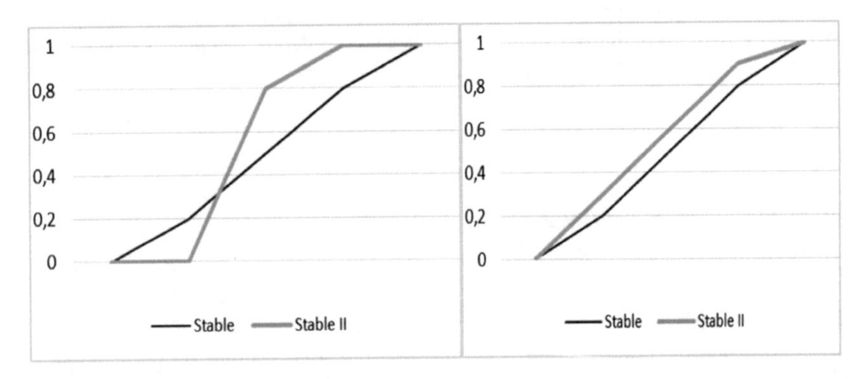

Fig. 11. Stable Type 1 and Type 2 fuzzy distributions with comparison of perception between two experts of Readiness/Willingness/Character of the management of the company to repay the loan

Result of one expert estimation of Type 2 stable company based on Readiness/Willingness/Character of the management of the company to repay the loan (Fig. 11) show a right trapezoid membership function compared to Type 1 where the result indicate a Z-membership function. Type 2 estimation of the same variable of a different expert from the same bank show an upwards shift of the same type of possibility distribution.

Finally, example provided in Fig. 12 with the results of less stable Type 1 and Type 2 fuzzy distributions of Capability/Quality of the company's management, with comparison of perception between two experts, display different shapes of membership functions. However, in case of Type 2 evaluation they demonstrate a shift upwards along the membership grade in the perception of both experts.

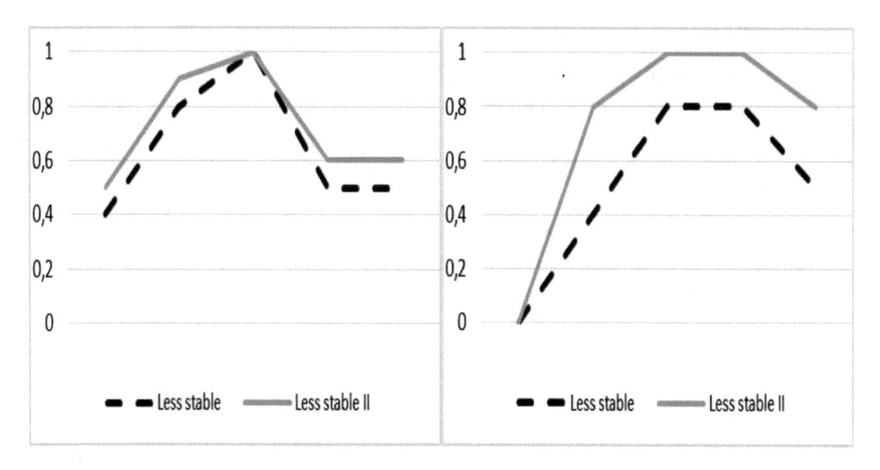

Fig. 12. Less stable Type 1 and Type 2 fuzzy distributions with comparison of perception between two experts of Capability/Quality of the company's management

4 Conclusion

In this paper, Type 2 fuzzy logic possibility distribution modeling of main soft variables, used for corporate client credit risk assessment in commercial banking. Banking experts were interviewed about the types of soft variables used for credit risk assessment of corporate clients, as well as for providing the inputs for generating Type 2 fuzzy logic membership functions of these soft variables. Similar to our analysis with Type 1 fuzzy distributions, all identified soft variables can be grouped into a number of segments, which may depend on the specific bank case. In this paper we looked into the following segments: (i) stability, (ii) capability and (iii) readiness/willingness of the bank client to repay a loan.

Our aim in this paper is to address inherent limitations of Type 1 fuzzy distributions. As demonstrated in our results, the Type 2 fuzzy distributions incorporate second-order uncertainties and thus include a broader range of banking data uncertainties which can be handled and combined with the corresponding hard data, in order to finally improve the loan decision making process.

With this we have built a thorough soft database with Fuzzy type 1 and 2 fuzzy logic possibility distributions based on expert interviews. In order to further improve the analysis we are currently in the process of expanding the soft database with inputs from other banks and experts. This database presents a new methodology for transforming linguistic and intuitive (soft) information about bank credit risk data into a series of mathematical fuzzy (possibility) distributions which can be handled quantitatively and combined (fused) with related probabilistic data. The results of this work represent a new approach for corporate client soft data usage/assessment in commercial banking with an aim of finally being incorporated into a new and superior soft-hard data fusion model via the Uncertainty Balance Principle (Hodzic [6], pp. 58–66, Hodzic [7], pp. 17–32) for the purpose of client credit risk assessment and other similar assessments. The Uncertainty Balance Principle (UBP) was defined to express uncertain data vagueness as represented by a fuzzy data models, with a non-uniqueness of related random data distributions (Hodzic [8], pp. 785–809).

Type 2 fuzzy logic gives a better basis for applying the method of UBP. This method transfers the fuzzy distribution in equivalent random (hard) distribution which is then combined with the original hard distribution of probability of default via the process of soft hard data fusion. This will be reported in our future papers. Finally, soft hard data fusion offers improved data for credit risk assessment.

Our final aim is to be able to improve bank credit risk assessments by using exact and more precise mathematical methodology.

References

1. Bank for International Settlements: International Convergence of Capital Measurement of Capital Measurement. Basel Committee on Banking Supervision, A Revised Framework Comprehensive Version, June 2006
2. Bennett, J.C., Bohoris, G.A., Aspinwall, E.M., Hall, R.C.: Risk analysis techniques and their application to software development. Eur. J. Oper. Res. **95**(3), 467–475 (1996)
3. Brkic, S., Hodzic, M., Dzanic, E.: Fuzzy logic model of soft data analysis for corporate client credit risk assessment in commercial banking. In: Fifth Scientific Conference with International Participation "Economy of Integration" ICEI 2017. SSRN, November 2017. https://ssrn.com/abstract=3079471
4. Gupta, V.K., Celtek, S.: A fuzzy expert system for small business loan processing. J. Int. Inf. Manag. **10**, Article 2 (2001). http://scholarworks.lib.csusb.edu/jiim/vol10/iss1/2
5. Hayden, E., Porath, D.: Statistical methods to develop rating models. In: The Basel II Risk Paramenters. Springer, London, pp. 1–12 (2011)
6. Hodzic, M.: Fuzzy to random uncertainty alignment. Southeast Eur. J. Soft Comput. **5**, 58–66 (2016)
7. Hodzic, M.: Uncertainty balance principle. IUS Period. Eng. Nat. Sci. **4**(2), 17–32 (2016)
8. Hodzic, M.: Soft to hard data transformation using uncertainty balance principle. In: International Symposium on Innovative and Interdisciplinary Applications of Advanced Technologies, IAT 2017. Advanced Technologies, Systems, and Applications II. Springer, pp. 785–809 (2017)
9. Karnik, N.N., Mendel, J.M., Liang, Q.: Type-2 fuzzy logic systems. IEEE Trans. Fuzzy Syst. **7**(6), 643–658 (1999)
10. Khashei, M., Bijari, M., Hejazi, S.R.: Combining seasonal ARIMA models with computational intelligence techniques for time series forecasting. Soft Comput. **16**(6), 1091–1105 (2012)
11. Khashei, M., Mirahmadi, A.: A soft intelligent risk evaluation model for credit scoring classification. Int. J. Financ. Stud. **3**, 411–422 (2015)
12. Lando, D.: Credit Risk Modeling: Theory and Applications. Princeton Series in Finance. Princeton University Press, Princeton (2004)
13. Mendel, J.M.: Type-2 fuzzy sets: some questions and answers. In: IEEE Connections, Newsletter of the IEEE Neural Networks Society 1, pp. 10–13 (2003)
14. Mendel, J.M.: Type-2 fuzzy sets and systems: an overview. IEEE Comput. Intell. Mag. **2**, 20–29 (2007)
15. Shang, K., Hossen, Z.: Applying fuzzy logic to risk assessment and decision-making, casualty actuarial society. Canadian Institute of Actuaries, Society of Actuaries (2013)

16. Thomas, L.C., Edelman, D.B., Crook, J.N.: Credit Scoring and its Applications. SIAM Monographs on Mathematical Modeling and Computation. SIAM, Philadelphia (2002)
17. Zadeh, L.A.: Fuzzy sets. Inf. Control **8**, 338–353 (1965)
18. Zadeh, L.A.: The concept of a linguistic variable and its application to approximate reasoning–1. Inf. Sci. **8**, 199–249 (1975)
19. Zadeh, L.A.: Fuzzy sets as a basis for a theory of possibility. Fuzzy Sets Syst. **1**, 3–28 (1978)
20. Zadeh, L.A.: Possibility theory and soft data analysis, Selected papers by Lotfi Zadeh, pp. 515–541 (1981)
21. Zirakja, M.H., Samizadeh, R.: Risk analysis in e-commerce via fuzzy logic. Int. J. Manag. Bus. Res. **1**(3), 99–112 (2011)
22. Wu, D.: On the fundamental differences between interval type-2 and type-1 fuzzy logic controllers. IEEE Trans. Fuzzy Syst. **20**(5), 832–848 (2012)
23. Zimmermann, H.-J.: Fuzzy Set Theory – and Its Applications, 4th edn., pp. 158–241, 369–404. Kluwer Academic Publishers (2001)

Design and Experimental Analysis of the Mobile System Based on the Android Platform

Anida Đuzelić(✉)

Faculty of Technical Engineering, University of Bihac,
Bihac, Bosnia and Herzegovina
anidad7@gmail.com

Abstract. A remote control vehicle is typically defined as any mobile device that is controlled by a means that does not restrict its motion with an origin external to the device. This is often a cable between control and vehicle, an infrared controller or nowadays control using smartphone. A remote control vehicle (RCV) differs from a robot in that the RCV is always controlled by a human and takes no positive action autonomously. Based on the Wi-Fi network of remote control vehicle use is mainly through the combination of single-chip and wireless network data, through the mobile terminal to send instructions to achieve vehicle according to the instructions to exercise, and can achieve data transmission. This paper discusses and analyzes the control system of remote control vehicle based on Wi-Fi network technology using smartphone. This control system is based on two technology, microcontroller and control using web server. The system uses data transmition from smartphone to microcontroller which is connected on H-bridge motor. This small DC motor allows movement of vehicle in any directions and courses. It is vital that a vehicle should be capable of proceeding accurately to a target area; maneuvering within that area to fulfill its mission and returning equally accurately and safely to base.

Keywords: Remote control vehicle · Microcontroller · HTTP server
H-bridge · ESP 8266 module · DC motor

1 Introduction

In todays world people expect to be able to do everything with their smartphone or tablet which is always within arm's reach [1]. Due to this it is a big advantage for a product to have an easily accessible interface that works with such devices. In a larger perspective more and more devices get the ability for network connectivity which opens many new possibilities [2]. There is a lot of talk about the internet of things to describe this phenomena where people no longer have only one special device for connectivity but have it conveniently built into most of the devices around them instead [3]. It is in this light Ericsson have predicted that by 2020 over 50 billion devices will be available online which can only be achieved if people have more than one device [4].

This creates a demand for inexpensive and easy to use solutions for adding wireless connectivity to products. Currently there is a high competition in many consumer

S. Avdaković (Ed.): IAT 2018, LNNS 59, pp. 470–479, 2019.
https://doi.org/10.1007/978-3-030-02574-8_37

electronics with many products seen as commodities [3]. This leads to price sensitive customers so the solution has to be cheap and also simple to keep the development cost low.

Finding such a solution is at the core of this project. The solution has to be cheap and have a reasonable complexity [2]. It should also be able to integrate with most embedded applications meaning it cannot take up too many resources. Although here we use it for remote control, the application could be anything from a sensor sending statistics to a server as well as streaming audio. In this paper is represented a prototype by choosing one chip among many available on the market and developed the software needed to get the remote control to work, the prototype has been verified and works.

The Wi-Fi module solution was developed by researching what Wi-Fi chip solutions were available on the market and then choosing one solution that is optimised for low price, ease of use and bandwidth. The Wi-Fi chip ESP8266 was chosen has been developed and constructed [8]. The software part of the paper consist of low level serial peripheral interface (SPI) drivers developed in MikroC code for an PIC microcontroller used in the communication with the Wi-Fi module [6]. To remote control of this project, server and client software have also been implemented. The server side software consists of a HTTP (HyperText Markup Language) server programmed in C code which lets the client access a web interface written in HTTP and Javascript that controls the mobile platform [5]. The web interface layout is automatically generated by a script written in HTML (HyperText Markup Language) and CSS (Cascading Style Sheets) that parses an Extensible Markup Language (XML) file provide.

2 Hardver Components

This chapter introduces the main hardware components used in this project. List of components used in the realization of the assembly are described in this chapter.

2.1 Microcontroller

The microcontroller on control board is an PIC 16F886 made by Microship. PIC is a family of microcontrollers made by MicrochipTechnology. The name PIC initially referred to *Peripheral Interface Controller*. The first parts of the family were available in 1976, by 2013 the company had shipped more than twelve billion individual parts, used in a wide variety of embedded systems [6].

This Microchip PIC16F886 microcontroller series is based on the RISC-Harvard architecture [7]. With the RISC controller there is a shorter instruction execution time than CISC and if the controller is based on RISC architecture, this is necessarily stated in the technical characteristics. The controller is built into high speed CMOS technology that consumes significant current only in transitions from log 0 to log 1 and log 1 to log 0, so it allows less power in RUN mode, i.e. normal mode, extremely low in IDLE and SLEEP modes when the controller runs at a reduced rate of 32 kHz.

2.2 Wi-Fi Module ESP8266

The ESP8266 is low-cost Wi-Fi microchip with full TCP/IP stack and microcontroller capability produced by Sanghai-based Chinese manufacturer, EspressifSystems [5].

The Wi-Fi chip is an off the shelf chip that handles the wireless communication. First the requirements for the Wi-Fi chip is specified before the different alternatives are presented and then comes relevant information about the chosen chip. The data rate column states the data rate for the Wi-Fi part of the communication while the data rates stated in the interface column are the speed of the communication protocol between the Wi-Fi chip and the microcontroller. The data rate stated in the interface column gives a more realistic view of the throughput of the Wi-Fi chip but could not be used to compare the different chips since only a handful of the chips stated the interface data rate in their data sheets.

2.3 Printed Circuit Board (PCB)

A Printed Circuit Board was designed to hold the Wi-Fi module, microcontroller and some supporting components. Connection to the host microcontroller which is already located on the main control board of this platform is done through a 28-pin connector. How everything is connected was specified in a schematic. With help of the schematic a PCB layout was designed and later assembled (Fig. 1).

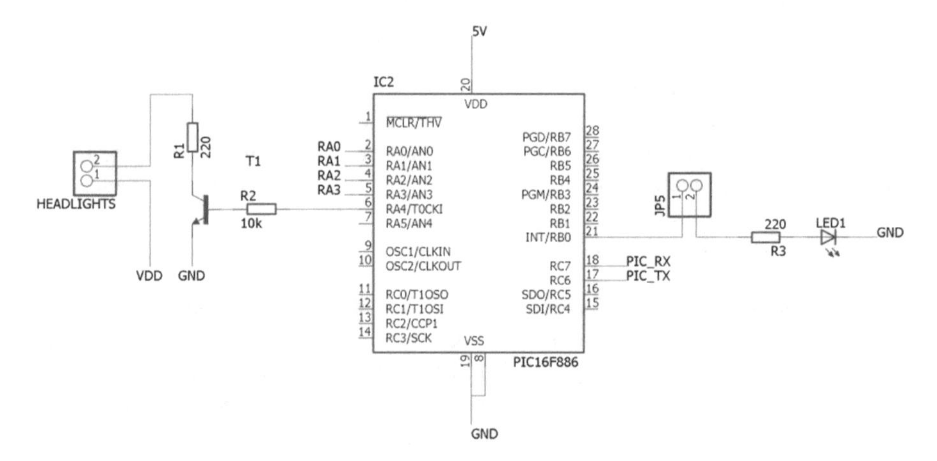

Fig. 1. Electric controller scheme

In general an H-bridge is a rather simple circuit, containing four switching element, with the load at the center, in an H-like configuration (Fig. 2):

The switching elements (Q1–Q4) are usually bi-polar or FET transistors. Integrated solutions also exist but whether the switching elements are integrated with their control circuits or not is not relevant for the most part for this discussion. The diodes (D1–D4) are called catch diodes and are usually of a Schottky type. The top-end of the bridge is connected to a power supply (battery for example) and the bottom-end is grounded. In

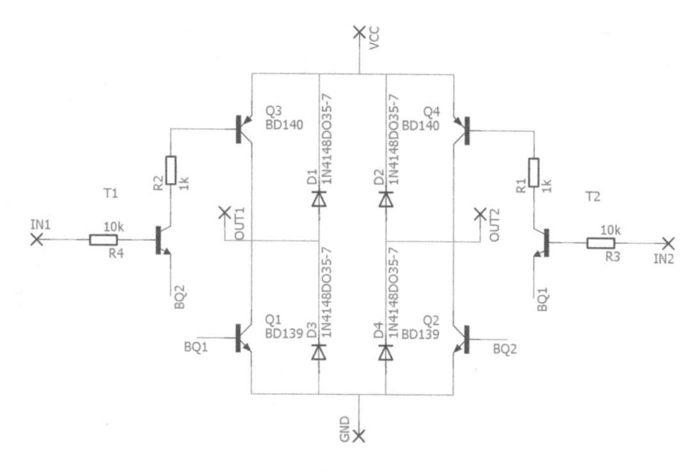

Fig. 2. H-bridge

general all four switching elements can be turned on and off independently, though there are some obvious restrictions [9].

2.4 Software

HTTP Server and Wi-Fi Control interface contains the software for the HTTP server that is used to get access to the remote control functions from an external device. The block also contains the commands that are used to control how the Wi-Fi module connects to the network and which network it chooses. The lower box called the Wi-Fi thread is the part that is run on the microcontroller connected to the Wi-Fi module. As the name implies it runs as a thread on the real time OS of the microcontroller with a low priority. The interface builder generates the web page used by the HTTP server and only has to be executed once.

3 Implementation

This chapter describes design of the system based on the android platform. The system is created using hardware components described in the previous chapter.

Figure 3 shows a high level overview of the Wi-Fi remote control system developed in this project. The figure illustrates how the communication goes from the device, a smartphone, via the Wi-Fi module and finally to the DC motor implemented in a vehicle. Smartphone device has to be connected to the network provided by the Wi-Fi module to send messages between them and enable the remote control features. In this system the smartphone will be referred to as the server and the device as the client.

Figure 4 shows the inside of the client system with the part added in this project labeled as implemented hardware. For the Wi-Fi module there is the constraint of needing to use the existing 28-pin connector and the selection of communication

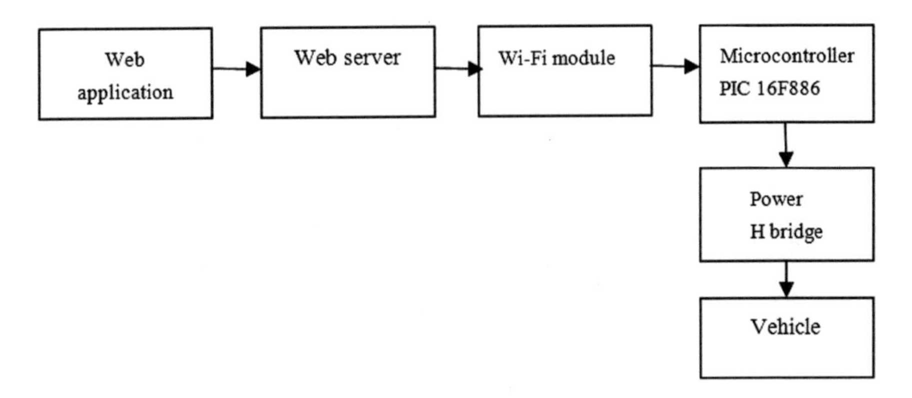

Fig. 3. Functional block diagram of Wi-Fi remote control system that has been developed in this project

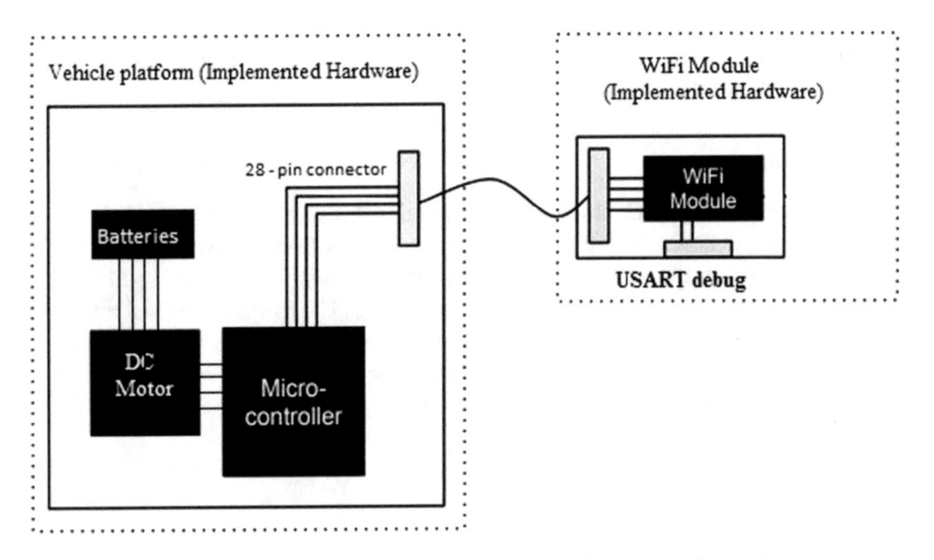

Fig. 4. Low level overview showing how implemented hardware fits in the system.

protocols that are available on the pins. The Wi-Fi module handles receiving and transmitting of the wireless data as well as the network stack. It also handles connecting to the network, connection information is saved in the internal non volatile memory. The microcontroller handles the communication with the Wi-Fi module and runs a server that clients can connect to and remote control through. The microcontroller also handles the rest of the vehicle system including monitoring DC motor. In order to handle all the tasks efficiently the microcontroller runs a real time operating system. Compared to the other tasks running the Wi-Fi solution have a low priority which means it can not use to much resources on the microcontroller.

After soldering the elements and placing all the necessary components in the car body, i got the ultimate look of one part of the circuit (Fig. 5).

Fig. 5. Ultimate look of one part of the circuit

The electronics built into the vehicle are divided into two parts, one of them is intended for receiving and processing control orders from a mobile device, while the other part is responsible for receiving logical signals from the first part and, on the basis of the received logic signals, drives the motors of the vehicle.

Vehicle control is achieved through a simultaneous power source, a 9 V battery. Inside the vehicle is a PIC16F886 microcontroller whose role is to act as a mediator between ESP8266 controllers and control electronics for motors. The vehicle's motors are controlled by a microcontroller, and in order to do this, it was necessary to realize the electro-assembly, the H-bridge.

The speed of the rotor rotation is proportional to the voltage applied to the engine outputs (Fig. 6).

The role of the PIC16F886 microcontroller is to be a mediator between ESP8266 controllers and control electronics for motors. The role of the ESP8266 controller is to receive control commands from the mobile device. These two controllers communicate with each other through the USART protocol. When supplying power to the electronics, the ESP8266 is a true Wi-Fi network called "WiCar" (Fig. 7) and internally at IP address 192.168.13.37 runs an HTTP server that can be accessed by any web browser of the mobile device connected to the WiCar network. This Wi-Fi network is locked for security reasons so that only the person who writes the correct code can only manage the vehicle. So, after switching on the vehicle, the power is brought to the vehicle electronics, the mobile device needs to be found and connected to the Wi-Fi protected Wi-Fi network. The next step after connecting to the WiCar network is to

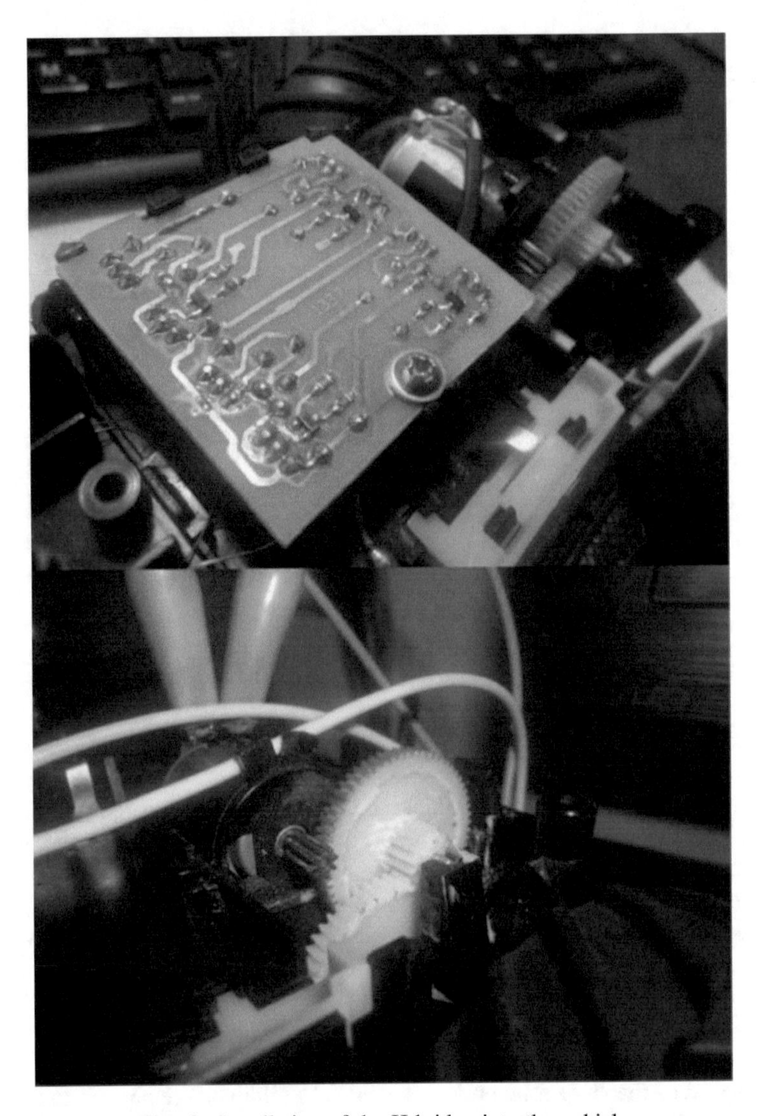

Fig. 6. Installation of the H bridge into the vehicle

enter the web address (chrome, firefox or other) of the mobile device into the address field and open the web page at the IP address 192.168.13.37.

The touch panel application consists of a total of eight touch points. In Fig. 8. is a display of a touch panel of touch panels on the touch points used to control the vehicle. Touchpoint features are:

- The vehicle moves in the direction of the north-west
- The vehicle moves in the direction of the north
- The vehicle moves in the direction of the north-east
- The vehicle is in the place with the front wheels curved to the left
- The vehicle is standing in the place with the front wheels shifted to the right

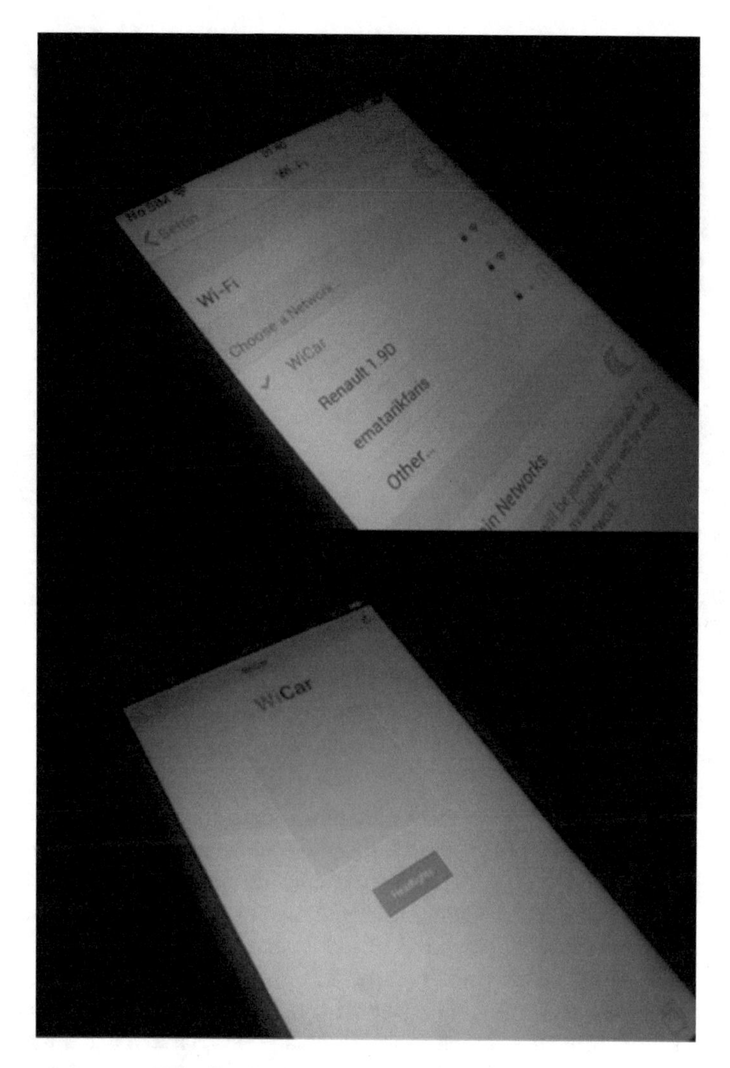

Fig. 7. Connection to the WiCar network

- The vehicle moves in the direction of the south-west
- The vehicle moves in the direction of the south
- The vehicle moves in the direction of the south-east.

Stopping when changing the touch point is not necessary, i.e. it is possible to move through the touch points on a touch-and-hold mode, for example, from point 2 it is possible to move to point 1 without lifting the finger from the screen of the mobile device. Application has one more button, the head lights, that serves for turn on or off head lights.

This web application is one real time application. User sends commands due Web server to a microcontroller that give commands to the vehicle's motor and eneble movement in real time.

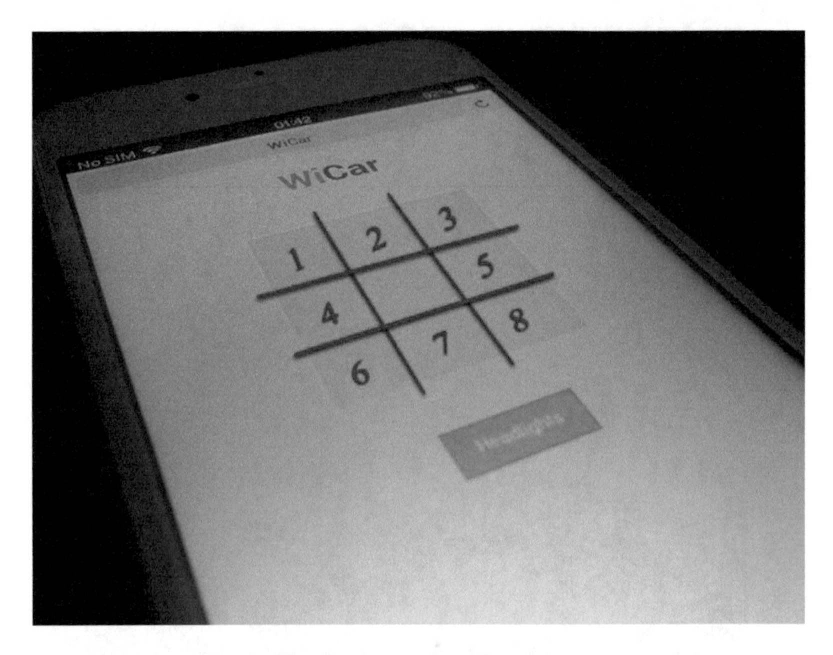

Fig. 8. Areas of touch panels

4 Conclusion

In this paper is described way to design mobile system for remote control of vehicle based on the android platform. Control functions of vehicle are controlled by Web server. Inside the vehicle is a PIC16F886 microcontroller whose role is to act as a mediator between ESP8266 module and electronics of motors. Vehicle movement is controlled due touch pannel. This kind of work serves as a example of what can be achieved by using one small wi-fi module, and that wi-fi is the transmission of the future of remote management. The small wi-fi module creates its own network and works in real time through a server. It would be a good idea to expand the reach, for now it's within a few meters, but it could be considerably expanded. Remote control has a great use in today's application, and has great advantages over manual operation.

References

1. Skoog, A., Stolt, F.: Wi-Fi based remote control of an audio-platform. Lund University, Department of Electrical and Information Technology, Master of Science thesis, Lund 2013. Accessed 12 Feb 2018
2. Zheng, Z., Cheng, J., Peng, J.: Design and implementation of teaching system for mobile crossplatform. College of Information Science and Technology, Hainan University, Haikou, China. Accessed 10 July 2017

3. Ugwu, C., Mesigo, T.: A novel mobile wallet based on Android OS and quick response code technology. Department of Computer Science, University of Port Harcourt, Rivers, Nigeria. Accessed 10 July 2017
4. Zhang, Y.-L., Wang, X.-K.: LBS mobile learning system based on android platform. Henan Mechanical and Electrical Vocational College, Zhengzhou. Accessed 10 July 2017
5. Ericsson, more than 50 billion connected devices, February 2011. Accessed 12 Feb 2018
6. http://www.esp8266.com. Accessed 12 Feb 2018
7. https://en.wikipedia.org/wiki/PIC_microcontroller. Accessed 12 Feb 2018
8. http://www.microchip.com/wwwproducts/en/PIC16F886. Accessed 12 Feb 2018
9. https://www.espressif.com/en/products/hardware/esp8266ex/overview. Accessed 12 Feb 2018
10. http://www.modularcircuits.com/blog/articles/h-bridge-secrets/h-bridges-the-basics. Accessed 12 Feb 2018

Last Mile at FTTH Networks: Challenges in Building Part of the Optical Network from the Distribution Point to the Users in Bosnia and Herzegovina

Anis Maslo[1(✉)], Mujo Hohzic[1], Aljo Mujcic[2], and Edvin Skaljo[1]

[1] BH Telecom Bosnia and Herzegovina, Sarajevo, Bosnia and Herzegovina
{anis.maslo,mujo.hodziv,edvin.skaljo}@bhtelecom.ba
[2] Faculty of Electrical Engineering Tuzla, Tuzla, Bosnia and Herzegovina
aljo.mujcic@untz.ba

Abstract. Since revenues in the telecom sector are falling, It's getting more and more important to reduce the cost of building of new optical networks. The technical-economic analysis based on practical experience is conducted in this paper. This analysis includes consideration of the costs (CAPEX-Capital Expenditure, OPEX-Operating Expenses) of building a part of the network from the distribution point in a street cabinet to the user. Different types of optical networks as urban area, suburban, rural area 1 and rural area 2 are considered.

1 Introduction

Strong competition in the telecommunications market sets the requirements for seeking cost effective solutions in the construction of optical dedicated network [1, 2]. In this goal the techno economic analyses of the part of network known as a last mile is conducted. In this analysis authors include the practical experience that gained through the construction of optical access networks on the territory of Bosnia and Herzegovina in urban, suburban and rural settlements.

Since the analysis relates to the last mile, the choice of active equipment is not considered. There are many definitions of the term Last mile.

2 Telecommunication Market in B&H

To be clear on which part of network is analysed the Figs. 1 and 2 are given [3].

Our analysis covers a part of the network from the street cabinet to the end user. Street cabinet can be self-standing or mounted on poles and represents the starting point of the analysis, the end point is the socket in the apartment at the end user.

There are three dominant telecom operators in BiH and more than a dozen regional alternative operators. Taking into account that the average salary of employees in BiH is about 450 Euros, each operator must adjust its cost of services to the purchasing power of subscribers. Therefore, prior to the construction of new telecommunication

S. Avdaković (Ed.): IAT 2018, LNNS 59, pp. 480–486, 2019.
https://doi.org/10.1007/978-3-030-02574-8_38

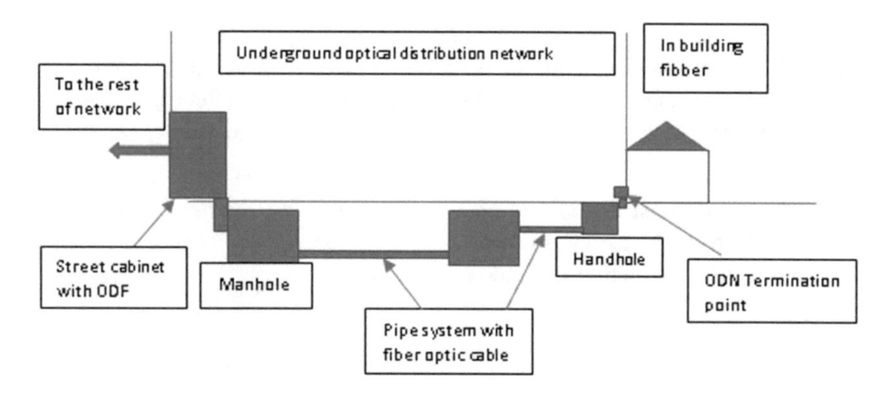

Fig. 1. Underground optical network organization

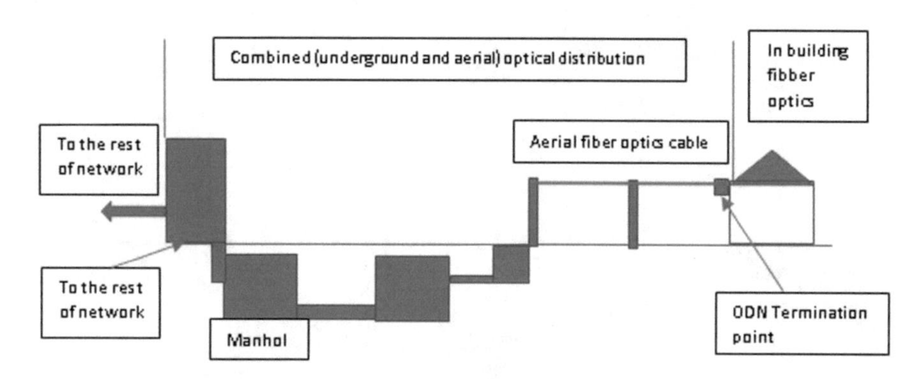

Fig. 2. Combined (underground and aerial) optical network organisation

optical access networks, a quality techno-economic analysis from the aspect: building speed, cost of investment (CAPEX), maintenance cost (OPEX) and the resolution of municipal permits (building permits for construction) should be done.

Market competition in BIH telecommunications grows from month to month. In addition to the three dominant telecom operators, the fourth cable operator has appeared, which can parry to existing operators.

The quality of the service and the speed of the Internet are crucial in market competition. In order to achieve the highest rates of data transmission over the Internet and the quality of the video service, the optical telecommunication access networks (OAN-Optical Access Network) are being built. The data transfer medium in these OANs is the optical fiber [4]. When building the OAN in the so-called Greenfield, it is planned to install an optical fiber as a transfer medium. However, when optimizing access networks, the copper pair is partially or completely changed by optical fiber. Therefore, the construction of an OAN with fiber optics as the transmission medium is a basic condition for the provision of quality service and reliability and OAN. The speed of construction depends primarily on the funds invested.

In addition to the speed of construction, the technology used is essential. Certainly, the speed of construction must not be achieved (realized) on the quality bill.

Analysis of the construction of the OAN (part from street cabinets to a telephone socket) in the last mile that the work being treated can be divided into two parts (Fig. 3):

1. Part relating to construction and erection from the last distribution point (cable connection) to the house installation.
2. Part related to construction of home installations.

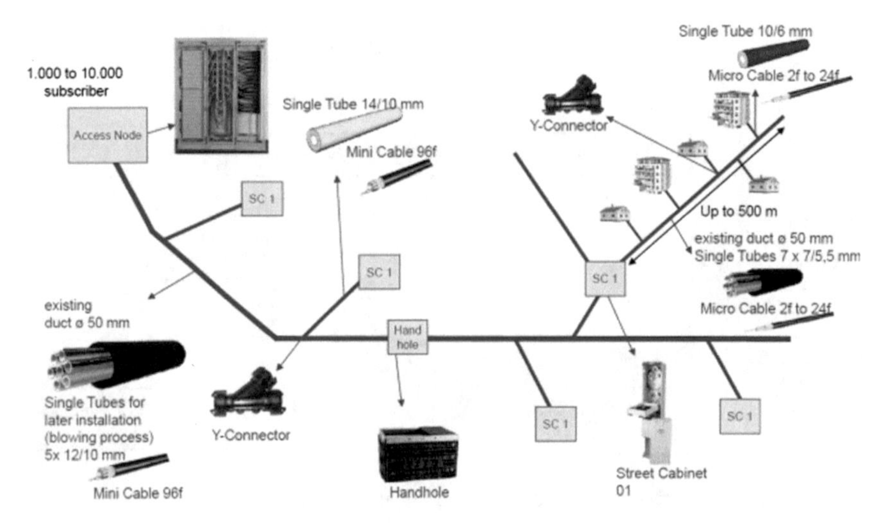

Fig. 3. Optical distribution network in with more details [4]

3 Analysis of Construction and Erection Works from the Last Distribution Point to the Beginning of the Home

These construction works can be realized in three ways:

1. Construction of underground cable duct (CD) through which the optical cables are fed to the user.
2. By building a wooden poles system by means of which the withdrawal of the aerial cable assembly is made to the user.
3. Combined method, where part of the construction work is performed underground and partly by air.

3.1 Construction of Underground Cable Duct

The construction of underground CD implies a complex process of obtaining utility and building permits of different levels of government in BiH. In addition to the complex process of obtaining building permits, it is time consuming and expensive, compared to the aerial network [5].

By analysing OAN in BIH cities: Sarajevo-city network, Bihac- combined urban/suburban network, Gradacac-suburban network, Gorazde-suburban/rural network, received the following parameters:

The average optical cable length from the last street cabinet to the home installation is:

1. City OAN - 93 m
2. Suburban OAN - 62 m
3. Rural OAN - 101 m

So the cost per user (works only in the observed share) for these types of networks is:

1. Urban OAN - 476 Euro
2. Suburban OAN - 526 Euro
3. Rural OAN - 772 Euro

Up to the cost of construction of the connection was made on the basis of Table 1. which gives an overview of the calculation of the individual stages of the works. The final cost of building connection for the user determines (Fig. 4):

- cable length - distance from the distribution point (street cabinet) to the splice box on the house
- length of home installation - distance from the splice box to the socket in the user's apartment,

Table 1. Calculation of the value of the construction of part OAN

Work phases	Method of calculation	Urban OAN	Suburban OAN	Rural OAN	Aerial OAN
Cable length of cable pulling	Price Euro/m	71,5	51	83	47
Drilling price concrete walls	Numbers of drill X 8 Euro/numbers user	3	16	16	1
Numbers of splices	4 Euro/splice	4	4	4	4
Length of PVC ducts made	1,5 Euro/m	37,5	15	15	6
Cost of civil work on the socket	10 Euro/piece	10	10	10	10
Measurement price	6 Euro/piece	6	6	6	6
Price of civil works	8,5 Euro/piece	340	425	637	100
Total sum [Euro]		**426**	**527**	**721**	**174**

- the amount of construction work inside the home - drilling and repair of concrete walls,
- the number of fiber optic connectors at analysed part of the last mile OAN,
- measurement the parameters of the installed fiber optic from the socket to the street cabinet.

Urban area - uses the same track on 75% of the length of construction the network, therefore the connection for users in buildings is cheaper.

Rural area - for single end users, 75% of the routes are used for one user only, so these connections are the most expensive.

The time of construction of the underground CD (Fig. 2) of these works for different types of OAN lasted:

1. City OAN - 1.66 h/m
2. Suburban OAN - 2.5 h/m
3. Rural OAN - 2.5 h/m

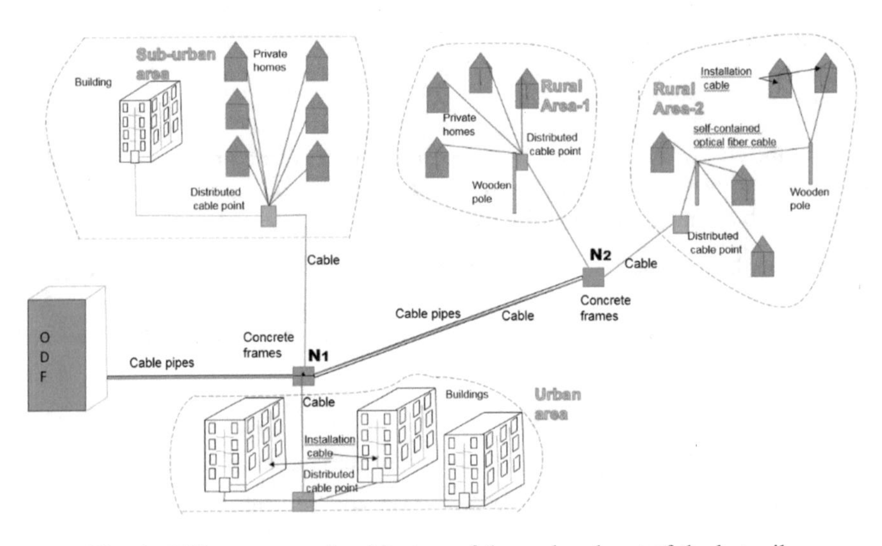

Fig. 4. Different types of architecture of the analyzed part of the last mile

3.2 Analysis of the Network Based on the Application of Aerial Cables (Self-supporting)

In this case, we have analysed the part of the aerial OAN from the point of street cabinet to the home installation on the example of the networks in Sarajevo and Mostar. So, the part of OAN distribution network was built using wooden poles and aerial fiber optic cables. The wooden poles are designed in a way that the stars of the network topology are realized. This means that in the observed distribution parts of the OAN, the optical cable to the end users is supplied directly from the cable outlet (street cabinet) or the wooden poles. In the earlier period cascading connections were used over the roof of private houses. In the event of a malfunction or a request to expand the network, it was impossible to do so due to the prohibition of access to private property/roofs. The time to build these networks is noticeably shorter, as well as the time for obtaining a building permit for utility approval.

The average length of cable from the last street cable outlet to the home installation (Home spice box) is:

1. Suburban OAN - 57 m
2. Rural OAN - 60 m

So the cost per user for parts of these types of networks is:

1. Suburban OAN - 173 Euro
2. Rural OAN - 185 Euro

The time of construction of the underground CD of these works for different types of OAN lasted:

1. Suburban OAN - 1.5 h/m
2. Rural OAN - 1.6 h/m.

However maintenance costs (OPEX) are considerably higher. None of the analyzed networks has reached the age foreseen for investment maintenance. But it is clear that the wooden pillars will have to be replaced in most cases after the age of 20. The fact that the estimated lifespan of the underground built-in CD is about 40 years suggests that these networks are in the best interests of the investor.

4 Analysis of the Home Installations According to Experience from BiH Is as Follows

Various techniques or materials were used when constructing house installations are presented in next [4]:

1. Classical technique using PVC ducts to which cables are subsequently placed.
2. Technique using micro ducts for vertical corridors and PVC ducts for horizontal individual joints with pre-assembled optical cables.
3. Technique with the use of fiber optic cable installations in hardcover (jargon riser cables).

We have preliminarily analyzed the construction of home installations. These three techniques installations (we concluded) are approximately the same for all three techniques. However, the v of installing optical cables in built-in home installations is the slowest and most demanding in the first case. The third case is a practical cable already installed and pull the fibers out of it as we needed. A Second case is optimized to quickly install optical fiber into a micro tube, but not to waste optical fiber without any need.

5 Conclusion

By analysing the cost of constructing various OANs in this paper, we have concluded that the price of construction decreases if the density of population is increased. We also showed that the fiber optic length for urban OAN is the shortest and increases for suburban and rural OAN. By analysing the construction time, it has also been shown that it is being built more quickly with a higher density of population, but it is crucial that the underground CD or the air network is under development. The combined underground/airborne construction method is most effective, especially if the existing infrastructure of the poles of electricity distribution is used. The analysed examples of OAN maintenance costs are slightly higher for air networks, and for underground CD maintenance costs are approximately the same for all types of OAN. We also concluded that the star is a technology that pierces from the columns to the end user instead of using a cascade connection through private objects the most efficient model. Therefore, the conclusions of the paper are logical and well known, the advantage is that this analysis has confirmed the conclusions on concrete examples and give concrete numerical relationships for different types of networks from the aspect of population density.

References

1. Škaljo, E., Hodžić, M., Mujčić, A.: A cost effective topology in fiber to the home point to point networks based on single wavelength bi-directional multiplex. In: International Work Shop on Fiber Optics in Access Networks FOAN, Brno-Ches Republik (2015)
2. Hodžić, M., Maslo, A.: Analysis techno-economic profitability of FTTH suburban network on the example of construction access network (AN) Srebrenica. In: Conference BHAAAS, International Symposium on Information and Communication Tehnologies-ISICT, Teslić (2017)
3. Von Bogaert, J., et al.: FTTH Buisness Guide. 5 edn. FTTH Council Europe. http://www.ftthcouncil.eu/
4. Bauer, K.: Installation, last mile and last meter challenges and issues. LEONI NBG Fiber Optics GmbH, Leoni (2007)
5. Olsen, B.T.: Techno-economic evaluation of narrowband and broadband access network alternatives and evolution scenario assessment. IEEE J. Sel. Areas Commun. **14**, 1184–1203 (1996)

Which Container Should I Use?

Esmira Muslija(✉) and Edin Pjanić

Faculty of Electrical Engineering, University of Tuzla,
Tuzla, Bosnia and Herzegovina
esmira.muslija@fet.ba, edin.pjanic@untz.ba

Abstract. In this paper we present results of performance testing we performed on three containers from the standard library: `vector`, `map` and `unordered_map`. The goal of this work is to make practical guidelines in order to help programmers choose the right container for a particular use case. We made a series of tests that resembles different circumstances that can come in real world applications. In order to make better conclusions, tests were performed on four different computer configurations and processors. Our initial hypothesis is that the performance of containers not only depends on their general complexity that is a consequence of their internal design and algorithms as such, but should additionally be connected to cache utilization. After analysis of the results obtained, we made several conclusions in terms of better use of C++ containers in particular use cases.

1 Introduction

Many computer applications utilize a large set of data, but many of them do not. However, most of applications require fast data retrieval and processing. In order to optimize memory usage and speed of data access and processing, quite a few data structures are developed, but only several of them are standard ones and built for general use. Those standard data structures are designed to be collections of other objects of arbitrary type and are called containers [1]. Each type of container has its own internal structure and organization, which has consequences for asymptotic complexity [2] of its operations. Developers tend to use the standard containers for storing data in random access memory (RAM) having in mind their well known theoretical asymptotic complexities.

Although processor clock speed has been constantly increasing in the last several decades, RAM access time has not been taking the same pace and is usually a magnitude slower than the processor. However, modern processors are very complex and utilize very fast memory called cache that is on the die near the processor core. Purpose of the cache memory is to make data from random access memory (RAM) available for the central processing unit (CPU) when it is required. Instead of wasting up to 30 clock cycles for loading a piece of data from RAM, the data is loaded from cache memory in only several cycles [3]. In order to accomplish that task, cache systems utilize sophisticated algorithms to predict what data to load from RAM and what data to keep in cache for later. When the data the CPU needs is already loaded in cache, we call that situation a cache hit. Otherwise, when the data that is needed by the CPU is not in the

S. Avdaković (Ed.): IAT 2018, LNNS 59, pp. 487–504, 2019.
https://doi.org/10.1007/978-3-030-02574-8_39

cache, we call it a cache miss. In that case, the data has to be loaded from somewhere else. Cache memory that is next to the CPU core, and the fastest as well, is called level 1 (L1) cache. Usually, cache is organized in 3 or 4 levels, so in case of a cache miss at level 1, data is searched in upper cache levels (L2, L3 and L4), each slower than the previous level. In the worst case, data is loaded from RAM into cache and made available to the CPU. During that process, not only data from memory location the CPU needs is loaded from RAM into the cache but data from several other memory locations are loaded to the cache as well. What data is loaded from RAM to the cache depends on cache's sophisticated algorithms, but usually it is a block of data next to the requested data. Obviously, increase in cache miss rate of an application is directly related to its poor performance.

Having that in mind, it would be beneficial if we could use data structures that make many cache hits and few cache misses. There are techniques to keep the data in specially organized data structures in order to improve cache behavior such as in [4–7], but these techniques are too complicated and there are many limitations in their use. In common, all techniques agree that data should be kept in consecutive memory locations in order to have less cache misses. So, what to do when faced with a decision what container to use and how to organize your data?

In this article we try to answer to a common question of a usual programmer: "Which container should I use in a certain scenario?", and we try to give some guidelines to common use cases.

Our focus is on the standard C++ containers [8] that allow access to underlying data by searching a given key: map (`std::map`), hash map (`std::unordered_map`) and vector (`std::vector`). They are the most utilized containers of all. We conducted a series of experiments in order to simulate common scenarios and calculate average access time. Based on results of the experiments, we give several guidelines.

The paper is organized as follows. After Introduction we briefly review selected containers from C++ Standard Library. In Sect. 3 we give an overview and methodology of conducted experiments. Section 4 contains results of the experiments with general guidelines for containers usage. Finally, conclusions and future work are presented in Sect. 5.

2 Common Data Structures

2.1 Vector (std::vector)

Main feature of the vector container is that its elements are stored contiguously, in one chunk of heap memory. Elements of this container are stored in an underlying dynamically allocated array. It means that maximal number of elements in vector as a container is exactly the size of the underlying array, which is a containers capacity, as depicted in Fig. 1. Vector as a container maintains required space for its elements. From a perspective of a container's user, capacity of the vector changes automatically whenever we need extra space for new elements, which means whenever the size of the vector could become larger than its capacity. In that case, new memory is allocated on the heap, then all existing elements are copied/moved to the new array and finally new element is inserted. New capacity is larger than the previous one by some factor. For vector in Standard Library, that factor is 2 which means that every time a vector grows, its capacity doubles.

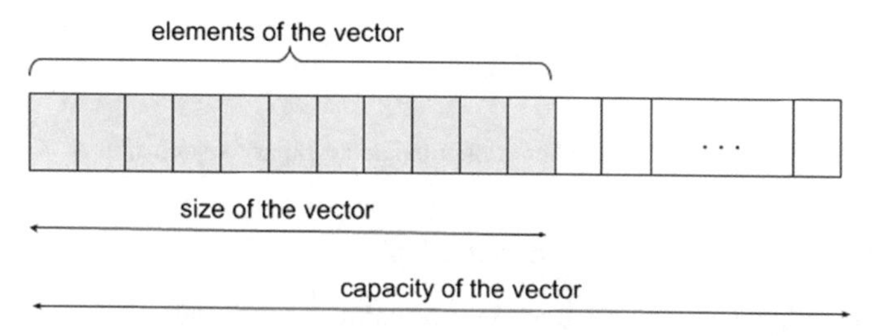

Fig. 1. Simplified vector container

Asymptotic complexity of access to an element of vector with known sequence number is O(1). However, when searching by given value, the complexity is greater. Generally, sequential search is used, where we approach each element of the vector and compare it with the one we search. Complexity of sequential search algorithm is O(n).

Complexity of searching in a vector is significantly improved when vector is sorted. In that case it is possible to use binary search with logarithmic complexity O(log n).

Removing an element from a vector container requires moving all elements placed after the removed element up until the last element for one place to the left. It is linear complexity.

2.2 Map (std::map)

Map is an associative container of pairs in form <key, value>. Key is used in making criteria for map sorting. Internal structure of map container is balanced binary search tree (BST) as simplified depicted in Fig. 2. Elements are stored in nodes sorted by keys. Map does not take contiguous portion of memory, as vector do. Instead, its elements are scattered all over the heap memory. With every insertion of an element, a new node with a key-value pair is dynamically allocated somewhere in the heap and its pointer is saved at its parent's node in the BST.

Search complexity of map container is logarithmic O(log n) because of binary search tree organization. However, due to independent dynamic allocation of its nodes, pieces of data belonging to this container are placed in separate parts of memory which potentially leads to many cache misses.

Deletion of an element with a given key happens in several steps. First, it is necessary to find the element with a given key in the tree, then delete it from the map and perform balancing of the tree if necessary. Complexity of this operation is O(log n).

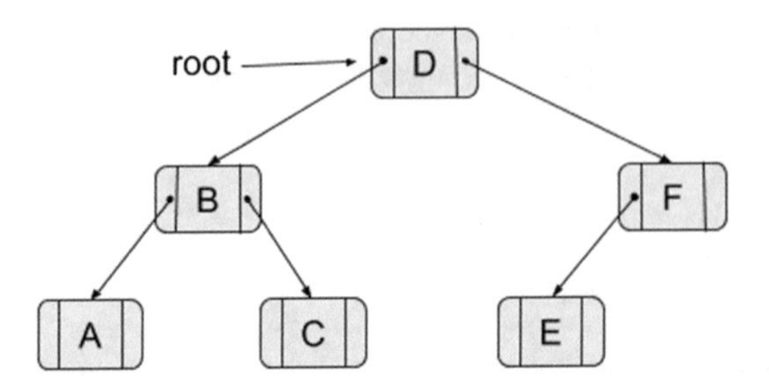

Fig. 2. Map container as a balanced binary search tree

2.3 Unordered Map (std::unordered_map)

Unordered map is an associative container of pairs <key, value>, too. Just like in map, uniqueness of its elements is determined by key. However, elements in this container are not ordered by key and they don't form a binary tree. Instead, they are grouped in buckets which are organized with hash table using a hash function, as depicted in Fig. 3. Hash function calculates an index of element in the hash table. Every key with the same hashed value goes into the same bucket. Buckets are basically lists with few elements.

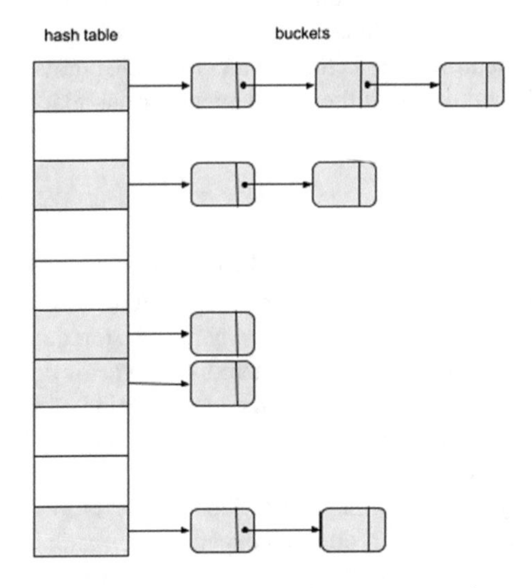

Fig. 3. Illustration of unordered map container

Searching takes place in two steps, first the given key is hashed and then associated bucket is sequentially searched. It has near constant complexity, because sequential search in a bucket overall has very little impact. Same applies on deleting a single element.

3 Performance Testing

3.1 Testing Setup and Methodology

Performance testing is performed on three types of containers described above. Vector container is used because of its sequential memory layout and constant time access by index. If we search element by a given key using binary search in previously sorted vector we can expect logarithmic complexity with good cache efficiency. However, although search in map container is of logarithmic complexity and search in unordered map is of constant time complexity, they utilize dynamic memory allocation for every element, so it is not clear which container of all will behave the most efficiently. Other containers, such as list, are excluded because of their obvious linear or greater asymptotic complexity.

Purpose of these tests is to seek out efficiency of each container in various circumstances. We performed experiments following two scenarios.

3.1.1 Scenario 1: Isolation

In the first scenario, we measure performance of every type of container alone, without interference with each other or any other data. We measure average time for insertion of elements, searching for a given key and deletion of every element by a given key. Performances are measured for several data types independently: int, string (size = 8, 16, 32, 64), char arrays (size = 8, 16, 32, 64), and a custom structure `struct Element` with size of 125 ints. (`struct Element{int id; int[124]}`). In the case of `struct Element`, its member `id` is used as a key for the hash function. For other data types, value itself is used as a key.

Each container is tested with the same randomly generated unique values. Vector is previously sorted so search is performed utilizing binary search algorithm. Additionally, we perform sequential search on the vector container because we expect it to have good cache efficiency on small vector sizes and hence good performance.

Besides default constructed vector that automatically expands its capacity we create another vector container object with preallocated required space and test its insert performance. Operations on default constructed and preallocated vectors in terms of searching and deletion does not differ so only one of them (default constructed) is tested for performance of those two operations.

For each data type in tests we create testing data which has randomly generated unique values, and then we perform independent experiments on each type of container with steps as follows:

1. Create a container object (vector, preallocated vector, map or unordered map).
2. For each value of testing data insert a copy into the container object. Save total insert time for the current container.
3. If the container is a vector, sort it and add the sorting time to the insertion time of the vector.
4. For each value of testing data perform search by key selected from testing data.
5. For each value of testing data delete corresponding elements from the container. Save total insert time for the current container.

6. Repeat all steps multiple times (300 times) and find an average insertion, search and deletion time.

3.1.2 Scenario 2: Pseudo-Real world

In summary, in this scenario we try to make allocations random as much as possible in order to see which container will have better performance that is directly related to cache efficiency. In order to do that, we fill all containers with random but the same data, just as we did in the first scenario. At the same time we dynamically allocate a large amount of dummy data in order to make gaps in heap, simulating some random data that could possibly be allocated in a real world application.

Specifically, for each data type used, as in Scenario 1, we perform independent experiments with steps as follows:

1. Create testing data which has randomly generated unique values.
2. Create all containers (vector, map, unordered map).
3. For each value of testing data insert a copy into all containers and dynamically allocate 30 kB of dummy data. Save total insert time for every container.
4. Sort vector. Add the sorting time to the insertion time of the vector.
5. For each value of testing data perform search based on a key and dynamically allocate 30 kB of dummy data in order to invalidate the cache.
6. For each value of testing data delete corresponding elements from all containers and dynamically allocate 30 kB of dummy data for cache invalidation. Save total insert time for every container.
7. Repeat all steps multiple times (300 times) and find an average insertion, search and deletion time.

Setup in this scenario should be a simulation of insertion, deletion and searching data in containers alternated with other activities that overwrites containers' data in cache. It is similar to a process with occasional search in a container.

In order to test performances of searching and processing of all data from a container in one batch without interruption, we made a sub case of this scenario. In this sub case we do not perform dynamic allocation of dummy data during searching.

3.2 Performance Test Results

For better comparison of the containers' performances, the tests were performed on four computers with different processors and different clock and cache configurations. The configurations are as follows:

Configuration 1

CPU: Intel i7-3610QM @ 2.40 GHz [9]
Level 1 cache size:
  4×32 KB 8-way set associative instruction caches
  4×32 KB 8-way set associative data caches
  Level 2 cache size: 4×256 KB 8-way set associative caches
Level 3 cache size: 6 MB 12-way set associative shared cache
RAM: 8 GB

Configuration 2

CPU: Intel i5-3337U @ 1.80 GHz [10]
Level 1 cache size:
  2 × 32 KB 8-way set associative instruction caches
  2 × 32 KB 8-way set associative data caches
Level 2 cache size: 2 × 256 KB 8-way set associative caches
Level 3 cache size: 3 MB 12-way set associative shared cache
RAM: 8 GB

Configuration 3

CPU: Intel Pentium M 740 @ 1.73 GHz [11]
Level 1 cache size:
  32 KB 8-way set associative instruction cache
  32 KB 8-way set associative write-back data cache
Level 2 cache size: 2 MB 8-way set associative cache
RAM: 1.25 GB

Configuration 4

CPU: AMD Athlon 64 TF-20 @ 1.60 GHz [12]
Level 1 cache size:
  64 KB 2-way set associative instruction cache
  64 KB 2-way set associative data cache
Level 2 cache size: 512 KB 16-way set associative cache
RAM: 2 GB

3.2.1 Results of Performance Testing Following the Scenario 1

We performed tests for selected data types and input sets with sizes in range from 1 to 1.000.000 elements. Figures 4, 5, 6, 7, 8, 9 and 10 depicts a part of results that significantly describes performances of all the operations we tested.

The results show that containers perform as expected in case of large number of elements. For searching, time complexity of vector and map containers is logarithmic and unordered map has near constant time complexity for all types of data. In case of searching for a given int value, binary search in vector has a certain advantage over search in map container. In all other cases map container is better than vector. However, unordered map dominates over other containers in all operations, including operations of insert and delete. Exception is the insertion operation of ints into a vector container. However, in insert/delete intensive processes vector requires sorting, so complexity of its combined operations would be much higher and performance would be very poor. In that case unordered map is a container of choice.

Figures 11, 12, 13, 14, 15, 16 and 17 shows performances of containers with small input sets.

When we have small input sets vector shows better performance in searching over other containers. Vector even more dominates in case of insertion operation and this is more evident when inserting elements into preallocated vector container. Deletion operation is fastest in case of unordered map. Although all operations are important,

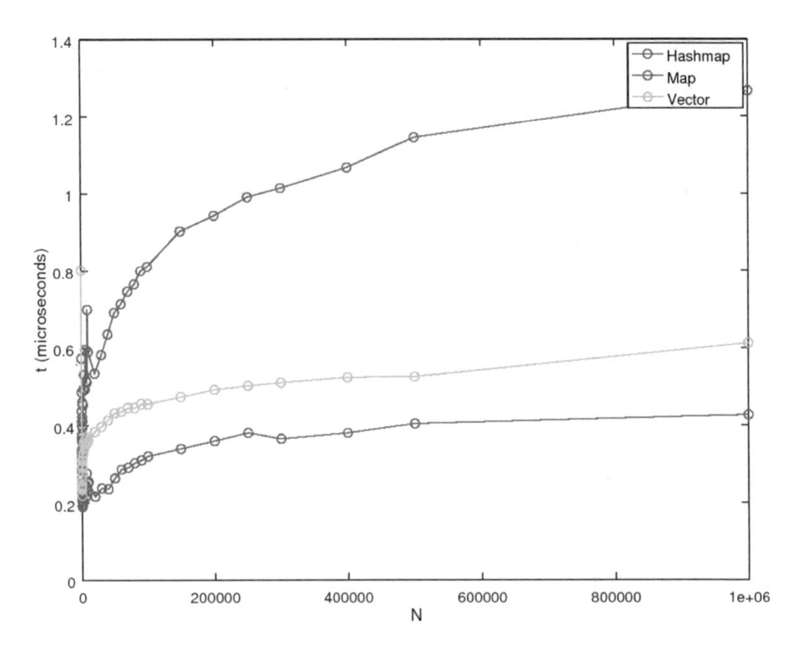

Fig. 4. Average search time in containers with elements of type int – i5

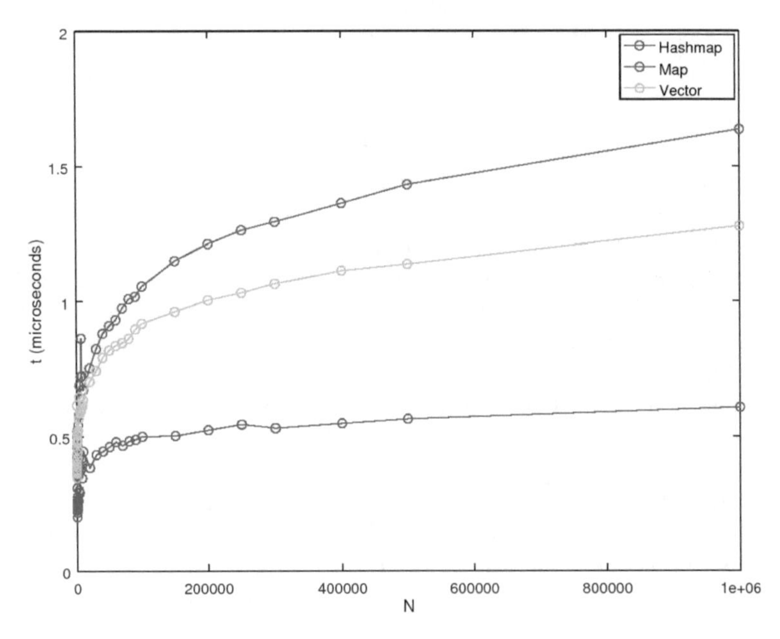

Fig. 5. Average search time in containers with elements of a custom type (search by given id, size 500 bytes) – i5

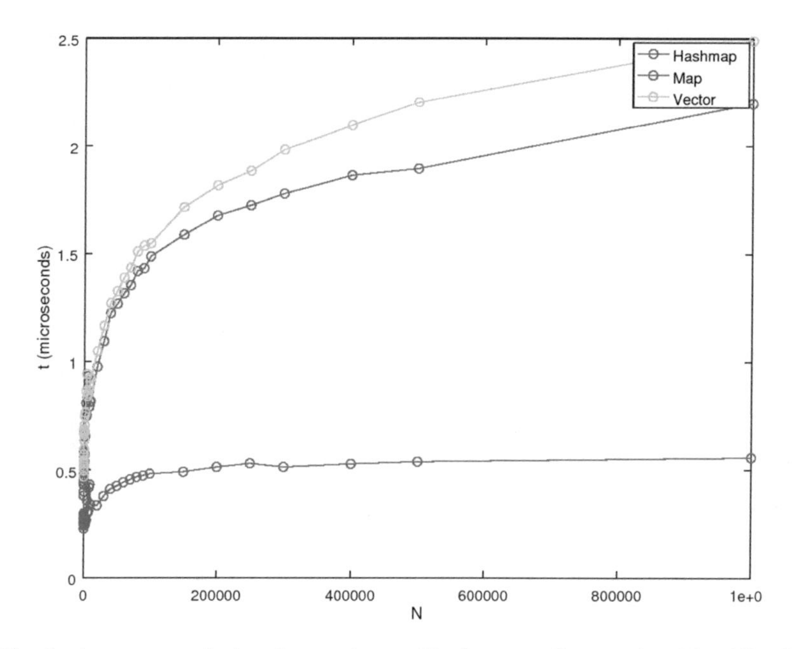

Fig. 6. Average search time in containers with elements of type string (size 32) - i5

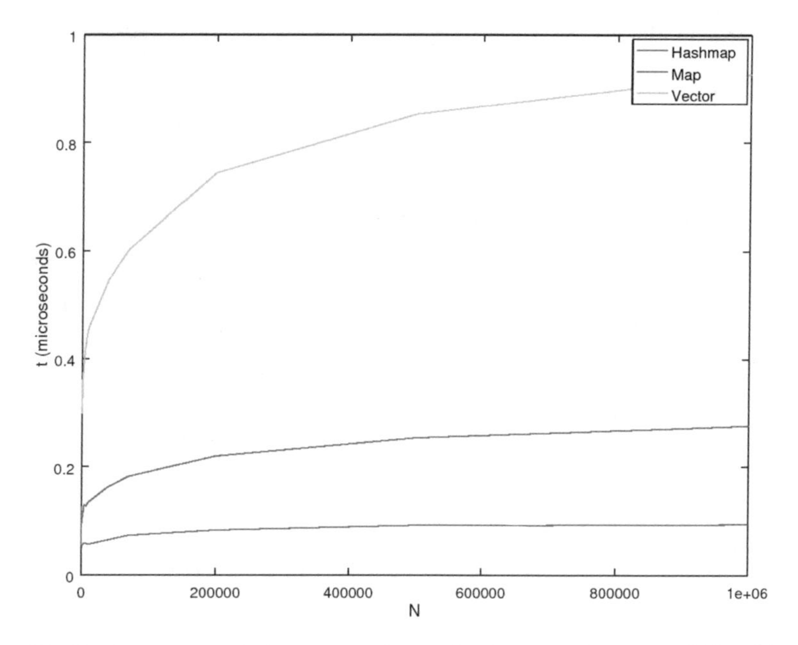

Fig. 7. Average search time in containers with elements of type char[64] – i7

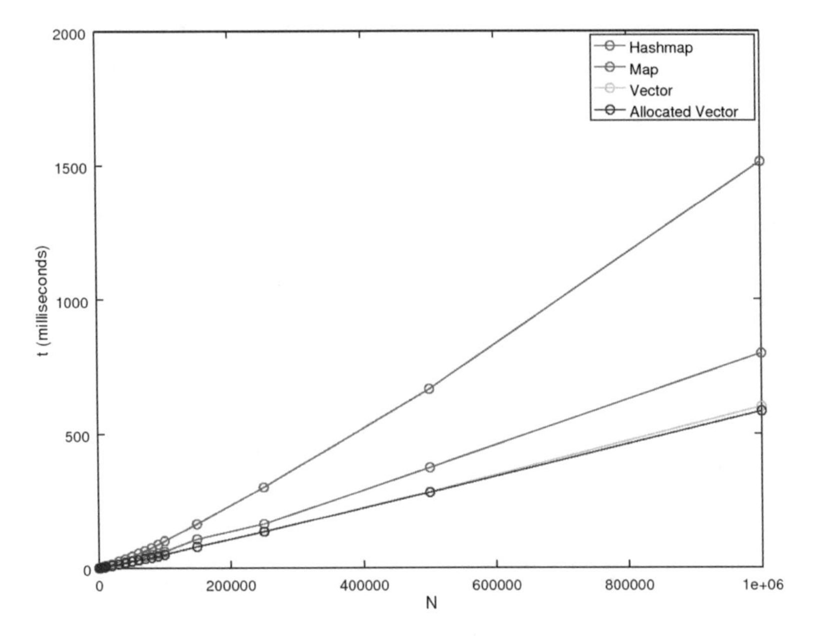

Fig. 8. Cumulative insertion time of elements of type int – i5

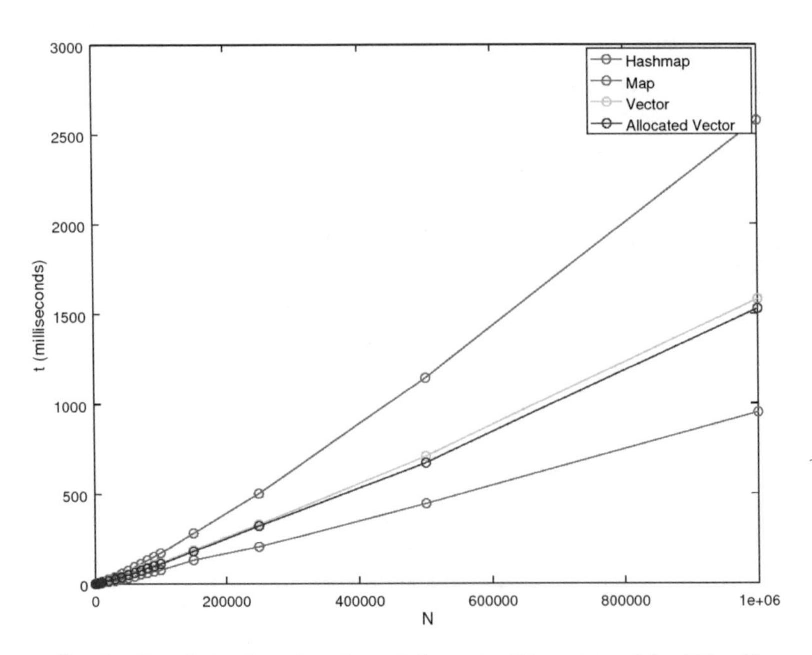

Fig. 9. Cumulative insertion time of elements of type string (size 32) – i5

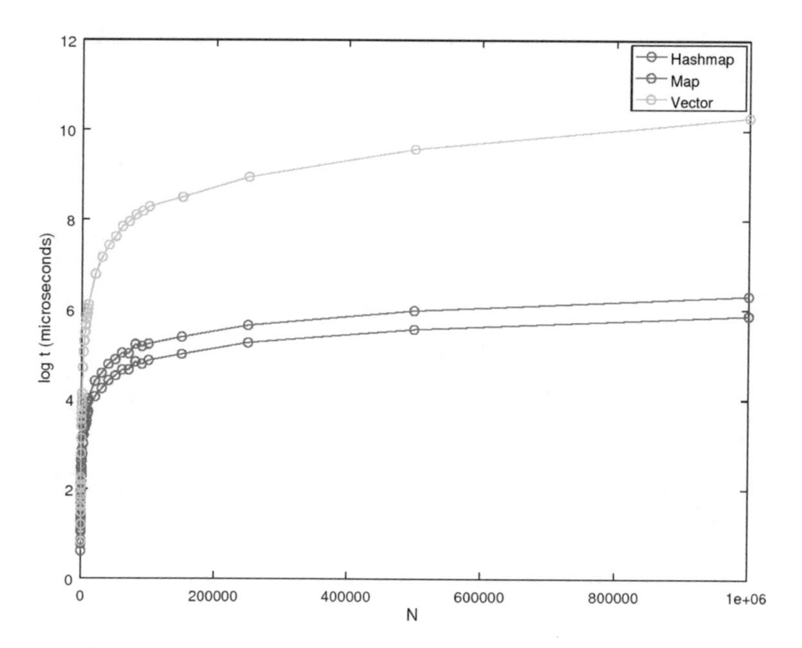

Fig. 10. Average deletion time of custom element (size 500 b) – i5

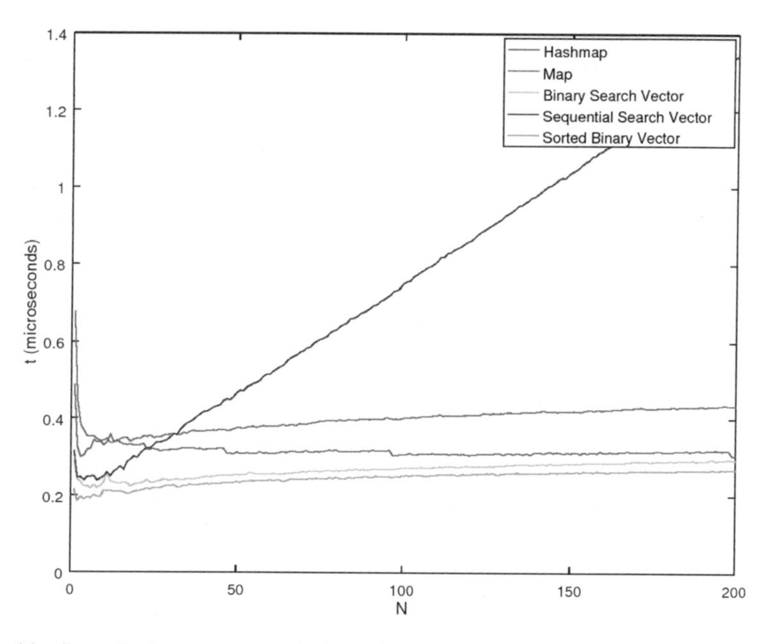

Fig. 11. Scenario 1: average search time of data type char[64] for small input sets – i7

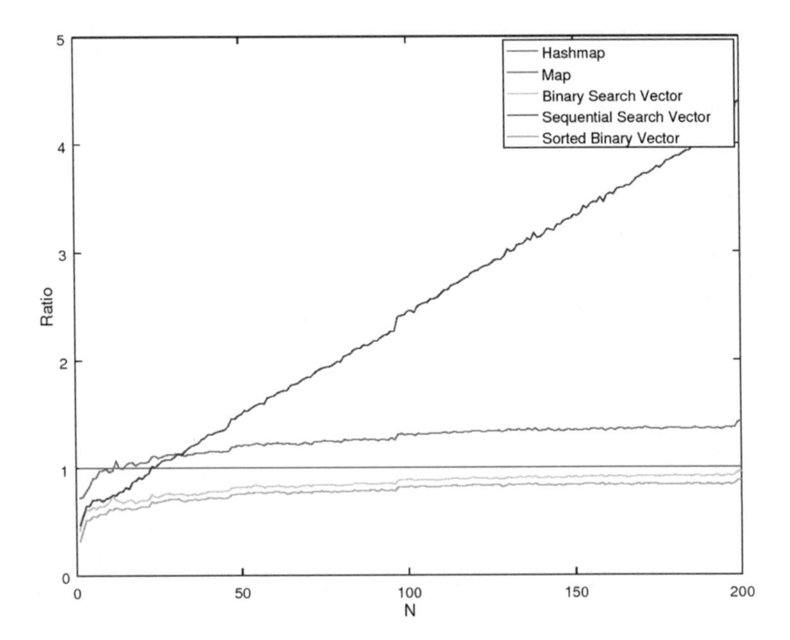

Fig. 12. Scenario 1: average search time relative to unordered map container search time (same data as in Fig. 11)

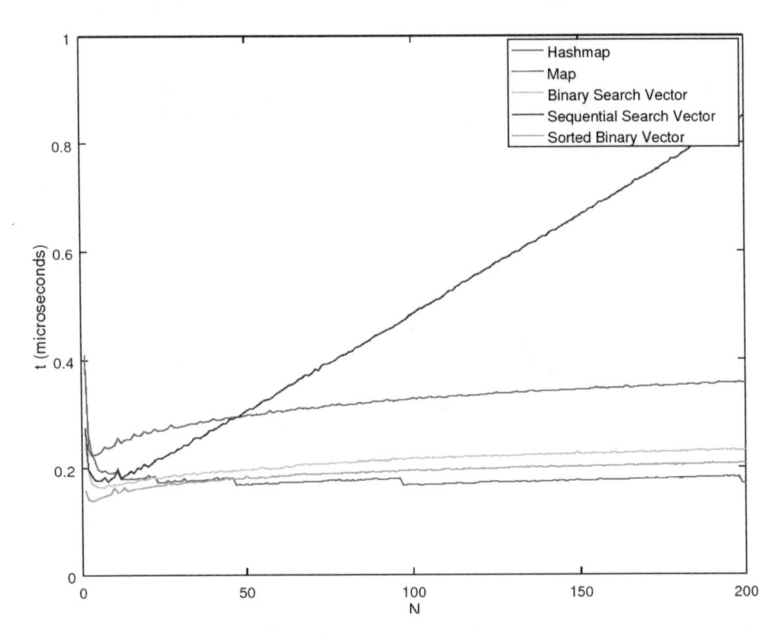

Fig. 13. Scenario 1: average search time of custom data type (size 500 bytes) for small input sets – i7

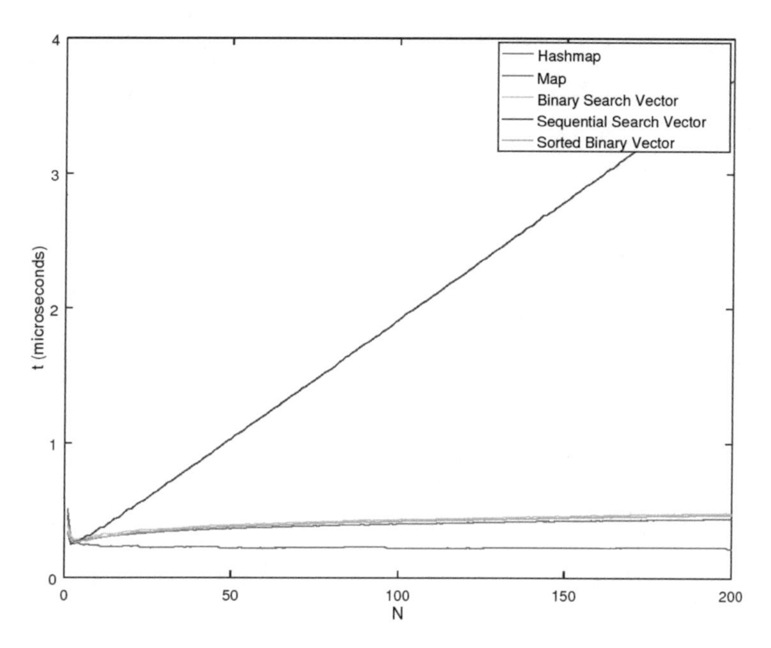

Fig. 14. Scenario 1: average search time in of string (size 32) for small input sets – i7

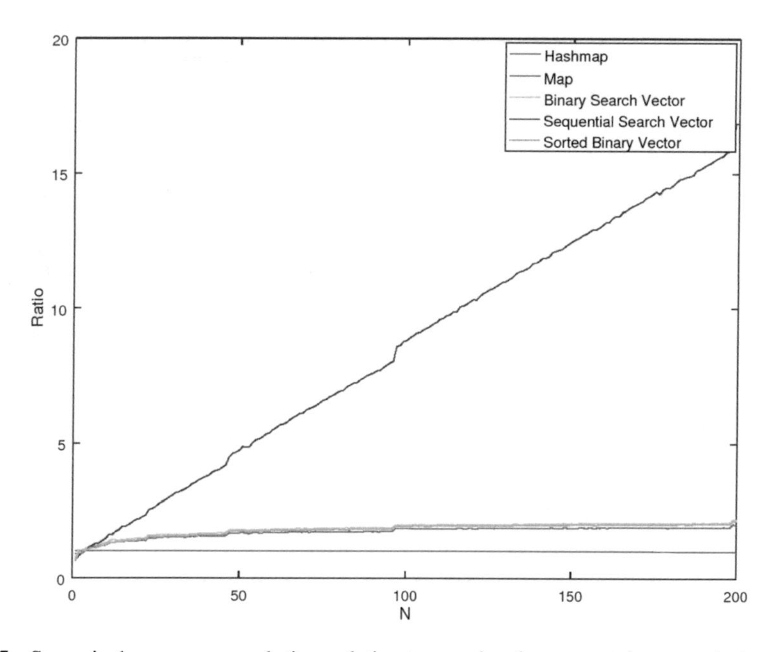

Fig. 15. Scenario 1: average search time relative to unordered map container search time (same data as in Fig. 14)

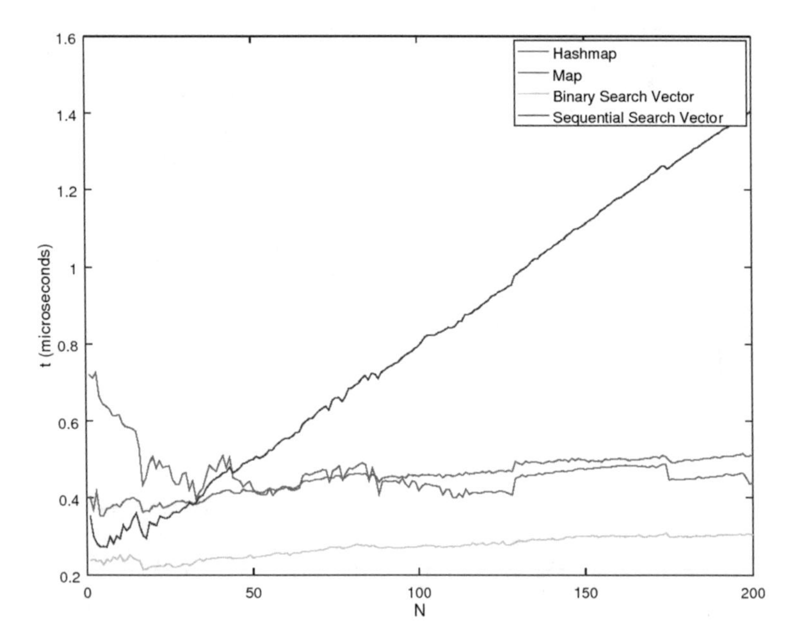

Fig. 16. Scenario 2: average search time of data type char[64] for small input sets – i7

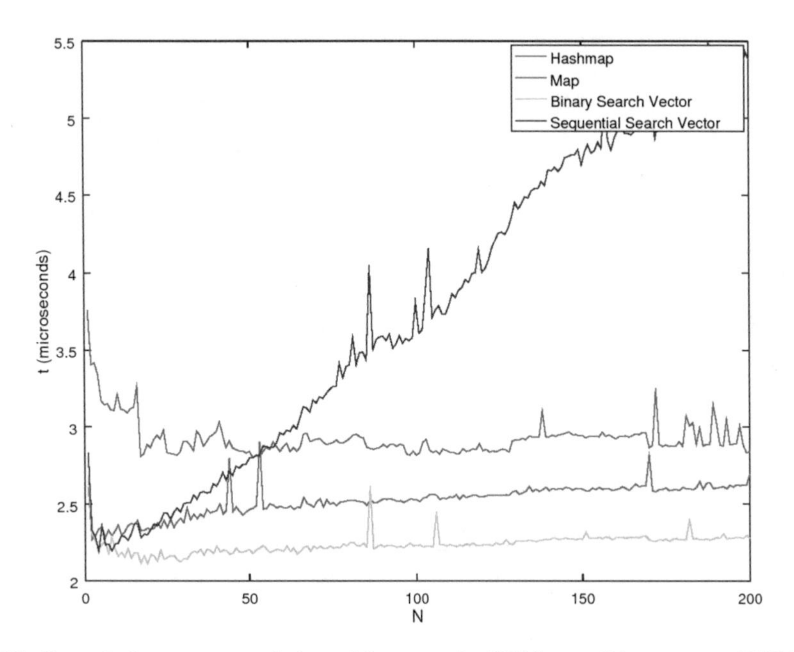

Fig. 17. Scenario 2: average search time of data type char[32] for small input sets – AMD TF-20

still in most cases searching is the most important operation on containers. After the analysis of the results for all containers and all computer configurations in scenario 1 for small input sizes, we concluded that vector has better performance with objects that do not utilize dynamic memory allocation so that they can be placed directly inside container's underlying array. Even vector with sequential search (does not require sorting) can be considered if input set is not larger than 20 elements. In case of objects that utilize dynamic allocation, such as `std::string`, advantage of using vector ends with input sets of size about 10, as depicted in Fig. 14 and 15. For larger input sets unordered map shows far better performance. As the number of elements in container grows, an advantage of unordered map increases.

3.2.2 Results of Performance Testing Following the Scenario 2

Scenario 2 was intended to make nodes of map and unordered map more scattered all over the heap, hence making searching more memory and time intensive due to increased cache misses. Indeed, time diagrams are really funny and toothed but after many repeated experiments we concluded that the results are really as they are depicted in diagrams in Figs. 16, 17, 18, 19 and 20, which show part of the results we obtained. We believe the reason for that is connected to many cache misses during searching, but further investigation is beyond these tests.

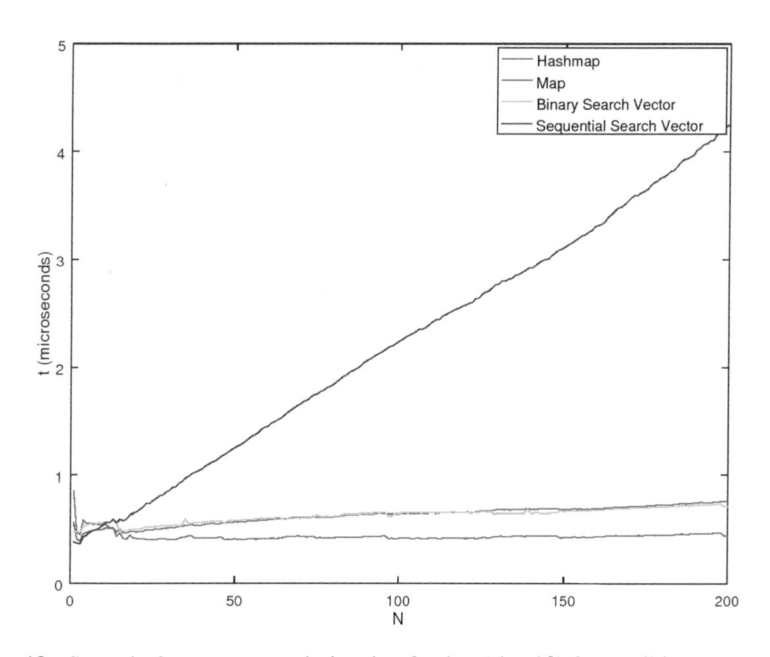

Fig. 18. Scenario 2: average search time in of string (size 32) for small input sets – i5

However, the results and all diagrams in this scenario are not significantly different than the results in scenario 1 for small input sizes. As an illustration we can make comparison of diagrams in Figs. 11 and 16.

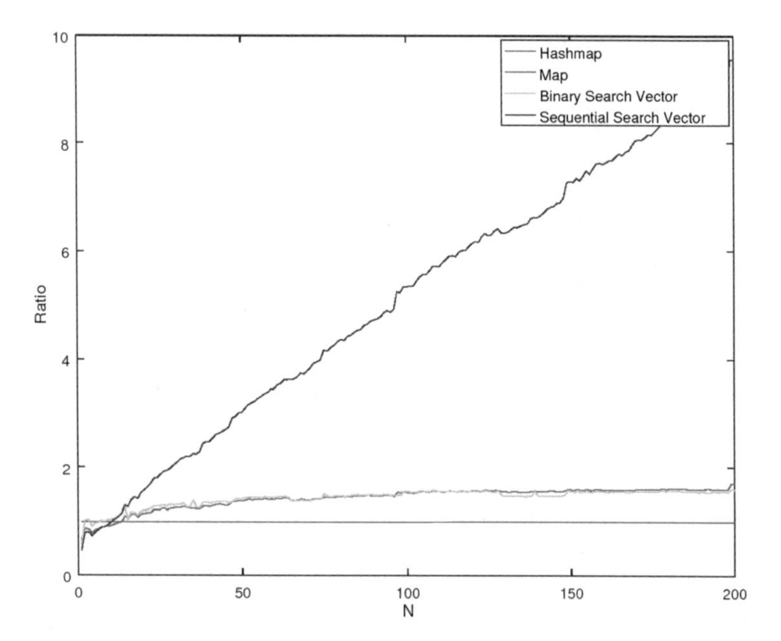

Fig. 19. Scenario 2: average search time relative to unordered map container search time (same data as in Fig. 18)

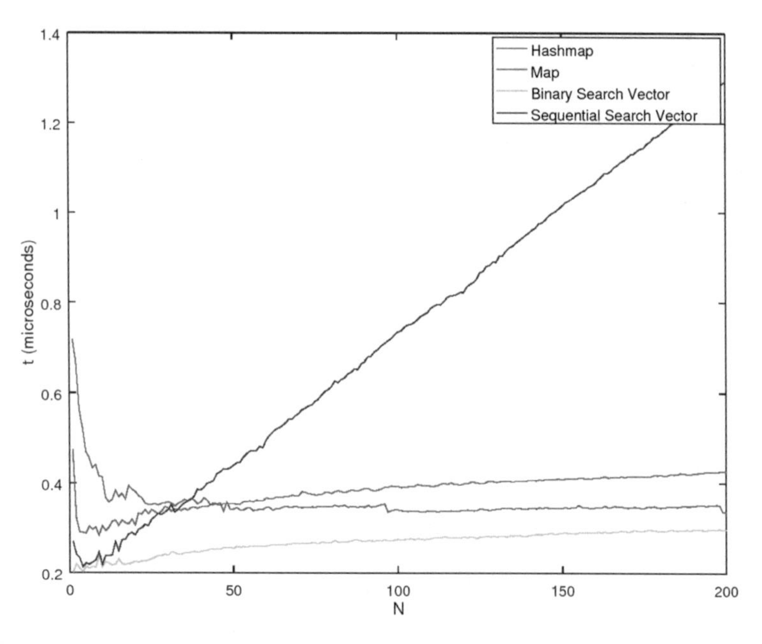

Fig. 20. Scenario 2 – sub case 2: average search time of data type char[64] for small input sets – i7

Additionally, all conclusions regarding the searching of certain data types we made in scenario 1 stand here too.

We are always interested to the searching results of one of the most used types in C ++: `std::string`. An example of diagrams that shows time performance of the searching for a string key is depicted on Figs. 18 and 19.

After experiments following the second sub case of the scenario 2 we concluded that there is no significant difference here neither. The only difference compared to the first sub case is in better smoothness of these diagrams. Figure 20 depicts one of the experiment result set. The results are comparable with results depicted in Figs. 16 and 11.

4 Conclusion

In this paper we performed performance testing of three containers from the standard library: `vector`, `map` and `unordered_map`. We made a series of tests that resembles different circumstances that can come in real world. In order to make better conclusions, tests were performed on four different computer configurations and processors. Our initial hypothesis was that the performance of containers not only depends on their general complexity that is consequence of their internal design and algorithms but should additionally be connected to cache utilization. We expected for vector container to have far better performance for small number of input data. We didn't know what would be practical guidelines.

After considering the results obtained, we can make several conclusions in terms of a better choice of C++ containers in a certain use case.

For large input sets one should still use a vector container if it is possible to use index for element's position, which gives us complexity O(1). If we cannot use index and have to find elements by a given key, then unordered map, which has near constant time complexity, is a container of choice. The only requirement is that we have to have a good hash function for key type. Fortunately, there are default implementations for many standard types. If we have to use some custom type for the key we must implement a corresponding hash function. Since `std::unordered_map` is a feature of C++11 and later, in case we have only old compiler we can use `std::map` which has logarithmic complexity for all operations.

For small input sets the situation is not so straightforward. First, we should always use vector if we can identify an element by index, that's obvious. If we have to search for an element by a given key then we could make the following reasoning. Integral types always perform better in vector of size up to 200 elements in which case we should utilize binary search by a given int key in previously sorted vector. If the number of elements in vector is less than 20 we can use sequential search too, sorting is not required then. If objects in container are of a plain struct-like type, meaning all members are in-place and not dynamically allocated then we can still use binary searched vector container, and it must be sorted. In that case upper boundary of a good performance depends of the objects' size. If the objects are not bigger than several tenths of bytes upper boundary is about 200, but if objects are several hundred bytes or larger than the upper bound is about 20 elements. This includes all primitive types and all structs or arrays of such types. If the objects have members that are pointers to dynamically allocated objects or are objects that utilize dynamic memory allocation then upper bound of number of elements for using vector is 10 or less. This leads to an important fact concerning `std::string` type that is very obvious from these experiments: always consider using char array instead of `std::string`. Additionally, one should consider partitioning of complicated objects into two or more parts, so that parts that are most likely to be used are placed together in the same container. These experiments also show that there are few arguments to use `std::map` container in new C++ code. We should use unordered map or vector container instead.

Future work could be based on further detailed investigation of cache performance of the tested containers in order to better design objects that are going to be placed into containers. That could lead to a new variation in design of an existing containers or designing a completely new container.

References

1. Malik, D.S.: Data structures using C. Course Technology/Cengage Learning India (2012)
2. Cormen, T.H., Leiserson, C.E.: Introduction to Algorithms, 3rd edn. The MIT Press, Cambridge (2009)
3. Irazoqui, G., Eisenbarth, T., Sunar, B.: Systematic reverse engineering of cache slice selection in Intel processors. In: 2015 Euromicro Conference on Digital System Design (2015)
4. Truong, D., Bodin, F., Seznec, A.: Improving cache behavior of dynamically allocated data structures. In: Proceedings of 1998 International Conference on Parallel Architectures and Compilation Techniques (Cat. No. 98EX192) (1998)
5. Adcock, C.M.S.: Improving cache performance by runtime data movement. University of Cambridge (2009)
6. Dudziak, T., Herter, J.: Cache analysis in presence of pointer-based data structures. ACM SIGBED Rev. **8**, 7–10 (2011)
7. Larus, J., Hill, M., Chilimbi, T.: Making pointer-based data structures cache conscious. Computer **33**, 67–74 (2000)
8. Josuttis, N.M.: The C Standard Library: A Tutorial and Reference. Addison-Wesley, Boston (2015)
9. CPU-World: Intel Core i7-3610QM specifications. http://www.cpu-world.com/CPUs/Core_i7/Intel-Core%20i7-3610QM%20Mobile%20processor.html
10. CPU-World: Intel Core i5-3337U specifications. http://www.cpu-world.com/CPUs/Core_i5/Intel-Core%20i5-3337U%20Mobile%20processor.html
11. CPU-World: Intel Pentium M 740 (Socket 479) specifications. http://www.cpu-world.com/CPUs/Pentium_M/Intel-Pentium%20M%20740%20RH80536GE0302M%20(BX80536GE1733FJ).html
12. CPU-World: AMD Athlon 64 TF-20 specifications. http://www.cpu-world.com/CPUs/K8/AMD-Athlon%2064%20TF-20%20-%20AMGTF20HAX4DN.html

Author Index

S. Avdaković (Ed.): IAT 2018, LNNS 59, pp. 505–506, 2019.
https://doi.org/10.1007/978-3-030-02574-8

Printed in the United States
By Bookmasters